Progress in Botany

Volume 80

Series Editors
Ulrich Lüttge, TU Darmstadt,
FB Biologie (10), Schnittspahnstraße 3–5,
64287 Darmstadt, Germany

Francisco M. Cánovas
Universidad de Málaga, Depto. Biología
Molecular y Campus de Teatinos, 29071
Málaga, Spain

Rainer Matyssek
Technische Universität München
Hans-Carl-von-Carlowitz-Platz 2
Wissenschaftszentrum Weihenstephan
85354 Freising, Germany

Hans Pretzsch
Technische Universität München
Hans-Carl-von-Carlowitz-Platz 2
Lehrstuhl für Waldwachstumskunde
85354 Freising, Germany

More information about this series at http://www.springer.com/series/124

Francisco M. Cánovas • Ulrich Lüttge •
Rainer Matyssek • Hans Pretzsch
Editors

Progress in Botany Vol. 80

Springer

Editors
Francisco M. Cánovas
Department of Molecular Biology
and Biochemistry
University of Málaga
Málaga, Spain

Ulrich Lüttge
Department of Biology
Technical University of Darmstadt
Darmstadt, Germany

Rainer Matyssek
Ecophysiology of Plants
School of Life Sciences Weihenstephan
Technical University of Munich
Freising, Bayern, Germany

Hans Pretzsch
Chair for Forest Growth and Yield Science
School of Life Sciences Weihenstephan
Technical University of Munich
Freising, Bayern, Germany

ISSN 0340-4773 ISSN 2197-8492 (electronic)
Progress in Botany
ISBN 978-3-030-10760-4 ISBN 978-3-030-10761-1 (eBook)
https://doi.org/10.1007/978-3-030-10761-1

Library of Congress Control Number: 2017959590

This Springer imprint is published by the registered company Springer Nature Switzerland AG
The registered company address is: Gewerbestrasse 11, 6330 Cham, Switzerland

Contents

List of Contributors

Mark Arick II Institute for Genomics, Biocomputing & Biotechnology (IGBB), Mississippi State University, Mississippi State, MS, USA

Concepción Ávila Departamento de Biología Molecular y Bioquímica, Facultad de Ciencias, Universidad de Málaga, Campus Universitario de Teatinos, Málaga, Spain

Eugenio Benvenuto Department of Sustainability, Division of Biotechnology and Agroindustry, Laboratory of Biotechnology, ENEA - Italian National Agency for New Technologies, Energy and the Environment, Rome, Italy

Siegmar-W. Breckle Formerly Department of Plant Ecology, University of Bielefeld, Bielefeld, Germany

Rafael A. Cañas Departamento de Biología Molecular y Bioquímica, Facultad de Ciencias, Universidad de Málaga, Campus Universitario de Teatinos, Málaga, Spain

Francisco M. Cánovas Department of Molecular Biology and Biochemistry, University of Málaga, Málaga, Spain

Francisco Javier Cejudo Instituto de Bioquímica Vegetal y Fotosíntesis, Universidad de Sevilla and CSIC, Sevilla, Spain

Fernando de la Torre Departamento de Biología Molecular y Bioquímica, Facultad de Ciencias, Universidad de Málaga, Campus Universitario de Teatinos, Málaga, Spain

Peter Geigenberger Fakultät für Biologie – Pflanzenmetabolismus, Ludwig-Maximilians-Universität München, Munich, Germany

Markku Keinänen Department of Environmental and Biological Sciences, University of Eastern Finland, Joensuu, Finland

Sarita Keski-Saari Department of Environmental and Biological Sciences, University of Eastern Finland, Joensuu, Finland

Sari Kontunen-Soppela Department of Environmental and Biological Sciences, University of Eastern Finland, Joensuu, Finland

E. Le Deunff Normandie Université, UNICAEN, ICORE, Caen, France

Jenna Lihavainen Department of Environmental and Biological Sciences, University of Eastern Finland, Joensuu, Finland

Silvia Massa Department of Sustainability, Division of Biotechnology and Agroindustry, Laboratory of Biotechnology, ENEA - Italian National Agency for New Technologies, Energy and the Environment, Rome, Italy

Belén Naranjo Fakultät für Biologie – Molekularbiologie der Pflanzen, Ludwig-Maximilians-Universität München, Munich, Germany

Elina Oksanen Department of Environmental and Biological Sciences, University of Eastern Finland, Joensuu, Finland

Mª Belén Pascual Departamento de Biología Molecular y Bioquímica, Facultad de Ciencias, Universidad de Málaga, Campus Universitario de Teatinos, Málaga, Spain

Daniel G. Peterson Institute for Genomics, Biocomputing & Biotechnology (IGBB), Mississippi State University, Mississippi State, MS, USA

Ombretta Presenti Department of Sustainability, Division of Biotechnology and Agroindustry, Laboratory of Biotechnology, ENEA - Italian National Agency for New Technologies, Energy and the Environment, Rome, Italy

Jean-Philippe Reichheld Laboratoire Génome et Développement des Plantes, Université de Perpignan Via Domitia, Perpignan, France

Laboratoire Génome et Développement des Plantes, Centre National de la Recherche Scientifique, Paris, France

Robert Schneider Chaire de recherche sur la forêt habitée, Département de Biologie, Université du Québec à Rimouski, Rimouski, QC, Canada

Arne Sellin Institute of Ecology and Earth Sciences, University of Tartu, Tartu, Estonia

Anu Sõber Institute of Ecology and Earth Sciences, University of Tartu, Tartu, Estonia

Ina Thormählen Fakultät für Biologie – Pflanzenmetabolismus, Ludwig-Maximilians-Universität München, Munich, Germany

José Abraham Trujillo-Hernandez Laboratoire Génome et Développement des Plantes, Université de Perpignan Via Domitia, Perpignan, France

Laboratoire Génome et Développement des Plantes, Centre National de la Recherche Scientifique, Paris, France

Curriculum Vitae

Siegmar-W. Breckle born on 27 February 1938 in Stuttgart, studied chemistry, biology, geology and geography at Stuttgart, Innsbruck and Hohenheim. In 1966 he got his Dr. rer. nat. with a dissertation on the ecology of *Quercus suber* in Catalunya (with Heinrich Walter) at the University of Hohenheim. After 3 years as a lecturer at the Faculty of Science, University of Kabul/Afghanistan he was Assistant Professor at the Institute of Pharmaceutical Biology at the University of Bonn. After his habilitation in 1976 (halophyte ecology in Utah/USA), he became Associate Professor in Bonn. At the same time, he had a guest lecturer-ship at the University of Marburg. In 1979, he was appointed Professor and Head of the Department of Ecology at the new University of Bielefeld. After retiring in 2003, he was guest professor at the University of Utsunomiya/Japan.His scientific interests are plant ecology and geo-botany, vegetation and stress ecology. Projects in deserts (Negev, Namib, Aralkum) and in tropical mountain rain forests (Costa Rica, Ecuador) were carried out in close co-operation with colleagues in the host countries. The projects were funded by DFG (German Research Foundation), BMBF (German Ministry of Education and Research), BMBL (German Ministry of Agriculture), DAAD (German Academic Exchange Service), VW-Foundation, Schimper-Foundation, Alexander-von-Humboldt Foundation and the University of Bielefeld.His publication record lists more than 300 papers, as well as publication, edition and co-edition of many scientific books. He served the Faculty as Academic Dean in 1984–1995 and from 1990 to 2003 he was responsible for the environmental council of the University of Bielefeld. He supervised many doctoral theses (c. 20), diploma theses (c. 130) and teachers' examination (c. 25). He was an external examiner at the University of Cairo/Egypt, Aswan/Egypt, Karachi/Pakistan, Tel Aviv/Israel, Coimbra/Portugal and Vienna/Austria. He served as examiner in many evaluation panels and on editorial boards of scientific journals. He founded four affiliation treaties of the Faculty of Biology/University of Bielefeld, with the University of Coimbra, with the Live Science Department at Tel Aviv University, with the Jardín Botánico Nacional at Santo Domingo/Dominican Republic and with the Sede Occidente in San Ramón/University of Costa Rica. For more than 10 years, he was speaker of the Afghanistan Research Group; he is, to date, vice president of the Swiss Afghanistan Foundation and the Schimper-Foundation (Stuttgart-Hohenheim). He is a member of the Academy of Science and Arts Peter I. St. Petersburg. *Astragalus brecklei* I. Deml, *Acantholimon brecklei* Rech. f. & Schiman-Czeika and *Anemone X brecklei* Rech. fil. were named after collections in his herbarium; some 15–20 new species and subspecies were described from his Afghan specimens.

Vegetation, Climate and Soil: 50 Years of Global Ecology

Siegmar-W. Breckle

Contents

Abstract In this essay, more than 50 years of studies on a broad spectrum of ecological topics is reviewed. Many research questions in desert ecology, in tropical ecology and in stress ecophysiology were studied, mainly in the field but often completed by analysis in the lab.

To keep in mind the complexity of ecological systems, it was necessary to go into detail with very specific approaches but always keeping in mind a synthetic view. More and more it became clear that the biodiversity in many ecosystems plays an important role for the future. Vegetation, climate and soil are interwoven in

Communicated by Ulrich Lüttge

S.-W. Breckle (✉)
Formerly Department of Plant Ecology, University of Bielefeld, Bielefeld, Germany
e-mail: sbreckle@gmx.de

Progress in Botany (2019) 80: 1–64, DOI 10.1007/124_2017_14,
Published online: 11 March 2018

ecosystems; the rapid developments in science – as well as the fast development of methods, techniques and data pools – are a big challenge. Many aspects of global ecology have been studied in the last 50 years. Some examples are demonstrated here. This essay will remain a rather historical treatise. It will give examples of experimental and field work. It will be a look back with great gratitude on what was able to be achieved and experienced.

1 Introductory Remarks and Early Background at Stuttgart, Innsbruck, Hohenheim, Kabul, Bonn, Logan (Utah) and Bielefeld

The following review gives a personal overview of some of the scientific topics tackled over the last 50 years and characterizes a somewhat unusual career, which asked many questions on many very different ecological topics. The scientific depth was sometimes limited, and the scientific breadth was immense and always satisfying.

Breadth or depth? I always have had the feeling that, as a scientist, I was lucky not to do research for 25 or 40 years on *Sinapis* seedlings or on one gene of *Arabidopsis*, although it is certainly very important too.

I had the privilege to study three subjects at the University (Technische Hochschule) of Stuttgart (chemistry, biology and geology) for my teacher's exam. Later, geology had to be replaced by geography. But ethnology, astrophysics, engineering, philosophy or even Beethoven piano sonatas could come into lectures. This was fascinating, as was studying for 1 year at the University of Innsbruck with a touch of high mountain ecology. After two obligatory practical courses at schools (each for 4 weeks), the three classes in chemistry were so different that even the teacher had problems in keeping the classes on the same parallel level.

The thesis for the 1963 teacher's exam was in inorganic chemistry at Stuttgart (supervised by Joseph Goubeau). The title was "Preparation of a Compound Between Perchloric Acid and Hydrogen Peroxide" (Breckle 1963). Those who know chemistry would judge that this has something to do with rocket explosives. The questions were: How stable is the ionic compound? Is a salt existent that could be called "perhydronium perchlorate"? (Fig. 1). In fact, this is a very explosive salt, only stable below $-50°C$. By infrared and Raman spectroscopy, the cohesive force and the bond angles could be calculated. However, the monocrystalline sodium chloride (NaCl) cuvette, after several tests, exploded once, and all of the windows of the lab – and also, for some days, my auditory sense and hearing – were damaged. Thus I started writing the thesis and finished my teacher's studies in 1963 with the state examination (inorganic and organic chemistry, zoology, botany, geography and the Philosophicum).

Then I switched back to botany, my old hobby. Instead of becoming a teacher or having a well-paid job at one of the large chemical companies (Bayer, BASF), I preferred to tackle the question: What are the competitive conditions between

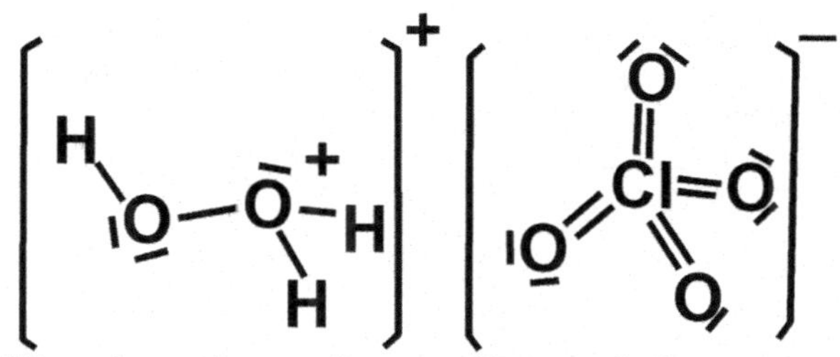

Perhydronium-Perchlorate

Scheme on force constants of dimeric hydrid-configurations as bond descriptor						
Free electron pairs-	Force constants [mdyn / Å = 100 N / m]					
	$k_{C\text{-}C}$		$k_{N\text{-}N}$		$k_{O\text{-}O}$	
0	C_2H_2	15.7	$N_2H_6^{2+}$	5.37	.....	
1	C_2H^-	13.1	$N_2H_5^+$	4.65	.....	
2	C_2^{2-}	??	N_2H_4	3.77	$H_4O_2^{2+}$	??
3	.....		$N_2H_3^-$	3.35	$H_3O_2^+$	??
4	.....		$N_2H_2^{2-}$	??	H_2O_2	3.99
5	.....		.....		HO^-	3.7
6	.....		.....		O_2^{2-}	??

Fig. 1 Structure of perhydronium perchlorate and scheme of force constants of dimeric hydrid configurations as a bond descriptor (Breckle 1963)

Quercus suber and *Quercus ilex* in Catalunya in north-eastern Spain? From the University of Hohenheim with an old Volkswagen Beetle and a 1,000 DM fellowship from the Carlos Faust Foundation in Barcelona – the only external funding – I started at the famous Mar I Murta Botanical Garden in Blanes, where they wanted me at once as a gardener! Advice came by letters from my supervisor, Heinrich Walter, in South America, where he was on his sabbatical. It turned out that the cork oak, *Quercus suber*, uses twice as much water as *Quercus ilex* (Fig. 2), with a very large root system (Breckle 1966). And, growing in Germany, young trees of *Quercus suber* were more frost resistant than those of *Quercus ilex* – in contrast to many observations in Spain or in Italy at Lake Maggiore (Sakai and Larcher 1987).

There was some pressure to publish the work, since Heinrich Walter was near retirement. On the other hand, a doctoral thesis at that time was acknowledged as a publication providing more details than a publication in a journal, as Beck (2017) also indicated, and a cumulative dissertation was not known or allowed at that time. The Rigorosum (viva) at the University of Stuttgart-Hohenheim was very comprehensive (one long oral of 90 min), covering botany, ecology and seed science.

My studies at Stuttgart and Hohenheim were a very helpful basis for future research. Teaching at the Chemieschule and during practicals and field trips around Hohenheim made me learn much about plants, their behaviour and ecology. Behind that were chemistry and chemical reactions and pathways. But doing studies and

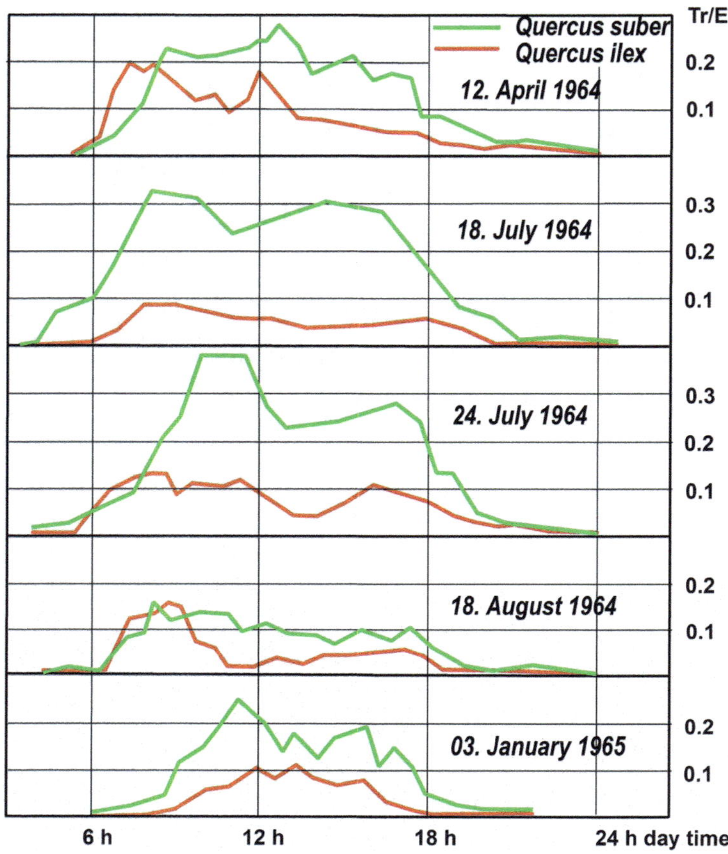

Fig. 2 Daily courses of the transpiration-to-evaporation ratio of *Quercus suber* and *Quercus ilex* in Catalunya (Breckle 1966)

measurements in nature proved to be more complicated; it took many replications to understand diurnal and annual fluctuations.

Parallel to my stays in north-eastern Spain, Helmut Freitag used his Carlos Faust Fellowship for research in the semi-desertic south-eastern Spain. He once mentioned to me the affiliation between the universities of Bonn and Kabul and that there was an existing German team of lecturers in Kabul at its Faculty of Science. We tried and then succeeded in getting a new job in Afghanistan – for us an unknown country. After Freitag's habilitation and my doctorate, we started at the end of 1966 to build up a curriculum for botany, giving lectures, courses, seminars and excursions, jointly for male *and* female students at the Faculty of Science, Kabul. Uta Breckle organized the lab work and the establishment of a herbarium.

Again, there were high mountains. During the summer vacation, when all of the steppe and semi-deserts in the lowlands were dry and arid, the Hindu Kush was an ideal place. One of the many highlights was an expedition to the Wakhan (Fig. 3a),

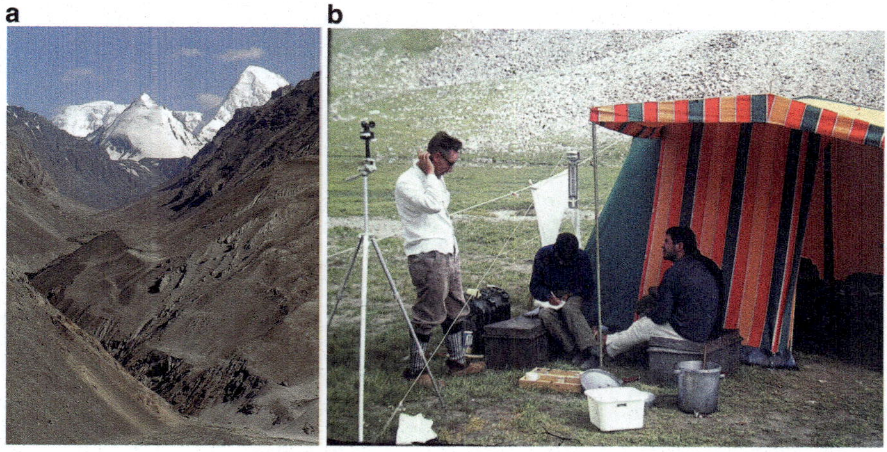

Fig. 3 (**a**) Noshaq (7,492 m), the highest peak in Afghanistan, in the eastern Hindu Kush, near Qazi Deh, Wakhan (photo courtesy of C. Naumann, 2002). (**b**) Base camp at Kotale-Wazit (4,620 m) for microclimatic measurements; data collecting and evaluations with Afghan counterparts (photo courtesy of P. Schneider, August 1968)

with the permission of the Afghan King Mohammad Zahir Shah (deposed in 1973); in August 1968 we spent 1 month in the Wazit area (at an altitude of 4,600 m) as our base camp. Some microclimatic measurements (Fig. 4), not repeatable until now – indicating the specific high-altitude conditions, vegetation surveys and many herbarium collections, etc. – were made (Breckle 1973). Advanced Afghan students who later got fellowships to study in Germany were involved (Fig. 3b).

Other mountain ranges (Koh-e Baba, Safed Koh, etc.) were studied up to the nival region (Breckle 1975; Breckle and Frey 1974, 1976). The number of vascular plants known from above 5,000 m increased from 12 to 40 species. Decades later, many of the data and the photographs (see Sect. 7.2) were used for inventories.

As a member of the German team within the affiliation between Bonn University and the University of Kabul, I had the opportunity to start as an assistant at the Institute of Pharmaceutical Botany at the University of Bonn; the first textbook for students was published in 1978 (Kating and Breckle 1978) and the eighth edition came out in 2014 (Leistner and Breckle 2014).

In Afghanistan, it was very obvious that in dry areas the role of halophytic plant species is very important in many different vegetation types. Evaluation of material and data from Afghanistan, as well as a project comparing two competing halophytic species in the Great Salt Lake area and at the Utah State University in Logan, finally led to my habilitation (Breckle 1976). The question was: Do recreting halophytes have an advantage in the salt desert of Utah, and under what conditions? Mainly, ecophysiological responses of *Atriplex confertifolia* (C_4) and *Ceratoides lanata* (C_3) were compared, and later in Germany cultivation experiments were performed. Sodium (Na) turnover on both sites was calculated from respective analytical data (Fig. 5). *Atriplex confertifolia* exhibited a much higher turnover by litter and salt

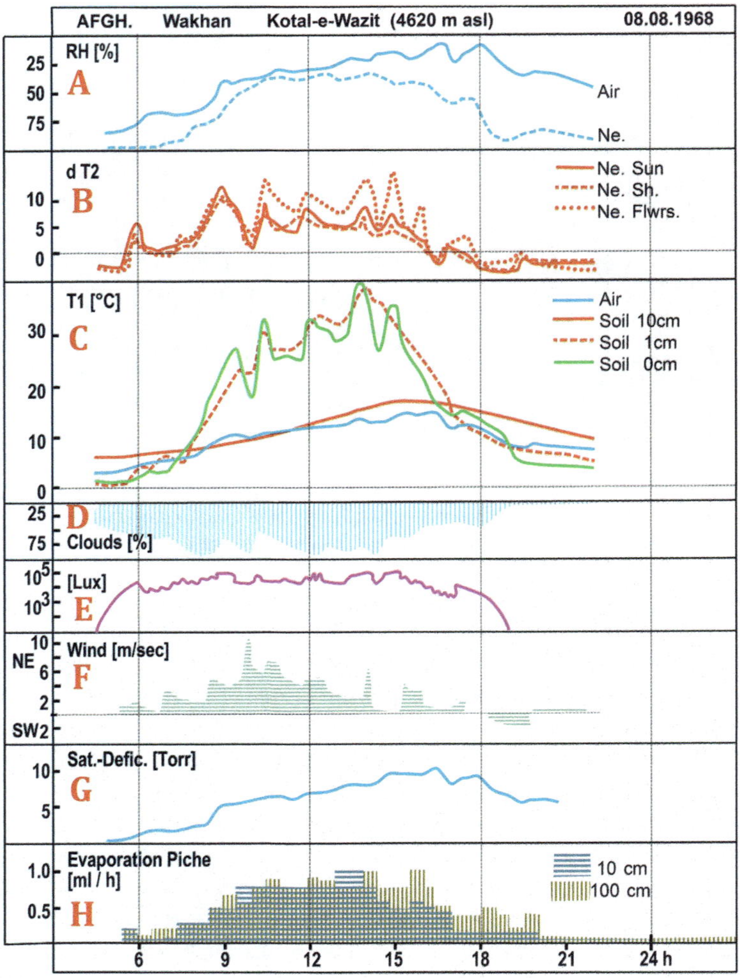

Fig. 4 Daily courses (measured on 8 August 1968) of some microclimatic and ecophysiological parameters at an alpine site at Kotale-Wazit (Wazit pass, 4,620 m; Wakhan) (Breckle 1973). (**a**) Relative humidity (air, 1.5 m; Ne, between *Nepeta pamirensis* [10 cm]). (**b**) Temperature difference (K) to air temperature of *Nepeta* sun leaves, shade leaves and flowers. (**c**) Air temperature (1.5 m) and soil temperatures (surface, 1 cm, 10 cm depth). (**d**) Cloudiness (% of whole sky). (**e**) Radiation (lux). (**f**) Wind speed (north-east or south-west). (**g**) Saturation deficit of air (torr). (**h**) Evaporation (Piche 3 cm) (ml h^{-1}), 10 and 100 cm

bladder hairs than *Ceratoides*, which, however, had a rather high salt content in its roots. Salt is accumulated in the roots of *Ceratoides*; in *Atriplex*, it is in the shoot. The dominant species of the salt deserts are shown in Fig. 6, according to their ecological requirements.

In Nov. 1979, I went to Bielefeld and started to establish the new Department of Plant Ecology. Together with a small team (some co-workers and a technician who came from Bonn), we had plenty of room but almost no equipment at the new

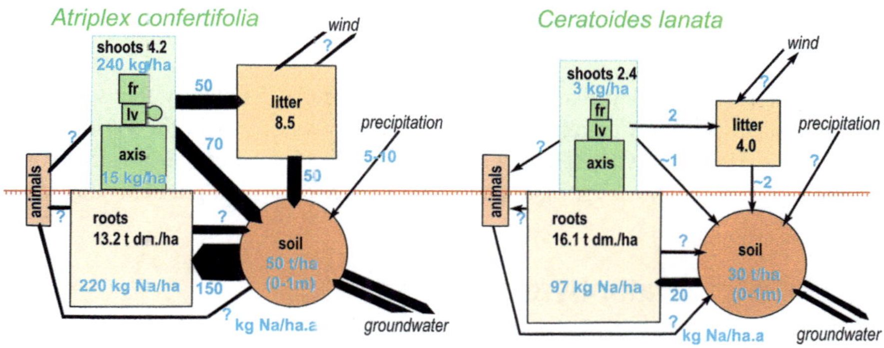

Fig. 5 Estimated sodium (Na^+) turnover (kg ha^{-1} a^{-1}) (*blue figures*) at the *Atriplex confertifolia* site (*left*) and at the *Ceratoides lanata* site (*right*) in the Curlew Valley, Utah; compartmental storage amount of sodium (kg ha^{-1}) and biomass (*black figures*) (total dry mass [tdm] ha^{-1}) (Breckle 1976)

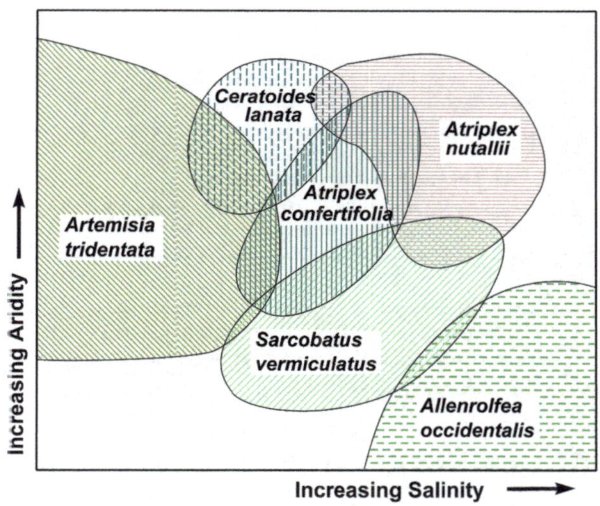

Fig. 6 Aridity/salinity ecogram of the main dominant species in the Great Salt Lake area, Utah (Breckle 1976)

university. Much new material had to be purchased; teaching started with new lectures, seminars, practical lab and field courses, and excursions.

Most of the first studies in Bielefeld had an autecological basis and focused on the effect of plants on distinct stress factors such as salt or heavy metals. Synecology, geobotany and vegetation surveys were later included in several research projects.

During the following years in Bielefeld, many projects and studies were undertaken in various countries abroad. Many candidates asked for a thesis, but we could not take all. We were able to publish some of the results in *Bielefelder Ökologische Beiträge* (BÖB), starting with volume 1 in 1985. After I became a pensioner in 2003, regular editions stopped with volume 20 in 2004, but volume 21 was published in 2017.

2 Halophytes and Salinity

Halophytic research in Afghanistan (1976) and Iran (1977) mainly focused on the question of mineral ion uptake and expression of physiotypes in the various halophytic taxa. The main types (Henkel and Shakhov 1945; Walter 1961) were defined according to salt uptake (Fig. 7), and their occurrence along salt gradients turned out

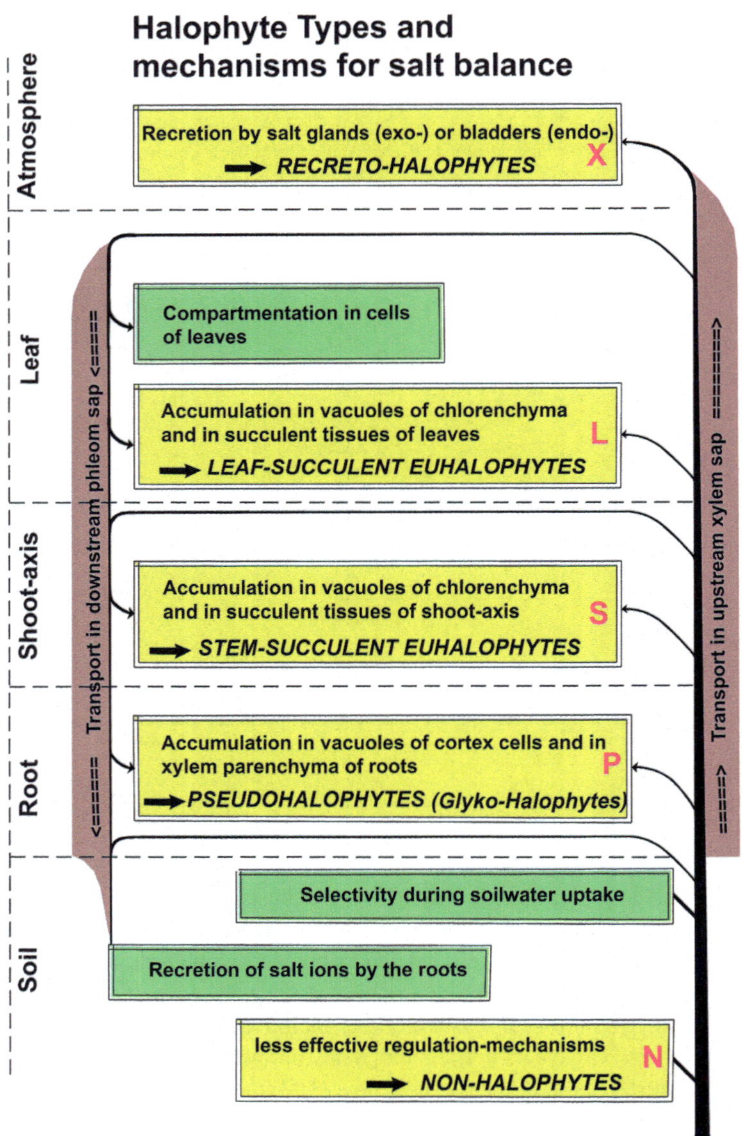

Fig. 7 Main halophyte types and applied mechanisms of salt balance in higher plants (Breckle 1976)

Halophyte-Type ← **% NaCl in soil**

15—10—5—3—2—1—0.5—0.2—0.1—0.05—0.02—0.01—<0.01

S	71	63	51	18	3	12	7	4	11	10	0	0
L	29	32	30	72	30	56	51	18	35	18	10	0
X	0	5	19	0	60	27	28	41	4	9	0	2
P	0	0	0	5	7	5	15	23	24	45	68	10
N	0	0	0	0	0	0	0	14	26	24	22	88

Abundance of halophyte types (percentages) along a salinity gradient

Fig. 8 Percentage values of calculated abundances of halophytic plant types (S, L, X, P, N; see Fig. 7) arranged according to the salinity gradient (logarithmic intervals) averaged from eight halophytic zonations in Iran and Afghanistan (Breckle 1986)

to be rather similar at many saline sites. The most salt tolerant are in general the stem-succulent halophytes (Fig. 8), often performing C_4 photosynthesis.

Already in Bonn, one of my first candidates had proved the salt tolerance of whole plants in comparison with tissue cultures in *Phaseolus* and *Suaeda* (Hedenström and Breckle 1974). The big difference that young intact plants exhibit is not mirrored in tissue cultures. To be a successful halophyte requires an intact complex network of storage space and transport systems.

Along ocean coasts, the quality of seawater is rather identical, with about 3.5% NaCl. Other ions such as potassium (K^+) or magnesium (Mg^{2+}) have only low concentrations. Coastal halophytes have thus evolved some kind of salt tolerance by different means. The dominant ecophysiological factor NaCl modifies many processes at all ecological and physiological levels (Fig. 9). It is an excellent example of how many biological processes are interwoven at the various levels of complexity (Breckle 2002a).

In arid desert regions, in contrast to saline coastal areas, the potential evaporation exceeds the water input by precipitation and thus enrichment of various soluble ions takes place (Breckle 2002b). The chemistry is more varied; the pH can be very alkaline, and sulphate (SO_4^{2-}) or bicarbonate (HCO_3^{-}) can be accumulated. Sometimes K^+ or Mg^{2+} or lithium (Li^+) is enriched. The uptake of those ions is often very specific for different species (thus physiotypes can be defined) (Albert 2005).

Studies on the ecology of halophytes were also done by several of my students over many years. The sodium-to-potassium (Na:K) ratio (Fig. 10), as well as the storage of anions and cations, showed great differences between species (Fig. 11). This has been shown for Iranian halophytes too (Matinzadeh et al. 2013).

In recreting halophytes such as *Limonium* species, one can distinguish between two processes for selecting ions: one is the discrimination of ions by the transport process from the root to the leaf, and the other from the leaves to the salt glands (Fig. 12). Obviously, other species behave rather differently (Wiehe 1986; Wiehe and Breckle 1990).

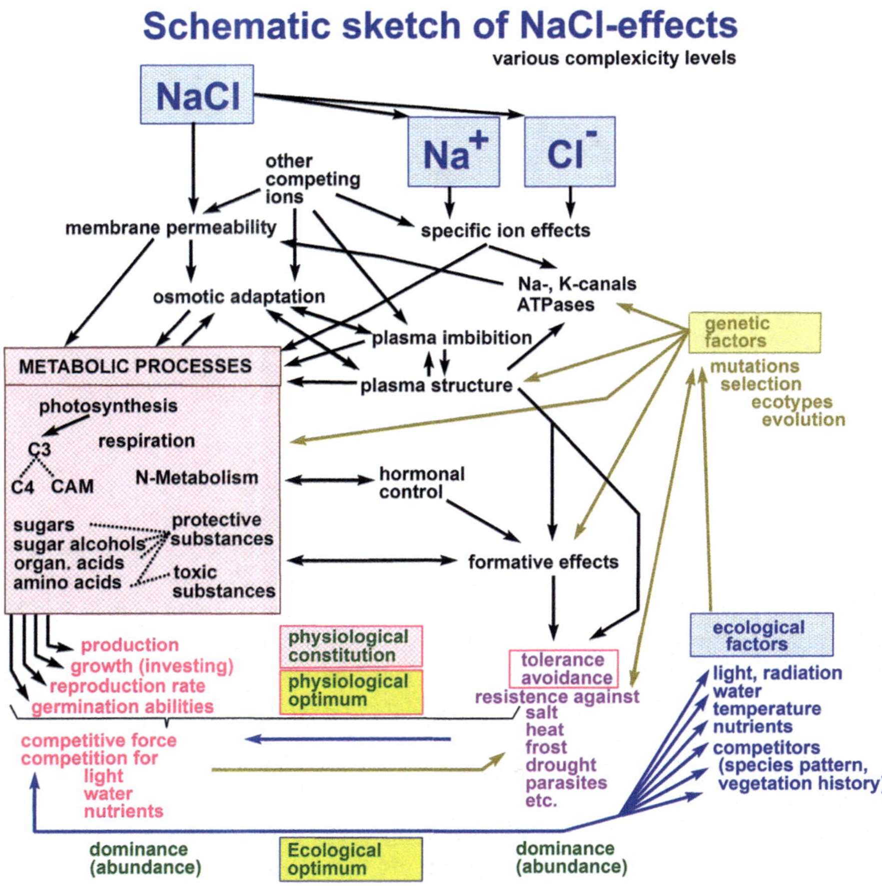

Fig. 9 Scheme of sodium chloride (NaCl) effects on plants at various complexity levels between genetic and ecological factors (Breckle 2005a)

In recreting halophytes such as *Atriplex*, only a few strongly halophytic species exhibit intensive NaCl turnover by bladders (Jones 1970; Lüttge 1971, 2016; Waisel 1972; Breckle 2002a; Osmond 1974; Reimann and Breckle 1988; Schirmer and Breckle 1982). The ratio of total bladder volume to leaf lamina volume in *Atriplex confertifolia* in very young leaves is above 2, keeping young growing tissues low in salt content; with ageing, it declines to about 0.4, and the bladders start to shrink or are blown away, thus eliminating NaCl from the leaves (Breckle 1976). In particular, *Atriplex* was studied very intensively (Tiedemann et al. 1984). Yuan et al. (2016) reported on how recretion in recreto-halophytes may occur. But salt tolerance and the genetic background in halophytes are still not really understood (Song and Wang 2015). Despite major efforts in analysing single genes responsible for salinity tolerance, some old questions (Breckle 1990, 1995) are still unresolved.

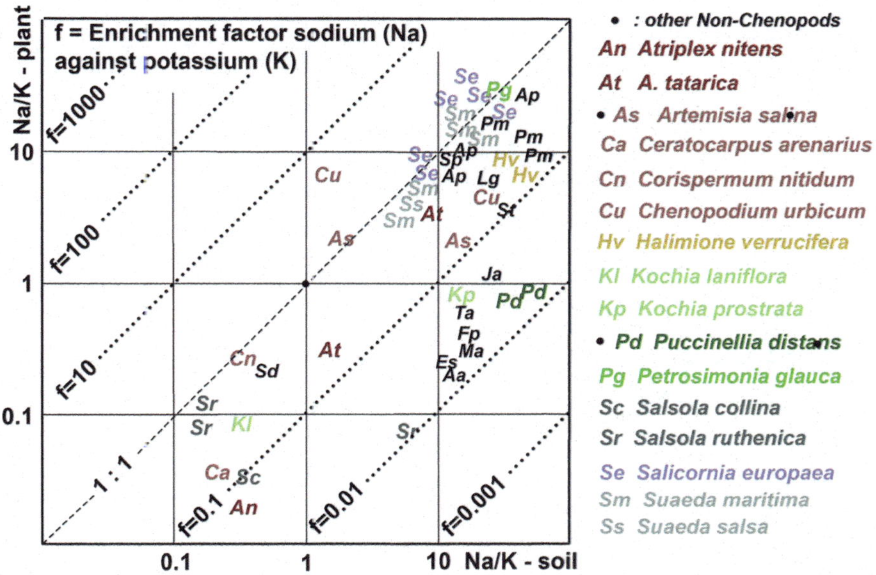

Fig. 10 The sodium-to-potassium (Na:K) ratio in soil is not mirrored in plants (Mirazai and Breckle 1978)

Roots of plants – the invisible half – have also been studied. In collaboration with Yoav Waisel from Tel Aviv [author of volumes on *Plant Roots: the Hidden Half* (Waisel et al. 1991, 1996, 2002; Eshel and Beeckman 2013)], some methods with hydroponics and aeroponics have been tested (Waisel and Breckle 1987). The question was: How do roots behave under salt stress? Large aeroponic chambers (Breckle et al. 2001a), more than 2 m high, were constructed in a new greenhouse in Bielefeld, similar to the Racine lab in Tel Aviv. Yoav Waisel stayed very often with us in Bielefeld. The sophisticated technique for keeping a constant pH and constant temperature of the nutrient solution could be managed for a period of several weeks. Tomato plants grown under different salt stress developed huge root systems in Bielefeld, as well as in Tel Aviv (Fig. 13); cutting of 90% of the roots did not have any effect on the transpiration of the upper plant parts. The architecture of root systems is very adaptive; the ratio of primary and secondary roots exhibits a strong shift to secondary roots with salt stress (Fig. 14). Field studies with mini-rhizotrons (glass tubes of 5 cm diameter) and TV cameras on roots of desert plants in the Negev Desert gave interesting results, but evaluation with automatic processing software is still not sufficiently resolved (Breckle et al. 2001c; Erz et al. 2005; Veste et al. 2005). It is a challenge for all root studies.

An interesting example of salinity effects was studied in the parasitic *Loranthus* growing on various halophytic hosts (Todt et al. 2000). In the southern Arava Valley (Negev Desert), *L. acaciae* was checked on five halophytic and ten non-halophytic host plants. Water content and succulence of mistletoes increased on halophytic

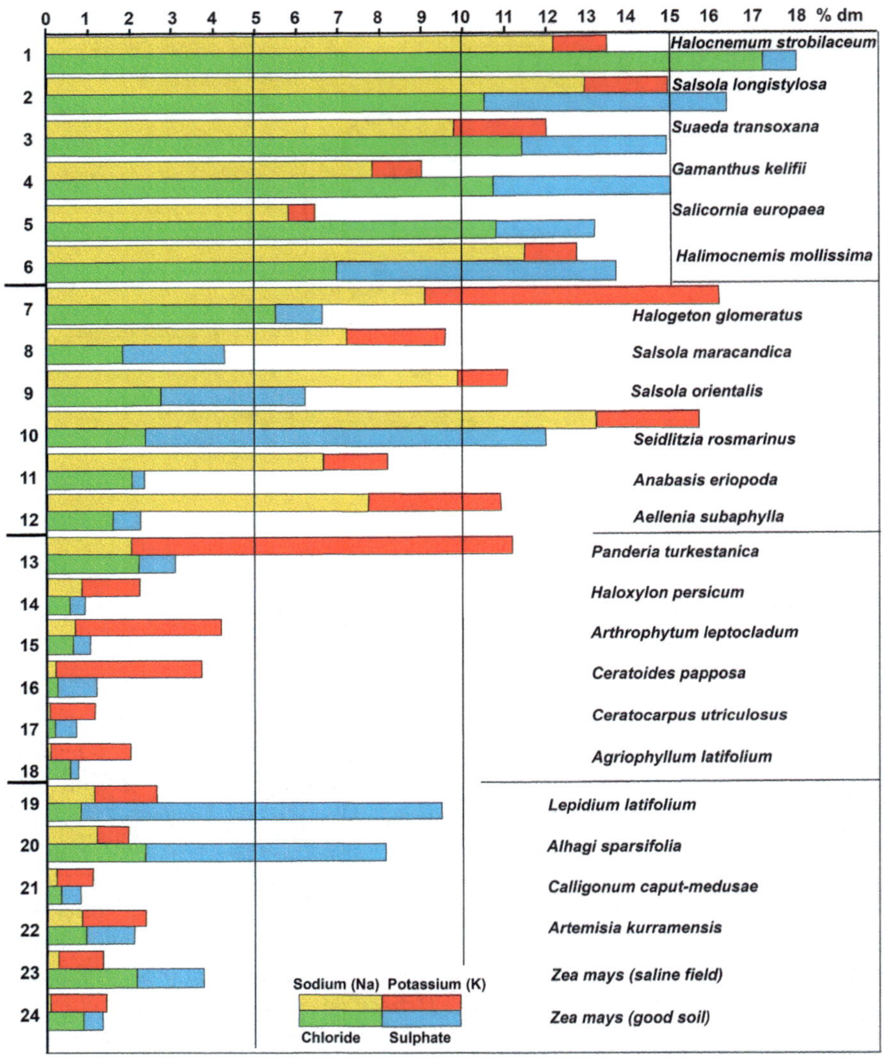

Fig. 11 Ion content in various halophytic and non-halophytic species from northern Afghanistan (after Mirazai and Breckle 1978)

hosts, and the leaf volume increased four to five times in comparison with those on non-halophytic hosts exhibiting typical halo-succulence. It behaves like a typical facultative eu-halophyte (Veste et al. 2014).

Of the world's land surface area, at least 6% is salt-affected land (Flowers and Yeo 1995; Munns 2005). Large areas in drylands and along coasts are naturally salt affected (Breckle 2002c), but with irrigation under arid climatic conditions, secondary salinization (Breckle 1989) is a big threat. Since the work of Boyko (1966), and then Waisel (1972) and Gallagher (1985), the use of seawater for agriculture has very

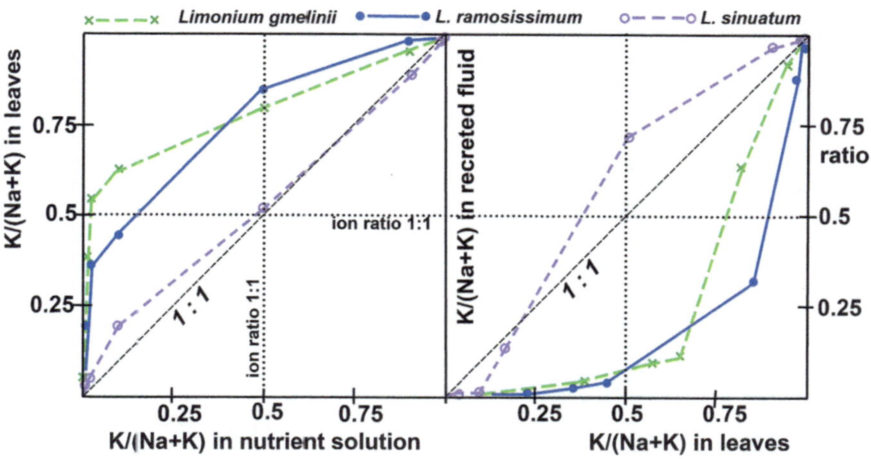

Fig. 12 Potassium-to-alkali ion ratio in leaves and in recreted fluid in three *Limonium* species under different salinities (Wiehe 1986; Wiehe and Breckle 1990)

Fig. 13 Tomato roots grown in aeroponic chambers in the root lab of the Tel Aviv Botanical Garden (photo: S. W. Breckle, 1997) (photo: S. W. Breckle)

often been discussed and studied. But despite intensive research and many projects, only a few organisms have been found that can be grown with seawater. Thus the question was asked: How can we grow plants with seawater? (Breckle 2009). Full-strength seawater can be used to sustainably irrigate only a few special plants, e.g. *Sesuvium*, *Aster*, *Salicornia* (Ventura and Sagi 2013) – maybe now that *Salicornia* salad is becoming quite popular – and *Spartina* along sandy seashores, without impairing soil quality provided that sufficient drainage exists. Intensive agriculture with seawater with high production per surface area is utopian. It was

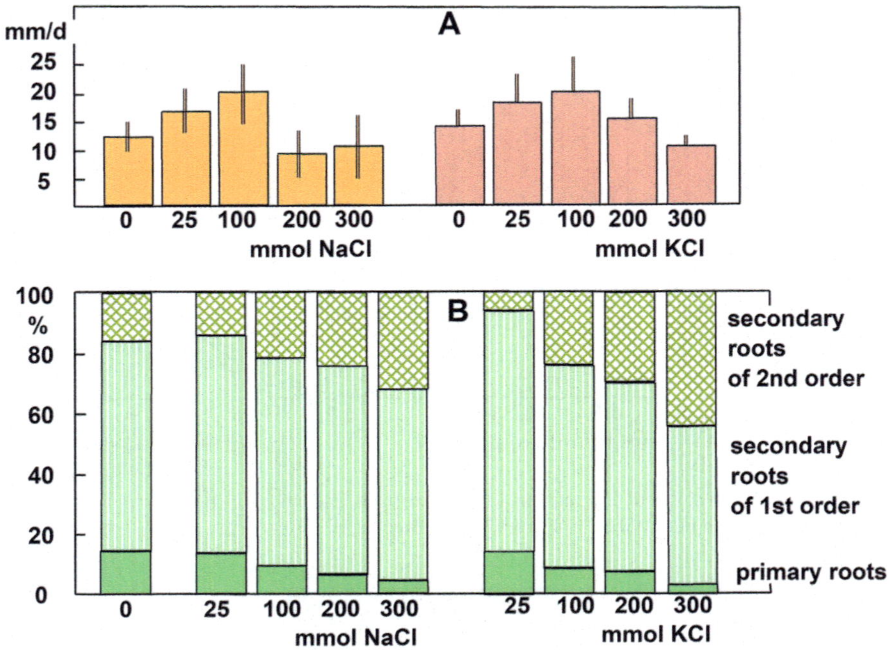

Fig. 14 Root structure of *Chenopodium album* under different salinities of sodium chloride (NaCl) and potassium chloride (KCl) treatments. (**a**) Mean daily growth of main root (mm per day). (**b**) Ratios of lengths of primary and secondary roots after 22 days (Seidel 1988; Breckle 1996b)

often said that greening of deserts, using halophytes, could partially help increase the biomass on the land surface and thus reduce global warming. But it would be much wiser to stop further deforestation in all of the humid climatic zones, especially in the tropics and in the Siberian and Canadian taiga, and to let all secondary forests grow, instead of transforming them from highly biodiverse primary forests into poor McDonald's beef sources (Breckle 2009). Additional costs to maintain sustainable irrigation and leaching systems to keep the salinity of the soil low pay more in the long run. The take-home message "no irrigation without drainage" means that it pays more to invest in desalination systems. This is especially important under very dry climatic conditions (Breckle et al. 2003) and when using urban water, as is done in southern Israel in the Yotvata farming area with brackish sewage water from Eilat.

3 Heavy Metals, Root Growth and Forest Dieback

Stress in plants is not only a result of salinity; many other factors cause stress. One is heavy metals. One of the first studies on lead (Pb) effects started in Bonn; along highways, the accumulation of Pb was checked (Lerche and Breckle 1975). When do trace metals have a measurable effect on the ecological behaviour of plants and trees? Along German highways, the Pb content in and on leaves exceeded the background content in the adjacent forests by 20–40 times. Later, we checked the accumulation of Pb, cadmium (Cd), zinc (Zn) and nickel (Ni) in the annual rings of beech and fir trees; it revealed a rather complex picture. The distribution of heavy metals in annual rings does not give a retrospective picture of the heavy metal stress over time. The physiological mobility of heavy metals within wood along radial pathways in most cases manifested in maximum concentrations of Pb and Cd along the sapwood–heartwood border and the lowest concentrations in the outermost rings, near the cambium (Hagemeyer and Breckle 1986; Hagemeyer et al. 1993; Brackhage et al. 1996). The findings were similar for potassium (K) and calcium (Ca), in contrast to Mg, Ni and Zn, for which the concentrations were lowest at the sapwood–heartwood transition. But both the distribution pattern and the concentrations of minerals in trunk wood were subject to seasonal variations.

Root growth is very sensitive to heavy metals. Reduced root growth, however, can be even overcompensated for when roots reach soil layers with low concentrations (control soil) (Weisser et al. 1990, Fig. 15). There, artificially enriched horizons in root boxes turned out to be very useful. Later also mini-rhizotrons (see Sect. 5.1) were found to be very suitable for experimental sets.

In many papers (Breckle and Kahle 1992; Kahle and Breckle 1992; Breckle 1996a, b), the effects of Pb and Cd on beech were shown. In the early 1990s, there

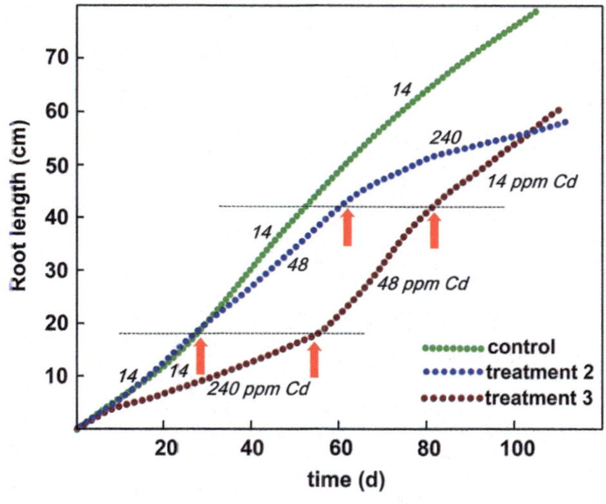

Fig. 15 Root length growth of beech seedlings in root chambers within 3 months. The root chambers were filled with three horizons of soil, differing in heavy metal content: *green*, all three horizons with low cadmium (Cd) (14 ppm); *blue*, increasing Cd content, lower 14, middle 48, upper horizon 240 ppm Cd; *brown*, decreasing Cd content, lower 240, middle 48, upper 14 ppm Cd. *Red arrows* indicate horizon limits (Weisser et al. 1990; Breckle and Kahle 1992; Hagemeyer and Breckle 1996)

was much research on forest dieback, especially in spruce. We started with beech. Later it became clear that deciduous trees also exhibited distinct signs of stress, with lowered soil pH, soluble aluminium (Al) in soil, and ozone (O_3) on leaves. With young beech saplings, it could be shown that in a rather short time of Cd stress, transpiration rates were much reduced (Hagemeyer et al. 1986, Fig. 16).

Only slight synergistic effects of Pb and Cd in beech saplings could be detected (Bertels et al. 1989). Changes in uptake of essential elements were, however, obvious but strongly dependent on soil pH (Kahle et al. 1989a, b). Mg and Ca uptake was antagonistic to Cd uptake, but this was not so with Pb. In general, it became obvious that the concentrations of Pb in exposed acid forest soils in Central Europe were sufficiently high to affect the germination and growth of saplings of beech. The "critical concentrations" could be defined as about 2 ppm of Cd and 25 ppm of Pb (1 N NH_4-acetate extractable fraction, pH 7). The reduction of wood formation seen with heavy metal contamination was distinct in some places of North Rhine–Westphalia (Lammersdorf, Stolberg), causing economic losses (Breckle and Hagemeyer 1992).

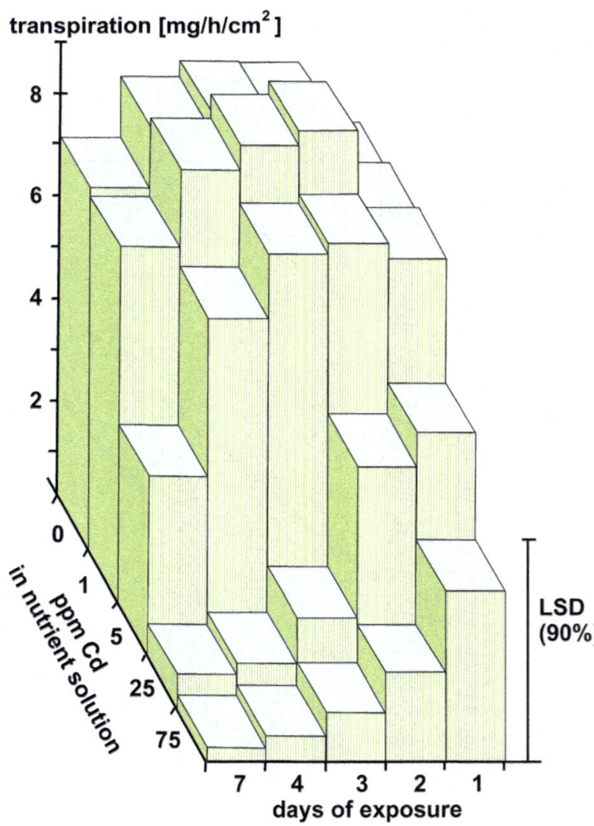

Fig. 16 Rapid effect of cadmium (Cd) stress on transpiration rates (mg h^{-1} cm^{-2}) of 2-year-old beech saplings (*Fagus sylvatica*) cultivated in nutrient solution, with indication of the least significant difference (LSD) of 90% (Hagemeyer et al. 1986)

Fig. 17 *Viola guestphalica* (the *blue* calamine *violet*) – the only endemic plant of Westphalia – at the Bleikuhle near Blankenrode (photo: S. W. Breckle)

Metallophytes – plants that are able to withstand high levels of heavy metals in the soil – often accumulate heavy metals in the plant body (Breckle 1997b). They are often very specific and very rare (Fig. 17). As with the idea of desalinating soils by halophytes (Breckle 2009), quite often one finds papers discussing the idea of decontaminating soils containing heavy metals. By simply calculating realistic turnover rates, we often disillusion people, since the desalination times are about 50–500 years; the times required for decontamination of heavy metals (Pb, Cd and Zn) are in the same range. Another purpose would be to use plants for mining (e.g. Ni mining by *Alyssum*) but then again the labour costs would make it inefficient. In the future, maybe better methods can be developed.

Later, during studies in the tropics (see Sect. 6), some heavy metal tests were carried out. We wanted to know in what concentrations essential trace metals are present. As one example, the manganese (Mn) content in tree fern fronds in a montane rainforest in Costa Rica was studied (Weber and Breckle 1994). Volcanic soils are rich in Mn; at the end of the rainy season (December–January), the content is lower than during the dry season (February–March) or at the end of the dry season (end of March) (Fig. 18). Additionally, it is remarkable how the Mn content differs between species.

4 High Mountains and Altitudinal Stress

In the summer of 2002, an expedition to the Pamir area was organized. After 34 years, a view to Afghanistan across the Tajik–Afghan border at the Abe Panj, the upper tributary of the Amu Darya, was again possible (Fig. 19a). This expedition – in the footsteps of my close friend Okmir Agakhanjanz (who had died a few months earlier) – was the first German expedition since the 1935 German Pamir expedition,

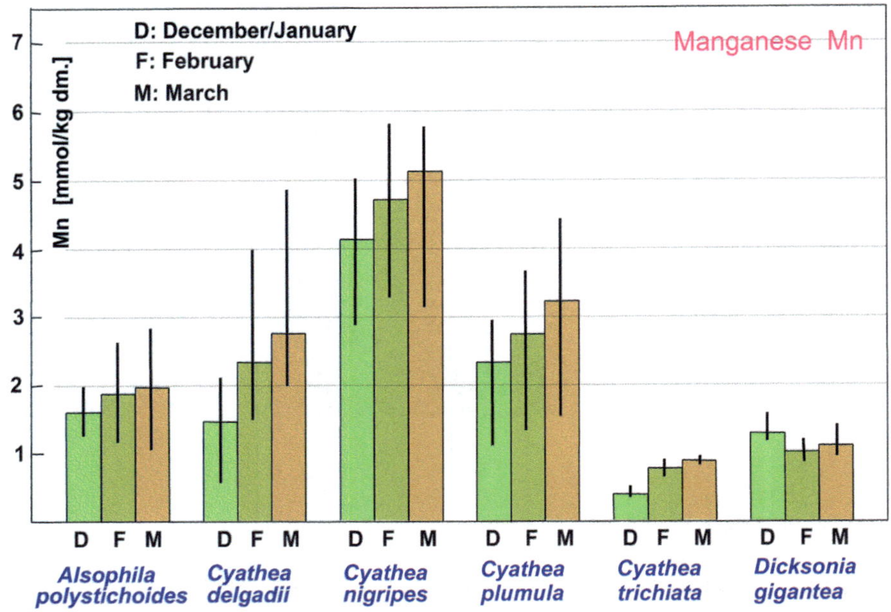

Fig. 18 Manganese (Mn) content (mmol kg^{-1} dm) in six tree fern fronds in the montane tropical forest near Rio Lorenzito in the Sierra de Tilarán, Costa Rica, during the wet season (December) and dry season (March) (Weber 1994; Weber and Breckle 1994)

along the upper Abe Panj and Pamir River, and reaching Zorkul (at an altitude of 4,200 m), the lake being one of the sources of the Amu Darya. The contrasts between western and eastern Pamir in terms of landscape morphology, climate and vegetation were striking. Also the comparison of old vegetation maps (Vanselow et al. 2016) and the present-day situation showed strong desertification.

The expedition to the high mountain deserts of eastern Pamir enabled a survey and comparison of mountain ecological conditions between eastern and western Pamir. This was the last research trip of Clas Naumann, during which he finally clarified the biology of the moth *Zygaena pamira* (Fig. 19b, c); at 4,250 m the eggs were found developing on the lower side of yak cow dung, close to our campsite.

An almost symbolic plant of eastern Pamir is *teresken* (*Ceratoides papposa*, Fig. 19d). This species has an enormous geographical distribution (from the Ebro basin in Spain, the Vienna area and Pannonia to Mongolia and western China) in deserts and semi-deserts. *Teresken* has been the main energy source for the people since the break-up of the Soviet Union and the civil wars in Tajikistan, and is also the main fodder for sheep and goats, for donkeys and camels, and for the famous Marco Polo sheep (*Ovis ammon polii*). But, interestingly, *teresken* is a rather slow-growing dwarf shrub with a shoot-to-root ratio of up to 1:40, thus it is very drought resistant. For fuel, the woody root is used. Old plants have a thick rootstock and can reach 130 years of age (Walter and Breckle 1994).

Fig. 19 (**a**) The Pamir River, Tajikistan, with meandering parts and natural gallery forests, joining the Wakhan River in the background (Afghanistan) (photo courtesy of C. Naumann, 2002). (**b**) *Zygaena pamira* on *Acantholimon diapensioides* at Turumtaikul (4,200 m), Pamir (photo courtesy of C. Naumann, 2002). (**c**) *Zygaena pamira* egg layings on the lower side of yak cow dung at Turumtaikul, Pamir (photo courtesy of C. Naumann, 2002). (**d**) *Krascheninnikovia ceratoides* (*teresken*), a middle-aged dwarf shrub, with thick woody rootstock, from eastern Pamir (photo: S. W. Breckle, 2002)

The "*teresken* syndrome" is one of the major problems in the eastern Pamir region and urgently needs a solution through replacement of *teresken* with other energy sources (projects from non-governmental organizations [NGOs] have started). A very similar and astonishingly parallel problem is the "tola syndrome" in the dry altiplano of Bolivia, where, again, people depend on almost only one life form. Again, slow-growing dwarf shrubs of the composite *Parastrephia* are the main source of fuel and fodder for lamas, alpacas and also the wild camelid vicuñas (Breckle and Wucherer 2006; Ahmadov et al. 2006).

Flora of high mountains is derived from floristic ancestors on the plains and in adjacent lower hill areas. In the Central Asian mountains, the migration of species, evolution of new species and extinction of species are leading to new mountain flora (Agakhanjanz and Breckle 1995). The character and speed of orogenesis, the climatic situation, the stages of forest belts and, finally, glaciation are the governing factors. A comparative survey of the phyto-diversity and known species numbers in

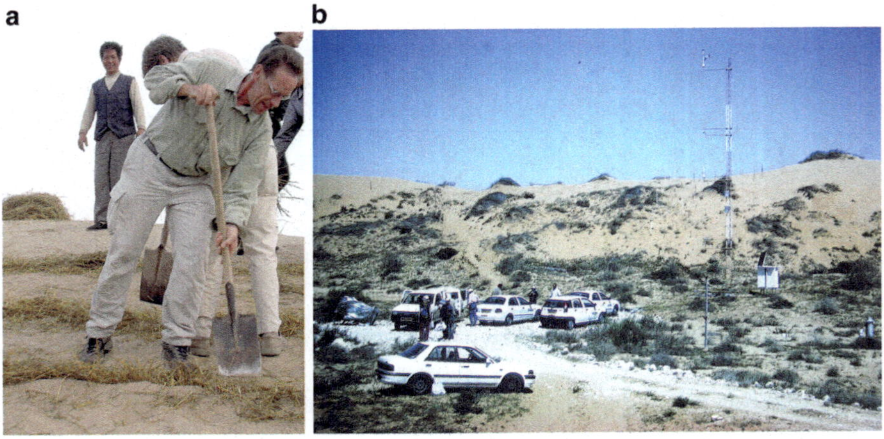

Fig. 20 (**a**) Combating desertification in Inner Mongolia. By a checker-straw technique, sand movement of mobile dunes is retarded, and diaspores of pioneer plants are able to invade within the straw squares (photo courtesy of C. Wissel, September 2004). (**b**) Nizzana sand dune area with longitudinal dunes (western Negev Desert); meeting of project groups (photo: S. W. Breckle, 1998)

the Eurasian mountains clearly indicates the need for small-scale inventories, especially for the orophytic belt (above the timberline). There endemics reach a high proportion in the subalpine and alpine belts but are almost absent in the subnival and nival belts, where some widespread boreal and even arctic species are present (Agachanjanz and Breckle 2002; Breckle 2004). In contrast, in arid mountains lacking a forest belt, endemics are more common in the lower belts. Concerning the Hindu Kush, the main Afghan mountain range, see Sect. 7.2.

5 Desert Ecology

Stress ecology is certainly very relevant in deserts. Heat stress and drought stress are the prominent ecological factors that organisms have to be adapted to or avoid. All of those questions of desert research have also been part of desertification studies. Only a few examples can be mentioned here. Desertification on a large scale has also been seen in several parts of China (Veste et al. 2006; Gao et al. 2007), with a severe dust situation in Beijing. Large-scale afforestation projects with millions of trees being planted in the loess area and sand-binding actions (Fig. 20a) are under way.

In 1999, a competent network of German scientific institutions was founded in Bielefeld (Desert☼Net) for exchange and planning of innovative joint projects. Close contacts with relevant political institutions were built up. Especially with the Secretary of the United Nations Convention to Combat Desertification (UNCCD) in Bonn, fruitful discussions took place, as well as participation at the Expo 2000 in Hanover. Participating in political discussions on the means necessary for combating desertification was, however, often rather disappointing, e.g. at the Fourth and Fifth Conferences of the Parties to the Convention to Combat Desertification (CCD-COP4 [held in Bonn] and CCD-COP5 [held in Geneva]). So, such political activities lasted only a few years. Then the European Desert☼Net was founded elsewhere, but it seems now just to be one among many other not especially creative or innovative organizations. Scientific desert ecology is more attractive and an important background for applications.

5.1 Negev (Nizzana Sand Dunes)

How do plants survive in deserts? Water availability is reduced in all deserts, but there are many different types of deserts, and generalizations are often misleading. Several projects concerning desert ecology were undertaken. The Namib was studied only randomly [three diploma theses (Veste and Mohr 2005)].

The Atacama had been visited during a 3-month stay in Chile, including study of several *salares* (salt flats) and halophytic sites, and checking of salinity gradients; however, before the return to Germany, all material, notes and samples – about 30 coloured films from all vegetation types between Arica and Chiloe – were lost after being stolen in Santiago just in front of the university library. It happens.

Certainly it was not as tragic as the loss of Alfred Russel Wallace's collection. He was in Amazonia, charting the Rio Negro for 4 years, collecting specimens and making notes on the peoples and languages he encountered, as well as the geography, flora and fauna. The ship's cargo caught fire, and the crew was forced to abandon the ship in August 1852 on the Atlantic Ocean. All of the specimens Wallace had on the ship, mostly collected during the last two – and most interesting – years of his trip, were lost. He managed to save a few notes and pencil sketches, and little else (Raby 2002, note by Lüttge).

The special longitudinal sand dune system of the western Negev, near Nizzana, was studied by a large joint project (Fig. 20b). The area is characterized by an arid typical desert climate with rather strong variability from year to year (Fig. 21). A synthesis of all studied topics was given by Breckle et al. (2008). Those longitudinal dunes were an ideal study system. The very regular ecotopic mosaic of dunes is characterized by the formation of a microbiotic crust (Fig. 22a) (Veste and Breckle 2003; Littmann and Veste 2005; Veste et al. 2011) despite the very arid climate. Within a few years, even on mobile sand (totally protected by the military close to the Egyptian border), it has a rather thick, strong appearance. This organic crust of cyanobacteria, lichens and mosses (Fig. 22b) greatly reduces the infiltration of the scarce rains in winter (Fig. 21) and results in a new pattern of water availability and a vegetation mosaic on formerly mobile sand dunes. This mosaic strongly influences the establishment of seed plants and is also a source of atmospheric nitrogen binding (Russow et al. 2008). The thickness and chlorophyll content along the gradient are shown in Fig. 23. Crust development strongly depends on the water regime, but dew and fog often are the main water sources. Here, just 20–25 fogs per year is one water

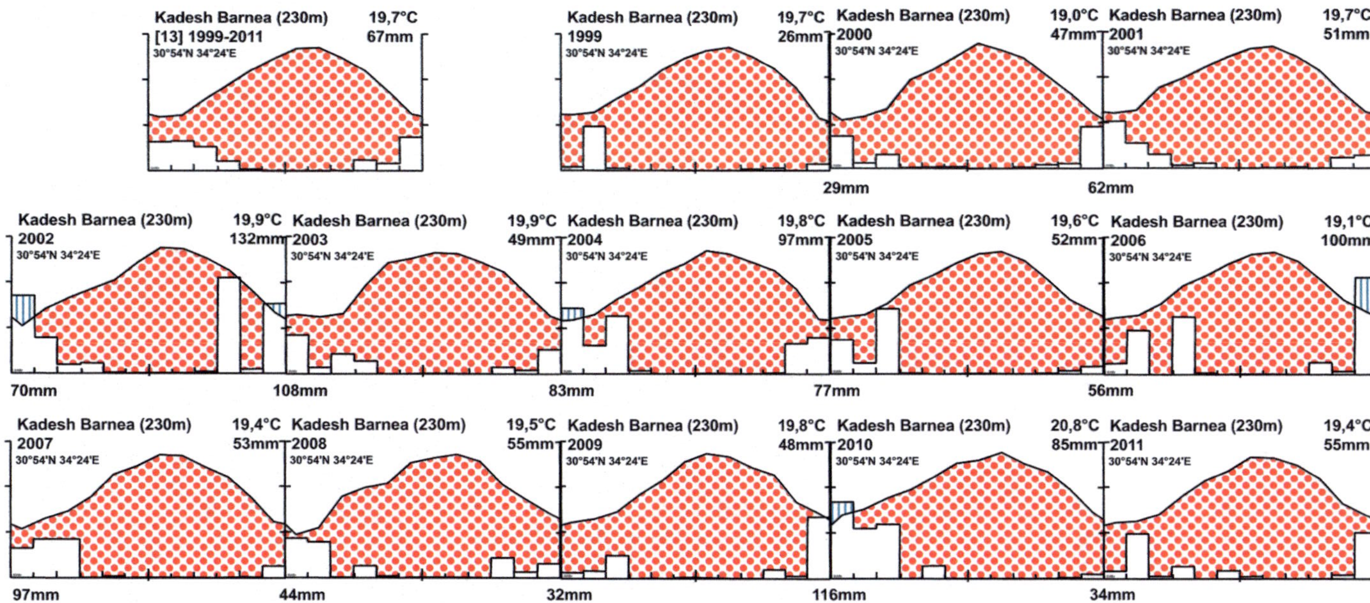

Fig. 21 Ecological climate diagram (*upper left corner*) indicating desert conditions at the Nizzana test site, and climatograms (indicating the conditions of 1 year), thus exhibiting the variability from year to year (during the period from 1999 to 2011). Data from the Barnea Kadesh meteorological station at Nizzana (courtesy of S. Berkowicz)

Fig. 22 (**a**) Biotic crusts on mobile sand dunes start with a cyanobacteria cover; this early stage can easily be destroyed by trampling (photo: S. W. Breckle, 1997). (**b**) Close-up view of the well-developed old biological crust on mobile sand dunes, Nizzana, Negev Desert, with cyanobacteria, lichens and mosses (photo: S. W. Breckle, 1998). (**c**) Large lysimeters at the Yotvata test site (southern Negev) with sophisticated nutrient solution supplies (photo: S. W. Breckle, 1996). (**d**) Rhizotron tubes around desert dwarf shrubs in the Nizzana sand dune area, suitable for a small television camera for semiautomatic picture processing (photo: S. W. Breckle, 1994)

source, and dew in the morning is the other – although it provides only about 0.06–0.14 mm of water input per night (Veste et al. 2008) – which could add up to about 26–33 m per year.

In many sand deserts, those biotic crusts are formed, but with strong winds, and even more with grazing and trampling, a closed thick crust normally cannot develop. On rocks, biotic crusts are present everywhere worldwide.

The interdisciplinary approach to the study of this highly complex and dynamic ecosystem mosaic revealed many astonishing details of the processes that form such a dune mosaic. The occurrence of surface run-off on sand dunes is not irrelevant, as is often said, and is seen on mobile sands. The accumulation of a thin cover of fine-grained particles and the existence of the widespread biological topsoil crust change

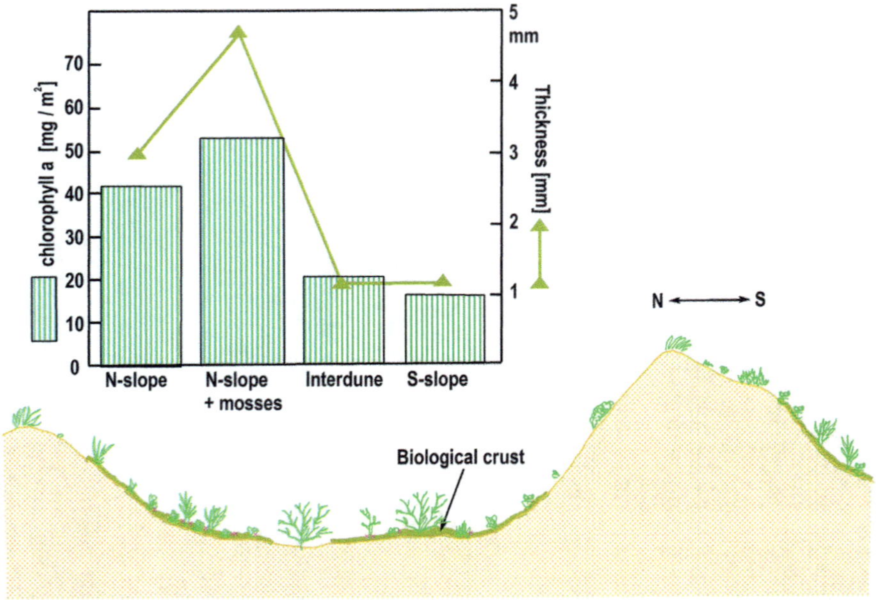

Fig. 23 Profile across dune slopes of the longitudinal dune system of Nizzana (Negev Desert), with formation of a biological crust, their thickness and their chlorophyll content (mg m^{-2}) (Veste et al. 2011)

the whole run-off and erosion processes and thus also the mosaic of vegetation cover (Kidron and Yair 2008; Yair 2008). These are long-term processes and warrant long-term observations. They are, of course, totally dependent on the various human impacts (former land use, grazing, collection, wood cutting, trampling, off-road traffic) and thus the special sand dune area at Nizzana was very suitable for study because it had been protected by the military. The contrast of the different land use systems is seen in satellite images, where the Israeli–Egyptian border exhibits a straight line of differing albedo. On the Egyptian side, the number of perennial shrubs was reduced from about 400 per km^2 (in 1945) to about 100 (in 1989), whereas on the Israeli side, with fluctuations, it increased to more than 600, making the sand dune area spotty green (Tsoar 2008). Another contrasting pattern is given by the steep gradient of aridity from south to north, from about 100 mm to less than 50 mm within a 50 km distance, but local rainfalls are very irregular. These sand dune mosaics are different in the north, where dune movements of the dune crests are lacking, in contrast to the south where more specially adapted sand dune plants are found. Many experiments could be performed to see the effects and recovery, as well as intensive measurements of the main parameters of desert plant life (Breckle et al. 2008), including root growth, by regular checking of rhizotron tubes around dwarf shrubs (Fig. 22c). Small television cameras provided pictures, which were processed by automatic image analysis (Jankowski et al. 1995; Breckle et al. 2001c; Erz et al. 2005).

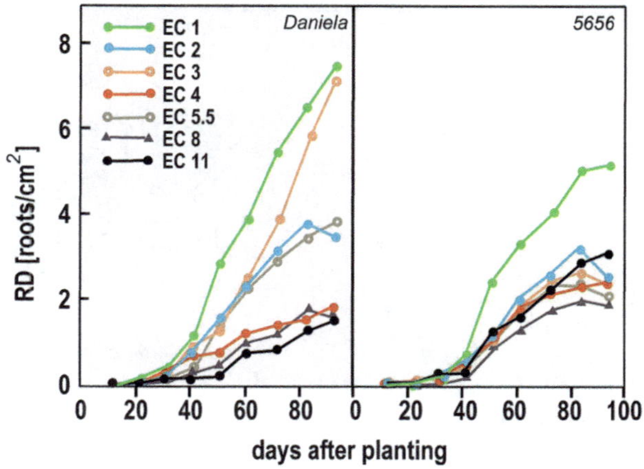

Fig. 24 Root density of tomato plants (breed Daniela and 5656) during 100 days of cultivation under salinity stress (electrical conductivity 1–11 mS) (Veste et al. 2005)

In the southern Negev (Kibbutz Yotvata), the large experimental fields and infrastructure, as well as large lysimeters (Fig. 22d), were used for several experiments to check water consumption under strongly arid and hot conditions. This project, titled "Optimization of Irrigation and Water Use under Desert Conditions", also checked root system behaviour, especially in tomatoes. The root density of tomato plants under irrigation with saline water and grown in large rhizotrons is shown in Fig. 24. Salt reduced the root growth rates, as well as below- and aboveground biomass, in both cultivars (Veste et al. 2005).

5.2 *Aralkum (The Desiccated Sea Floor)*

At the start of the long-term project titled "Aralkum", there were two main questions. First, what is the present situation on the desiccated sea floor, what flora and vegetation can be found, and what is starting to develop? Second, is there a chance to intervene in the process of primary succession to accelerate the vegetation cover in order to minimize salt and dust storms? The latter was a very important applied approach and was raised by the Federal Ministry of Education and Research (BMBF).

In 1960, the Aral Sea started to disappear because of an unequal hydrological equilibrium of the water balance of this endorheic system, caused by drastically enlarged irrigation areas to produce more than 7 million tons of cotton in Uzbekistan alone. Having been the world's fourth largest lake, today the Aral Sea no longer exists (Fig. 25). The whole larger area around the former Aral Sea became a catastrophic area with a strong social crisis (Fig. 26). Only slowly within the last decade have there been improvements in Kazakhstan by the construction of a dam, which, since 2005, has partly restored the northern Aral Sea. Still in dispute is how the situation can be improved (Breckle and Wucherer 2011; Breckle et al. 2012).

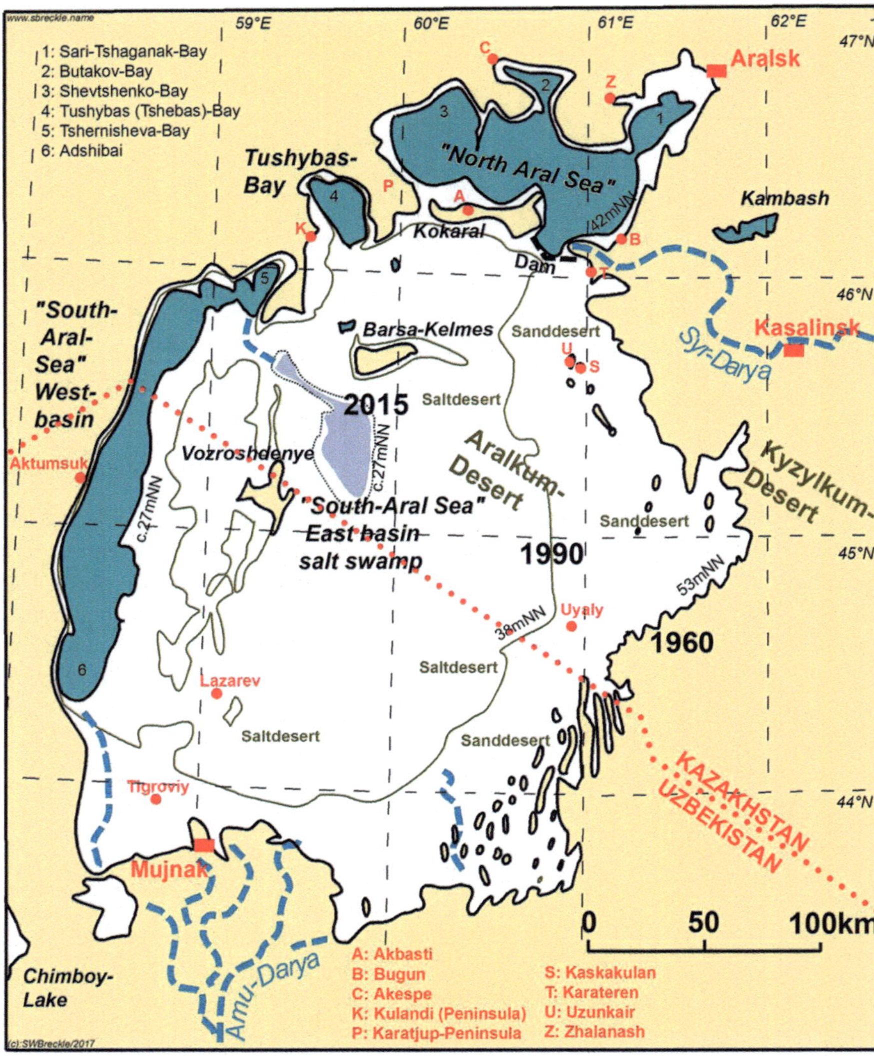

Fig. 25 Aral Sea area: geographical situation and extent of the Aralkum from 1960 to 1990 and 2015

A long-term study on the desiccated sea floor – the Aralkum – as a new desert (Fig. 25) revealed the primary succession (Breckle 2013; Breckle et al. 2012; Wucherer and Breckle 2001, 2012; Groeneveld et al. 2012). This succession often is called mankind's biggest succession experiment. The large new flat area – the desiccated sea floor – was invaded from the surroundings by diaspores. Depending on the substrate (older areas are sandy; younger areas are clayey and more saline), the succession was very dynamic, with species composition greatly varying from year to year. A general scheme for loamy soils is given in Fig. 27, indicating the

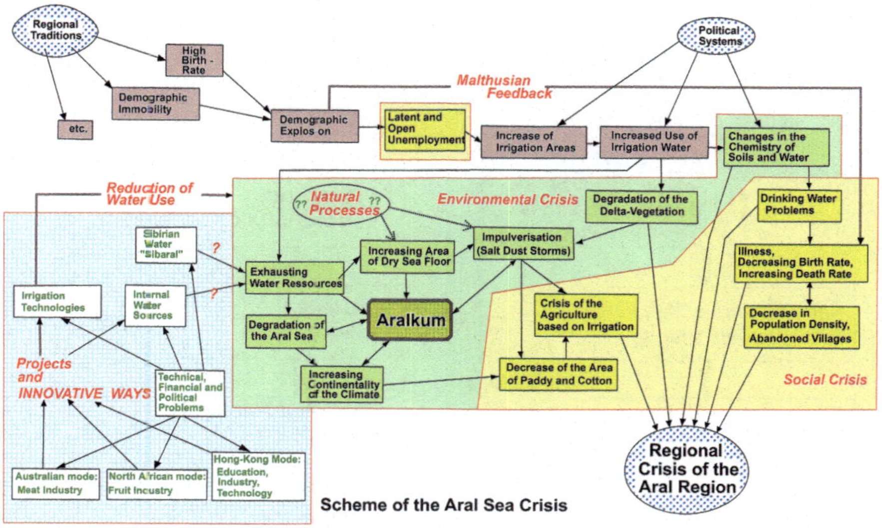

Fig. 26 Scheme of the Aral Sea crisis: environmental and social aspects, and possible future projects (after Breckle et al. 1998)

dynamics between several associations and bare salt desert patches. For the older sandy sea floor, the succession differs considerably because of the better water availability on sand. Small sand dunes can form with various *Calligonum* species but lately also with *Haloxylon aphyllum*.

The desiccation history is not easy to reveal. The changing shoreline from year to year left seeds and saplings, mainly of fast-germinating *Tamarix* species. After a few years, the former shorelines could be seen easily from a distance (Fig. 28a, b). At four transects, Breckle et al. (2014a) did a retrospective dating of the vegetation belts. Also, in the northern Aral Sea basin, the restart of inundation – as a result of the new dam – was visible.

Breckle et al. (2012) brought together the results of international and interdisciplinary studies on the new desert ecosystem. The physical characteristics of the area, mainly geological and climatological aspects and their dynamics, especially the dust storm dynamics, were discussed. There was a strong need to find ways to minimize the destructive salt–dust storms in the area. Phyto-melioration experiments (Breckle 2003) on 250 ha of sea floor with the help of Kazak villagers (financed by the BMBF and the German Society for Technical Cooperation [GTZ]) showed the best potential of *Haloxylon aphyllum* (saxaul) to grow and reproduce under these extreme conditions of continentality (−40°C in winter, +43°C in summer) and salinity. Under natural conditions, *Tamarix* species were the main invaders along the desiccation belts (Fig. 28a) but were not successful in planting experiments; they were always totally destroyed by rodents.

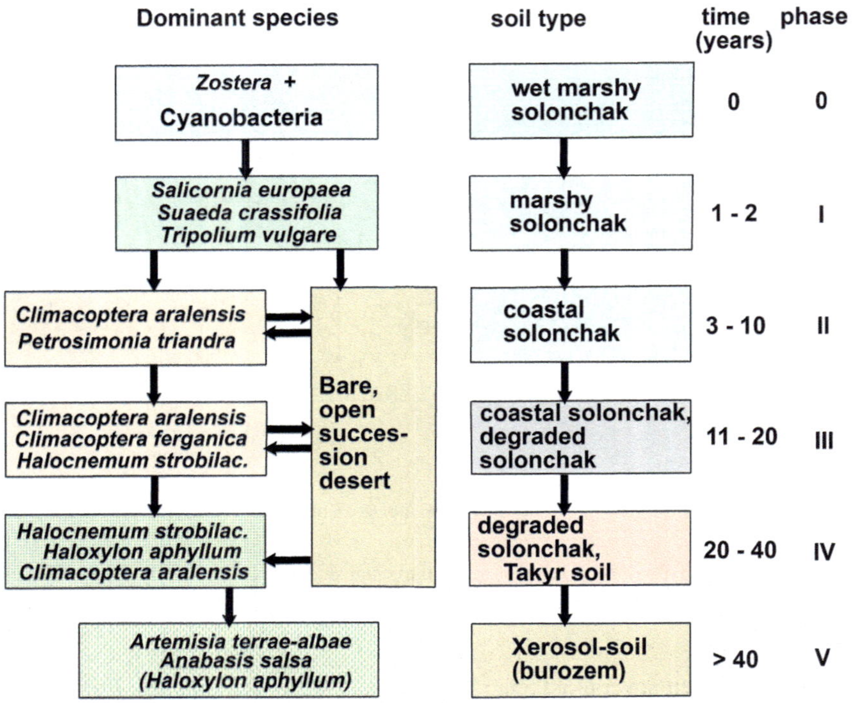

Fig. 27 Primary succession on loamy and clayey soils during the desiccation of the Aral Sea floor (modified from Breckle 2002b; Breckle and Wucherer 2012)

A distinct proportion of all 373 vascular plant species found on the desiccated sea floor are, so far, more or less halophytic. Table 1 gives an overview of the halophyte types and their salinity indicator value *S* (with *S* values from 1 to 9) (Breckle 1985; Ellenberg et al. 1991). Leaf-succulent halophytes play a major role, but non-halophytes and pseudo-halophytes often conquer where the upper soil is already desalinated and sand and dust have been accumulated; thus geomorphological processes are strongly linked with vegetation development.

Several other projects in Uzbekistan used various but similar methods. In Kazakhstan, a huge area was then declared a biosphere reserve (Breckle 2011a, b, 2013), enlarging the former Barsa-Kelmes Island area, partly according to our earlier proposals, parallel to the now open space as far as Kazka Kulan on the former east coast, where there are now herds of onagers (wild asses) and gazelles, often close to the warm springs.

a

b

Fig. 28 (**a**) View from the northern chinks [the steep terraces; see also (**b**)] to the northern Aral Sea basin at the desiccated Chevshenko Bay, with distinct *Tamarix* belts, indicating the desiccation

Table 1 Number of vascular plant species that have spontaneously invaded the Aralkum, their halophytic strategy types and related salinity indicator values (*S*) for the flora of the Aralkum (Breckle and Wucherer 2012)

Halophytic strategy type	$S = 1$	2	3	4	5	6	7	8	9	X	$\sum$
Non-halophytes	42	18	0	0	0	0	0	0	0	0	60
Pseudo-halophytes	4	30	42	15	8	4	1	0	0	4	108
Xero-succulents	0	0	0	0	0	0	0	0	0	1	1
Leaf-succulent eu-halophytes	0	2	2	5	14	7	22	9	1	0	62
Stem-succulent eu-halophytes	0	0	0	1	1	1	1	0	2	0	6
Endocrino-halophytes	0	0	2	5	1	2	0	0	0	0	10
Exocrino-halophytes	0	0	0	0	0	9	7	4	0	0	20
Hydro-halophytes	0	0	0	1	0	0	0	0	1	0	2
$\sum$	46	50	46	27	24	23	31	13	4	5	269
Undetermined strategy type											104
$\sum\sum$											373

6 Tropical Ecology, Canopy Structure and Growth Dynamics

6.1 Costa Rica

Similar to deserts but very different in appearance, untouched tropical rainforests certainly pose a wealth of open ecological questions. A small research station in the lower montane rainforest on the Sierra de Tilarán at the Rio Lorencito in Costa Rica, as emphasized by Stefan Vogel [in Vienna, formerly in Mainz], was a good starting point for studies on the following questions. How is the high biodiversity in tropical forests maintained? Which habitat structure do we have, and what are its dynamics?

Later, also jointly with other German groups, we used the San Francisco station in the montane rainforest in southern Ecuador, though there were fewer virgin forests than in Costa Rica. We first wanted to know the basic stand structure, the dynamics of the main trees and their growth behaviour. With 1 ha inventories and a large number of dendrometer bands, important basic data on the forest ecosystem could be collected (Homeier 2004; Homeier et al. 2002a, b, 2008; Homeier and Breckle 2004, 2008). Later, members of the department also worked in Kenya (Dalitz et al. 2005, 2011; Gliniars et al. 2011; Table 2).

Fig. 28 (continued) process (photo: S. W. Breckle, 2003). (**b**) The steep chinks provide an excellent view of the remnant of the northern Aral Sea (photo courtesy of W. Wucherer, 24 May 2003)

Table 2 Examples of research and teaching topics at the Department of Ecology at Bielefeld (Breckle 2005a)

Autecology etc.
Halophyte ecology (Breckle, Reimann, Freitas, Sanchez, Herrera, Waisel et al.)
Heavy metal stress (Breckle, Engenhart, Nabais, Pieczonka, Kahle, Dalitz et al.)
High mountain ecology (Breckle, Agakhanjanz)
Dendro-ecology and dendro-analysis (Hagemeyer et al.)
Root dynamics, aeroponics (Anlauf, Breckle, Waisel et al.)
Primary successions (Breckle)
Photosynthesis, biotic crusts (Veste, Breckle et al.)
Tropical ecology and global ecology
Costa Rica (Breckle, Dalitz, Homeier, Wattenberg et al.)
Ecuador (Breckle, Dalitz, Homeier, Oesker et al.)
Kenya (Dalitz, Todt, Uster et al.)
Vegetation and zono-biomes of the globe (Breckle et al.)
Afghanistan flora, ecology, vegetation (Breckle, Hedge, Rafiqpoor, Keusgen et al.)
Desert ecology
Central Asia: Aralkum project (Wucherer, Breckle et al.)
Inner Mongolia: combating desertification (Breckle, Veste et al.)
Sand dune ecosystems in the Negev: Nizzana project (Breckle, Veste, Yair et al.)
Namib (Breckle, Veste et al.), Atacama
Eastern Pamir and the *teresken* syndrome (Breckle, Wucherer, Trux)
Succulents, Southern Africa, Richtersveld (Veste)
Vegetation and landscape ecology
Riverine vegetation (Stockey et al.)
Nature protection (Breckle et al.)
Applied ecology
Nature sports, e.g. orienteering, ice skating (Breckle)
Desertification (Breckle, Veste, Wucherer et al.)

Gap dynamics play an important role in the future stand structure (Fig. 29). Gaps are the result of natural disturbances in forests, ranging from large branches breaking off, to a tree dying then falling over and bringing its roots to the surface of the ground, to landslides bringing down groups of trees. In the premontane tropical rainforests of the Sierra de Tilarán, the slopes are very steep; thus tree falls and small landslides are common; the same is also true in Ecuador (see Fig. 30b). Because of the range of causes, gaps have a wide range of sizes, including small and large gaps. Gaps allow an increase in light as well as changes in moisture and wind levels, leading to differences in microclimate conditions compared with those below the closed canopy, which are generally cooler and more shaded. Thus, gaps are more or less a special site with altered conditions (Fig. 29); these smaller or bigger islands can be the reason for a dynamic mosaic canopy structure. Gaps provide the ideal location and conditions for rapid plant reproduction and growth (Fig. 39) (Wattenberg 1996).

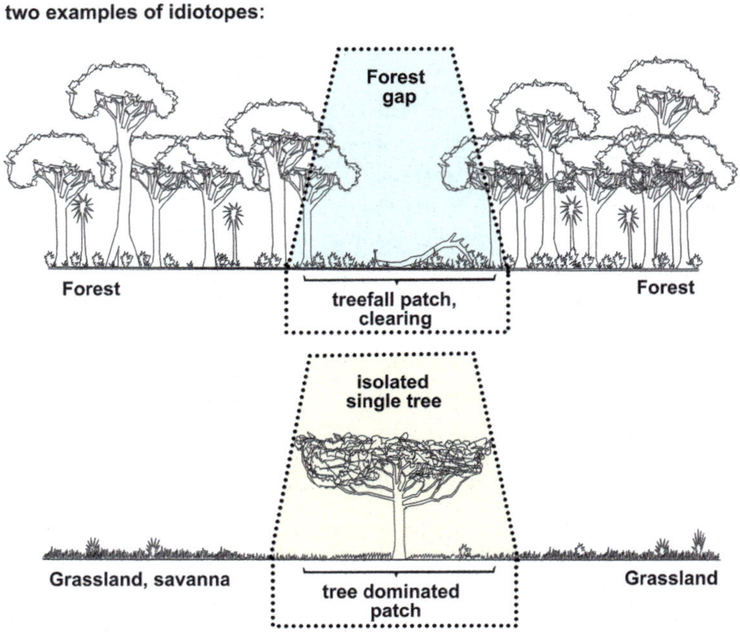

Fig. 29 Comparative structure and effects of idiotopes in a forest area and in savannah (Belsky and Canham 1994; Breckle 1999), indicating the specific island effect of shading from a single tree in the savannah and the opposite effect in a forest by gap formation

Fig. 30 (**a**) During the rainy season, the rain gauge is full every day and has to be emptied very regularly (photo courtesy of J. Homeier, May 2004). (**b**) Typical landslides on steep slopes of the Rio San Francisco valley, southern Ecuador (photo: S. W. Breckle, 2004)

Working in the tropics, the difficulties were major from the start. Inventories of a beech forest in Germany can take an afternoon, and it is then clear which species dominate and what their cover percentages are. In the tropics, even by restriction to 5 or 10 cm dbh (diameter at breast height), it takes weeks and months of intensive identification work to get reliable species lists. And it is obvious that phytosociological methods and the definition of syn-taxonomic units are senseless, since the main precondition of an existing "minimum area" is not fulfilled (Fig. 31). The very artificial syn-taxonomic nomenclature is useless in the tropics at least, though many syn-taxonomic papers are still published.

In Costa Rica, about 94 tree species per hectare (dbh $\geq$10 cm) were inventoried (Wattenberg and Breckle 1995). If the area was doubled, the number of tree species would be about 130, indicating the lack of an asymptotic minimum area (Fig. 31). The number of tree species in southern Ecuador (dbh $\geq$5 cm) is extremely high and almost a world record (Homeier 2004). At both study sites, the climate is typically a diurnal climate all year round (Fig. 32). Sometimes very large amounts of rain can fall within 1 day (Fig. 30a), regularly causing landslides on the steep slopes and thus dynamic reproduction and establishment processes.

In Costa Rica's 1 ha plot, more than one third of the tree species were present with only one individual tree, and one fifth with only two individuals (Fig. 33). It is still almost a miracle how, under these circumstances, successful pollination is possible. One probable explanation is the process of bird pollination.

Only in species that occur relatively frequently is it possible to delimit an altitudinal belt structure, as was shown for tree ferns. Figure 34 covers only an altitudinal distance of 500 m, but the tree ferns apparently exhibit strong niches (Bittner and Breckle 1995).

What is the growth dynamic of the trees? The growth of tree stems can be checked by study of the annual rings from drilled bore cores, but this method is destructive; furthermore, many tropical trees do not produce periodic annual rings. We used dendrometer bands. Biweekly measurements of 948 trees over several years gave good hints about the growth conditions, the variability within and between species, and the influence of site conditions. This could be shown in Costa Rica (premontane forest between 850 and 1,210 m asl [above sea level]), as well as in the montane tropical forest (1,850–2,450 m) in Ecuador (Homeier 2004; Homeier and Breckle 2004; Bräuning et al. 2008). Diameter growth was highly variable between the 22 investigated species, with annual increments from 0.6 to 5.7 mm in Ecuador and from 1.6 to 12.3 mm in Costa Rica. The maximum growth rates were in the lower diameter classes. Seasonality in growth was seen in the only deciduous tree, *Tabebuia chrysantha*.

The tropical forests in Costa Rica are rich in palm species. Figure 35 shows the growth of three palm species in time, derived from long-term measurements (J. Homeier, personal communication, 2017: Stattegger 2017). Also, the thickness of the palm stems depends on size and thus on age (Fig. 36), despite the fact that monocots are normally described in textbooks as not having a secondary growth of the stem. The two palms reached dbh values of 21 cm (*Euterpe*) and 26 cm (*Iriartea*). Both species start growing with a slender stem and show a limited

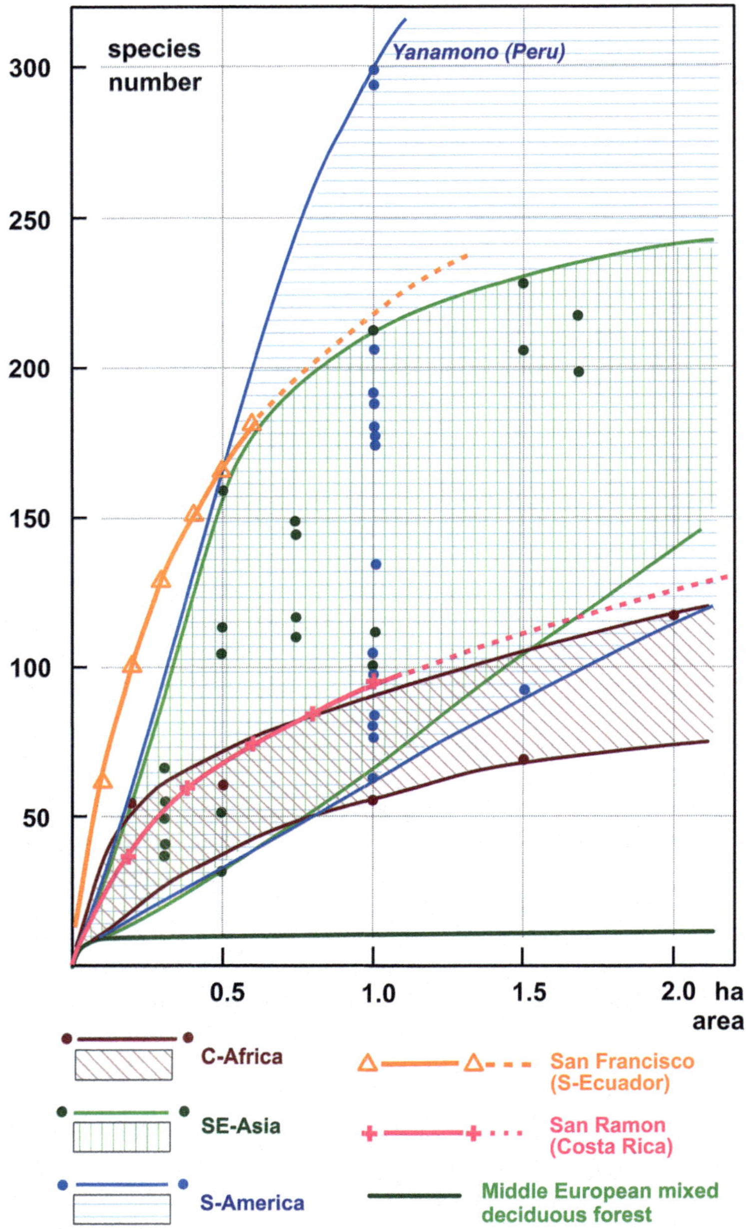

Fig. 31 Species/area curve of the montane tropical forest of San Ramón at Sierra de Tilarán, Costa Rica (diameter at breast height [dbh] $\geq$10 cm) and San Francisco, southern Ecuador (dbh $\geq$5 cm) in comparison with some other tropical forest regions and with the middle European mixed deciduous forest (data from Wattenberg and Breckle 1995; Homeier 2004)

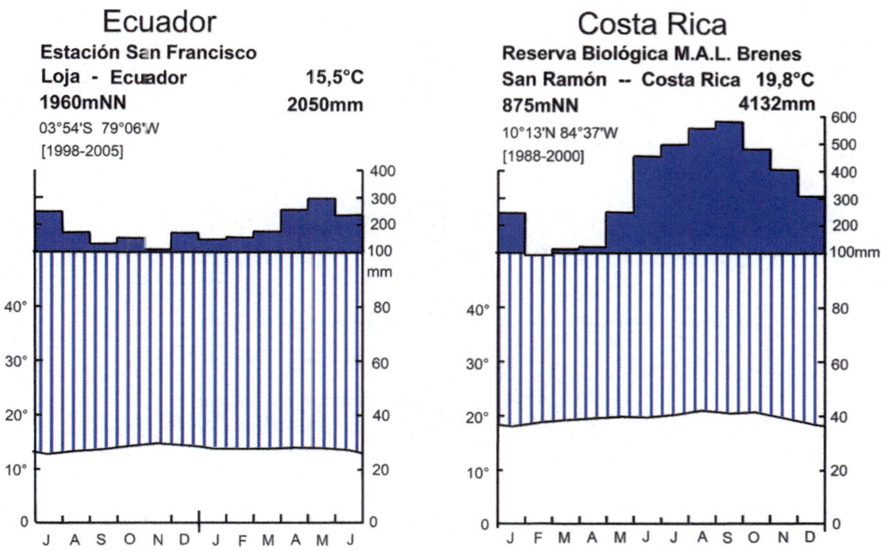

Fig. 32 Ecological climatic diagrams from southern Ecuador (San Francisco Biological Station) and from Costa Rica (San Ramón Biological Station, Rio Lorencito). The climate is typically humid and tropical all year round, with almost no annual temperature oscillation; most months are hyperhumid, with above 100 mm precipitation/month

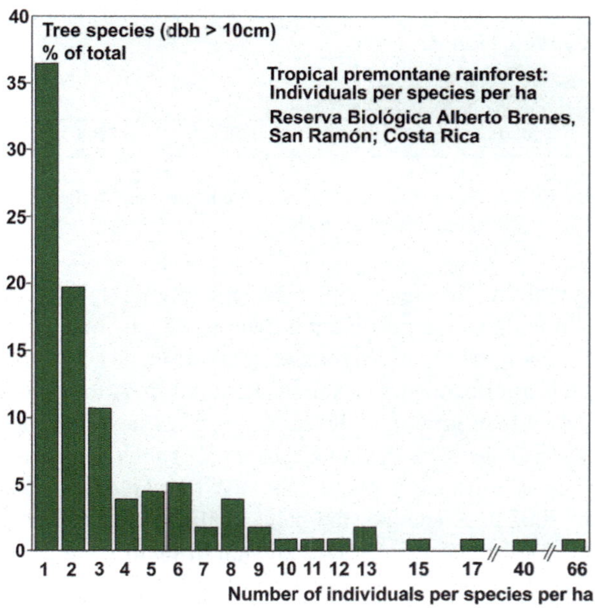

Fig. 33 Number of individuals per hectare of tree species (% of total) in the tropical montane rainforest in the Sierra de Tilarán at Reserva San Ramón, Costa Rica (Wattenberg and Breckle 1995); 37% of tree species are represented by only one individual, 20% of species have two individuals

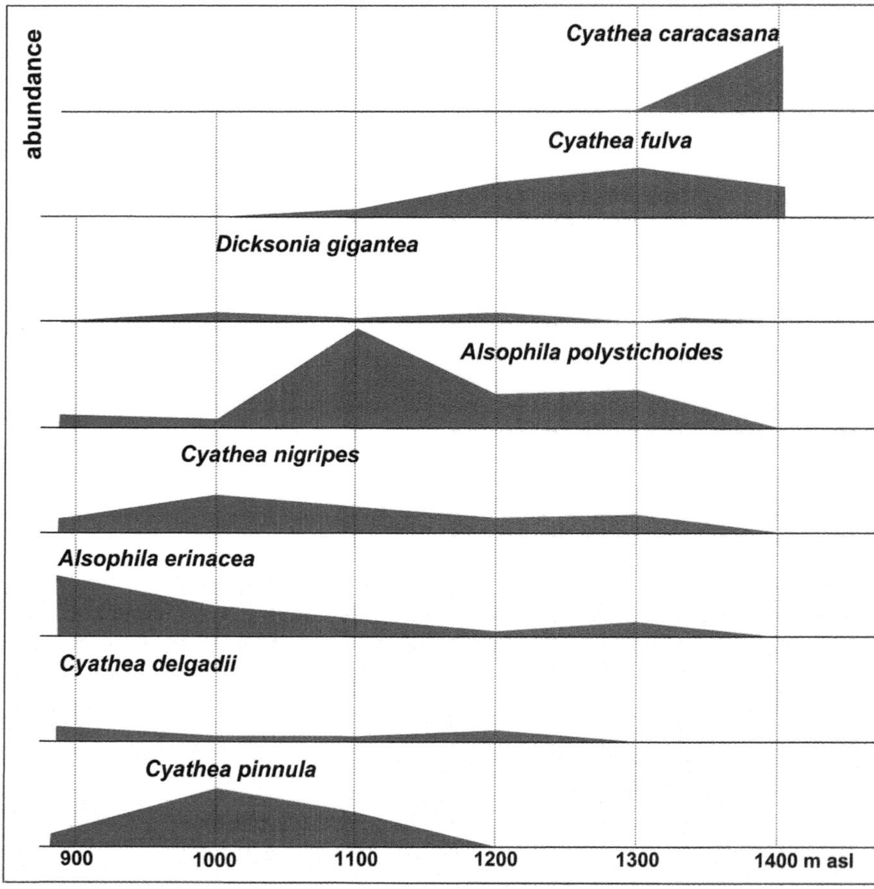

Fig. 34 Abundance of tree fern species along the altitudinal gradient in Sierra de Tilarán, near Rio Lorencito, Costa Rica (Bittner and Breckle 1995)

diameter increment by sustained cell expansion during their development. The relation of height to dbh shows different growth strategies for *Euterpe* and *Iriartea*. *Euterpe* palms grow regularly in height and in dbh (Fig. 35). In *Euterpe* palms, first diameter growth is prioritized up to a dbh of about 20 cm. After almost reaching their final diameter, they then grow mainly vertically. This means competitive strength changes according to age. The smaller *Cryosophila* palms increase their number of leaves during an establishment phase, and the stem reaches the final diameter (Homeier et al. 2002a). After that, the palm continues its vertical growth, with a diameter sufficient to support a maximum height of up to 10 m.

The reproduction and establishment of the various species is a rather complex phenomenon. It was only checked with very few species but is very important for future sustainable soft management of tropical forests, since introduced trees often suffer nutrient deficiencies (Breckle et al. 2005). The population structures of the

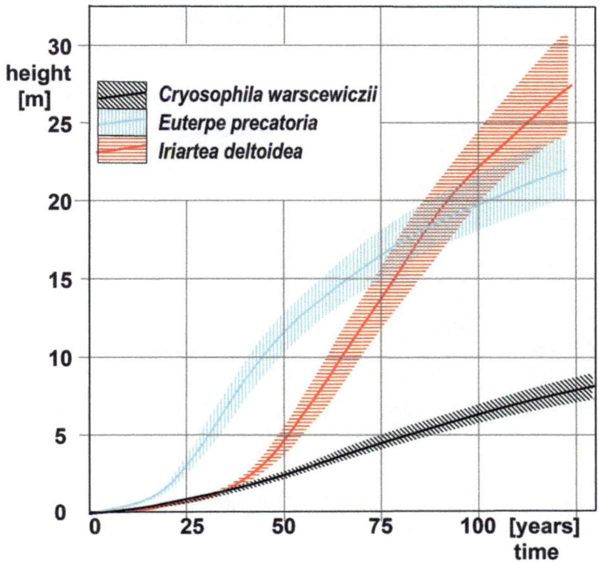

Fig. 35 Height growth of three palm species in the montane tropical forest of Sierra de Tilarán, Reserva Biológica, near San Ramón, Costa Rica, as a function of time (data from Homeier et al. 2002a)

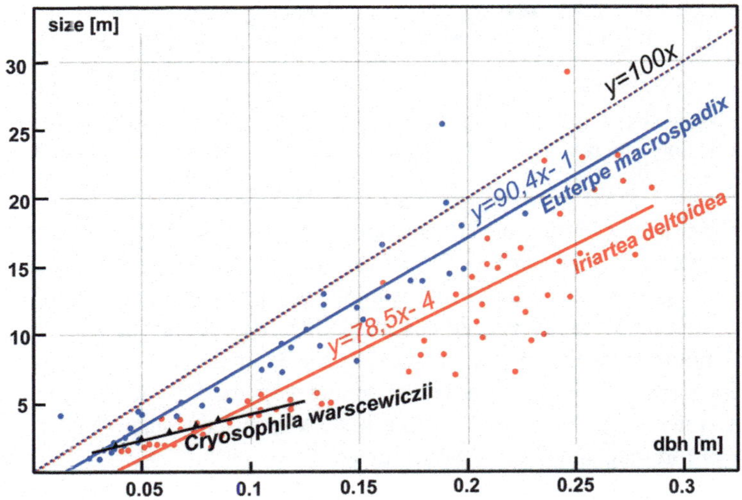

Fig. 36 Correlation of height (m) and diameter at breast height [dbh (m)] of three palm tree species in the montane tropical forest of Sierra de Tilarán, Costa Rica (data from Homeier 2004; Homeier et al. 2002a)

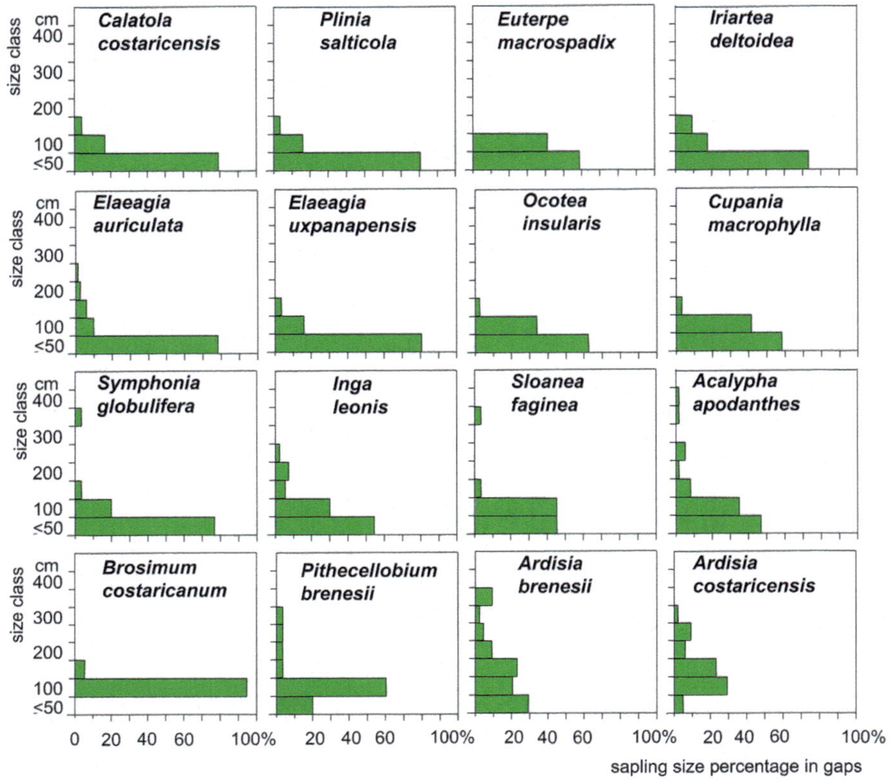

Fig. 37 Comparison of distribution of size classes of saplings and young trees of 16 species from the montane tropical forest of Sierra de Tilarán, Costa Rica (Wattenberg 1997; Breckle 1997a)

various tree species in the Costa Rican mountain rainforest differ from each other. A shortened version (given in Fig. 37) of the size classes of 16 species can also be interpreted as a kind of sapling bank, waiting for an eventual gap to happen. In any case, the number of small individuals in saplings and young trees is much higher than in adult trees, which is demonstrated by the example of *Pterocarpus hayesii* in Costa Rica (Fig. 38). The turnover of young gap trees is very high, with high mortality rates (Fig. 39). In conclusion, it is clear that the shape and composition of tropical forests change rather fast in time.

6.2 *Ecuador*

The study area covered the moist tropical mountain forests from the valley bottom of Rio San Francisco (1,800 m) up to the treeline at about 2,800 m. The whole area covers mainly northern exposed slopes around the biological station, forming the

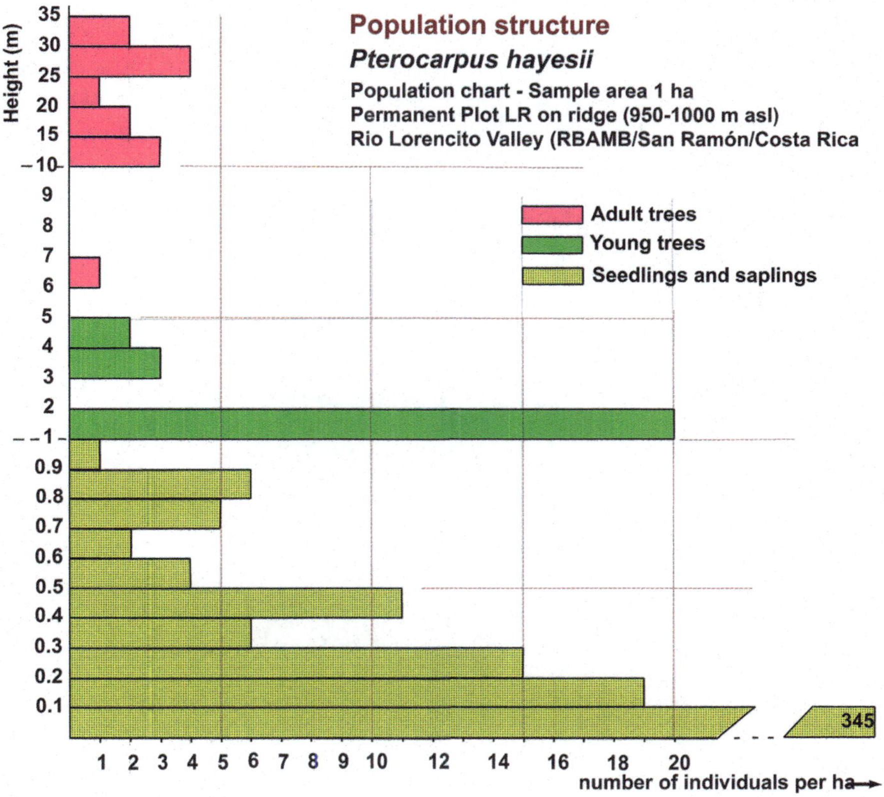

Fig. 38 Population structure of *Pterocarpus hayesii* in the montane tropical forest of Sierra de Tilarán, Costa Rica (Breckle 1997a)

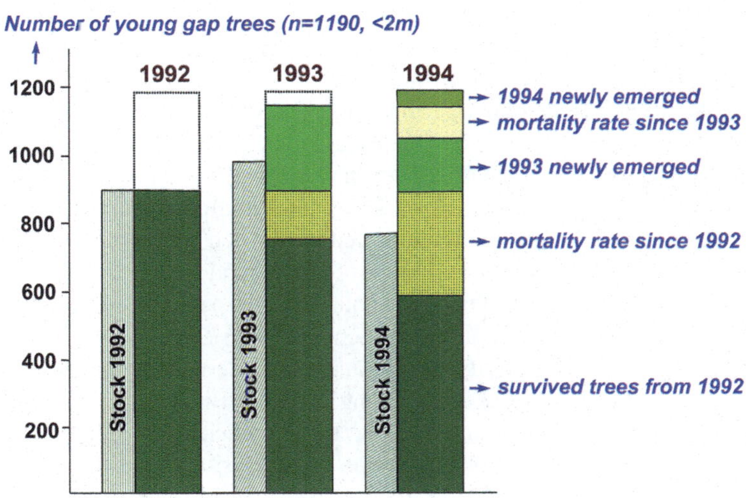

Fig. 39 Mortality dynamics in the montane tropical forest of Sierra de Tilarán, Costa Rica (after data from Wattenberg 1996); number of trees <2 m, per ha

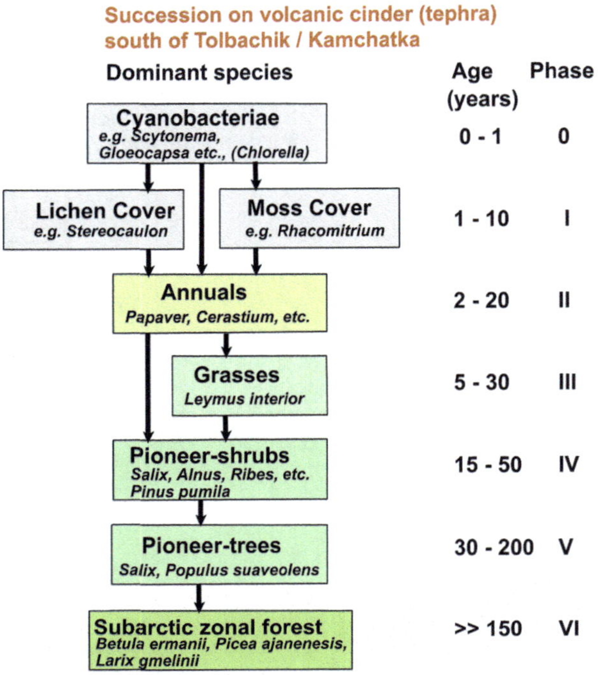

Fig. 40 Primary succession on volcanic cinder (tephra) south of Tolbachik, Kamchatka (Breckle 2002b)

Reserva Biológica San Francisco (RBSF) adjacent to the Podocarpus National Park. On the opposite side of the valley, the forests were cleared about 30 years ago for pine plantations and pastures. Thus, this whole valley – being a hotspot of biodiversity (see Sect. 7.1) – offers great opportunities for comparative studies.

The succession is apparently very dynamic and different in each place, especially in southern Ecuador. There, the long open and steep landslides (Fig. 30b) revealed orchids and ferns as members of the first successional stages (Gross 1998), together with annuals, certainly favoured by their small diaspores. This is rather similar to many examples of primary succession (Breckle 2002b), especially on open cinder slopes of volcanos, e.g. Tolbachik, Kamchatka (Fig. 40), and also in Fogo, Cap Verde; Kilauea, Hawaii; Arenal, Costa Rica; and Piton de la Fournesse, Réunion. But later successional stages seem to proceed very much by chance.

Many basic questions remain. How is the high biodiversity dependent on site conditions? Dalitz et al. (2005) tried to relate it to the spatial heterogeneity, mainly by canopy throughfall of rainwater and leaching of nutrients.

We participated in the priority programme of the German Research Foundation (DFG) on "The Mechanism of Maintaining Tropical Biodiversity". For a shorter period, we worked at the same time in Costa Rica and Ecuador. Homeier (2004) had carried out extensive comparative studies on tree diversity, forest structure and tree growth in both montane primary forests. In Ecuador on 15 permanent plots (400 m^2

each; in Costa Rica, 13 plots), 1,260 trees (dbh $\geq$5 cm) were inventoried (in Costa Rica, 605 stems). Those trees were identified as belonging to 183 species (in Costa Rica, 147 species) from 53 plant families (in Costa Rica, 55 families). For the whole reserve area in Ecuador, at least 230 tree species were known; four of them were new to Ecuador and nine of them were new to science. At both sites, Lauraceae, Melastomataceae and Rubiaceae were the most common plant families.

The census of biodiversity is a rather full-time job. First, Ingrid Wattenberg and then Jürgen Homeier did basic data collection (Wattenberg and Breckle 1995; Wattenberg 1996; Homeier et al. 2008; Liede-Schumann and Breckle 2008).

Several co-workers continued with their own projects. It turned out that, as in Costa Rica, botanical inventories in Ecuador faced great difficulties. This was because of our ignorance of tropical flora, the frequent lack of flowers or fruits in herbarium collections, or simply because the species have yet to be described (Brehm et al. 2008).

Which are the limiting resources? Are these resources always the available space for growth and species number area-wise? What about the available water in ecosystems? In the Atacama Desert, there are about 50 species and in the Negev Desert about 5–10, but in the tropical rainforests of Costa Rica or Ecuador, only about 1–2 species are found if you calculate them in relation to 1 mm of the available annual precipitation (Breckle 2006).

Nutrient availability is certainly a major factor in plant growth and competitive ability between species (Beck et al. 2008). Jürgen Homeier was able to pursue this question in Ecuador. Especially, the availability of nitrogen (N) and phosphorus (P) to a great degree governs the competitiveness of trees along the elevational gradient, as well as on ridges in contrast to valleys (Homeier et al. 2010).

The question of whether some kind of seasonality can be detected was also tackled. Close to the equator, the weather all year long is more or less rainy and humid, and high annual temperatures fluctuate only between day and night. As has already been mentioned, the wood increment was periodically active annually only for the deciduous *Tabebuia*, but flowering and fruiting also exhibited some annual periods in other species (Cueva Ortiz et al. 2006).

Many groups were also working at the same sites in southern Ecuador. There, strong ties developed between botanists, zoologists, soil scientists, forestry and sociology from both the German and Ecuadorian sides, and many joint studies and projects were carried out and many publications produced (Beck et al. 2008; Bendix et al. 2006, 2013).

7 Wealth of Biodiversity

7.1 Tropics

The question of wealth – what is the price for biodiversity – is a totally anthropocentric question. Do we need to know it? What can we conclude? If biodiversity is lost, it is irreversible; if climate changes, it is reversible. Natural biodiversity cannot easily be judged monetarily; it is an indefinite wealth. The global initiative TEEB (The Economics of Ecosystems and Biodiversity; teebweb.org) is focused on "making nature's values visible" in economic terms. It helps decision makers to recognize the wide range of benefits provided by ecosystems and biodiversity. In anthropogenic agro(forestry) systems, biodiversity stabilizes and sustains productivity. In such cases, via productivity, biodiversity can be better judged monetarily (note by Beck).

Though we were mainly tackling ecological questions, the problem of extremely high biodiversity was always present. In both regions (Costa Rica and Ecuador), the documentation of biodiversity was a necessary part, parallel to all other studies, and time consuming.

In southern Ecuador, one of the hotspots of biodiversity was investigated by many individual projects and research units. From this Ecuador group, as mentioned above, books (Beck et al. 2008; Bendix et al. 2013) and more than 600 papers were published, covering many aspects of natural and social sciences.

The checklists for various groups of organisms (Liede-Schumann and Breckle 2008) bring together earlier knowledge.

The checklist for spermatophytes has 131 plant families, 425 genera and 1,208 species in the RBSF. The pteridophytes have 250 species; the bryophytes have 320 liverworts, 204 mosses and three hornworts (altogether 527 species); the lichens list has 323 species, but many specimens are still unidentified. A very special list is that of the mycorrhizal fungi, where the only way of sure identification is direct sequencing of the associated fungi. Those ribosomal genotypes can rarely be related to morphological or biological species. The mammals are very incompletely known, and only the list of bats, with 35 species, may be rather complete. The birds list has 271 species in the RBSF, but there may well be close to 800, half of the 1,600 in Ecuador. The list of papilionoid butterflies has 245 species, which is estimated to be less than 50% of the existing fauna. The moths are extremely rich; 2,547 species have been recorded in the RBSF and in the close vicinity, including 42 Saturniidae and 36 Sphingidae. The checklist of Tettigoniidae has about 100 species. One hundred and ninety-three species from Oribatidae have been recorded, again including several ones new to science, as in other organismic groups. The checklist of amoebas has 135 species. Finally, also included is a list of useful plants and weeds, with about 360 plants in the Shuar and other indigenous communities.

It was, at that time, a wonderful experience to bring together so many different lists of organisms. Though several groups are still lacking, those lists are very important for future studies. Certainly the inventories may soon be outdated by the tremendous wealth of biodiversity, guaranteeing new finds for many years to come.

Documenting high plant biodiversity certainly always requires good herbarium material, but with new sophisticated digital cameras, photographs of plant portraits can also be used for image-based plant databases (Homeier 2004). With Visual Plants (http://www.visualplants.de/), one has access to records of digitized plant images, herbarium specimens and illustrations of plants from Kenya, Uganda, Costa Rica and Ecuador, and many additional data. This is a helpful tool for the determination of plants; it can be used in the field on laptop computers and also for the training of students before travel in tropical regions. In 2017, the collection included 27,100 images of 288 families, 1,855 genera and 4,215 species.

In 1990, we could study the elevational gradient not only on the tropical high volcano peaks of Mt Kenya and Mt Elgon but also in one of the rather untouched rainforests at Kakamega. Former co-workers of the Department of Ecology later participated in the large BIOTA consortium, where not only the dry vegetation gradients in Namibia and South Africa were studied but also (BIOTA-East) the area of the Kakamega Forest (1,460–1,760 m) in Western Kenya. Dalitz et al. (2011) provided basic data and a very valuable field guide on the woody plants of the area, with about 300 species of trees, shrubs, lianas and palms, indicating the high biodiversity but also the threats of logging and firewood collecting.

7.2 *Afghanistan*

Studying biodiversity in an arid country is another endeavour. After many years, in 2005, work on Afghan flora and ecology could be resumed after more than 30 years of war and destruction. It became my second home after 3 years of work, studies and teaching there. In Bielefeld, together with Clas Naumann (also infected by the "Afghanistan virus", but as a zoologist), we were able to organize an interesting symposium (Breckle and Naumann 1982) at a really bad time for the country after the Soviet occupation.

In the continental semi-desert of Afghanistan, the cold winters and the hot summers force plants and animals to use the underground. In the Dashte-Kushi, south of Kabul, we had an interesting phenomenon to observe (Breckle 1971). There were small anthills, sometimes only an opening and an underground nest of ants (*Cataglyphis bicolor*). Around them there were large circles devoid of vegetation but surrounded by a ring, about 4–6 m in diameter, of a therophytic flora rich in species and with distinctly favoured growth. In this belt, N and P content was distinctly higher than in samples from outside the belt. The additional source of nutrients is the underground waste chambers of the ants (garbage zone). Another ring, located outside the garbage zone, poor in small stones, shows vegetation more negatively influenced by the plant- and seed-collecting activity of the ants. This myrmecophilous dissemination varies within different plant species. In the course of several years, there evolves a typical weedy plant community, exhibiting the changed availability of nutrients and water. We saw a rather similar mosaic of plant-free rings later in Arizona, caused by ants (*Pogonomyrmex*). From Namibia, recently several papers described a similar pattern (called "fairy circles"), most probably caused by underground termites (Tarnita et al. 2017) as a multi-scale patterning.

Exactly at Christmas 1979, the Soviet troop invasion in Afghanistan started. Field research had been blocked for a few years even before then because of internal unrest and turmoil. After many years of war and destruction in Afghanistan, and after terrible civil war times, many of our old coloured slides of landscapes and plants

Fig. 41 (**a**) Title page of the *Afghan Plant Atlas* (Breckle and Rafiqpoor 2010). (**b**) Title page of the *Afghan Checklist* (Breckle et al. 2013)

became valuable historical documents. How were we to use them? By developing a "snowball system", after 2005 we were able to use photographs from many colleagues all over the world. This was the starting point for collecting photos for an illustrated flora. An easy-to-use botanical field guide was, for a long time, much needed. The *Field Guide Afghanistan* (Breckle and Rafiqpoor 2010, Fig. 41a) fulfilled that need for school teachers and university lecturers. It is also a basic reference source for forestry, horticultural and agricultural projects, for nature conservation and anyone interested in the very rich flora of Afghanistan. The idea of bringing together old slides – photographed by both the German team and others from Austria, Britain and Scandinavia working there during the 1960s and 1970s – had been discussed during the 6th Plant Life of South West Asia Symposium (PLoSWA) in Van, Turkey, in 2002. Modern communication facilities, especially e-mail, enabled us to complete it in a relatively short time, as well as intensive cooperation in a perfect "symbiosis" between the authors and contributors.

Many colleagues from all over the world contributed photographs, old slides (which we scanned) and digital photographs. Five thousand copies of the book were printed, with German Academic Exchange Service (DAAD) funding, to be distributed all over Afghanistan, and 500 additional copies, funded by the Sibbald Trust, Edinburgh, were distributed cost free to libraries and herbaria all over the world. The main part of this pictorial flora includes 2,000 colour photographs of approximately 1,200 plant species with short descriptions, all bilingual (in English and Dari). The general chapters give an introduction to the overall physical geography of Afghanistan (see also Breckle 2007; Djamali et al. 2012), including its geology, soils, climate and vegetation, but also the plants and their structure, morphology, anatomy, taxonomy and systematics are described.

During the acquisition of photographs, we got (from a private development project by Marga Flader) some digital photographs from a tree in a village in north-western Afghanistan and, after a short time, another one from the German Embassy in Kabul. It was immediately clear: this tree was *Ailanthus altissima*. The plant family Simaroubaceae, it turned out, had been forgotten in the Flora Iranica (Rechinger 1963 ff). Thus, after we contacted Iranian and other colleagues in the whole area, it became obvious; *Ailanthus* is very widespread (Breckle et al. 2015). An additional delivery for Flora Iranica could be prepared to make the series with 181 deliveries complete (still lacking only the pteridophytes).

Germany, especially, has long-lasting ties with Afghanistan (Breckle 2008). In 1966, the Scientific Working Group on Afghanistan (Arbeitsgemeinschaft Afghanistan [AGA]) was founded in Heidelberg by Karl Jettmar, Willi Kraus and Carl Rathjens. Coming back from Afghanistan at the end of 1969, I became a member of that group. This group organized symposia, edited books on Afghanistan and was an ideal information platform between both countries. From 1990 until 2001, I was the chairman of the group, which had more than 250 members from about 12 countries. From the beginning, there was close cooperation with the excellent Swiss archive Bibliotheca Afghanica in Bubendorf and the Swiss Afghanistan Foundation.

The next step was to give an answer to the simple-looking question: How many vascular plant species are known in Afghanistan? What proportion do we have illustrated in the *Field Guide Afghanistan* (Breckle and Rafiqpoor 2010)? It was necessary to produce a definitive checklist. All available floras and papers on taxonomy had to be evaluated. A checklist of a region enables one to quickly become familiar with the richness of its flora. Dr Souza Dias (executive secretary of the Convention on Biological Diversity, Montreal) wrote in the preface: "The book 'Vascular Plants of Afghanistan – An Augmented Checklist' (Breckle et al. 2013, Fig. 41b) provides the most accurate and up-to-date account of the plant biodiversity of this region. It complements the colourful 'Field Guide Afghanistan – Flora and Vegetation' by Breckle and Rafiqpoor (2010) during the International Year of Biodiversity".

The estimations of species numbers were about 3,500 species and 30% endemism. The more precise answer from the checklist is now approximately 5,000 species and about 25% endemism. The genus *Astragalus* has the most species, and the family Asteraceae has the most genera and species. And we know that the Hindu Kush has distinctly higher elevational belts than the Alps or the Caucasus (Breckle et al. 2017, Fig. 42).

Both books are scientifically as correct as possible on one side but also are an important tool for education on biodiversity topics for a wider audience (Breckle and

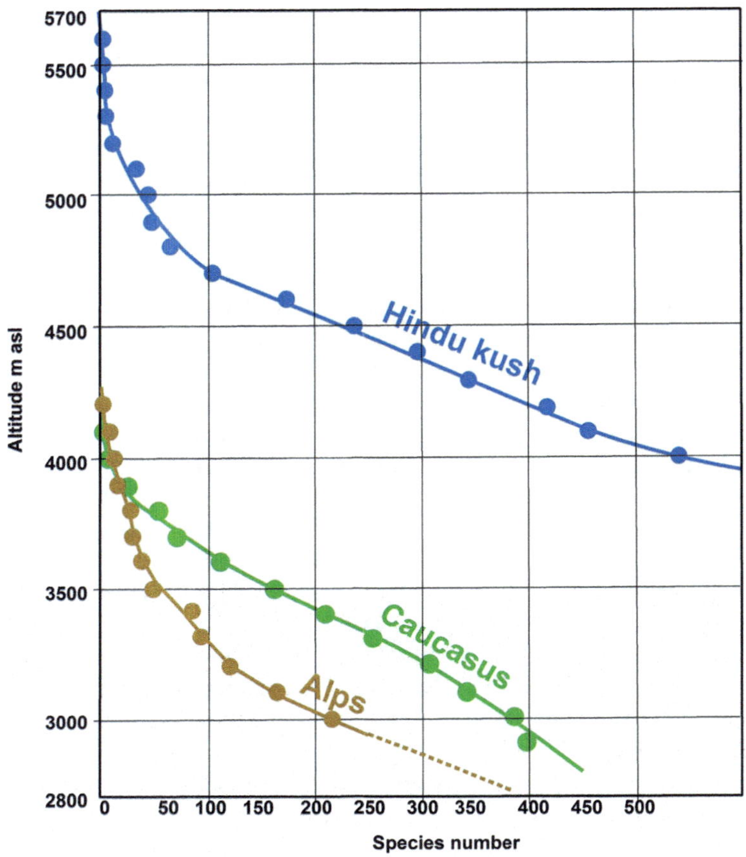

Fig. 42 Floristic "drainage" (decrease) in species numbers with increasing altitude in the Hindu Kush, Caucasus, and the Alps (Breckle and Rafiqpoor 2010; Breckle et al. 2017)

Rafiqpoor 2016; Lozan et al. 2016). Older studies on useful plants (Breckle and Unger 1977; Breckle 1979; Breckle and Koch 1982) were a good starting point.

A continuation is planned. The *Field Guide on Mosses and Liverworts of Afghanistan* is in press (Kürschner et al. 2017), a field guide on lichens is planned (Mayrhofer et al. 2019) and an updated treatise on Afghan medicinal plants is also planned (Keusgen et al. 2015), as well as a general treatise of Afghan biodiversity (Breckle et al. 2018).

8 Ecosystems of the Earth

The two volumes on the vegetation of the earth (Walter 1968, 1973, short version: Walter 1986, 1990) give, in approximately 2,000 pages, a survey of the vegetation and climate of the globe. A totally new edition, including ecological processes in the various ecosystems and ecozones, was much needed, so Heinrich Walter invited me to be a co-author. A new rather comprehensive and strictly hierarchical content of all of the necessary ecologically ranked units, and including relevant recent literature, led to the four volumes *Ökologie der Erde* (*Ecological Systems of the Geo-biosphere*) (Walter and Breckle 1983, 1984, 1986a, b, c, 1989, 1991a, b, c, 1994, 2004), including revised new editions. Mainly the third volume, with all of the Russian literature, needed another co-author after Heinrich Walter's death on 15 October 1989. Independently, we had the same solution: Okmir Agakhanjanz was the ideal partner, having been a DFG guest professor in Bielefeld in 1992 and a future close friend. His long-term studies in the Pamirs and many other Asian mountains were the incentive for the Pamir expedition in 2002 [with the help of GTZ-UNCCD (Bonn) and DAAD (Bonn) funding (Figs. 3a and 19a–d)], with all aspects of biological and geographical highlights (including Bavarian Television). All of the curious conditions and adventures of scientific work were illustrated by Agakhanjanz (2002).

In order to describe all of the various ecosystems of the climatic belts of the whole geo-biosphere, it seemed necessary not only to check all relevant literature but also to become personally familiar with most ecosystems. Examples from the tropics, lowlands, mountains and islands (see Fig. 44c) and the various hot deserts could be enlarged by experience and examples from savannah ecosystems, from the various Mediterranean biomes, from high mountains on most continents and from taiga and tundra. There are still many blank areas (on the global map) concerning our own experience. However, I am sad to say, there are increasing difficulties in studying ecological systems abroad, for political reasons and because of bureaucracy, civil wars and the ongoing destruction of natural ecosystems. Despite better possibilities to travel across continents, it has become more complicated to study natural ecosystems – often isolated in national parks – as a basis for comparison of the natural processes of ecosystems of the various zono-biomes.

Heinrich Walter had initiated a book series titled *Vegetationsmonographien der einzelnen Großräume* (*Vegetation Monographs of the Major World Regions*). The series was funded in 1965. In his survey (Walter 1976), he explained the whole concept of hierarchical ecosystems of zono-biomes, biomes and ecotones. After Heinrich Walter's death in 1989, I continued as the editor of this monograph series. The concept includes nine world regions. After volume 1 (North America, Knapp 1965), volume 2 (South American forests, Hueck 1966), volume 3 (Africa, Knapp 1973), volume 7 (North and Central Asia, Walter 1974) and volume 10 (climate diagrams of all continents, Walter et al. 1975), additional volumes to be published were volume 4 (Australia, Beadle 1981), volume 9 (Pacific Islands, Mueller-Dombois and Fosberg 1998) and volume 8 (not indicated as such: Europe, only open vegetation types, Klötzli et al. 2010). I was only partly successful in continuing this monograph series, despite many attempts to convince possible authors. Volumes 5 and 6 (East Asia, South Asia) are still lacking. The assembling of the huge existing information into an integrated comprehensive book is a major magisterial synthesis and a challenge to attempt. It requires long experience (Walter 1989).

The continuation of the aims of the AFW Schimper Foundation, founded by H. and E. Walter in 1968 to promote ecological research outside of Europe, was documented by the proceedings of the first symposium in October 1998 and a second in October 2002. From 1968 till now, almost 100 fellowships have been given. The fellows (the "Schimper Family") have presented many interesting results of their research, which quite often were a starting point for more detailed studies. The proceedings volumes (Breckle et al. 2000, 2004) document a broad wealth of worldwide ecological studies and, in part, up-to-date research data on specific subjects.

9 Applied Biology and Ecology

On my return from Afghanistan at the end of 1969 to the Institute of Pharmacognosy in Bonn (later the Institute of Pharmaceutical Biology) it became clear that my ecological interests should be combined with the chemistry of secondary compounds and also studies on nutrients and trace elements. A heavy load of teaching on the morphology and anatomy of plants, as well as on biochemical pathways of active principles in medicinal plants, led to the edition of a textbook on *Pharmazeutische Biologie* (Kating and Breckle 1978, 1981), with several subsequent new editions – the third in 1988 and now, in collaboration with Eckhard Leistner, the 8th edition (Leistner and Breckle 1988, 1992, 1997, 2000a, b, 2008, 2014) with new authors (Drews et al.).

When I started in Bielefeld in 1979, botanical ecology was becoming a dominant topic. There were also applied projects on nature conservation, mainly documentation of specific sites for future nature protection. They were the topic of many diploma theses. Numerous phytosociological and floristic studies were carried out around Bielefeld. Often, the documentation of the ecology of a distinct area led, sometimes many years later, to the declaration of such an area as a nature reserve – sometimes with unnecessary prohibitions on entering the area, since not all official flora–fauna habitat (FFH) regions (as declared by the European Union [EU]) were really worth nature protection. Many former diploma candidates are now working at biological stations and in communal environmental administration, or have founded private consulting agencies.

This scientific curiosity was not directed only towards purely ecological problems but also towards other scientific topics. One interesting question has been how much influence on – or damage to – nature has been caused by nature sports such as orienteering, which is very common in Scandinavia.

An anecdotal event should be narrated. In 1984, the German university championships in orienteering (O) were organized by the University of Bielefeld. The sandy Senne area, with a pine forest and fossil glacial sand dune hills, was an ideal competing ground. But what would be the impacts of such an event, where runners would go all over the forests, to find controls as fast as possible with the help of a special O-map? Three control sites were mapped and photographed before and after the event (Breckle et al. 1989, 2014b).

Some weeks later, because I was responsible for the event, I got a letter, which accused me of having partly destroyed the small Ölbach Valley. Apparently this had been a small nature reserve, though it was not marked as such. Several letters went to and fro; as an ecologist I was being charged with destroying a nature reserve. Finally, it could be clarified that there was not one official indicator plate in the area that indicated the borders of the nature reserve (as is necessary according to the law), that there was no damage (the few visible steps had disappeared after a few weeks) and that in such a sandy area, disturbances from time to time enhance the biodiversity and recovery of plant species; however, they were much less than the large military tanks were doing nearby. Thus, the penalty of a 3-month jail sentence (or 50,000 DM fine) for such an offence was not imposed. The area is in close proximity to the British and German military exercising area in the Senne. It has been a military area for about 120 years and now exhibits rare flora and fauna maintained by continuous special disturbances (Breckle 1993) – maybe in future it will be part of a national park?

Long-term studies of observation squares at former control sites since then revealed that the natural dynamics are very obvious but have nothing to do with any anthropogenic influences. The number of shoots of *Trientalis* and of *Maianthemum* oscillated tremendously (Fig. 43) according to ice/rain, growth of shading *Frangula* or the natural death and decay of pine trees.

10 Teaching

The topics mentioned in Table 2 illustrate the breadth of teaching but also various research projects, which, in most cases, were characterized by an interdisciplinary approach. The philosophy behind it was the rather important feedback between field research and lab studies and controlled experiments where students could participate. The importance of such an approach was also stressed by Beck (2017).

Starting in Bonn and continuing in Bielefeld, I was always engaged in preparing textbooks and their new editions, not only in ecology (see above, Breckle 1999, 2002a, b, c, d) but also in pharmaceutical biology (Kating and Breckle 1978, 1981; Leistner and Breckle 1988, 1992, 1997, 2000a, b, 2008, 2014).

The ongoing research projects were done in close cooperation with local people and institutions. We used all opportunities to inform relevant stakeholders, to raise capacity building and to improve resource management by organizing small meetings and symposia and, of course, by participating in talks and posters at relevant

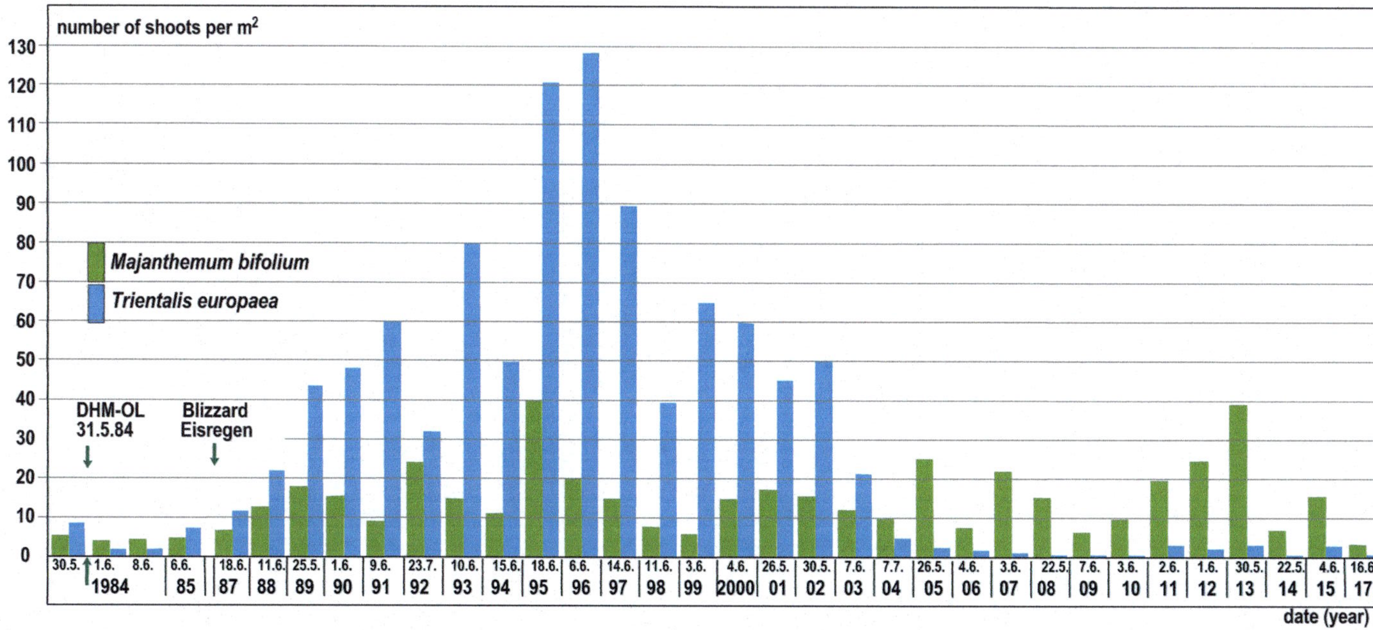

Fig. 43 Number of shoots during the last 33 years from *Maianthemum bifolium* and *Trientalis europaea* in sandy pine forest in the Senne area, Augustdorf, south of Bielefeld (Breckle and Breckle 2014) – the site of the German championships in orienteering (DHM-OL) in 1984

regional, national and international workshops and meetings (Breckle et al. 2001a, b, c, 2008, 2012; Breckle 2005b).

Since I retired in 2003, I have had more time to work constantly on specific interests. There were not only the books on Afghan flora and vegetation or new editions of textbooks, but also many older studies mainly on dry land ecology and desertification (e.g. Breckle et al. 2001a, b, c, 2008, 2012; Khan et al. 2016) and again on the effects of orienteering (Breckle et al. 2014b, Breckle and Breckle 2014) or even several editions of a basic treatise on ice skating (Breckle 2016).

11 Some Final Remarks

Ecology deals with the environment of organisms. Scientific ecology has many aspects; it depends on the scale. At the level of cells and tissues, many processes are important for the whole plant. Structure and function are closely related. One incredible type of structure is bladders in Chenopodiaceae (Fig. 44a).

At the level of the whole plant, there are fantastic adaptations to hostile environments but also aesthetic structures and an incredible variety of flowers, governing pollination and preparing for the next generation. Some plants are very common weeds, and others are extremely rare; one example is shown in Fig. 44b.

At the next structural level, very complex plant associations and their basic ecological factors can be comparatively studied from deserts to tropical rainforests (Fig. 44c).

At the next structural level, landscapes and vegetation belts are a striking feature, again demonstrating effects of ecological factors, climate, soil, etc. The treeline in mountains – under natural conditions – gives us an idea of the importance of forests worldwide. The highest treeline on the globe – that of the rosaceous *Polylepis tarapacana* on the Sajama slopes in Bolivia – is an incredible example of the struggle of plants (Fig. 44d).

Figure 44a–d is used here in a symbolic way, indicating the broad amplitude of ecology (along scales) and its fascination and beauty.

It is the privilege of university professors to work on topics that they find especially interesting and that are satisfying (at least part of the time). It is not only a job but also like a hobby. However, administration and ministries sometimes seem to have another opinion. Administration and management become increasingly dominant.

Good and basic knowledge, good education and outstanding science have to be offered to students. Information is not equal to knowledge; the latter must be followed by understanding. What can we do against the atomization of scientific subjects? Should we not always try to stay on the diagonal line in Fig. 45? Should students not be offered a broad spectrum of science in future too? University faculties are not small Max Planck Institutes with specialized research topics; they should not have only two or three main research centres. What is the important role of biology and ecology? Ecology is green; it is almost a religion, if you follow

Fig. 44 (**a**) Cell level: scanning electron micrograph (SEM) from the surface of a mature leaf of *Atriplex hastata*, showing various stages of development from fully turgid bladders to totally shrunken remnants (Schirmer and Breckle 1982). (**b**) Plant level: *Rhododendron afghanicum*, an extremely rare plant, formerly known only in two localities at the south-eastern Hindu Kush treeline, now most probably extinct; the collecting history from our excursion with Ian Hedge and Per Wendelbo can be found in Larsen (2009) (photo: S. W. Breckle, 1969). (**c**) Ecosystems biocoenosis level: tropical mountain canopy at Réunion with volcanic boulders (photo: U. Breckle, 2017). (**d**) Ecosystems landscape level: arid tropics in Bolivia, altiplano on the Sajama peak (6,542 m) and the very high forest belt with *Polylepis tarapacana* between 4,350 and 4,800 m, with the highest individuals at up to 5,300 m (photo: S. W. Breckle, 2003)

statements on television and in politics. At least now everybody knows: “Trees produce oxygen!” But the stoichiometry of carbon dioxide (CO_2) and global warming is something none of the public understand. Basic knowledge of natural science topics (MINT) urgently needs to be pushed in schools.

What is kept in mind by former students? These are mainly the “big excursions” of 1 or 2 weeks, the many coloured slides in lectures and the joint efforts in ice skating. What else? Why are almost no excursions offered for students today? The advice I gave students often included the following:

> Have clear aims about what you want to achieve. Specialize as late as possible. Have at least one or two hobbies. Biologists need good knowledge and understanding of chemistry. Use the internet and software critically; use it to make work easier. Always try to optimize applied methods. If something works well, keep it; be conservative and innovative. Set scientific rationality against irrationality and esotericism, fake news and fundamentalism; it

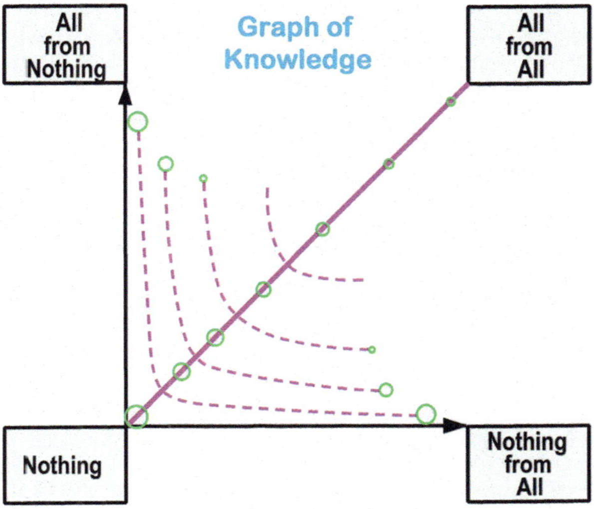

Fig. 45 Knowing and understanding everything from nothing and nothing from everything

is better to think before talking. Use international contacts and try to be able to communicate in more than one foreign language. Visit foreign countries and realize how small Germany is. Be aware of the short lifetime and the limited size of humans – in comparison with our planet, our sun, and our galaxy.

Acknowledgements Scientific curiosity has to be animated, by parents and by teachers. To both, I say thanks.

If university faculties offer excursions for students, they firstly have to be attractive and secondly worth the price. If, during the time of study, it is possible to join about ten 2-week excursions (soil science, botany, geography, geology, limnology, palaeontology, ethnology, zoology), then this is the most intensive learning process and best way of scientific mediation – though also a strain. I say thanks.

Personally, I would like to mention and acknowledge a few scientists and university teachers: Prof. Hermann Aldinger, Stuttgart (geology); Prof. Max Bense, Stuttgart (philosophy); Prof. Hellmut Bredereck, Stuttgart (organic chemistry); Prof. Heinz Ellenberg, Göttingen (ecology); Prof. Theodor Förster, Stuttgart (physical chemistry); Prof. Helmut Gams, Innsbruck (botany); Prof. Josef Goubeau, Stuttgart (inorganic chemistry); Prof. Emil Greiner, Ludwigsburg (physics); Prof. Helmut Heuberger, Innsbruck (geography); Prof. Johannes-Erich Hiller, Stuttgart (crystal chemistry); Prof. Herfried Hoinkes, Innsbruck (climatology); Prof. Karl Ilg, Innsbruck (ethnology); Prof. Helmut Kinzel, Vienna (ecophysiology) Prof. Hans Kinzl, Innsbruck (geography); Prof. Hermann Lautensach, Stuttgart (geography); Dr Werner Marten, Blanes (entomology); Prof. Wolfgang Meckelein, Stuttgart (geography); Prof. Otto Pflugfelder, Stuttgart-Hohenheim (zoology); Prof. Arthur Pisek, Innsbruck (ecophysiology of plants); Prof. Hans Pitschmann, Innsbruck (botany); Prof. Herbert Reisigl, Innsbruck (botany); Prof. Ernst Schlichting, Stuttgart-Hohenheim (pedology); Prof. Maximilian Steiner, Bonn (pharmacognosy); and Prof. Heinrich Walter, Stuttgart-Hohenheim (geobotany/ecology). I say thanks.

Additionally, I have to thank all of those colleagues who have been very cooperative in joint projects and other joint ventures or were visiting guests for some time; not all can be named: Prof. Otar Abdalaze (Tbilisi); Prof. Okmir Agakhanjanz (Minsk); Prof. Roland Albert (Vienna); Prof. Wilhelm Barthlott (Bonn); Prof. Kamal H. Batanouny (Cairo); Dr Rasik Bhadresa (Tehran, London); Prof. Martyn Caldwell (Logan); Dr Liliya Dimeyeva (Almati); Dr Eskandar Firouz (Tehran); Prof. Helmut Freitag (Kassel); Prof. Helena Freitas (Coimbra); Prof. Jan Greszta

(Kraków); Ian C. Hedge (Edinburgh); Prof. Paul Keßler (Leiden); Prof. Michael Keusgen (Marburg); Prof. Ajmal Khan (Karachi); Prof. Christian Körner (Basel); Prof. Ulrich Kull (Stuttgart); PD Dr Thomas Littmann (Halle); Prof. Gia Nakhutsrishvili (Tbilisi); Prof. Clas Naumann (Bonn); Dr Takajoshi Nishio (Utsunomiya); Prof. Nathalia Ogar (Almaty); Dr Tatsuhiro Ohkubo (Utsunomiya); Prof. Muazzez Önal (Istanbul); Dr Rodolpho Ortiz (San Ramón); Prof. Barry Osmond (Canberra); Prof. Münir Öztürk (Izmir); Prof. Fabricio Queirolo (Antofagasta); Dr Daud Rafiqpoor (Bonn); Dr Hamer Salazar (Costa Rica); Prof. Ernst-Detlef Schulze (Bayreuth/Jena); Prof. Zoldan Varga (Debrecen); Prof. Yoav Waisel (Tel Aviv); Prof. Per Wendelbo (Gothenburg); Prof. Christian Wissel (Leipzig); Dr Walter Wucherer (Greifswald); and Prof. Aaron Yair (Jerusalem).

Special thanks are due to Ian C. Hedge for long years of intense cooperation and for his help in smoothing this text.

Without the help of funding agencies – mainly the DFG (German Research Foundation), BMBF (German Ministry of Education and Research), AERC (Arid Ecosystem Research Center, Jerusalem), GTZ-CCD (Bonn), BMBL (German Ministry of Agriculture), DAAD (German Academic Exchange Service), MURL-NRW (Ministry for the Environment, State of North Rhine–Westphalia), VW Foundation, Schimper Foundation and University of Bielefeld – it would have been impossible to have carried out these projects and studies. To them I say many thanks.

I never tried to publish in nature or science or sought impact factors. But often, over the years, we received complimentary comments. Senckenberg's director (Dr Mosbrugger) wrote concerning our *Field Guide Afghanistan* book: "a marvellous book, even if there are no impact- or Hirsch-factors improved, this book has certainly a much longer value than most of the Nature or Science papers". I say thanks.

All co-workers in the (relatively small) department of ecology; all assistants; all candidates for doctoral theses (approximately 20), for diploma theses (approximately 130) and for the teacher's examination (approximately 25); and all Arbeits-Beschaffungs-Maßnahme (ABM; "job creation scheme" in English) people and technical assistants (especially Irmingard Meier and Anja Scheffer) were a good and reliable team. I say thanks.

This scientific curiosity is shared with my wife, Uta Breckle. She formed the framework of the family and was always of indispensable help in every decade, together with my three children (Hans, Rolf and Margit). To them I say the most important thanks.

References

Agachanjanz OE, Breckle S-W (2002) Plant diversity and endemism in high mountains of Central Asia, the Caucasus and Siberia. In: Mountain biodiversity – a global assessment. Körner C, Spehn E (eds) Parthenon Publ Group, Boca Raton, New York etc., chapter 9, pp 117–127

Agakhanjanz O (2002) Der Wind, der heißt Afghane. Forschungen auf dem Pamir im Jahr der Schlange. (Hg: Breckle S-W, U) Shaker, Aachen. 236 pp

Agakhanjanz OE, Breckle S-W (1995) Origin and evolution of the mountain flora in middle and central Asia. In: Chapin FS III, Körner C (eds) Arctic and alpine biodiversity: patterns, causes and ecosystem consequences. Ecological Studies, vol 113, pp 63–80

Ahmadov KM, Breckle S-W, Breckle U (2006) Effects of grazing on biodiversity, productivity, and soil erosion of alpine pastures in Tajik mountains. In: Spehn E, Körner C, Liberman M (eds) Land-use change and mountain biodiversity. CRC Taylor & Francis, Boca Raton, FL, pp 253–261

Albert R (2005) Das Physiotypen-Konzept – ein Modell zur Erklärung ökophysiologischer Anpassungen. In Veste M, Wucherer W, Homeier J (eds) Ökologische Forschung im globalen Kontext. Festschrift zum 65. Geburtstag von Prof. Dr S.-W. Breckle, pp 1–23

Beadle NCW (1981) In: Walter H, Breckle S-W (eds) The vegetation of Australia. Vegetationsmonographien der einzelnen Großräume, vol 4, Fischer, Stuttgart, 690 pp

Beck EH (2017) Torn between nature and lab: a dying breed of plant scientists? Prog Bot 78:3–41
Beck E, Bendix J, Kottke I, Makeschin F, Mosandl R (2008) Gradients in a tropical mountain ecosystem of Ecuador. Ecol Stud 198. 525 pp
Belsky AJ, Canham CD (1994) Forest gaps and isolated savanna trees. Bioscience 44:77–84
Bendix J, Homeier J, Cueva Ortiz E, Emck P, Breckle S-W, Richter M, Beck E (2006) Seasonality of weather and tree phenology in a tropical evergreen mountain rain forest. Int J Biometeorol 50:370–384
Bendix J, Beck E, Bräuning A, Makeschin F, Mosandl R, Scheu S, Wilcke W (eds) (2013) Ecosystem services, biodiversity and environmental change in a tropical mountain ecosystem of South Ecuador, Ecological Studies, vol 221. Springer, Heidelberg. 480 pp
Bertels C, Rüther P, Kahle H, Breckle S-W (1989) Die Entwicklung des Wurzelsystems von Buchenkeimlingen bei Cadmium- und kombinierter Cadmium-/Blei-Belastung. Verh Ges Ökol 18:367–371
Bittner J, Breckle S-W (1995) The growth rate and age of tree fern trunks in relation to habitats. Am Fern J 85(2):37–42
Boyko H (1966) Basic ecological principles of plant growing by irrigation with highly saline or seawater. In: Boyko H (ed) Salinity and aridity. Junk, The Hague
Brackhage C, Hagemeyer J, Breckle S-W, Greszta I (1996) Radial distribution patterns of Cd and Zn in stems of Scots pine (*Pinus sylvestris* L.) trees analyzed 12 years after a contamination event. Water Air Soil Pollut 90:417–428
Bräuning A, Homeier J, Cueva E, Beck E, Günter S (2008) Growth dynamics of trees in tropical mountain ecosystems. Ecol Stud 198:291–302
Breckle S-W (1963) Darstellung einer Additionsverbindung zwischen Wasserstoffperoxid und Perchlorsäure Teacher Thesis TH Stuttgart, 53 pp
Breckle S-W (1966) Ökologische Untersuchungen im Korkeichenwald Kataloniens PhD Diss Stuttgart-Hohenheim, 190 pp
Breckle S-W (1971) Die Beeinflussung der Vegetation durch hügelbauende Ameisen (*Cataglyphis bicolor* Fabricius) auf der Dasht-i-Khoshi (Ost-Afghanistan). Ber Dtsch Bot Ges 84:1–18
Breckle S-W (1973) Mikroklimatische Messungen und ökologische Beobachtungen in der alpinen Stufe des afghanischen Hindukusch. Bot Jahrb Syst 93:25–55
Breckle S-W (1975) Ökologische Beobachtungen oberhalb der Waldgrenze des Safed Koh (Ost-Afghanistan). Vegetatio (Acta Geobotanica) 30:89–97
Breckle S-W (1976) Zur Ökologie und den Mineralstoffverhältnissen alsalzender und nicht absalzender Xerohylophyten (unter besonderer Berücksichtigung von *Atriplex confertifolia* und *Ceratoides lanata* in Utah/USA). Diss Bot 35. 169 pp
Breckle S-W (1979) Afghanische Drogen und ihre Stammpflanzen II. Süßholz (shirin buia). Afghan J (Graz) 6:87–91
Breckle S-W (1985) Die Siebenbürgische Halophyten-Flora – Ökologie und ihre pflanzengeographische Einordnung. Siebenbürgisches Archiv 3. Folge 20:53–105
Breckle S-W (1986) Studies on halophytes from Iran and Afghanistan. II. Ecology of halophytes along salt gradients. Proc Roy Bot Soc Edinb 89B:203–215
Breckle S-W (1989) Role of salinity and alkalinity in the pollution of developed and developing countries. In: Öztürk MA (ed) International Symposium on the Effect of Pollutants to Plants in Developed and Developing Countries in Izmir/Turkey 22. 28.8.1988, pp 389–409
Breckle S-W (1990) Salinity tolerance of different halophyte types. In: Bassam NEl et al (eds) Genetic aspects of plant mineral nutrition. Plant Soil, vol 148, Kluwer, Dordrecht, pp 167–175
Breckle S-W (1993) Gedanken zur Konzipierung eines Großschutzgebietes am Beispiel Nationalpark Senne. In: Truppenübungsplätze und Naturschutz. Schriftenreihe Dtsch Rates für Landespflege 62(4):72–82
Breckle S-W (1995) How do halophytes overcome salinity? In: Khan MA, Ungar IA (eds) Biology of salt tolerant plants. Karachi, pp 199–213
Breckle S-W (1996a) Growth of tree roots under heavy metal (Pb-) stress. Acta Phytogeogr Suec 81:39–43

Breckle S-W (1996b) Root growth and root architecture of non-halophytes under saline soil conditions. Acta Phytogeogr Suec 81:44–47
Breckle S-W (1997a) Population studies on dominant and abundant tree species in the montane tropical rain forests of the biological reserve north of San Ramón (Sra. de Tilarán, Costa Rica). Tropical Ecol 38:259–272
Breckle S-W (1997b) Ecotoxicology of trace metals in forest ecosystems. In: Adriano DC, Chen Z-S, Yang S-S, Iskandar IK (eds) Biogeochemistry of trace metals. Advances in environmental sciences, Science Reviews, Northwood, UK, pp 357–379. Proceedings of the 2nd International Conference on the Biogeochemistry of Trace Elements, Sept. 1993, Taipeh, Taiwan
Breckle S-W (1999) Vegetation und Klimazonen (begr v H Walter), UTB 14, 7. Aufl., Ulmer-Verlag Stuttgart, 544 pp
Breckle S-W (2002a) Salinity, halophytes and salt affected natural ecosystems. In: Läuchli A, Lüttge U (eds) Salinity: environment – plants – molecules. Kluwer Academic Publishers, Dordrecht, The Netherlands, pp 53–77
Breckle S-W (2002b) Primärsukzessionen – Beispiele von Sanddünen, Vulkanaschen, Salzböden und anderen Substraten. Ber d Reinh-Tüxen-Ges (Hannover) 14:5–16
Breckle S-W (2002c) Walter's vegetation of the earth. Ecological systems of the geo-biosphere. Springer, Heidelberg, 527 pp
Breckle S-W (2002d) Salt deserts in Iran and Afghanistan. In: Böer B, Barth H-J (eds) Sabkha ecosystems. Kluwer, Dordrecht, The Netherlands, pp 109–122
Breckle S-W (2003): Rehabilitation of the Aral Sea environment, Kazakhstan. Proceedings of the International Workshop (Aleppo, May 2002): Combating Desertification – Rehabilitation of Degraded drylands and Biosphere Reserves. UNESCO-MAB Dryland Series No 2:47–57
Breckle S-W (2004) Flora, vegetation und Ökologie der alpin-nivalen Stufe des Hindukusch (Afghanistan). In: Breckle S-W, Schweizer B, Fangmeier A (eds) Proceedings of 2nd Symposium AFW Schimper-Foundation: Results of Worldwide Ecological Studies. Stuttgart-Hohenheim, pp 7–117
Breckle S-W (2005a) Naturwissenschaftliche Neugierde, 40 Jahre ökologisch-geobotanische Forschung – Rückblick und Ausblick. In: Veste M, Wucherer W, Homeier J (eds) Ökologische Forschung im globalen Kontext. Festschrift zum 65. Geburtstag von Prof. Dr S.-W. Breckle, pp 273–295
Breckle S-W (2005b) Möglicher Einfluss des Klimawandels auf die Waldvegetation Nordwestdeutschlands? LÖBF-Mitteilungen 2/05:12–17
Breckle S-W (2006) Deserts and biodiversity – is it area- or resource-related? J Arid Land Stud 16:61–74
Breckle S-W (2007) Flora and vegetation of Afghanistan. Basic Appl Dryland Res 1(2):155–194
Breckle S-W (2008) Der Stellenwert des Deutschen als Wissenschaftssprache in Afghanistan. In: Zerwinsky S (ed) Lessing in Kabul. Deutsche Sprache, Literatur und Germanistik in Afghanistan. Iudicium-Verlag, München, pp 191–199
Breckle S-W (2009) Is sustainable agriculture with seawater realistic? In: Ashraf M, Öztürk M, Athar HR (eds) Tasks for Vegetation Science, vol 44, pp 187–196
Breckle S-W (2011a) Der verlorene Aralsee – aber 60 000 km^2 Neuland. In: Anhuf D, Fickert T, Grüninger F (eds) Ökozonen im Wandel, Passauer Kontaktstudium Geographie, vol 11, pp 91–100
Breckle S-W (2011b) Vegetationszonen und Klima – gestern, heute und morgen. In: Anhuf D, Fickert T, Grüninger F (eds) Ökozonen im Wandel, Passauer Kontaktstudium Geographie, vol 11, pp 27–36
Breckle S-W (2013) From Aral Sea to Aralkum – an ecological disaster or halophytes' paradise. Prog Bot 74:351–398
Breckle S-W (2016) Alles zum Eislaufen. 10. Auflage. Sport-Kurs Sport-Fakultät Univ. Bielefeld "Eislaufen" mit Eiskunstlaufen, Eistanz, Synchronlaufen, Eisschnellauf, Eishockey, Spiele auf dem Eis. Info-Buch WS 2016/17, 110 pp

Breckle S-W, Breckle U (2014) Orienteering and environmental impacts, 30 years of longterm vegetation observations of control sites. Sci J Orienteering 19:3–24

Breckle S-W, Frey W (1974) Die Vegetationsstufen im Zentralen Hindukusch. Afghanistan J (Graz) 1:75–80

Breckle S-W, Frey W (1976) Beobachtungen zur heutigen Vergletscherung der Hauptkette des Zentralen Hindukusch (Afghanistan). Afghanistan J (Graz) 3:95–100

Breckle S-W, Hagemeyer J (1992) Untersuchung des Einflusses von Schwermetallen (insbes. Cd, Zn) auf die Kambiumaktivität mitteleuropäischer Waldbäume (Buche, Fichte). Forschungsberichte Nordrhein-Westfalen Luftverunreinigungen und Waldschäden, pp 1–142

Breckle S-W, Kahle H (1992) Effects of toxic heavy metals (Cd, Pb) on growth and mineral nutrition of beech (*Fagus sylvatica*). Vegetatio 101:43–53

Breckle S-W, Koch M (1982) Afghan. Drogen und ihre Stammpflanzen III Hanf und Haschisch. Afghan J (Graz) 9:115–123

Breckle S-W, Naumann C (eds) (1982) Forschungen in und über Afghanistan, vol 22. Dtsch Orient-Institut, Mitteilungen, 230 pp

Breckle S-W, Rafiqpoor DM (2010) Field guide Afghanistan – flora and vegetation. Scientia Bonnensis, 864 pp

Breckle S-W, Rafiqpoor MD (2016) Bildung und Öffentlichkeitsarbeit zur Artenvielfalt – Bilderfloren und Internetgalerien, Beispiel Afghanistan. In: Lozan JL, Breckle S-W, Müller R, Rachor E (eds): Warnsignal Klima: Die Biodiversität. Wissensch Auswertungen in Koop. mit GEO. pp 338–343

Breckle S-W, Unger W (1977) Afghanische Drogen und ihre Stammpflanzen. I Gummiharze von Umbelliferen. Afghan J (Graz) 4:86–95

Breckle S-W, Wucherer W (2006) Vegetation of the Pamir (Tajikistan): land use and desertification problems. In: Spehn E, Körner C, Liberman M (eds) Land-use change and mountain biodiversity. CRC Taylor & Francis, Boca Raton, pp 239–251

Breckle S-W, Wucherer W (2011) Hat der Aralsee eine Zukunft? In: Lozan JL et al. (eds) Warnsignal Klima: Genug Wasser für alle? Wissenschaftliche Auswertungen, GEO/Hamburg 3. Aufl., pp 210–219

Breckle S-W, Wucherer W (2012) Halophytes and salt desertification in the Aralkum area. In: Breckle, S-W, Dimeyeva L, Wucherer W, Ogar NP (eds) Aralkum – a man-made desert. The desiccated floor of the Aral Sea (Central Asia). Ecological Studies, vol 218, pp 271–299

Breckle S-W, Agachanjanz OE, Wucherer W (1998) Der Aralsee; Geoökologische Probleme. Naturwiss Rundschau 51:347–355

Breckle S-W, Breckle H, Breckle U (1989) Vegetation impact by orienteering? A phytosociological long-term study. Sci J Orienteering 5:25–36

Breckle S-W, Schweizer B, Arndt U (Hg) (2000) Ergebnisse weltweiter ökologischer Forschungen (results of worldwide ecological studies, Proceedings of the 1st Symposium of the AFW. Schimper-Foundation, established by H & E Walter, Hohenheim) Verlag Günter Heimbach, Stuttgart, 498 pp

Breckle S-W, Richardt U, Scheffer A, Veste M, Waisel Y (2001a) Wurzeln in der Luft – ein Werkzeug der Wurzelforschung. In: Leuschner C, Hertel C, Hölscher D (Hg) Ökologie des Wurzelraumes, 6. Jahrestagung des AK Experimentelle Ökologie der Gesellschaft für Ökologie. 1–3. April 2001, Göttingen, p 17

Breckle S-W, Veste M, Wucherer W (eds) (2001b) Sustainable land-use in deserts. Proceedings of the International Königswinter Workshop. Springer, Berlin, 465 pp

Breckle S-W, Jankowski M, König C, Minnich M, Posch S, Sagerer G, Veste M (2001c) Anwendung der Minirhizotron-Technik und automatische Bildauswertung. In: Leuschner C, Hertel C, Hölscher, D (Hg) Ökologie des Wurzelraumes, 6. Jahrestagung des AK Experimentelle Ökologie der Gesellschaft für Ökologie. 1–3. April 2001, Göttingen, p 16

Breckle S-W, Haverkamp M, Scheffer A, Veste M (2003) Ökologische Optimierung der Wassernutzung bei Bewässerungsverfahren mit salzhaltigem Wasser in ariden Gebieten. Bielefelder Ökologische Beiträge Band 16. 108 pp

Breckle S-W, Schweizer B, Fangmeier A (Hg) (2004) Ergebnisse weltweiter ökologischer Forschungen (results of worldwide ecological studies, Proceedings of the 2nd Symposium of the AFW. Schimper-Foundation, established by H & E Walter, Hohenheim) Verlag Günter Heimbach, Stuttgart, 397 pp

Breckle S-W, Breckle U, Homeier J, Scheffer A (2005) Mineral deficiencies in a pine-plantation in southern Ecuador. Ecotropica 11:79–85

Breckle S-W, Veste M, Yair A (eds) (2008) Arid dune ecosystems – the Nizzana sands in the Negev Desert. Ecol Stud, 200. 475 pp

Breckle S-W, Dimeyeva L, Wucherer W, Ogar NP (eds) (2012) Aralkum – a man-made desert. Ecol Stud 218:484 pp

Breckle S-W, Hedge IC, Rafiqpoor MD (2013) Vascular plants of Afghanistan – an augmented checklist. Scientia Bonnensis, Bonn, 598 pp

Breckle S-W, Breckle U, Wucherer W (2014a) Gradient studies and desiccation history of the Aral Sea floor. Proceedings 75 Years Nature Reserve Barsa Kelmes. Aralsk, pp 112–117

Breckle S-W, Breckle H, Breckle R, Breckle U (2014b) Orienteering and environmental impacts, observations of controls before and after an O-event and generalizations. Sci J Orienteering 19:25–58

Breckle S-W, Vitek E, Akhani H, Gharemaninejad F, Joharchi MR, Mehrabian AR, Mozzaffarian V, Naderi R, Ranjabr M (2015) Simaroubaceae. Flora Iranica, NHM Wien Lfg 181:15 pp

Breckle S-W, Fayvusch G, Murtazaliev R, Nakhutsrishvili G (2017) Floristic analysis of the subnival-nival vegetation of Hindu Kush and Caucasus (in preparation)

Breckle S-W, Hedge IC, Rafiqpoor MD (2018) Biodiversity in Afghanistan. In: Pullaiah T (ed) Global biodiversity, vol. 1 Asia and Australia. Apple Academic Press (in press)

Brehm G, Homeier J, Fiedler K, Kottke I, Illig J, Nöske NM, Werner F, Breckle S-W (2008) Are mountain rain forests in southern Ecuador a hotspot of biodiversity? Limited knowledge and diverging patterns. In: Beck E et al (eds) Gradients in a tropical mountain ecosystem of Ecuador. Ecological Studies, vol 198, pp 15–23

Cueva Ortiz E, Homeier J, Breckle S-W, Bendix J, Emck P, Richter M, Beck E (2006) Seasonality in an evergreen tropical mountain rain forest in southern Ecuador. Ecotropica 12:69–85

Dalitz H, Oesker M, Todt H, Wolter A (2005) Spatial heterogeneity of bulk precipitation in tropical rain forests of Costa Rica, Ecuador and Kenya. In: Veste M, Wucherer W, Homeier J (eds) Ökologische Forschung im globalen Kontext. Festschrift zum 65. Geburtstag von Prof. Dr S.-W. Breckle, pp 163–174

Dalitz C, Dalitz H, Musila W, Masinde S (2011) Illustrated field guide to the common woody plants of Kakamega forest. Ber Inst Landschafts u Pflanzenökologie Univ. Hohenheim Beiheft 24. 615 pp

Djamali M, Brewer S, Breckle S-W, Jackson ST (2012) Climatic determinism in phytogeographic regionalization: a test from the Irano-Turanian region, SW and Central Asia. Flora 207:237–249

Ellenberg H, Weber HE, Düll R, Wirth V, Werner W, Paulißen D (1991) Zeigerwerte von Pflanzen in Mitteleuropa. Scripta Geobot 18:1–248

Erz G, Veste M, Anlauf H, Breckle S-W, Posch S (2005) A region and contour based technique for automatic detection of roots of tomatoes in minirhizotron images. Appl Bot Food Qual 79:83–88

Eshel A, Beeckman T (2013) Plant roots: the hidden half, 4th edn. CRC Press, Boca Raton, FL. 848 pp

Flowers TJ, Yeo AR (1995) Breeding for salinity resistance in crop plants: where next? Aust J Plant Physiol 22:875–884

Gallagher J (1985) Halophytic crops for cultivation at seawater salinity. Plant Soil 89:323–336

Gao J, Nishio T, Ichizen N, Veste M, Breckle S-W (2007) Desertification and rehabilitation in China – an overview. J Arid Land Stud 17:101–112

Gliniars R, Becker GS, Braun D, Dalitz H (2011) Monthly stem increment in relation to climatic variables during 7 years in an East African rainforest. Trees Struct Funct 27:1129–1138

Groeneveld J, Wucherer W, Dimeyeva LA, Jeltsch F, Wissel C, Breckle S-W (2012) Spatial distribution of plant functional types along stress gradients – a simulation study orientated towards the plant succession on the desiccating Aral Sea floor. In: Breckle S-W, Dimeyeva L, Wucherer W, Ogar NP (eds): Aralkum – a man-made desert. The desiccated floor of the Aral Sea (Central Asia). Ecol studies, vol 218, pp 301–311

Gross A (1998) Terrestrische Orchideen einer Hangrutschung im Bergwald Süd-Ecuadors: Verteilung, Phytomasse, Phänologie und Blütenmerkmale. Thesis Univ. Ulm

Hagemeyer J, Breckle S-W (1986) Cadmium in den Jahrringen von Eichen: Untersuchungen zur Aufstellung einer Chronologie der Immissionen. Angew Botanik 60:161–174

Hagemeyer J, Breckle S-W (1996) Growth under trace element stress. In: Waisel Y, Eshel A, Kafkafi U (eds) Plant roots: the hidden half. 2nd edn. pp 415–433

Hagemeyer J, Kahle H, Breckle S-W, Waisel Y (1986) Cadmium in *Fagus sylvatica* L. trees: leaching, uptake and effect on transpiration. Water Air Soil Pollut 29:347–359

Hagemeyer J, Lohrmann D, Breckle S-W (1993) Development of annual xylem rings and shoot growth of young beech (*Fagus sylvatica* L.) grown in soil with various Cd and Zn levels. Water Air Soil Pollut 69:351–361

Hedenström HV, Breckle S-W (1974) Obligate halophytes? A test with tissue culture methods. Z Pflanzenphysiol 74:183–185

Henkel PA, Shakhov AA (1945) The ecological significance of the water regime of certain halophytes (Russ). Bot Zh 30:154–166

Homeier J (2004) Baumdiversität, Waldstruktur und Wachstumsdynamik zweier tropischer Bergregenwälder in Ecuador und Costa Rica. Diss Bot 391:1–207

Homeier J, Breckle S-W (2004) Growth of some frequent tree species from an Ecuadorian montane forest. In: Breckle S-W, Schweizer B, Fangmeier A (eds) Proceedings of 2nd Symposium AFW Schimper-Foundation: Results of Worldwide Ecological Studies. Stuttgart-Hohenheim, pp 375–379

Homeier J, Breckle S-W (2008) Gap-dynamics in the tropical montane forest in South Ecuador. In: Beck E et al (eds) Gradients in a tropical mountain ecosystem of Ecuador, Ecol studies, vol 198, pp 311–317

Homeier J, Breckle S-W, Dalitz H, Leyers C, Ortiz R (2002a) Demography, spatial distribution and growth of three arborescent palm species in a tropical premontane rain forest in Costa Rica. Ecotropica 8:239–247

Homeier J, Dalitz H, Breckle S-W (2002b) Waldstruktur und Baumartendiversität im montanen Regenwald der Estación Científica San Francisco in Südecuador. Ber Reinh-Tüxen-Ges (Hannover) 14:109–118

Homeier J, Werner FA, Gradstein R, Breckle S-W, Richter M (2008) Potential vegetation and floristic composition of Andean forests in South Ecuador, with a focus on the reserve San Francisco. In: Beck E et al (eds) Gradients in a tropical mountain ecosystem of Ecuador, Ecol studies, vol 198, pp 87–100

Homeier J, Breckle S-W, Günter S, Rollenbeck RT, Leuschner C (2010) Tree diversity, forest structure and productivity along altitudinal and topographical gradients in a species-rich Ecuadorian montane rain forest. Biotropica 42:140–148

Hueck K (1966) In: Walter H (ed) Die Wälder Südamerikas. Veget.-Monographien der einzelnen Großräume, vol 2. Fischer, Stuttgart, 422 pp

Jankowski M, Posch S, Sagerer G, Breckle S-W, Veste M (1995) Automatische Detektion von Wurzelsystemen in Minirhizotron-Bildern. In: Sagerer G, Posch S, Kummert F (eds) Mustererkennung. DAGM-Tagung Informatik aktuell. Springer, pp 176–185

Jones R (ed) (1970) The biology of *Atriplex* Canberra. 128 pp

Kahle H, Breckle S-W (1992) Blei und Cadmium – Zeitbombe in unseren Waldböden. Biol Unserer Zeit 22:21–27

Kahle H, Bertels C, Noack G, Röder U, Rüther P, Breckle S-W (1989a) Untersuchungen zur Toxizität von Blei und VCadmium für Jungwuchs der Buche (*Fagus sylvatica* L.) Bielefelder Ökologische Beiträge 5:99–126

Kahle H, Bertels C, Noack G, Röder U, Rüther P, Breckle S-W (1989b) Wirkungen von Bei und Cadioum auf Wachstum und Mineralstoffhaushalt von Buchenjungwuchs. Allg Forst Zeitschr 29/30:783–788

Kating H, Breckle S-W (1978) Pharmazeutische Biologie I. Grundlagen, Stellung der Arzneipflanzen im System. 1/2.Aufl. Thieme, Stuttgart, 440 pp

Kating H, Breckle S-W (1981) Pharmazeutische Biologie I. Grundlagen, Stellung der Arzneipflanzen im System. 1/2.Aufl. Thieme, Stuttgart, 496 pp

Keusgen M, Breckle S-W, Rafiqpoor MD (2015) Medicinal plants of Afghanistan. Univ. Marburg, 247 pp

Khan AM, Böer B, Őzturk M, Clüsener-Godt M, Gul B, Breckle S-W (eds) (2016) Sabkha ecosystems vol. V: The Americas. 431 S. Springer, Heidelberg

Kidron GJ, Yair A (2008) Runoff and erosion processes within a dune system. In: Breckle S-W, Veste M, Yair A (eds) Arid dune ecosystems – the Nizzana sands in the Negev Desert. Ecological Studies, vol 200, pp 239–249

Klötzli F, Dietl W, Marti K, Schubiger C, Walther G-R (2010) Vegetation Europas. Ott, Hep/Bern, 1190 pp

Knapp R (1965) In: Walter H (ed) Die Vegetation von Nord- und Mittelamerika. Veget.-Monographien der einzelnen Großräume, vol 1, Fischer, Stuttgart, 373 pp

Knapp R (1973) In: Walter H (ed) Die Vegetation von Afrika. Veget.-Monographien der einzelnen Großräume, vol 3, Fischer, Stuttgart, 626 pp

Kürschner H, Lüth M, Frey W (2017) In: Breckle S-W, Hedge IC, Rafiqpoor MD (eds) Field guide Afghanistan: bryophytes. Flora and augmented checklist. UNDP/Kabul, Genf. Ca 350 pp

Larsen OL (2009) Per Wendelbo – en norsk plantejeger – og reintroduksjonen av to sjeldne rhododendronarter. Lapprosen 12:11–13

Leistner E, Breckle S-W (1988) Pharmazeutische Biologie. Grundlagen, Systematik, Humanbiologie. 3. – 8. Aufl. WVA, Stuttgart, 538 pp

Leistner E, Breckle S-W (1992) Pharmazeutische Biologie. Grundlagen, Systematik, Humanbiologie. 3. – 8. Aufl. WVA, Stuttgart, 538 pp

Leistner E, Breckle S-W (1997) Pharmazeutische Biologie. Grundlagen, Systematik, Humanbiologie. 3. – 8. Aufl. WVA, Stuttgart, 550 pp

Leistner E, Breckle S-W (2000a) Pharmazeutische Biologie. Grundlagen, Systematik, Humanbiologie. 3. – 8. Aufl. WVA, Stuttgart, 512 pp

Leistner E, Breckle S-W (2000b) Pharmazeutische Biologie. Prüfungsfragen. WVA, Stuttgart, 294 pp

Leistner E, Breckle S-W (2008) Pharmazeutische Biologie. Grundlagen, Systematik, Humanbiologie. 3. – 8. Aufl. WVA, Stuttgart, 713 pp

Leistner E, Breckle S-W (2014) Pharmazeutische Biologie. Grundlagen, Systematik, Humanbiologie. 3. – 8. Aufl. WVA, Stuttgart, 740 pp

Lerche H, Breckle S-W (1975) Untersuchungen zum Bleigehalt von Baumblättern im Bonner Raum. Angew Bot 48:309–330

Liede-Schumann S, Breckle S-W (eds) (2008) Provisional checklists of flora and fauna of the San Francisco Valley and its surroundings (Reserva Biológica San Francisco, Prov. Zamora-Chinchipe, Southern Ecuador). Ecotrop Monogr 4:256 pp

Littmann T, Veste M (2005) Klima und Vegetaion in Trockengebieten – eine geographische Betrachtung. In: Veste M, Wucherer W, Homeier J (eds) Ökologische Forschung im globalen Kontext. Festschrift zum 65. Geburtstag von Prof. Dr S-W Breckle, pp 71–80

Lozan JL, Breckle S-W, Müller R, Rachor E (eds) (2016) Warnsignal Klima: Die Biodiversität – unter Berücksichtigung von Habitatveränderungen, Umweltverschmutzung, Globalisierung. Wissenschaftliche Fakten. Wissensch. Auswertungen in Koop. mit GEO. 350 pp

Lüttge U (1971) Structure and function of salt glands. Annu Rev Plant Physiol 22:23–44

Lüttge U (2016) Transport processes: the key integrators in plant biology. Prog Bot 77:3–65

Matinzadeh Z, Breckle S-W, Mirmassoumi M, Akhani H (2013) Ionic relationships in some halophytic Iranian Chenopodiaceae and their rhizospheres. Plant Soil. Doi: https://doi.org/10.1007/s11104-013-1744-7

Mayrhofer M, NN (2019) In: Breckle S-W, Hedge IC, Rafiqpoor MD (eds) Field guide Afghanistan: lichens. Flora and augmented checklist. Planned, in preparation. c. 300 pp
Mirazai NA, Breckle S-W (1978) Untersuchungen an afghanischen Halophyten. I. Salzverhältnisse in Chenopodiaceen Nord-Afghanistans. Bot Jahrb Syst 99:565–578
Mueller-Dombois D, Fosberg FR (1998) Vegetation of the tropical Pacific islands. Veget.-Monographien der einzelnen Großräume, vol 9. Springer, NewYork 733 pp
Munns R (2005) Genes and salt tolerance: bringing them together. New Phytol 167:645–663
Osmond CB (1974) Leaf anatomy of Australian saltbushes in relation to photosynthetic pathways. Aust J Bot 22:39–44
Raby P (2002) Alfred Russel Wallace: a life. Princeton University Press, Princeton
Rechinger KH (1963 ff. Flora Iranica ADEVA/Graz; NHM Wien. Parts, pp 1–181
Reimann C, Breckle S-W (1988) Anatomie und Entwicklung der Blasenhaare von *Chenopodium*-Arten. Flora 180:275–288
Russow R, Veste M, Breckle S-W, Littmann T, Böhme F (2008) Nitrogen input pathways into the sand dunes: biological fixation and atmospheric nitrogen deposition. In: Breckle S-W, Veste M, Yair A (eds) Arid dune ecosystems – the Nizzana sands in the Negev Desert, Ecological Studies, vol 200, pp 319–336
Sakai A, Larcher W (1987) Frost survival of plants. Ecol Stud 62:198, 199
Schirmer U, Breckle S-W (1982) The role of bladders for salt removal in some Chenopodiaceae (mainly *Atriplex* species) In: Sen DN, Rajpurohit KS (eds) Contributions to the ecology of halophytes. Tasks for vegetation science, vol 2, Junk, Den Haag, pp 215–231
Seidel G (1988) Untersuchungen zum Wurzelwachstum bei *Chenopodium album* L. bei unterschiedlichem Kalium- und Natrium-Angebot. Dipl. Arb. Univ. Tübingen, Bielefeld, 76 pp
Song J, Wang B (2015) Using euhalophytes to understand salt tolerance and to develop saline agriculture: *Suaeda salsa* as a promising model. Ann Bot 115:541–553
Tarnita CE, Bonachela JA, Sheffer E, Guyton JA, Coverdale TC, Long I RA, Pringle RM (2017) A theoretical foundation for multi-scale regular vegetation patterns. Nature 541:398–401. https://doi.org/10.1038/nature20801
Tiedemann AR, McArthur ED, Stutz HC, Stevens R, Johnson KL (eds) (1984) Proceedings – Symposium on the Biology of *Atriplex* and Related Chenopods (Provo 1983); Ogden/Utah GTR-172, 309 pp
Todt H, Breckle S-W, Veste M (2000) The mistletoe *Loranthus acaciae* (Loranthaceae) on halophytic and non-halophytic hosts in the southern Arava-Valley (Israel). In: Breckle S-W, Schweizer B, Arndt U (eds) Ergebnisse weltweiter Forschung. Günter Heimbach, Stuttgart, pp 475–480
Tsoar H (2008) Land use and its effect on the mobilization and stabilization of the north-western Negev sand dunes. In: Breckle S-W, Veste M, Yair A (eds) Arid dune ecosystems – the Nizzana sands in the Negev Desert, Ecol studies, vol 200, pp 79–89
Vanselow K, Samimi C, Breckle S-W (2016) Preserving a comprehensive vegetation knowledge base – an evaluation of four historical Soviet vegetation maps of the western Pamirs (Tajikistan). PLoS One 11(2). https://doi.org/10.1371/journal.pone.0148930
Ventura Y, Sagi M (2013) Halophyte crop cultivation: the case for *Salicornia* and *Sarcocornia*. Environ Exp Bot 92:144–153
Veste M, Breckle S-W (2003) Untersuchungen zur Populationsökologie und zur Ökophysiologie halophiler Arten in den Sanddünen der nordwestlichen Negev. Bielefelder Ökologische Beiträge Band 17. 98 pp
Veste M, Mohr M (2005) Vegetation der Lineardünen der zentralen Namib und deren Ionenhaushalt. In: Veste M, Wucherer W, Homeier J (eds) Ökologische Forschung im globalen Kontext. Festschrift zum 65. Geburtstag von Prof. Dr S.-W. Breckle, pp 93–107
Veste M, König C, Minnich M (2005) Anwendung der Minirhizotron-Technik zur Untersuchung des Wurzelwachstums und der Wurzelbiomasse von Tomaten (*Lycopersicum esculentum* Mill.) bei unterschiedlicher Bewässerung. In: Veste M, Wucherer W, Homeier J (eds) Ökologische

Forschung im globalen Kontext. Festschrift zum 65. Geburtstag von Prof. Dr S.-W. Breckle, pp 55–69
Veste M, Gao J, Baoping S, Breckle S-W (2006) The green Great Wall – combatting desertification in China. Geogr Rdschau Int Edit 2(3):14–20
Veste M, Heusinkfeld BG, Berkowicz SM, Breckle S-W, Littmann T, Javobs AFG (2008) Dew formation and activity of biological soil crusts. In: Breckle S-W, Veste M, Yair A (eds) Arid dune ecosystems – the Nizzana sands in the Negev Desert, Ecological Studies, vol 200, pp 305–318
Veste M, Breckle S-W, Eggert K, Littmann T (2011) Vegetation pattern in arid sand dunes controlled by biological soil crusts along a climatic gradient in the northern Negev Desert. Basic Appl Dryland Res 5:1–16
Veste M, Todt H, Breckle S-W (2014) Influence of halophytic hosts on their parasites – the case of *Plicosepalus acaciae*. Ann Bot Plants 7:plu084. https://doi.org/10.1093/aobpla/plu084
Waisel Y (1972) Biology of halophytes. Academic Press, New York, London, 395 pp
Waisel Y, Breckle S-W (1987) Differences in responses of various radish roots to salinity. Plant Soil 104:191–194
Waisel Y, Eshel A, Kafkafi U (1991) Plant roots: the hidden half. 1st edn. CRC Press, Marcel Dekker, New York, 948 pp
Waisel Y, Eshel A, Kafkafi U (1996) Plant roots: the hidden half. 2nd edn. CRC Press, Marcel Dekker, New York, 1002 pp
Waisel Y, Eshel A, Kafkafi U (2002) Plant roots: the hidden half. 3rd edn. CRC Press, Marcel Dekker, New York, 1136 pp
Walter H (1961) Salinity problems in the arid zones. The adaptations of plants to saline soils. Arid Zone Res 14:65–68
Walter H (1968) Die Vegetation der Erde in öko-physiologischer Betrachtung Band II: Die gemäßigten und arktischen Zonen. Fischer, Stuttgart, 1001 pp
Walter H (1973) Die Vegetation der Erde in öko-physiologischer Betrachtung. Band I: die tropischen und subtropischen Zonen. 3 Aufl. Fischer, Stuttgart, 743 pp
Walter H (1974) In: Walter (ed) Die Vegetation Osteuropas, Nord- und Zentralasiens. Veget.-Monographien der einzelnen Großräume, vol 7, Fischer, Stuttgart, 452 pp
Walter H (1976) Die ökologischen Systeme der Kontinente (Biogeosphäre). Fischer, Stuttgart, 131 pp
Walter H (1986) Allgemeine Geobotanik. 3 Aufl. Ulmer, Stuttgart, 279 pp
Walter H (1989) Bekenntnisse eines Ökologen. Erlebtes in 8 Jahrzehnten und auf Forschungsreisen in allen Erdteilen mit Schlußfolgerungen. 6. Aufl. Fischer, Stuttgart, 365 pp
Walter H (1990) Vegetation und Klimazonen. 6 Aufl. Ulmer, Stuttgart 382 pp
Walter H, Breckle S-W (1983) Ökologie der Erde, Band 1, 2. Aufl, Grundlagen. Fischer, Stuttgart, 240 pp
Walter H, Breckle S-W (1984) Ökologie der Erde. Band 2. Spezielle Ökologie der Tropischen und Subtropischen Zonen. Breckle S-W: 3 Aufl, UTB Große Reihe, Spektrum, Heidelberg, 764 pp
Walter H, Breckle S-W (1986a) Ecological systems of the geo-biosphere. Vol. 1: ecological principles in global perspective. Springer, 242 pp
Walter H, Breckle S-W (1986b) Ecological systems of the geo-biosphere. Vol. 2: tropical and subtropical zonobiomes. Springer, 465 pp
Walter H, Breckle S-W (1986c) Ökologie der Erde, Band 3. In: Breckle S-W, Agakhanjanz OE (eds) 2. Auflage. Spezielle Ökologie der gemäßigten und arktischen Zonen Euro-Nordasiens. Fischer, Stuttgart, 726 pp
Walter H, Breckle S-W (1989) Ecological systems of the geo-biosphere. Vol 3: temperate and polar zonobiomes of Northern Eurasia. Springer, 581 pp
Walter H, Breckle S-W (1991a) Ökologie der Erde, Band 1, 2. Aufl, Grundlagen. Fischer, Stuttgart, 240 pp
Walter H, Breckle S-W (1991b) Ökologie der Erde, Band 4, Gemäßigte und arktische Zonen außerhalb Euro-Nordasiens. Fischer, Stuttgart, 586 p

Walter H, Breckle S-W (1991c) Ökologie der Erde. Band 2. Spezielle Ökologie der Tropischen und Subtropischen Zonen. Breckle S-W: 3 Aufl, UTB Große Reihe, Spektrum, Heidelberg, 764 pp

Walter H, Breckle S-W (1994) Ökologie der Erde, Band 3. In: Breckle S-W, Agakhanjanz OE (eds) 2. Auflage. Spezielle Ökologie der gemäßigten und arktischen Zonen Euro-Nordasiens. Fischer, Stuttgart, 726 pp

Walter H, Breckle S-W (2004) Ökologie der Erde. Band 2. Spezielle Ökologie der Tropischen und Subtropischen Zonen. Breckle S-W: 3 Aufl, UTB Große Reihe, Spektrum, Heidelberg, 764 pp

Walter H, Harnickell E, Mueller-Dombois D (1975) In: Walter H (ed) Klimadiagramm-Karten der einzelnen Kontinente und die ökologische Klimagliederung der Erde. Veget.-Monographien der einzelnen Großräume, vol 10, Fischer, Stuttgart, 36 pp + 10 Karten

Wattenberg I (1996) Struktur, Diversität und Verjüngungsdynamik eines prämontanen Regenwaldes in der Cordillera de Tilarán in Costa Rica. Bielef Ökol Beiträge 10:1–195

Wattenberg I (1997) Comparative studies on population ecology of *Iriartea deltoidea*, *Elaeagia uxpanapensis* and *Calatola costaricensis* in a tropical premontane rain forest in the Cordillera de Tilarán, Costa Rica. Bielef Ökol Beiträge 11:47–59

Wattenberg I, Breckle S-W (1995) Tree species diversity of a premontane rain forest in the Cordillera de Tilarán, Costa Rica. Ecotropica 1:21–30

Weber C (1994) Mineralstoffverhältnisse bei Baumfarnen in einem prämontanen Regenwald Costa Ricas Dipl Arb Bielefeld, 115 pp

Weber C, Breckle S-W (1994) Mineralstoffverhältnisse bei Baumfarnen in der Reserva Forestal de San Ramón, Costa Rica. Poster 7. Jahrestagung der GTÖ, Febr 1994, Bremen

Weisser F, Kahle H, Breckle S-W (1990) Regenerationsvermögen von Buchenwurzeln in unterschiedlich bleibelasteten Wurzelbodenzonen. Verh Ges Ökol 20:178–179

Wiehe W (1986) Untersuchungen zur Salzrekretion bei ausgewählten *Limonium*-Arten Dipl Arb Bielefeld. 135 pp

Wiehe W, Breckle S-W (1990) Die Ontogenese der Salzdrüsen von *Limonium* (Plumbaginaceae). Bot Acta 103:107–110

Wucherer W, Breckle S-W (2001) Vegetation dynamics on the dry sea floor of the Aral Sea. In: Breckle S-W, Veste M, Wucherer W (eds) Sustainable land use in deserts. Springer, Berlin, pp 52–68

Yair A (2008) Succession stages in the recovery processes of the topsoil crust in a disturbed sand arid area. In: Breckle S-W, Veste M, Yair A (eds) Arid dune ecosystems – the Nizzana sands in the Negev Desert, Ecological Studies, vol 200, pp 285–303

Yuan F, Leng B, Wand B (2016) Progress in studying salt secretion from plant glands in recretohalophytes: how do plants secrete salts? Front Plant Sci 7:977. https://doi.org/10.3389/fpls.2016.00977

Engineering Plants for the Future: Farming with Value-Added Harvest

Silvia Massa, Ombretta Presenti, and Eugenio Benvenuto

Contents

Abstract Plants and their rich variety of natural compounds are used to maintain and to improve health since the earliest stages of civilization. Despite great advances in synthetic organic chemistry, one fourth of present-day drugs have still a botanical origin, and we are currently living a revival of interest in new pharmaceuticals from plant sources.

Modern biotechnology has defined the potential of plants to be systems able to manufacture not only molecules naturally occurring in plants but also newly engineered compounds, from small to complex protein molecules, which may originate even from non-plant sources. Among these compounds, pharmaceuticals

---

Communicated by Francisco M. Cánovas

S. Massa, O. Presenti, and E. Benvenuto (✉)
Department of Sustainability, Division of Biotechnology and Agroindustry, Laboratory of Biotechnology, ENEA - Italian National Agency for New Technologies, Energy and the Environment, Rome, Italy
e-mail: eugenio.benvenuto@enea.it

Progress in Botany (2019) 80: 65–108, DOI 10.1007/124_2018_20,
Published online: 30 June 2018

such as vaccines, antibodies and other therapeutic or prophylactic entities can be listed. For this technology, the term plant molecular farming has been coined with reference to agricultural applications due to the use of crops as biofactories for the production of high-added value molecules. In this perspective, edible plants have also been thought as a tool to deliver by the oral route recombinant compounds of medical significance for new therapeutic strategies. Despite many hurdles in establishing regulatory paths for this "novel" biotechnology, plants as bioreactors deserve more attention when considering their intrinsic advantages, such as the quality and safety of the recombinant molecules that can be produced and their potential for large-scale and low-cost production, despite worrying issues (e.g. amplification and diffusion of transgenes) that are mainly addressed by regulations, if not already tackled by the plant-made products already commercialized. The huge benefits generated by these valuable products, synthesized through one of the safest, cheapest and most efficient method, speak for themselves.

Milestone for plant-based recombinant protein production for human health use was the approval in 2012 by the US Food and Drug Administration of plant-made taliglucerase alfa, a therapeutic enzyme for the treatment of Gaucher's disease, synthesized in carrot suspension cultures by Protalix BioTherapeutics.

In this review, we will go through the various approaches and results for plant-based production of proteins and recent progress in the development of plant-made pharmaceuticals (PMPs) for the prevention and treatment of human diseases. An analysis on acceptance of these products by public opinion is also tempted.

Keywords Plant molecular farming, Plant-derived antibodies, Plant-derived vaccines, Responsible research and innovation

1 Introduction

Plants have provided mankind with useful molecules for centuries. The idea to produce heterologous (exogenous, non-plant) protein in plants to get plant-made pharmaceuticals (PMPs) has been available and rapidly advanced, for just over 30 years (Ma et al. 2003). The first plant-made pharmaceutically relevant protein was the human growth hormone, expressed in transgenic tobacco in 1986 (Barta et al. 1986). Other human proteins have been produced in an increasing variety of plant species and of techniques. In 1989, the expression of the first antibody in tobacco (Hiatt et al. 1989) and, subsequently, the production of the first experimental plant-made vaccine (e.g. hepatitis B virus surface antigen) (Mason et al. 1992) clearly showed that plants had the capability to assemble functional glycoproteins with complex structure. The structural authenticity and preservation of function of plant-produced recombinant proteins was confirmed by the production of more complex antibody derivatives such as the secretory immunoglobulin A (Ma et al. 1995) and new immunogenic proteins (Haq et al. 1995). Also antigenic peptides, expressed as fusions to plant viral coat proteins, were made, and, in many cases, the assembled viral-like particles were demonstrated able to protect animals from the challenge with the pathogen from which the epitope was derived (Rybicki 2014; Kushnir et al. 2012). At the same time, fusions with plant-derived proteins or only plant-typical sequences were found to improve the immunogenicity of antigens for vaccination against tumours (Massa et al. 2011, 2017).

The evolution of molecular techniques has led to the development of systems like the Gateway-mediated cloning (reviewed by Dafny-Yelin and Tzfira 2007), Golden Gate (Binder et al. 2014) and GoldenBraid (Sarrion-Perdigones et al. 2011) just to name a few, thanks to which even the assembly of multi-modular constructs for the expression of multiple proteins or enzymes in plants is now possible.

Beside proteins for pharmacological use, also plant-made industrial products (PMIPs) have been produced, such as enzymes, proteins for research use, nutritional supplements, polymers, etc. (Davies 2010; Tschofen et al. 2016).

Unlike mammalian cell-derived drugs, plant-derived antibodies, vaccines and other pharmacologically relevant proteins are particularly advantageous in that they are free from mammalian viral vectors and/or human pathogens. Advantages offered by plants may include also low cost of cultivation and high biomass production, relatively fast "gene-to-protein" time, low investment costs, good scalability, eukaryotic post-translational modifications and a high protein yield (Table 1).

The availability of effective technical means of expressing proteins in plants offers the prospect of using them as bioreactors for the production of cost-effective pharmaceuticals for both human and animal health sectors.

Over the past four decades, a wealth of literature has demonstrated that the production of proteins in plants for health applications is a promising approach in the area of biologics manufacturing. It is hoped that together to what has been defined in the MIT Technology Review "the biggest biotech discovery of the century", the CRISPR technology, new frontiers of biomedical discoveries, ranging from the pharmaceutical sector to the agro-industry, will be opened and ready for this "next-generation" plant-based medicine.

In its intent, the present review is thought to make the point on the evident opportunities that arise from the "Farming for Pharming" of biologics through plants. If properly addressed, this technology is destined to tackle human and animal diseases breaking the limits of current standard production technologies of biopharmaceuticals. This in turn will help reduce disparities in health rights and guarantee better health protection in the name of the guiding principle of reduction of costs.

2 The Strategies and the Technological Platforms of Plant Transformation

The diversity of plant molecular farming systems available reflects the wealth of possibilities offered by plants. Recombinant proteins have been produced in many different plant species wherein there is a choice of whole plants or various cell/tissue

Table 1 Comparison of expression hosts for the production of proteins of pharmacological interest (content partially sourced from Ma et al. 2003; Franconi et al. 2010; Loh et al. 2017)

Host	Expression level	Production time	Production cost	Scale-up capacity	Glycosilation pattern (vs human)	Risk of contamination
Bacteria	Medium/ high	Short	Low/ moderate	High	None	High: endotoxins
Yeast	Low/high	Medium	Medium	High	Incorrect: high mannosylation	Low
Insect cells	Low/high	Medium	High	Medium	Incorrect: high mannosylation	High: baculoviruses and mammalian viruses
Mammalian cells	Low/ medium	High	High	Very low	Correct	High: mammalian viruses, prions, oncogenic DNA
Animal	Medium/ high	Very high	High	Low	Correct	High: mammalian pathogens, prions, oncogenic DNA
Plant cell cultures	Medium/ high	Short	Low-moderate	High	Minor differences, possibility to modify	Very low
Plant organs (e.g. hairy roots)	Medium/ high	Medium/ short	Low-moderate	High	Minor differences, possibility to modify	Very low
Whole-plant transient (agro-infiltration)	Medium/ high	Medium	Low	High	Minor differences, possibility to modify	Very low
Whole-plant stable (nuclear/plastidial)	Medium/ high	Long	Low	High	Minor differences, possibility to modify	Very low

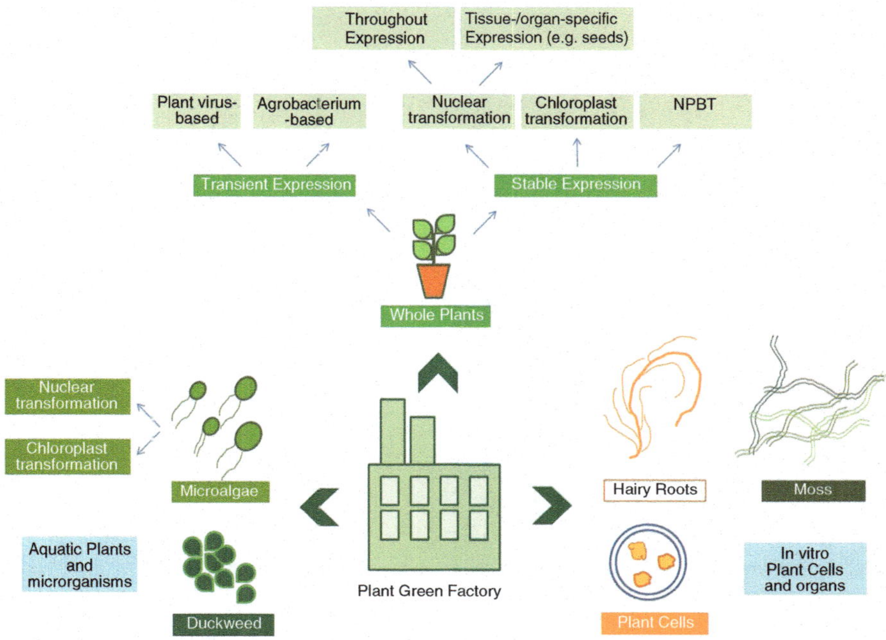

Fig. 1 The plant factory expression platforms (content partially sourced from Xu et al. 2012; NPBT: new plant breeding techniques)

culture formats (Fig. 1). Plants may be suitable for stable expression (including nuclear and plastid transformation in some species) and transient expression (which can be achieved using *Agrobacterium tumefaciens*, plant virus vectors or combinations of both). The various approaches for plant-based production of proteins are schematically illustrated in Fig. 2.

2.1 Stable Transformation

2.1.1 The "Whole-Plant" Approach

Stable nuclear and chloroplast transformations are the two approaches used to stably express heterologous (i.e. non plant) recombinant proteins in plants. *Agrobacterium*-mediated and particle bombardment-mediated stable transformation have a long history and are achieved by stable integration of DNA into a plant genome which results in the expression of transgenes (Tinland 1995).

In the field of whole-plant transformation, mainly tobacco and its relative *Nicotiana benthamiana*, which has a fast growth rate and an apparent defective RNA silencing system (Goodin et al. 2008), potato, tomato and lettuce are the main leafy crops used with the primary advantage of an abundant leaf biomass

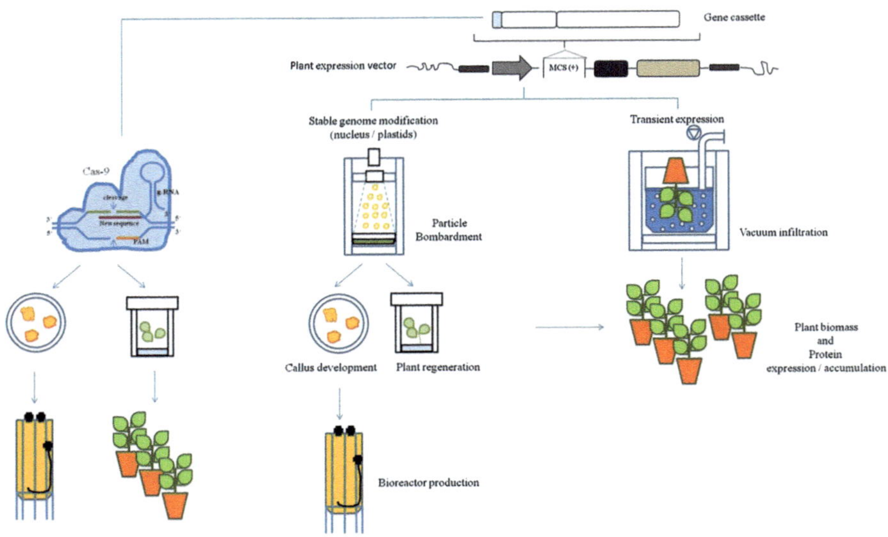

Fig. 2 Production of recombinant proteins in plants using stable or transient transformation methods (partially sourced from Loh et al. 2017)

accumulation accompanying the expression of the exogenous protein (Tremblay et al. 2010). Among cereals, maize, rice and barley (Nandi et al. 2005; Sabalza et al. 2013) have been tried along with soybean, among legumes. While the watery tissues of leaves impose to extract rapidly the target protein in order to minimize degradation, desiccated seeds allow protein to remain stable and are suitable for a possible oral administration.

Despite the large array of plant systems available and the great advantage of playing a seminal role in the development of "green" bioreactors, the "whole-plant" transgenic approach is labour-intensive and time-consuming due to the necessary steps for plant regeneration and selection, with an average lead time ranging from 12 to 18 months to engineer plants that often show non-competitive contents of the target protein (Gleba et al. 2005).

Stable introduction of target genes into chloroplast genome (transplastomics) allows for higher levels of expression if compared to nuclear transformation, due to the lack of gene silencing phenomena and high gene copy numbers. The technique is also advantageous since it prevents possible transgene escape (plastids are inherited through the maternal tissue in the majority of species) and absence of chloroplasts in pollen and improbability of their transfer, which quells environmental concerns (Meyers et al. 2010; Cardi et al. 2010). This method allows high concentration of recombinant protein (Oey et al. 2009). Nevertheless, it is technically difficult since several generations of plant regeneration are needed to select transformed plants with all chloroplasts bearing the gene of interest (homoplasmic state). In addition, this platform lacks most post-translational modifications being successful in a limited number of species.

Aquatic plants like duckweed (*Lemna minor*) and other non-vascular plants such as moss (*Physcomitrella patens*) have been considered as alternative expression platforms (Reski et al. 2015). Having properties in common with both terrestrial plants (in that they are differentiated whole plants) and cell suspension cultures (they can be grown in containment and in simple mediums), technologies applicable to aquatic plants and cell suspension cultures are similar with the difference being that aquatic plants require light, while cell suspension cultures are usually grown in the dark with the addition of a carbon source.

2.1.2 Cell Suspension Cultures

Removal of cell walls and gene transfer into the resulting protoplasts and suspension culture is an easier method in comparison to whole-plant genetic manipulation, since this might simplify purification and downstream processing (Santos et al. 2016). In addition, suspension culture allows homogeneity in cell proteins and sugars (*N*-glycans) due to the high uniformity of type and cell size (Lienard et al. 2007). Plant cells can be cultivated aseptically using classical fermentation technology, are easy to scale up and must comply with regulatory requirements similar to those established for consolidated systems based on both microbial and mammalian cells. In fact, plant suspension cultures, being made of single cells, allow comparisons with the mostly used mammalian cell production of recombinant pharmaceuticals, Chinese hamster ovary (CHO) cells (Table 2). Cell-specific production rates of 8 pg/cell/day have been reported for the monoclonal antibody M12 produced in tobacco BY-2 cells (Havenith et al. 2014) compared to typical production rates of 20–40 pg/cell/day for CHO cells (typically carrying thousands of gene copies), showing that the difference between these systems is less than an order of magnitude (Santos et al. 2016).

Tobacco cell lines BY-2 and NT-1 have been used to produce many recombinant proteins (Ullisch et al. 2012), but other platforms include carrot (*Daucus carota*) and rice (*Oryza sativa*) cell lines (Hellwig et al. 2004; Xu et al. 2012). Innovative methods have been also developed for the production of cell packs from suspension cultures to facilitate accumulation and purification of target proteins (Rademacher 2013).

Table 2 Differences in biopharmaceutical production between mammalian and plant cell cultures

Mammalian cell production	Plant cell production
High initial investment (>$250 million; e.g. expensive stainless steel bioreactors)	Low initial investment (<$250 million; e.g. inexpensive polyethylene bags)
Long timeline for capacity expansion	Rapid scale-up and capacity expansion
Growth and manufacturing under strict controlled environment	Growth and manufacturing at room temperature
Expensive maintenance	Less costly "hands-on" maintenance
Risk of human pathogen contamination	No risk of human pathogen contamination

It should be noticed that taliglucerase alfa (Elelyso®), the first plant-made licensed recombinant pharmaceutical protein, is produced by carrot cell suspension cultures (Shaaltiel et al. 2007) as well as the veterinary vaccine for poultry against Newcastle disease approved by the US Department of Agriculture that is produced in suspension-cultured tobacco cells.

Despite general concerns about a possible genetic instability, plant cell cultures are fundamentally a reliable system in which medium optimization, process engineering, experimental/process design and scale-up can be tuned on demand (reviewed in Santos et al. 2016).

2.1.3 Green Unicellular Microalgae

In the recent past, other systems have been devised, which can easily be contained, propagated and transformed to produce recombinant proteins. A protein expression system that is based on the unicellular green alga *Chlamydomonas reinhardtii* has been developed (Mayfield et al. 2003). This alga has nuclear, plastidial and mitochondrial genomes completely sequenced. In this system, chloroplast-targeted transgenes were used to express different recombinant, health-related proteins. Like bacteria, the chloroplast lacks the machinery to perform complex post-translational modifications such as glycosylation (the glycosylated proteins come from the endoplasmic reticulum), but, unlike *E. coli*, *Chlamydomonas* chloroplast allows the disulphide bond formation and is able to carry out some types of phosphorylation. Unlike higher plants, *C. reinhardtii* has a single chloroplast, with about 80 genome copies. Consequently, conversion of all copies of the chloroplast genome to recombinant homoplasmy is facilitated. Complex molecules such as fully functional antibodies, therapeutics (among which a candidate therapeutic vaccine against human papillomavirus-related tumours based on a soluble, immunogenic form of the E7 viral protein) and other biologics have been produced with various yields highlighting the potential of microalgae as alternative platforms for the production of biologics for human uses (Mayfield et al. 2003; Demurtas et al. 2013). This relatively novel platform offers advantages including short time from transformation to scaling-up, rapid growth (doubling time of few hours), ease of cultivation, safety (microalgae do not harbour human pathogens and many are generally regarded as safe (GRAS) organisms) facilitating production of biopharmaceuticals in GMP conditions and homogeneity of protein production. Technologies for molecular pharming in *C. reinhardtii* are still in the infancy, while efforts are made to bring productivity to levels comparable to those of well-established platforms.

2.1.4 The "Hairy Root" Culture System

Together with cell suspensions, organ cultures such as hairy root offer advantages including containment, defined cultivation conditions and product homogeneity (Schillberg et al. 2013). Hairy roots (HR) are particularly attractive for the industrial-scale production of secondary metabolites but also of pharmaceutical proteins. In fact, like undifferentiated cells, they grow in simple, defined media but are genetically stable and easy to handle, with protocols available to establish transgenic lines highly scalable, producing significant biomass accumulation (Guillon et al. 2006).

HR cultures are generated by *Agrobacterium rhizogenes* infection. This infection leads to the generation of root clones characterized by an extensive secondary branching that can be cultivated under contained sterile conditions in hormone-free media (Franconi et al. 2010). Importantly, recombinant proteins can be secreted in the culture medium facilitating the downstream purification processes (Guillon et al. 2006).

HR have been used to express a range of recombinant proteins, such as enzymes (Woods et al. 2008), vaccines (Skarjinskaia et al. 2013), monoclonal antibodies (Wongsamuth and Doran 1997) and anti-HIV microbicides (Drake et al. 2013).

2.1.5 The CRISPR-Cas9 System

Editing the genome of a plant without introducing foreign DNA into cells is now a reality. Programmable nucleases, like zinc-finger nucleases (ZFNs), transcription activator-like effector nucleases (TALENs) and RNA-guided endonucleases, have been used for genome editing in plant cells, amplifying the horizon of plant molecular farming (Li et al. 2013; Shan et al. 2013; Nekrasov et al. 2013). Recently, the CRISPR (clustered regularly interspaced short palindromic repeat)/Cas9 (CRISPR-associated) system has been successfully used in a wide range of plant species (Bortesi and Fischer 2015) but not in plant suspension cells. The system is based on a short RNA guide (sgRNA) which associates to the Cas9 endonuclease to create a double-stranded break (DSB) in the target genomic DNA. As a consequence, mutations are generated through either error-prone non-homologous end-joining (NHEJ) or homology-directed repair (HDR) of the intended cleavage site. NHEJ has been used to generate mutagenic insertions/deletions often leading to gene inactivation. The resulting changes at chromosomal target sites can be indistinguishable from naturally occurring genetic variation. However, the system can be used to introduce larger modifications and insertion of exogenous sequences. Moreover, similar to "traditional" gene insertion, if *A. tumefaciens* is used as a delivery vector of CRISPR-Cas9, the resulting genome-edited plants contain foreign DNA sequences, in addition to those that encode the programmable nucleases, in the host genome. On the other hand, the use of episomal plasmids to deliver these nucleases into plant cells has to take into account that transfected plasmids are rapidly degraded in cells by endogenous nucleases and that the resulting DNA fragments can be inserted at off-target sites in host cells (Kim et al. 2014).

The delivery of preassembled Cas9-gRNA ribonucleoproteins (RNPs) into plant cells has been shown to remove the possibility to introduce recombinant DNA into the host plant genome (Cho et al. 2013). Furthermore, it as has been shown that these premixed RNPs cleave target DNA sites immediately after transfection and are degraded rapidly by endogenous proteases, thus reducing the frequency of off-target effects in regenerated plants (Kim et al. 2014; Cho et al. 2013). Moreover, when using protein-and-RNA-only systems, there is no need to optimize codon usage or to find promoters that will express Cas9 and gRNAs. In 2015, the delivery of RNPs into protoplasts of *Arabidopsis thaliana*, *Nicotiana attenuata* and *Oryza sativa* and the induction of targeted genome modifications in regenerated plants were reported. Purified Cas9 protein was mixed in molar excess with gRNAs targeting genes from the three plant species in vitro to obtain preassembled RNPs. The RNPs were incubated with protoplasts in the presence of polyethylene glycol. Indels were detected at the expected positions, with various frequencies (Woo et al. 2015).

2.2 Transient Expression (Delivery of Genes to Somatic Tissues)

The alternative plant-based technology to stable transformation of genomes is transient expression. As mentioned, stable transformation is not suitable for the rapid production of pharmaceuticals. This feature would be an obstacle to the use of plant-based technologies particularly in specific fields like those related to the production of biologics to combat emerging diseases.

The transient technological platforms imply the introduction of episomal vectors into plant tissues. There are two main transient expression strategies. The first is agro-infiltration, where leaves are infiltrated with *A. tumefaciens* (by injection or by applying vacuum to transfect whole plants at a glance). The second implies the infection of plant tissues directly with recombinant plant viruses that infect cells, replicate and spread from cell to cell and systemically guide the expression of recombinant protein in every cell. The majority of plant viral vectors used are based on single-stranded RNA viruses, such as tobacco mosaic virus, potato virus X and cowpea mosaic virus (CPMV) (Loh et al. 2017).

The use of deconstructed virus genomes delivered by *A. tumefaciens* is a methodology that is in between agro-infiltration and viral infection. The target gene is cloned into a modified plant viral vector which can be integrated into the *Agrobacterium* vector and delivered into the plant tissues by infiltrating with the corresponding transformed *Agrobacterium*. This facilitates the transfer of T-DNA to a very high number of cells. With the use of a deconstructed virus, the issue of instability of viral genome due to the introduction of large target genes can be resolved. Moreover, a certain degree of containment can be obtained because it implies the transfection of many cells with the virus genome followed by its cell-to-cell movement but no systemic spreading (Peyret and Lomonossoff 2015).

Icon Genetics, a German plant biotechnology company, has adapted this technology as MagnICON™ for the manufacturing of various plant-based vaccines, including those based on the hepatitis B virus surface antigen (300 mg/kg *N. benthamiana* fresh leaves) in the form of VLP (Huang et al. 2006), norovirus capsid proteins (Scotti and Rybicki 2013; Rybicki 2014) and non-Hodgkin lymphoma vaccines, which proceeded to a phase I clinical trial (see also Tables 4 and 5) (McCormick et al. 2008).

Fraunhofer USA Center for Molecular Biotechnology (CMB) has developed a "launch vector", an advanced, hybrid, gene infiltration system that combines the elements of TMV vector and *A. tumefaciens* binary plasmids (Musiychuk et al. 2007). Among others developed, the launch vector pBID4 contains the 35S promoter from cauliflower mosaic virus (35S CaMV) that drives transcription of the viral genome, the nopaline synthase terminator, genes for virus replication and cell-to-cell movement and the target gene cloned under the transcriptional control of the coat protein subgenomic mRNA promoter. Following infiltration, primary transcripts produced in the nucleus are transported into the cytoplasm, resulting in robust protein production (such as 1 mg/g of fresh weight) (Musiychuk et al. 2007; Massa et al. 2007).

The pEAQ system is based on full-length or truncated versions of CPMV RNA-2 for efficient and rapid protein production without viral replication (Sainsbury et al. 2009). These vectors contain the 35S CaMV promoter, nos terminator, the p19 sequence encoding a suppressor of silencing and 5′- and 3′-UTRs from CPMV RNA-2. The gene of interest is inserted between the UTRs. The "HT" (hypertranslatable) variants of these family of vectors provide extremely high translational efficiency and, eventually, high level of the recombinant protein accumulation in plant biomass (Peyret and Lomonossoff 2015).

Transient expression of target proteins in plants is, therefore, considered a more feasible approach when compared to stable transformation, due to its rapid production capabilities often leading to the expression of large amounts of recombinant protein in a short time. The process is scalable just by virus-infecting or agro-infiltrating more plants.

Agro-infiltration has demonstrated the highest efficiency and highest levels of target protein expression with the potential for cost-effective production (Loh et al. 2017). *Nicotiana benthamiana* is particularly amenable to infiltration methods. The maximum of protein expression is generally observed within 7 days post-infiltration which is faster if compared to the mere virus strategy that requires at least 2 weeks to generate a systemic spreading and expression. As mentioned, yields using this approach can reach 1 g of product per kilogram of leaves, though the levels are obviously protein-dependent. Successful clinical trial has indicated safety and efficacy of the protein therapeutics and biologics made by agro-infiltrated plants (see Table 5). The main example is represented by the production in *Nicotiana benthamiana* of a vaccine candidate against H1N1 influenza pandemic that occurred in 2009: the Canadian company Medicago was able to produce a ready to be administered hemagglutinin-based vaccine by transient expression in 19 days (D'Aoust et al. 2010).

3 Plant-Made Antigens for Developing Vaccines

The production of plant-based pharmaceuticals is strongly motivated especially in the field of vaccines, which are the most effective tools for the prevention and treatment of infections and, in some cases, cancer. Unfortunately, the progress toward commercialization of vaccines takes the biggest effort and time compared to other biopharmaceuticals, and this is also the case of plant-based vaccine candidates.

The possibility of manufacturing recombinant biopharmaceutical proteins in plants initially gave the prospect of a possible vaccine delivery by edible plant material ("edible vaccines") (Haq et al. 1995). Nevertheless, the idea of vaccination through ingestion of raw plant material (e.g. fruits or tubers) turned out to be not feasible due to different issues (uppermost, uniformity in dosage). The concept of oral antigen delivery in processed plant material is now considered more feasible (Streatfield 2006). A limited number of clinical trials involving oral delivery of antigens have been performed. In all cases no major safety concerns were detected, and formulations turned out to be well tolerated by vaccinees.

The first trials were conducted with the non-toxic B subunit of heat-labile enterotoxin (LTB) of enterotoxigenic *E. coli* contained either in raw potato or maize administered orally to healthy volunteers for safety and immunogenicity testing. No adverse effects of vaccination were noticed. LTB-specific IgA-secreting cells in peripheral blood and increased levels of LTB-specific serum IgG and IgA were detected after vaccination (Tacket et al. 1998; Tacket 2007).

Norovirus capsid protein VP1 was produced in potato tubers. Vaccination with approximately 500 μg of recombinant VP1 was done. Despite the high infectivity of the virus resulting in pre-existing serum immunity in a portion of volunteers, vaccinees developed VP1-specific serum IgG titres (Tacket et al. 2000).

Also a pathogen like the hepatitis B virus (HBV), able to evolve chronic infection and cancer both in adults and in infants, was targeted by PMF. HBV surface antigen (HBsAg) transgenic lettuce leaves containing 0.1–0.5 μg of HBsAg/100 g of fresh tissue were given to three adult volunteers showing transient protective levels of specific IgG after vaccination, but no HBsAg-specific serum IgA (Kapusta et al. 1999). When HBsAg was produced in transgenic potato (Thanavala et al. 2005), a dose-dependent, elevated serum HBsAg antibody titre induction was observed after administration of 850 $\pm$ 210 μg of antigen/dose, in a clinical trial on volunteers over the 70-day follow-up period after the first immunization.

Fragments encoding a chimeric protein of G protein and N protein of the rabies virus fused with that of alfalfa mosaic virus coat protein were introduced into a TMV-derived plant expression vector. The vaccine transiently expressed in spinach was orally administered (three doses of spinach; 20 g corresponding to 84 μg of chimeric rabies peptide each) in a clinical trial revealing the induction of elevated rabies-specific IgG in a quote of both in individuals previously vaccinated with a commercial injection-type vaccine and in volunteers with no history of rabies vaccination (Yusibov et al. 2002).

Overall, these studies have indicated that an immune response can be mounted in individuals fed plant material expressing a disease antigen.

Several other antigens have been expressed through different methodologies to the purpose of manufacturing (nonedible) candidate vaccines intended for clinical (Tables 3 and 4) and veterinary use (Table 5). Plant-made antigens targeting various pathogens have been shown to be effective in animal models (Table 3). Many of these candidates have reached phases I–II human clinical trials (Table 4). Safety of these vaccines was demonstrated; nevertheless, at present, there are no plant-made vaccine antigens approved for clinical use.

Despite this, plant-based expression platforms demonstrated to be able to tackle important sanitary problems even in emergency contexts. As already explained, in emergency, the transient methods assure the shortest leading times to the final products. That's why the major examples of plant-made vaccines in clinical trial are from transient approaches (Table 4). As an example, influenza virus, a very complex pathogen with 16 different hemagglutinin (HA) subtypes and characterized by occurrence of antigenic shift abolishing cross-protective immunity of the host even against strains of the same subtype, has the potential to be pandemic as happened in 2009 for the H1N1-type influenza virus. To control its diffusion by vaccination, fast production of a new HA antigen was needed. Medicago Inc. was able to develop a technology obtaining the accumulation of HA virus-like particles (virus-like particles, VLPs) in the apoplast of *N. benthamiana* cells by transient expression (D'Aoust et al. 2010; US patent application number 20130183341). Phase I (5, 10 or 20 μg of H5-VLP subcutaneously injected twice with alum adjuvant) and II (20, 30 or 45 μg of H5-VLP) clinical trial of the VLP composed of HA protein of H5N1 influenza virus (A/Indonesia/5/05) has showed, respectively, the induction of hemagglutinin inhibition titres at all tested doses and cross-protective CD4+ T-cell responses, indicating strong induction of long-term cell-mediated immunity by plant-made H5-VLP after 6 months of vaccination (Landry et al. 2012). No detectable IgE responses (and, therefore, allergy or hypersensitivity) against plant-specific mannose residues were found, one of the main issues that are often risen against the production of protein-based pharmaceuticals in plant-based systems (Ward et al. 2014). Medicago Inc. performed also a phase I clinical trial with 5, 13 or 28 μg of H1N1 influenza (A/California/7/09) VLPs that demonstrated safety and induced immune response to the virus, including cell-mediated immunity (Landry et al. 2010).

Other plant-based influenza vaccines (HA from A/California/04/2009 H1N1 (HAC1) and A/Indonesia/05/05 H5N1 (HAI-05); Fraunhofer CMB USA) based on transient expression have successfully completed phase I clinical trial (Shoji et al. 2011, 2015).

Plant-based platforms have been identified also as promising approaches to tackle chikungunya virus, the emerging pathogen initially found in East Africa and currently spread in many regions of the world. No licensed vaccines are available, and thus the need for advancing in this research field is great. Beside the preclinical stage candidates in trial, also the implementation in parallel of low-cost production platforms will be determinant as chikungunya is mainly affecting developing countries where access to vaccines is limited due to the high cost of conventional formulations. Based on the previous experiences on influenza and other viral pathogens, plant-made VLPs seem the most efficacious approach to render immunogenic formulations also in this case (Salazar-González et al. 2015).

Table 3 Examples of plant-derived vaccines for the prevention and treatment of human diseases (sourced from Loh et al. 2017)

Target protein	Host/expression strategy	Functionality evaluation	Reference
Anthrax protective antigen 83 (PA83)	*Nicotiana benthamiana*/ transient	Detection of high-titre toxin-neutralizing antibodies. 100% survival of immunized rabbits (IM) against lethal anthrax challenge	Chichester et al. (2013)
	Brassica juncea (mustard)/transgenic (nuclear)	Detection of systemic and mucosal immune responses. 60% survival of orally immunized mice against lethal anthrax challenge	Gorantala et al. (2014)
	Nicotiana tabacum (tobacco)/transgenic (chloroplast)	Detection of systemic and mucosal immune responses. 80% survival of orally immunized mice against lethal anthrax challenge	Gorantala et al. (2014)
Dengue consensus domain III of envelope glycoprotein (cEDIII) with 6D8 anti-Ebola IgG	*N. benthamiana*/ transient	Detection of virus-neutralizing specific anti-cEDIII humoral immune response in immunized mice (SC)	Kim et al. (2015)
Ebola glycoprotein (GP) in fusion with 6D8 anti-Ebola IgG	*N. benthamiana*/ transient	Detection of humoral immune responses. 80% survival of immunized mice (SC) against lethal EBV challenge	Phoolcharoen et al. (2011)
EBOV GP1 in fusion with E. coli heat-labile enterotoxin B subunit	Tobacco/transgenic (nuclear)	Detection of serum IgG in immunized mice (SC) and faecal IgA in immunized mice via oral administration	Rios-Huerta et al. (2017)
Hepatitis B virus (HBV) small surface antigen	*Lactuca sativa* (lettuce)/transgenic (nuclear)	Detection of serum IgG in immunized mice via oral administration	Czyż et al. (2014)
HBV surface antigen (HBsAg) SUV against HBV	*Solanum tuberosum* (potato)/transgenic (nuclear)	Induction of serum antibodies and stable immunological memory in immunized mice fed with transgenic potato tubers	Rukavtsova et al. (2015)
Human immunodeficiency virus (HIV) gp120 multi-epitopic envelope protein	Lettuce/transgenic (nuclear)	Detection of cell-mediated and humoral immunities in immunized mice via oral administration	Govea-Alonso et al. (2013)
HIV gp120 and gp41 multi-epitopic envelope proteins	Tobacco/transgenic (chloroplast)	Detection of antibody and cellular responses as well as specific IFN-g production in immunized mice via oral administration	Rubio-Infante et al. (2015)

(continued)

Table 3 (continued)

Target protein	Host/expression strategy	Functionality evaluation	Reference
HIV-1 envelope proteins (Gag/Dgp41) VLPs	*N. benthamiana*/ transient	Induced Gag-specific serum antibody and CD4 and CD8 T-cell responses in mice via systemic (IP) and mucosal (IN) immunizations	Kessans et al. (2016)
Human papillomavirus type 16 (HPV-16) HPV-16L1	Tobacco/ transgenic (nuclear)	Detection of cell-mediated and humoral immunities in immunized mice via oral administration	Hongli et al. (2013)
HPV-16 E6 and E7 fusion (HPV-16L1 E6/E7) VLP	*Solanum lycopersicum* (tomato)/transgenic (nuclear)	Detection of persistent neutralizing antibodies and 57% tumour reduction in immunized mice via oral administration	Monroy-García et al. (2014)
HPV-16 E7	*N. benthamiana* (transient)	Detection of antibodies and cell-mediated responses, tumour reduction in immunized mice models via subcutaneous administration	Massa et al. (2007); Venuti et al. (2009)
Influenza H1N1 HA from A/California/04/09 strain (HAC-VLPs)	*N. benthamiana*/ transient	Detection of serum HI antibody responses in immunized mice (IM)	Shoji et al. (2013)
Influenza H3N2 nucleoprotein	*Zea mays* (maize)/ transgenic (nuclear)	Detection of humoral immune responses in immunized mice via oral administration	Nahampun et al. (2015)
Influenza H5N1 HA1 domain (HA1-MY)	*N. benthamiana*/ transient	Detection of serum HI antibody responses in immunized mice (IM)	Pua et al. (2017)
Rabies virus glycoprotein in fusion with ricin toxin B chain	*Solanum lycopersicum* (tomato) hairy roots/transgenic (nuclear)	Detection of serum IgG and Th2 lymphocyte responses in immunized mice via intra-mucosal administration	Singh et al. (2015)
Severe acute respiratory syndrome coronavirus (SARS-CoV) nucleocapsid (N) protein	*N. benthamiana*/ transient	Recognition of SARS patient sera by purified N protein	Demurtas et al. (2016)

After pioneering examples that we have already mentioned in the field of "edible vaccines" (e.g. against hepatitis B), and others undertaken against human immunodeficiency virus (in lettuce, tobacco) and human papillomavirus (in tomato) as main cases, stable transformation approaches (both nuclear and plastidial) are also currently being undertaken and evaluated in preclinical context to tackle highly relevant sanitary problems such as those related to deadly pathogens like anthrax (in mustard and tobacco) and Ebola virus (tobacco) (see Table 3).

Table 4 Recent examples of plant-based vaccine antigens at various stages of clinical trials (sourced from Loh et al. 2017)

Product/Application	Plant host	Status	Company (sponsor)
Pfs25 VLP-FhCMB Malaria transmission blocking vaccine	*N. benthamiana*	NCT02013687 Phase I (completed in 2015)	FhCMB, USA
PA83-FhCMB Against anthrax	*N. benthamiana*	NCT02239172 Phase I (completed in 2015)	FhCMB, USA
HAC1 H1N1 seasonal influenza virus	*N. benthamiana*	NCT01177202 Phase I (completed in 2012)	FhCMB, USA
HAI-05 H5N1 pandemic influenza virus	*N. benthamiana*	NCT01250795 Phase I (completed in 2011)	FhCMB, USA
H1 (VLP) H1N1 seasonal influenza virus	*N. benthamiana*	NCT01302990 Phase I (completed in 2011)	Medicago, Canada
Quadrivalent VLP H1N1, H3N2 seasonal influenza B viruses	*N. benthamiana*	NCT01991587 Phases I and II (completed in 2014)	Medicago, Canada
Quadrivalent VLP H1N1, H3N2, seasonal influenza B viruses	*N. benthamiana*	NCT02233816 Phase II	Medicago, Canada
Quadrivalent VLP H1N1, H3N2, seasonal influenza B viruses	*N. benthamiana*	NCT02236052 Phase II	Medicago, Canada
H5 VLP H5N1 pandemic influenza virus	*N. benthamiana*	NCT00984945 Phase I (completed in 2010)	Medicago, Canada
H5 VLP H5N1 pandemic influenza virus	*N. benthamiana*	NCT01244867 Phase II (completed in 2011)	Medicago, Canada
H5 VLP H5N1 pandemic influenza virus	*N. benthamiana*	NCT01991561 Phase II (completed in 2014)	Medicago, Canada
H5-VLP + GLA-AF H5N1 pandemic influenza virus	*N. benthamiana*	NCT01657929 Phase I (completed in 2014)	Medicago (IDRI), Canada
H7 VLP H7N9 pandemic influenza virus	*N. benthamiana*	NCT02022163 Phase I (completed in 2014)	Medicago, Canada

Unfortunately, at present, the only plant-made vaccine approved is a purified and injectable Newcastle disease vaccine for poultry produced in tobacco suspension cells (Rybicki 2010, Table 6). The approval of this veterinary plant-made vaccine represents a quite important milestone. It should be considered that plant-based technologies are very suitable for developing veterinary vaccines. Especially in the

Table 5 Plant-based vaccine antigens for veterinary use

Product/application	Plant host	Reference/status
Chicken Newcastle disease Haemagglutinin-neuraminidase	Tobacco suspension cells	Katsnelson et al. (2006) Approved by USDA
Chicken Newcastle disease F protein	Maize	Guerrero-Andrade and Loza-Rubio (2006)
Chicken infectious bronchitis virus (IBV) S1 glycoprotein	Potato	Zhou et al. (2004)
Chicken IBV VP2	Rice	Wu et al. (2007)
Pig enterotoxigenic *E. coli* (ETEC) Fimbriae (F4)	Tobacco (chloroplast)	Kolotilin et al. (2012)
Pig ETEC Fimbriae (F4)	Alfalfa	Joensuu et al. (2006)
Pig ETEC Cholera toxin B subunit	Rice	Takeyama et al. (2015a, b)
Pig foot-and-mouth disease virus VP1	*N. benthamiana*	Yang et al. (2007)
Pig porcine transmissible gastroenteritis virus Spike protein	Tobacco	Tuboly et al. (2000)
Cattle bovine herpesvirus gD protein	Tobacco	Pérez Filgueira et al. (2003)
Cattle bovine viral diarrhoea virus E2 protein	Alfalfa	Peréz Aguirreburualde et al. (2013)
Cattle rinderpest virus Hemagglutinin	Peanut	Khandelwal et al. (2003)

case of fisheries, since farming relies on ocean water and excessive use of antibiotics may have impact on the environment, the use of (edible) vaccines for disease prevention may mitigate this problem (Clarke et al. 2013).

4 Plantibodies

Antibodies represent a peculiar example of biopharmaceuticals that can be efficiently expressed utilizing plants as low-cost heterologous hosts. In fact, prokaryotic expression systems are generally inappropriate for complete antibodies, due to the lack of some post-translational modifications such as *N*-linked glycosylation, which is often required for therapeutic proteins to be effective, and the inconsistency of others like the stabilization of free sulphydryls (i.e. from cysteines) and the reduction of disulphide bonds necessary for peptides and proteins to fold correctly.

As already mentioned, in the past decades, after the first, impressive evidence of correct expression and assembly of a full immunoglobulin in plant cells (Hiatt et al. 1989), unimaginable before the early 1990s, plant-made antibodies have been

Table 6 Recent examples of plant-based antibodies against tumour malignancies and infectious diseases

Antibody	Host/expression strategy	Functionality evaluation	Reference
Trastuzumab (humanized)	*N. benthamiana* (transient)	Anti-proliferative effects, similarly to the current drug Herceptin	Komarova et al. (2011); Grohs et al. (2010)
Rituximab (BLX-300 mouse/human chimeric anti-CD20 antibody C2B, with human-like glycosylation)	Aquatic plant *Lemna minor* (LEX System)	In preclinical studies showed the same target cell-binding activity, improved antibody-dependent cellular cytotoxicity against B cells and superior B-cell depletion, induction of apoptosis but attenuated complement-dependent cytotoxicity	Cox et al. (2006); Gasdaska et al. (2012)
Rituximab immunocytokine (scFv-Fc fused to human IL-2 with homogeneous glycosylation devoid of xylose/fucose *N*-glycosylation)	*N. benthamiana* (transient)	In vitro CD20 binding activity comparable to Rituximab and efficient elicitation of antibody-dependent cell-mediated cytotoxicity (ADCC) hIL-2 in the context of 2B8-Fc-hIL-2 maintained the biological activity and subsequent triggering of T-cell proliferation	Marusic et al. (2016)
Obinutuzumab (humanized from mouse monoclonal antibody against a CD20 epitope partially overlapping that of rituximab and glycoengineered)	*N. benthamiana* (transient)	Substantially equivalent to that derived from conventional Chinese hamster ovary (CHO) cells in binding to CD20 and expanding the immune-mediated target cell death	Lee et al. (2018)
2G12 (human mAb against a neutralizing epitope on HIV-1 GP120 protein)	Tobacco, maize and *Arabidopsis* (transgenic)	Potent and broad HIV-1-neutralizing activity in vitro and in vivo	Hessell et al. (2009)
		Double-blind, placebo-controlled, randomized, dose-escalation, phase I clinical trial was performed	Ma et al. (2015)
	Tobacco (transient)	Binding and neutralization properties with the "humanized" *N*-glycans equivalent to that of the original mAb derived from CHO cells. More effective in virus neutralization	Sainsbury et al. (2010)
	N. tabacum and *N. benthamiana* (transgenic, sIgA version of 2G12)	Effective aggregation of HIV virions and enhanced stability in mucosal secretions when compared to the cognate IgG format	Paul et al. (2014)

(continued)

Table 6 (continued)

Antibody	Host/expression strategy	Functionality evaluation	Reference
Paramyxovirus respiratory syncytial virus humanized IgG1 mAb palivizumab (different isotypes and *N*-glycoform variants)	*N. benthamiana* (transient) also utilizing glico-engineered plant hosts	Compared to commercially available palivizumab with respect to both in vitro receptor and C1q binding and in vivo efficacy. Antigen binding and neutralization activity of each variant were indistinguishable from those of palivizumab. Fcγ receptor binding profiles demonstrated significant differences depending on isotype and glycan profile	Hiatt et al. (2014)
Human mAb against anthrax main virulence factor(PA)("humanize-d", glycosylated with plant-specific glycans or glycan-deprived)	*N. benthamiana* (transient)	Used in both prophylactic and therapeutic treatment of individuals exposed or infected with anthrax. Both formats were able to bind PA, neutralizing anthrax lethal toxin, and to protect mice against a lethal spore challenge. The glycan-deprived mAb demonstrated improved half-life providing full protection in non-human primates against anthrax spore inhalation	Mett et al. (2011)
Chimeric complete IgG and a IgG-like format (scFv-Fc) against opportunistic fungal pathogens derived from a murine mAb (2G8)	*N. benthamiana* (transient)	Retained the efficacy of the original mAb, being able to inhibit directly the growth of *Candida albicans* and to induce the protection in animal models mimicking a systemic or vaginal *C. albicans* infection	Capodicasa et al. (2011)
		Both Abs promoted the killing of fungal cells by human polymorphonuclear neutrophils in ex vivo assays In a further development, a recombinant IgA version was also expressed and correctly assembled in dimeric form in plants	Capodicasa et al. (2017)
IgG against Ebola virus GP1	*N. benthamiana* (transient)	Protective against infection	Huang et al. (2010)
ZMapp (MB-003 mAb cocktail consisting of three human and human-mouse chimeric mAbs c13C6, h13F6 and c6D8 produced plants)	*N. benthamiana* (transient)	Three times as effective as CHO cell-produced recombinant counterpart mAbs, most likely due to the absence of core fucose residues on the Fc region of the plant-derived mAbs, which results in enhanced antibody-dependent cellular cytotoxicity	Zeitlin et al. (2011)

(continued)

Table 6 (continued)

Antibody	Host/expression strategy	Functionality evaluation	Reference
ZMapp cocktail, consisting of selected components from MB-003 (c13C6) and ZMAb (humanized c2G4 and c4G7)	*N. benthamiana* line XT/FT (transient)	Compassionate use of experimental interventions to seven Ebola patients as a post-exposure therapy. Of these patients, five significantly improved and recovered from the disease even though the treatment was initiated at least 9 days after infection A randomized phase I/II clinical trial of ZMapp as a putative investigational therapeutic in the treatment of patients with known Ebola infection started in 2015	Qiu et al. (2014)

promoted and validated as one of the most promising plant cell-based biologics. In fact, preclinical and current clinical trials hold promise for immunotherapy based on these recombinant molecules with peerless safety and economic qualities.

The recent success gained against Ebola virus by ZMapp™, a cocktail of three mouse/human chimeric antibodies, developed and produced in *N. benthamiana* by Mapp Biopharmaceutical Inc. (Davey Jr et al. 2016), shed new light on the potential of these "green" recombinant molecules as novel products of the healthcare system, possibly in the context of the "biosimilars" or "biobetters" class, able to lower the barriers to patients' access to treatment. To this end, it should be considered that some of the original antibody-based drugs (i.e. those from the 1980s and 1990s) are or will soon be coming off patent, and, therefore, they may attract investments in the development of novel industrial facilities to meet the increasing demand of these drugs. Indeed, at the moment, all protein-based active pharmaceutical ingredients derived from plants are still outlandish drugs for most pharmaceutical industries, although these herbal reagents would be competitive thanks to flexible and very large-scale production capacity, especially when the profitability ratios of the mammalian cell counterpart falter (Whaley et al. 2014).

4.1 Plant-Made Antibodies in Cancer Diagnostics and Therapy

Monoclonal antibodies (mAbs) are highly effective pharmaceuticals in cancer therapy (Gaughan 2016) with an estimated market value of $150 billion by 2021 (https://www.researchandmarkets.com/research/b3gj4m/the_development) (Ecker et al. 2015). The demand for some major products is in the order of tonne-scale production in the developed countries. mAbs' consumption may even increase if the supply of these biopharmaceuticals is to be extended to developing countries, which currently have limited access to these biologics, due to the current costs of manufacturing in mammalian cell culture. This scenario would unlock the full potential of plants as a novel production platform.

mAbs directed to several tumour-associated antigens (Table 6) and involved in different mechanisms of interference/diagnosis of the relevant cancer type have been expressed in both plant leaves and seeds (Pujol et al. 2007). The genes for heavy and light chain of the base format of the immunoglobulin have been engineered for transient or stable expression through different techniques of transgenesis involving both nuclear and chloroplast genome (Ko et al. 2005; Komarova et al. 2010; Daniell 2003).

As mentioned elsewhere in this chapter, every expression technique (either stable or transient) presents advantages and drawbacks. Therefore, a careful evaluation of the system has to be adopted on a case-by-case basis. Nevertheless, while cumbersome procedures are required for the establishment of stable genome transgenic lines and literature highlights expression yields usually non-competitive (De Muynck et al. 2010), transient expression systems allow high yields of recombinant protein in few days. The transient approach has been prevalent in recent years in an attempt to make the plant system highly productive for the synthesis of biologic medicines. As an example, by this approach it is currently possible to produce ten million doses of influenza vaccine in the remarkable time of 6 weeks (Pillet et al. 2016).

In a pioneering work of agro-infiltration, a clinically relevant recombinant antibody (diabody) against the carcinoembryonic antigen (CEA) in tumour imaging was expressed in tobacco plants (Vaquero et al. 2002) validating the feasibility of the approach. In subsequent study, 500 mg/kg yields of a full-size tumour-specific human mAb A5 were achieved using the "MagnICON" system (magnifection) developed by Icon Genetics (Giritch et al. 2006; Marillonnet et al. 2004).

Since the discovery of competitive manufacturing yields, anticancer therapeutic antibodies produced as plant biosimilars occupy an area of active interest due to the potential to guarantee access to more affordable treatments also in view of patents' expiry dates for quite a number of antibodies by 2020. The following examples witness a fervent activity in this sense.

TheraCIM, a recombinant humanized antibody against the epidermal growth factor receptor (EGF-R) that had orphan drug status for glioma, starting from 2014 in the United States and EU, was one of the first antibodies of commercial value to be transiently expressed in plants. Engineered in an aglycosylated format in tobacco, this antibody, produced from agro-infiltrated leaves, maintained unchanged the recognition of the EGF-R on the surface of human tumour cultured cells (Rodriguez et al. 2005).

Trastuzumab, a humanized monoclonal approved by the US Food and Drug Administration for the treatment of metastatic breast cancer, recognizing the human epidermal growth factor receptor 2 (HER2/neu), has been transiently expressed in *N. benthamiana* plants (Komarova et al. 2011; Grohs et al. 2010). In these independent reports, this plant-made mAb, expressed using either the

MagnICON system or tobacco mosaic virus- and potato virus X-based vectors, revealed anti-proliferative effects, similarly to the current drug Herceptin.

The mouse/human chimeric anti-CD20 antibody C2B8 (rituximab) is the first antibody-based drug approved by regulatory authorities for the treatment of patients with recurrent B-cell lymphomas and is capable of depleting B cells in vivo (Reff et al. 1994). Although only about 48% of patients treated with rituximab respond to the therapy, with <10% showing a complete remission of the tumour (Davis et al. 2014), treatment with this drug associated with chemotherapy is a pillar for the treatment of non-Hodgkin's lymphomas (Plosker and Figgitt 2003).

In an effort to increase cost-effectiveness of rituximab, Biolex Therapeutics used the proprietary protein expression (LEX System) in the aquatic plant *Lemna minor* to produce a plant version of rituximab with fully human-like glycosylation pattern with a single major *N*-linked glycan devoid of xylose or fucose (Cox et al. 2006). This new product, named BLX-300, was shown to have the same target cell-binding activity, improved antibody-dependent cellular cytotoxicity against B cells and superior B-cell depletion, induction of apoptosis but attenuated complement-dependent cytotoxicity in preclinical studies (Gasdaska et al. 2012).

A further approach to improve efficacy of therapy of B-cell malignancies is based on the use of anti-CD20 antibodies to deliver cytokines to the tumour microenvironment. In particular, human interleukin 2 (IL-2)-based immunocytokines have shown enhanced antitumour activity in several preclinical studies (List and Neri 2013).

The first successful example of a recombinant immunocytokine produced in agro-infiltrated *N. benthamiana* plants was reported in an unconventional yet effective approach (Marusic et al. 2016). The immunocytokine was designed as a protein fusion between an antibody scaffold derived from rituximab and the human IL-2. In particular, a dimeric bivalent antibody format based on a scFv-Fc fused to IL-2 with highly homogeneous glycosylation devoid of the typical xylose/fucose *N*-glycosylation plant signature. In vitro studies showed that this antibody conjugate has a CD20 binding activity comparable to that of rituximab and an efficient elicitation of antibody-dependent cell-mediated cytotoxicity (ADCC). In addition, hIL-2 in the context of 2B8-Fc-hIL-2 maintained the biological activity, as validated by 2B8-Fc-hIL-2 binding to the hIL-2 receptor and subsequent triggering of T-cell proliferation. The final touch was the discovery that glycan engineering, performed utilizing RNAi-silenced transgenic plants devoid of xylosyltransferase and fucosyltransferase activity (Strasser et al. 2008), was the best way to achieve an overall improvement of the biological functions (Marusic et al. 2018).

Among antibodies targeting B-cell malignancies, obinutuzumab is a third generation fully humanized from mouse monoclonal antibody that binds a CD20 epitope partially overlapping that of rituximab (Klein et al. 2013). Obtained through a proprietary GlycoMAb® technology by GlycArt Biotechnology and developed by Roche as a therapeutic of chronic lymphocytic leukaemia and other B-cell non-Hodgkin's lymphomas (Marcus et al. 2017), this glycoengineered mAb, when produced in *N. benthamiana* plants, has been demonstrated substantially equivalent to that derived from conventional Chinese hamster ovary (CHO) cells in binding to CD20 and expanding the immune-mediated target cell death (Lee et al. 2018).

4.2 *Plant-Made Antibodies and Immunotherapy of Infectious Diseases*

Numerous mAbs have shown the potential to treat infectious diseases targeting microbial infections by directly interacting with key epitopes and/or harnessing the immune response.

Among anti-infectious mAbs, the human mAb 2G12, recognizing a distinctive neutralizing epitope on the GP120 envelope protein of HIV-1, has a potent and broad HIV-1-neutralizing activity in vitro and in vivo (Hessell et al. 2009). Due to these properties, mAb 2G12 has been chosen by Pharma-Planta Consortium (http://www.pharma-planta.net) as the best candidate to be produced in plants under GMP conditions for the first in-human, double-blind, placebo-controlled, randomized, dose-escalation, phase I clinical trial (Ma et al. 2015). This mAb, representing a significant milestone in the commercial development of plant-derived biologics, has been produced by several expression systems and transgenic hosts (i.e. tobacco, maize and *Arabidopsis*) as well as in plant cell systems. However, the production of this mAb in transient expression systems was far superior to that obtained by stable transgenics in terms of yield and speed of expression. In particular, the highest yield of purified mAb (105 mg/kg fresh weight tissue) was obtained through the CPMV-HT system and retention in the ER (Sainsbury et al. 2010). In addition, binding and neutralization properties of plant-produced 2G12 with the "humanized" *N*-glycans were equivalent to that of the original mAb derived from CHO cells, being even more effective in virus neutralization.

Finally, a sIgA version of 2G12 has been produced in both transgenic *N. tabacum* and transiently transformed *N. benthamiana* (Paul et al. 2014) demonstrating effective aggregation of HIV virions and enhanced stability in mucosal secretions, when compared to the cognate IgG format.

A further antiviral mAb expressed in plants is the Hu-E16, a humanized mAb that binds an epitope on the envelope protein (domain III of GP-E) of West Nile virus (WNV) (Oliphant et al. 2005). Using both the MagnICON system and a geminivirus-based vector, codon-optimized HC and LC encoding sequences, high level of production (~260/800 and mg per kilogram of fresh weight) has been achieved in *N. benthamiana* and lettuce (Lai et al. 2010, 2012).

Further efforts have been made to improve Hu-E16 mAb efficacy, engineering diverse modified formats to be transiently expressed in both wild-type and glyco-engineered XT/FT *N. benthamiana* (He et al. 2014; Lai et al. 2014). Overall data demonstrate that plant-made Hu-E16 show in mice a strong neutralizing activity and induced pre- and post-exposure protection, equivalently to what has been observed with Hu-E16 derived from mammalian cell, possibly transforming this preclinical candidate into a cost-effective therapeutic of medical and veterinary significance against WNV.

Most current experimental therapies against Ebola virus (EBOV) are addressed to the principal virulence factor GP, a transmembrane protein triggering virus host cell entry and exerting cytopathic effects in infected cells (Lee et al. 2008; Simmons et al. 2002) while inducing protective antibodies (Wilson et al. 2000). At the moment, the most advanced Ebola treatment is based on antibody-mediated therapy. In an early attempt, remarkable expression levels (0.5 mg per gram of fresh tissue) of a protective IgG against Ebola virus GP1, harnessing the geminivirus DNA replicon of bean yellow dwarf virus, were reported (Huang et al. 2010). At that time, the multi-replicon vector represented a significant advance in transient expression technology for antibody production in plants.

As previously illustrated, Mapp Biopharmaceutical, Inc., is involved in a research to find the best immunotherapy against Ebola, and the company has tested the drug MB-003, a mAb cocktail consisting of three human and human-mouse chimeric mAbs c13C6, h13F6 and c6D8 produced in CHO cells (Olinger Jr et al. 2012). When produced from plants, MB-003 resulted three times as effective as CHO cell-produced recombinant counterpart mAbs, most likely due to the absence of core fucose residues on the Fc region of the plant-derived mAbs, which results in enhanced antibody-dependent cellular cytotoxicity (Zeitlin et al. 2011). Nowadays the most advanced, optimized anti-Ebola mAb composition is the ZMapp cocktail, consisting of selected components from MB-003 (c13C6) and ZMAb (humanized c2G4 and c4G7) (Qiu et al. 2014). Using Magnifection, ZMapp was entirely produced in the glycoengineered *N. benthamiana* line ΔXT/FT (Strasser et al. 2008) as a product of LeafBio (San Diego, California, USA), a joint venture of Mapp Biopharmaceutical and Defyrus (Toronto, Canada) and was manufactured under cGMP at the Kentucky BioProcessing facility.

Although ZMapp safety had not been previously evaluated in humans, during the 2014–2015 Ebola outbreak in West Africa, it was administered outside clinical trials in compassionate use of experimental interventions to seven Ebola patients as a post-exposure therapy. Of these patients, five significantly improved and recovered from the disease even though the treatment was initiated at least 9 days after infection.

Nevertheless, a randomized phase I/II clinical trial of ZMapp as a putative investigational therapeutic in the treatment of patients with known Ebola infection is currently underway (ClinicalTrials.gov. 2015. Putative investigational therapeutics in the treatment of patients with known Ebola infection. Stud. Rec. NCT02363322, Natl. Inst. Health, Bethesda, MD. https://clinicaltrials.gov/ct2/show/NCT0236332237).

Today's state of the art indicates ZMapp as an effective drug for Ebola patients; however its supply is limited. Providing sufficient quantities of this biological is challenging due to the high dosage required for optimal therapy (Qiu et al. 2014) and to the limited capacity of the plant production system. Nevertheless, Mapp Biopharmaceutical has signed a contract with the US Department of Health and Human Services with the aim to accelerate the drug's development (McCarthy 2014).

Other plant biopharming companies are joining the global efforts against Ebola. Medicago, (Quebec, Canada), in the framework of a task order from the Biomedical Advanced Research and Development Authority at the US Department of Health and Human Services, is producing three anti-Ebola virus monoclonal antibodies (mAbs) with expected performance comparable to that of ZMapp™. In parallel, the Fraunhofer USA Center for Molecular Biotechnology is producing these anti-EBOV mAbs for the Biomedical Advanced Research and Development Authority, and

Caliber Biotherapeutics LLC declares itself ready to produce commercial quantities of ZMapp mAbs quickly and cost-effectively. Overall efforts will potentially boost production amounts and worldwide supply of these important drugs.

The paramyxovirus respiratory syncytial virus (RSV) can cause devastating lower respiratory tract infections in preterm infants and small children or elderly patients when other serious health problems are present. Passive immunization is an effective immunoprophylaxis against RSV, and treatment with a humanized IgG1 mAb, palivizumab (brand name Synagis, manufactured by MedImmune), is the current standard of care. In a recent report, plant-made different isotypes and *N*-glycoform variants were compared with commercially available palivizumab with respect to both in vitro receptor and C1q binding and in vivo efficacy. Whereas the antigen binding and neutralization activity of each variant were indistinguishable from those of palivizumab, their Fcγ receptor binding profiles demonstrated significant differences. Overall results indicate that interaction with FcγR, hence antibody-dependent cell-mediated cytotoxicity and not virus neutralization, plays the major role in palivizumab's efficacy. Therefore isotype and glycan profile engineering utilizing glycoengineered plant hosts have been put into play, paving the way for enhancing antibody-dependent cellular cytotoxicity and the efficacy of these anti-RSV mAbs (with yields of ~200 mg per kilogram of plant biomass) (Hiatt et al. 2014).

Anthrax is a zoonotic disease caused by the gram-positive, spore-forming bacterium *Bacillus anthracis.* Although the incidence of the disease has continually decreased since the late nineteenth century, anthrax is considered one of major concerns in the era of bioterrorism in that inhalation of aerosolized spores has very high mortality rates. *B. anthracis* possesses a main virulence factor, the 83-kDa form of protective antigen (PA), composed of two exotoxins, also known as lethal toxin and oedema toxin, which cause cell death following introduction into the cytoplasm (Young and Collier 2007). PA elicits specific neutralizing antibodies and is the major target for the development of both anthrax vaccines and mAbs. A human mAb against the *B. anthracis* PA has been overexpressed in plants to be used in both prophylactic and therapeutic treatments of individuals exposed or infected with anthrax. A humanized glycosylated (e.g. "decorated" with plant-specific glycans) and a glycan-deprived version of this mAb have been devised. Both formats were able to bind PA, neutralizing anthrax lethal toxin, and to protect mice against a lethal spore challenge. Surprisingly, the glycan-deprived mAb demonstrated improved half-life providing full protection in non-human primates against anthrax spore inhalation (Mett et al. 2011).

Infections caused by opportunistic fungal pathogens pose serious issues for both prevention and treatment especially in immune-compromised or hospitalized individuals. In the perspective of being used as a wide-spectrum antifungal therapeutics, a chimeric complete IgG and IgG-like format (scFv-Fc) have been produced in agro-infiltrated *N. benthamiana* plants (Capodicasa et al. 2011). These engineered Abs were derived from a murine mAb (2G8) that inhibits fungi growth conferring significant protection against different opportunistic pathogens such as *Candida albicans*, *Aspergillus fumigatus* and *Cryptococcus neoformans* in animal models (Torosantucci et al. 2009). By binding to a fungal cell wall polysaccharide (β-glucan), both 2G8 chimeric IgG and scFv-Fc produced in plants retained the efficacy of the original mAb, being able to inhibit directly the growth of *Candida albicans* and to induce protection in animal models mimicking a systemic or vaginal *C. albicans* infection. In addition, both Abs promoted the killing of fungal cells by human polymorphonuclear neutrophils in ex vivo assays. In a further development, a recombinant IgA version was also expressed and correctly assembled in dimeric form in plants (Capodicasa et al. 2017). These recombinant Abs may be exploited to tackle a broad range of human fungal infections, in alternative or in addition to classical chemotherapy (Table 6).

5 Plant-Made Proteins for Other Pharmacological Uses

Currently, only two products derived from this technology have been brought on the market. The first, Elelyso, is an enzyme replacement therapy (ERT) for humans, while the second, Interberry-alpha, has been developed for veterinary use (canine gingivitis).

In 2012, the US FDA approved ELELYSO™ (human recombinant taliglucerase alfa or glucocerebrosidase, GC), an enzyme produced in genetically engineered carrot cells for treating type 1 Gaucher's disease (GD) by Protalix BioTherapeutics (an Israeli enterprise established in 1993) and its partner, Pfizer (Fox 2012). GD is a lysosomal storage disorder caused by a hereditary deficiency of the enzyme glucocerebrosidase, which is involved in glycolipid metabolism. GD creates a disabling condition curable only with permanent ERT.

As a result of advances in stable plant transformation technology and upstream and downstream manufacturing process, GD can be currently treated by enzyme replacement therapy using this recombinant GC that is administered intravenously every 2 weeks (Shaaltiel et al. 2007) and represents an alternative, cheaper choice with respect to the CHO-based manufacturing (i.e. the recombinant human GC Cerezyme™ -Genzyme- and (VpriV®) -Shire Pharmaceuticals-). Moreover, since expression of taliglucerase alfa is targeted to the storage vacuoles of the cell using a plant-specific sorting signal, this determines the presence of mannose sugars. In fact, GC utilized for ERT must display mannose moieties for entry into the macrophage. The expression process in the plant cell results in mannose decoration without glycan enzymatic remodelling, which is, instead, a downstream process necessary for drug of mammalian cell origin elevating the cost of goods for this specific product. Protalix brilliantly took advantage of the plant secretory pathway to obtain homogeneous paucimannosidic glycans in a cost-effective glycoengineering intervention within the cells. A fortunate circumstance to speed up ELELYSO approval was the US FDA warning about the potential for foreign-particle contamination in some Genzyme products including Cerezyme™ (Drake et al. 2017). This event has been fundamental to raise awareness of the scarcity of approved safer alternatives to mammalian cell manufacturing. Moreover, it sets a positive example to be followed that could renew investors' interest in molecular farming companies, similar to what has occurred right after commercial approval for ELELYSO that determined a collaboration agreement Protalix BioTherapeutics-Pfizer. In addition, it should be positively noticed that 1-year treatment with taliglucerase allows patients and national sanitary systems for an estimated cost reduction by 25% with respect to administration of imiglucerase for the same period. Also the oral route of administration is currently being tested that might reduce discomfort or pain related to intravenous administration. The results demonstrate that the carrot cells protect the recombinant protein in the gastric environment and may enable absorption in vitro. Feeding rats and pigs with carrot cells containing GC proved that active recombinant enzyme can be found in the digestive tract and blood and can reach the target organs (Shaaltiel et al. 2015). These results demonstrate that the oral administration of proteins encapsulated in plant cells is feasible and may be a valuable alternative to intravenous administration of ERT.

Recombinant canine interferon-alpha (Interberry-alpha) was produced by Hokusan Co. Ltd. in the National Institute of Advanced Industrial Science and Technology (NAIST), Hokkaido, Japan. Interberry-alpha is manufactured in genetically modified strawberries in a contained biosafety level 2 facility for transgenic plants (avoidance of gene release into the environment). This product was approved for commercialization by the Japanese Ministry of Agriculture, Forestry and Fisheries, and processed strawberries containing were marketed from 2014 for the treatment of periodontal disease in dogs (Drake et al. 2017).

Production of proteinaceous drugs other than vaccines and antibodies has been proven in a variety of plant-based expression platforms (from plant cell suspension cultures to moss, to duckweed) for different applications (Table 7). The advancement status of these products reached phase I and II clinical trial and hopefully can reach commercial approval.

6 Bringing "Functional" Plants to the Marketplace

6.1 Good Manufacturing Practices

Importantly, and independently from the method used (stable/transient expression), current EMA regulatory guidelines (EMA 2008) indicate that all biopharmaceutical products intended for clinical trials should be manufactured according to GMP. Thus, GMP compliance appears to be a crucial point in developing plant-based pharmaceuticals for clinical use. In Table 8 the main GMP factories for the production of plant-made biopharmaceuticals are listed. Besides the GMP facility for the production of the only plant-made therapeutic enzyme approved for human use in Israel (Protalix/Pfizer), two big facilities are capable of biomanufacturing plant-derived pharmaceutical proteins under GMP conditions. Fraunhofer CMB in the United States deals with regulatory and clinical affairs by managing a GMP pilot plant equipped with complete processing cycles for transient expression (i.e. plant farming, bacterial cultures, plant infiltration, biomass harvest and protein purification). Also Medicago Inc., Québec CA, is into clinical trial operating in a similar facility.

Table 7 Examples of plant-based protein pharmacological entities at various stages of clinical trials

Product/application	Plant host	Status	Company (sponsor)
Taliglucerase alfa (Human glucocerebrosidase, prGCD) ERT for Gaucher's disease	*Daucus carota* (carrot) cell culture	NCT00376168 Phase III (completed in 2012) FDA-approved in 2012	Protalix BioTherapeutics, Israel/Pfizer
Moss-aGal (Human alpha-galactosidase A) ERT for Fabry disease	*Physcomitrella patens* (moss)	NCT02995993 Phase I	Greenovation Biotech GmbH, Germany
PRX-102 (Human alpha-galactosidase A) ERT for Fabry disease NCT01769001	Tobacco cell culture	Phases I and II	Protalix BioTherapeutics, Israel<aq2
Recombinant human intrinsic factor Dietary supplement for vitamin B12 deficiency	*Arabidopsis thaliana*	NCT00279552 Phase II (completed in 2006)	University in Aarhus, Denmark
Recombinant lactoferrin Anti-inflammation treatment for HIV patients	*Oryza sativa* (rice)	NCT01830595 Phase II (completed in 2006)	Jason Baker (MMRF), USA
rhLactoferrin Treatment for chronic inflammation in the elderly	Rice	NCT02968992 Phase II	Johns Hopkins University, USA
Locteron (controlled-release interferon alpha 2b) Antiviral treatment for hepatitis C virus infection	*Lemna minor* (duckweed)	NCT00593151 Phases I and II (completed in 2009)	Biolex Therapeutics, USA
Interberry-alpha Canine interferon-alpha treatment of periodontal disease in dogs	Strawberry (stable)	Manufacturing and marketing approval by the Japanese Ministry of Agriculture, Forestry and Fisheries	NIAIST Institute (Japan)

Table 8 The main GMP factories for the production of plant-made biopharmaceuticals

Company	Plant/expression system	Main biopharmaceutical product
Kentucky Bioprocessing (USA)	Potato (stable) Tobacco (transient)	Norovirus vaccine Ebola virus vaccine (ZMapp)
SIGMA Aldrich (USA)	Maize (stable)	Trypsin
Medicago Inc. (Canada)	Tobacco (transient)	Influenza vaccine
Protalix/Pfizer (Israel)	Carrot (stable)	Taliglucerase alfa
Fraunhofer (CMB-USA/IME-Germany)	Tobacco (transient, stable)	Influenza vaccine, anti-HIV antibodies
NIAIST Institute (Japan)	Strawberry (stable)	Canine interferon-alpha
Tokyo University (Japan)	Rice (stable)	Cholera toxin B subunit

6.2 Risk Analysis and Regulations

As any pharmaceutical molecules, plant-based biologics may be associated with some risks. In the particular case of plant-based pharmaceuticals, some specific features have to be considered.

It is obvious that these products should be free of classic impurities described in other recombinant product manufacturing and in the case of a biosimilar or biobetter requires demonstration of substantial equivalence to any legally marketed product. This means that the new product is at least as safe and effective as the predicate. In the area of plant-derived biopharmaceuticals, the main safety issues to be addressed could be the different post-translational modifications of the plant cell, such as the *N*-glycosylation that may induce allergic responses (mainly superable by glycoengineering of plant expression; see below).

The potential risk of impact on the environment is limited, since plant-based pharma production is strictly regulated and mainly envisaged under GAP (Good Agricultural Practices) and GMP indoor manufacturing. Transient expression systems seem to pose some risks in that, while enabling the rapid and robust production of target proteins, they require the large-scale use of *A. tumefaciens* or viral vectors, which are the only real GMO involved in the process. Therefore, agro-infiltration is covered by the existing legislation and risk assessment as GM micro-organisms in plants and therefore might not fall under the scope of the genetically modified plant legislation (Sparrow et al. 2013).

The use of CRISPR-Cas9 systems may alleviate regulatory concerns related to genetically modified plants. Beside CRISPR, also other new plant breeding techniques (NPBT) differ substantially from the transgenic techniques that have been used in the last two to three decades. For some of them, the regulatory framework in the EU may no longer be feasible. In fact, the nuclease techniques used for only point mutations or a few nucleotide changes might be considered not to determine GMO as no foreign DNA is integrated in the plant cell. NPBTs used, instead, for gene integration are covered by the existing guidance for risk assessment and GMO legislation as they do not differ from the integration of genes by transgenesis, even though greatly enhancing the accuracy of gene integration at the target locus and minimizing unattended side effects which might result in an easier risk assessment.

In the United States, the FDA ensures the safety of manufacturing (including approval of GMP facilities for plant production) and clinical use of plant-based biopharmaceuticals before licensing. The US Department of Agriculture (USDA) also plays a role in the introduction of plant-made pharmaceuticals, being responsible of the control and approval of veterinary biologics through the control of the genetic background of plants, the probability of cross-pollination and evaluating risk-management strategies.

In the European Union, the European Food Safety Authority (EFSA) and the European Medicines Agency (EMA) are in charge of the regulation concerning food safety, hence transgenic plants cultivation, and concerning medicines, respectively.

EMA published a draft guidance on the quality of the biologically active substances produced by stable transgene expression in higher plants, which only goes through plants with stable expression of transgenes and does not consider transiently transfected plants and/or plant cell cultures, the use of which still requires regulation. EMA states the importance of the establishment of master and working seed banks from the final transformant that have to be characterized in respect to the transgene (i.e. sequence, integrity, site of insertion, copy number, fate of the marker sequence), recombinant protein expression (i.e. tissue and organ specificity, regulation, expression level) and unintended changes in the levels of endogenous plant proteins and storage properties (storage conditions, shelf-life). Importantly, EMA regulatory guidelines indicate that all biopharmaceutical products in trials should be manufactured according to GMP (European Medicines Agency. Guideline on the Quality of Biological Active Substances Produced by Stable Transgene Expression in Higher Plants www.ema.europa.eu/docs/en_GB/document_library/Scientific_guideline/2009/09/WC500003154.pdf).

The Pharma-Planta Consortium (including 33 academic and companies in Europe and South Africa), funded by the CE Framework Programme 6, was the first entity to be engaged with regulatory authorities in this field, from 2004 to 2011. Great merit should be attributed to this consortium that significantly contributes to the maturation of the EMA guidelines in this research area (as summarized previously in the text). As a consequence, a GMP manufacturing license (i.e. for the above-mentioned anti-HIV antibody P2G12) was granted to Fraunhofer IME for plant-derived monoclonal antibodies by the Germany regulatory authority and subsequent clinical trials application approved by the UK regulatory authority. These two achievements demonstrated that a GMP-compliant process for transgenic plants and the administration of a plant-made recombinant biopharmaceutical were feasible and acceptable to regulators throughout Europe (Ma et al. 2015).

6.3 Acceptance and Inclusion of Citizens in the Decisional Processes

Today, the users of products are increasingly active and engaged consumers, especially when goods are relevant to their health. People are more and more aware of the consequences of health effects, and it is fundamental to be responding to people's expectations. The vast majority of consumers agree that development in science and technology could make changes to their lives too quickly and that they may hide negative side effects on both health and environment (Castle and Dalgleish 2005). Social and environmental impacts are increasingly complex, and it is necessary to demonstrate that any research is ethically, socially and environmentally responsible. Biotechnology innovations, certainly successful on technical merits, must also be accompanied by social acceptance, assuming that market demand is favourable for a commercial success. In other words, it is a fact that benefits derived from PMP are now well highlighted; still regulatory agencies need to cooperate with developers to ensure that there is strict regulatory oversight of these products, in order to gain trust, confidence and support of the public, concerned stakeholders and investors and then progress on the marketplace.

Responsibility in science and technology is a hot topic among policy-makers, researchers and innovators in Europe. Thanks to concerted efforts, in recent years, more people now understand changes that the European research and development system is going through. More citizens are involved in science through public engagement exercises. Biology and society scientists have established collaborations. Users are leading innovation. Open access trends are changing the publishing system. The so-called responsible research and innovation (RRI; Fig. 3) represents a cross-cutting issue that is strongly pursued through the Horizon 2020 funding scheme, in line with Europe 2020 – the EU's strategy for growth launched in 2010 – and following the 2009 Lund Declaration (updated in 2015), which called upon European nations and institutions to focus research on the "grand challenges" facing society, such as climate change, water shortages and ageing populations. Innovative solutions that improve people's lives and platforms that better communicate these solutions and address ethical, social and environmental issues are the pillars of RRI (von Schomberg 2013).

If there is an area where the benefits of responsible research activities can be easily proved, it is the health sector. There is no doubt that citizens more readily welcome the results of research if they affect the quality of life. Integrating principles of responsible research and innovation in research and development activities in the field of PMPs will probably help to improve social acceptance of plant biotechnology innovations. Social actors may be motivated to participate to research and

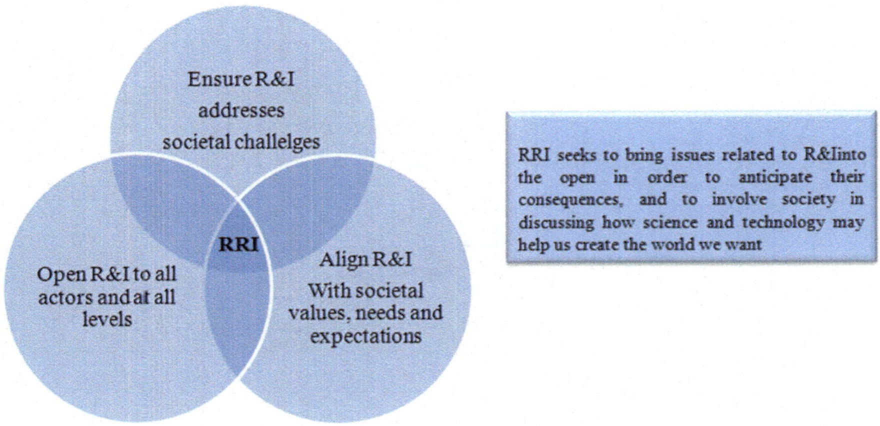

Fig. 3 Responsible research and innovation (RRI). Sourced from "A practical guide to responsible research and innovation key lessons from RRI tools" (https://www.rri-tools.eu/documents/10184/16301/RRI+Tools.+A+practical+guide+to+Responsible+Research+and+Innovation.+Key+Lessons+from+RRI+Tools)

innovation processes and may discover that outcomes of research and innovation in the field of PMPs meet the expectations of safety, desirability, acceptability and quality.

Civil society actors should be encouraged to work in contact with scientists and companies at a deeper level throughout the whole R&I (research and innovation) process. This will ensure that R&I results meet societal expectations.

If social acceptance is of sufficient importance, then gauging public attitudes toward any new technology becomes an important step in market assessment and justification of financial investment to conduct research and development. Plant-made biopharmaceuticals are on the intersection among the R&D areas of modern agriculture, medicinal plants/supplements and (plant) biotechnology (Fig. 4). The available studies show an acceptance of PMPs as a rare combination of medical and agricultural biotechnologies (Castle and Dalgleish 2005). As shown by some studies, there appears to be a positive public outlook about engineered pharmaceuticals, in general (Castle and Dalgleish 2005). The development of PMPs promises significant advantages for the production of vaccines and other therapeutic/prophylactic entities. However, the benefits and risks associated with new technologies must still be communicated and discussed with the public early, to maximize social acceptance. In fact, the above-mentioned study underlines that people are normally excluded from public debates and policy-making and normally do not have access to data and information about issues, if not through the media which most of the interviewees consider as giving misleading information. Studies have shown that even if people associate technology with relatively high risks and unknown consequences (especially genetic technologies), they still might not reject the technology (Castle and Dalgleish 2005). Oversight by regulatory agencies may give confidence to the general public and facilitate acceptance of new technologies, despite negative perceptions with regard to specific risks. More empirical research on public

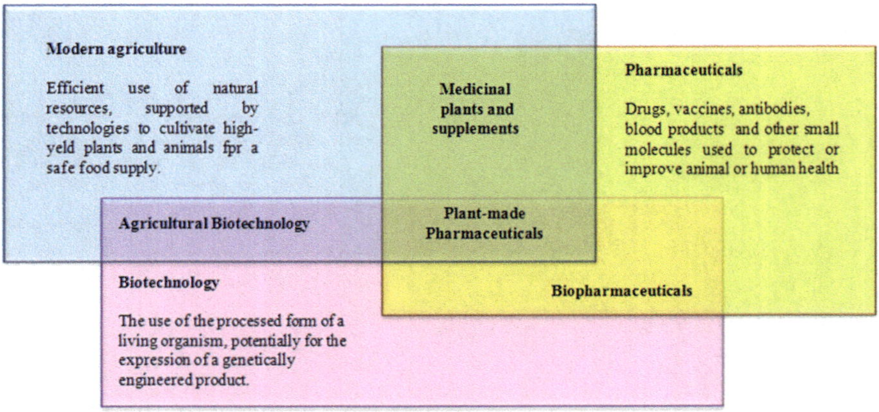

Fig. 4 Plant-made biopharmaceuticals on the intersection among the R&D areas of modern agriculture, medicinal plants/supplements and (plant) biotechnology (sourced from https://www.rri-tools.eu/it/how-to-stk-pm-implement-rri-at-national-level; R&I: Research and Innovation)

perceptions of agricultural biotechnology specific to producing novel vaccines and pharmaceuticals is needed before substantive generalizations can be made. Given that (1) oral pharmaceuticals are preferred, (2) people believe that most vaccines are genetically modified, and (3) public has expressed a high acceptance for PMPs, further development of this technology by commercial parties is desirable, if paralleled with appropriate market demand for specific products. Investment in clear communication by scientists and regulators will further enhance the public trust, optimism and ultimate acceptance for PMPs.

Even though the first players that tried to move the border forward were academia and public parties, due to the commercial uncertainties caused by production processes in the absence of certain regulations, nevertheless Elelyso is a positive example of how interest and engagement of a company can generate a big value for citizen's health and can reassure people on the possibility to get profit of some pharmaceuticals that are "biobetters by nature" due to the plant origin.

Even if the probability and severity of risks is different between open-field production for food and pharmaceutically based GMP production, depends on the plant species and has to be determined on a case-by-case basis for each plant-based product, the evidences of harmlessness of GM crops used for food supply may probably help to communicate to end-users the intrinsic safety of plant-produced biopharmaceuticals for both people and environment. As an example, meta-analysis of 21 years of open-field data (from 1996 to 2016) on the impact of genetically engineered maize clearly demonstrated the fulfilment of the desired effects on the target insect, no substantial effect on insect community diversity, reduced health risks for human consumption (due to relevant reduction of mycotoxin content in grains) and improved quality of the production (Pellegrino et al. 2018).

Reaching citizens and stakeholders with correct and peer-reviewed scientific information should finally clear both risks and benefits related to genetically engineered plants and reduce debate, hopefully. As gene technology moves forward, new traits and new health properties will be conferred to plants, and new experimental data should be available allowing researchers and regulators to draw further conclusions on the agro-environmental and health risks of GE plants.

7 Conclusions

Platforms for PMP are different and may involve the use of whole plants in greenhouses, vertical farming facilities or plant cell/organ cultures subjected to a transient or stable expression (targeted or constitutive). In addition, the unaltered version of the engineered plant may be a food or feed crop, or neither. Expression may involve transgene(s) vehiculation into nuclear or organelle genome resulting in protein monomers, multimers or virus-like particles. The PMP may be intended for purification, or administration as a crude extract or whole-plant tissue with a route of administration may be parenteral oral, nasal or topical. All these aspects may emphasize the advantages of the plant-based system but also influence the steps toward the commercialization process. The aim of molecular farming is to produce large quantities of active and secure cost-effective pharmaceutical proteins. These approaches, in comparison with other established microbial and animal expression systems, show various advantages in the biosynthesis of pharmaceutical biomolecules. So far, lots of valuable pharmaceutical proteins and antibodies have been produced by this method, which can help the treatment of patients remarkably in developing countries where the production and preservation costs of such medicines cannot be afforded. However, there are still some disputes, such as the acceptance by the general public and the biosecurity hurdles. On the other end, it is astonishing to consider that a technology that began with the hypothesis that the plant cells could behave as a reactor for newly induced biosynthesis has actually resulted in life-saving treatments like in the case of Z-Mapp and Elelyso. The recent past has seen the emergence of PMP-based companies and the beginning of commercial era of PMP products. The success of Protalix Biotherapeutics and Medicago (just to name one of the expanding companies in the field) and many others will probably sustain the future of the PMP field. Looking ahead, the progress of further products to market and the gaining of a wider acceptance of the biopharmaceutical industry are hoped.

Acknowledgments Silvia Massa is the recipient of the special Grant from MIUR (Italian Ministry of University and Research) "ENEA 5 x Mille" (Young investigator Project: New therapeutic strategies for the treatment of cancer).

References

Barta A, Sommergruber K, Thompson D, Hartmuth K, Matzke MA, Matzke AJ (1986) The expression of a nopaline synthase - human growth hormone chimaeric gene in transformed tobacco and sunflower callus tissue. Plant Mol Biol 6(5):347–357. https://doi.org/10.1007/BF00034942

Binder A, Lambert J, Morbitzer R, Popp C, Ott T, Lahaye T, Parniske M (2014) A modular plasmid assembly kit for multigene expression, gene silencing and silencing rescue in plants. PLoS One 9(2):e88218. https://doi.org/10.1371/journal.pone.0088218

Bortesi L, Fischer R (2015) The CRISPR/Cas9 system for plant genome editing and beyond. Biotechnol Adv 33:41–52. https://doi.org/10.1016/j.biotechadv.2014.12.006

Capodicasa C, Chiani P, Bromuro C, De Bernardis F, Catellani M, Palma AS, Liu Y, Feizi T, Cassone A, Benvenuto E, Torosantucci A (2011) Plant production of anti-β-glucan antibodies for immunotherapy of fungal infections in humans. Plant Biotechnol J 9:776–787. https://doi.org/10.1111/j.1467-7652.2010.00586.x

Capodicasa C, Catellani M, Moscetti I, Bromuro C, Chiani P, Torosantucci A, Benvenuto E (2017) Comparative analysis of plant-produced, recombinant dimeric IgA against cell wall β-glucan of pathogenic fungi. Biotechnol Bioeng 114(12):2729–2738. https://doi.org/10.1002/bit.26403

Cardi T, Lenzi P, Maliga P (2010) Chloroplasts as expression platforms for plant-produced vaccines. Expert Rev Vaccines 9(8):893–911. https://doi.org/10.1586/erv.10.78

Castle D, Dalgleish J (2005) Cultivating fertile ground for the introduction of plant-derived vaccines in developing countries. Vaccine 23(15):1881–1885. https://doi.org/10.1016/j.vaccine.2004.11.022

Chichester JA, Manceva SD, Rhee A, Coffin MV, Musiychuk K, Mett V, Shamloul M, Norikane J, Streatfield SJ, Yusibov V (2013) A plant-produced protective antigen vaccine confers protection in rabbits against a lethal aerosolized challenge with Bacillus anthracis Ames spores. Hum Vaccin Immunother 9(3):544–552

Cho SW, Lee J, Carroll D, Kim JS, Lee J (2013) Heritable gene knockout in Caenorhabditis elegans by direct injection of Cas9-sgRNA ribonucleoproteins. Genetics 195(3):1177–1180. https://doi.org/10.1534/genetics.113.155853

Clarke JL, Waheed MT, Lössl AG, Martinussen I, Daniell H (2013) How can plant genetic engineering contribute to cost-effective fish vaccine development for promoting sustainable aquaculture? Plant Mol Biol 83(1–2):33–40. https://doi.org/10.1007/s11103-013-0081-9

Cox KM, Sterling JD, Regan JT, Gasdaska JR, Frantz KK, Peele CG, Black A, Passmore D, Moldovan-Loomis C, Srinivasan M, Cuison S, Cardarelli PM, Dickey LF (2006) Glycan optimization of a human monoclonal antibody in the aquatic plant Lemna minor. Nat Biotechnol 24:1591–1597. https://doi.org/10.1038/nbt1260

Czyż M, Dembczyński R, Marecik R, Wojas-Turek J, Milczarek M, Pajtasz-Piasecka E, Wietrzyk J, Pniewski T (2014) Freeze-drying of plant tissue containing HBV surface antigen for the oral vaccine against hepatitis B. Biomed Res Int 2014:485689. https://doi.org/10.1155/2014/485689

Dafny-Yelin M, Tzfira T (2007) Delivery of multiple transgenes to plant cells. Plant Physiol 145(4):1118–1128

Daniell H (2003) Medical molecular pharming: expression of antibodies, biopharmaceuticals and edible vaccines via the chloroplast genome. In: Plant biotechnology and beyond 2002. Springer, Dordrecht, pp 371–376. https://doi.org/10.1007/978-94-017-2679-5_76

D'Aoust MA, Couture MM, Charland N, Trépanier S, Landry N, Ors F, Vézina LP (2010) The production of hemagglutinin-based virus-like particles in plants: a rapid, efficient and safe response to pandemic influenza. Plant Biotechnol J 8(5):607–619. https://doi.org/10.1111/j.1467-7652.2009.00496.x

Davey RT Jr, Dodd L, Proschan MA, Neaton J, Nordwall JN, Koopmeiners JS, Beigel J, Tierney J, Lane HC, Fauci AS, Massaquoi MBF, Sahr F, Malvy D (2016) A randomized, controlled trial of ZMapp for Ebola virus infection. N Engl J Med 375:1448–1456. https://doi.org/10.1056/NEJMoa1604330

Davies HM (2010) Commercialization of whole-plant systems for biomanufacturing of protein products: evolution and prospects. Plant Biotechnol J 8(8):845–861. https://doi.org/10.1111/j.1467-7652.2010.00550.x

Davis BTA, Grillo-Lopez AJ, White CA, Mclaughlin P, Czuczman MS, Link BK, Maloney DG, Weaver RL, Rosenberg J, Levy R (2014) Rituximab anti-CD20 monoclonal antibody therapy in non-Hodgkin's lymphoma: safety and efficacy of re-treatment. J Clin Oncol 18:3135–3143. https://doi.org/10.1200/JCO.2000.18.17.3135

De Muynck B, Navarre C, Boutry M (2010) Production of antibodies in plants: status after twenty years. Plant Biotechnol J 8:529–563. https://doi.org/10.1111/j.1467-7652.2009.00494.x

Demurtas OC, Massa S, Ferrante P, Venuti A, Franconi R, Giuliano G (2013) A Chlamydomonas-derived human papillomavirus 16 E7 vaccine induces specific tumor protection. PLoS One 8(4): e61473. https://doi.org/10.1371/journal.pone.0061473

Demurtas OC, Massa S, Illiano E, De Martinis D, Chan PK, Di Bonito P, Franconi R (2016) Antigen production in plant to tackle infectious diseases flare up: the case of SARS. Front Plant Sci 7:54–66. https://doi.org/10.3389/fpls.2016.00054

Drake PM, de Moraes ML, Szeto TH, Ma JK (2013) Transformation of Althaea officinalis L. by Agrobacterium rhizogenes for the production of transgenic roots expressing the anti-HIV microbicide cyanovirin-N. Transgenic Res 22(6):1225–1229. https://doi.org/10.1007/s11248-013-9730-7

Drake PM, Szeto TH, Paul MJ, Teh AY, Ma JK (2017) Recombinant biologic products versus nutraceuticals from plants - a regulatory choice? Br J Clin Pharmacol 83(1):82–87. https://doi.org/10.1111/bcp.13041

Ecker DM, Jones SD, Levine HL (2015) The therapeutic monoclonal antibody market. MAbs 7(1):9–14. https://doi.org/10.4161/19420862.2015.989042
European Medicines Agency (EMA) (2008) Guideline on the quality of biological active substances produced by stable transgene expression in higher plants (EMEA/CHMP/BWP/48316/2006). European Medicines Agency, London
Fox JL (2012) First plant-made biologic approved. Nat Biotechnol 30:472 https://www.nature.com/articles/nbt0612-472
Franconi R, Demurtas OC, Massa S (2010) Plant-derived vaccines and other therapeutics produced in contained systems. Expert Rev Vaccines 9(8):877–892. https://doi.org/10.1586/erv.10.91
Gasdaska JR, Sherwood S, Regan JT, Dickey LF (2012) An afucosylated anti-CD20 monoclonal antibody with greater antibody-dependent cellular cytotoxicity and B-cell depletion and lower complement dependent cytotoxicity than rituximab. Mol Immunol 50:134–141. https://doi.org/10.1016/j.molimm.2012.01.001
Gaughan CL (2016) The present state of the art in expression, production and characterization of monoclonal antibodies. Mol Divers 20(1):255–270. https://doi.org/10.1007/s11030-015-9625-z
Giritch A, Marillonnet S, Engler C, van Eldik G, Botterman J, Klimyuk V, Gleba Y (2006) Rapid high-yield expression of full-size IgG antibodies in plants coinfected with noncompeting viral vectors. PNAS 103(40):14701–14706. https://doi.org/10.1073/pnas.0606631103
Gleba Y, Klimyuk V, Marillonnet S (2005) Magnifection: a new platform for expressing recombinant vaccines in plants. Vaccine 23:2042–2048. https://doi.org/10.1016/j.vaccine.2005.01.006
Goodin MM, Zaitlin D, Naidu RA, Lommel SA (2008) Nicotiana benthamiana: its history and future as a model for plant-pathogen interactions. Mol Plant Microbe Interact 21(8):1015–1026. https://doi.org/10.1094/MPMI-21-8-1015
Gorantala J, Grover S, Rahi A, Chaudhary P, Rajwanshi R, Sarin NB, Bhatnagar R (2014) Generation of protective immune response against anthrax by oral immunization with protective antigen plant-based vaccine. J Biotechnol 176:1–10. https://doi.org/10.1016/j.jbiotec.2014.01.033
Govea-Alonso DO, Rubio-Infante N, García-Hernández AL, Varona-Santos JT, Korban SS, Moreno-Fierros L, Rosales-Mendoza S (2013) Immunogenic properties of a lettuce-derived C4(V3)6 multiepitopic HIV protein. Planta 238(4):785–792. https://doi.org/10.1007/s00425-013-1932-y
Grohs BM, Niu Y, Veldhuis LJ, Trabelsi S, Garabagi F, Hassell JA, McLean MD, Hall JC (2010) Plant-produced trastuzumab inhibits the growth of HER2 positive cancer cells. J Agric Food Chem 58(18):10056–10063. https://doi.org/10.1021/jf102284f
Guerrero-Andrade O, Loza-Rubio E, Olivera-Flores T, Fehérvári-Bone T, Gómez-Lim M (2006) Expression of the Newcastle disease virus fusion protein in transgenic maize and immunological studies. Transgenic Res 15:455–463. https://doi.org/10.1007/s11248-006-0017-0
Guillon S, Trémouillaux-Guiller J, Pati PK, Rideau M, Gantet P (2006) Hairy root research: recent scenario and exciting prospects. Curr Opin Plant Biol 9(3):341–346. https://doi.org/10.1016/j.pbi.2006.03.008
Haq TA, Mason HS, Clements JD, Arntzen CJ (1995) Oral immunization with a recombinant bacterial antigen produced in transgenic plants. Science 268(5211):714–716
Havenith H, Raven N, Di Fiore S, Fischer R, Schillberg S (2014) Image-based analysis of cell-specific productivity for plant cell suspension cultures. Plant Cell Tissue Organ Cult 117:393–399. https://doi.org/10.1007/s11240-014-0448-x
He J, Lai H, Engle M, Gorlatov S, Gruber C, Steinkellner H, Diamond MS, Chen Q (2014) Generation and analysis of novel plant-derived antibody-based therapeutic molecules against West Nile virus. PLoS One 9:e93541. https://doi.org/10.1371/journal.pone.0093541
Hellwig S, Drossard J, Twyman RM, Fischer R (2004) Plant cell cultures for the production of recombinant proteins. Nat Biotechnol 22(11):1415–1422. https://doi.org/10.1038/nbt1027
Hessell AJ, Rakasz EG, Poignard P, Hangartner L, Landucci G, Forthal DN, Koff WC, Watkins DI, Burton DR (2009) Broadly neutralizing human anti-HIV antibody 2G12 is effective in

protection against mucosal SHIV challenge even at low serum neutralizing titers. PLoS Pathog 5:e1000433. https://doi.org/10.1371/journal.ppat.1000433

Hiatt A, Cafferkey R, Bowdish K (1989) Production of antibodies in transgenic plants. Nature 342(6245):76–78. https://doi.org/10.1038/342076a0

Hiatt A, Bohorova N, Bohorov O, Goodman C, Kim D, Pauly MH, Velasco J, Whaley KJ, Piedra PA, Gilbert BE, Zeitlin L (2014) Glycan variants of a respiratory syncytial virus antibody with enhanced effector function and in vivo efficacy. PNAS 111:5992–5997. https://doi.org/10.1073/pnas.1402458111

Hongli L, Xukui L, Ting L, Wensheng L, Lusheng S, Jin Z (2013) Transgenic tobacco expressed HPV16-L1 and LT-B combined immunization induces strong mucosal and systemic immune responses in mice. Hum Vaccin Immunother 9:83–89. https://doi.org/10.4161/hv.22292

Huang Z, Santi L, LePore K, Kilbourne J, Arntzen CJ, Mason HS (2006) Rapid, high-level production of hepatitis B core antigen in plant leaf and its immunogenicity in mice. Vaccine 24(14):2506–2513. https://doi.org/10.1016/j.vaccine.2005.12.024

Huang Z, Phoolcharoen W, Lai H, Piensook K, Cardineau G, Zeitlin L, Whaley KJ, Arntzen CJ, Mason HS, Chen Q (2010) High-level rapid production of full-size monoclonal antibodies in plants by a single-vector DNA replicon system. Biotechnol Bioeng 106(1):9–17. https://doi.org/10.1002/bit.22652

Joensuu JJ, Verdonck F, Ehrström A, Peltola M, Siljander-Rasi H, Nuutila AM, Oksman-Caldentey KM, Teeri TH, Cox E, Goddeeris BM, Niklander-Teeri V (2006) F4 (K88) fimbrial adhesin FaeG expressed in alfalfa reduces F4+ enterotoxigenic Escherichia coli excretion in weaned piglets. Vaccine 24(13):2387–2394. https://doi.org/10.1016/j.vaccine.2005.11.056

Kapusta J, Modelska A, Figlerowicz M, Pniewski T, Letellier M, Lisowa O, Yusibov V, Koprowski H, Plucienniczak A, Legocki AB (1999) A plant-derived edible vaccine against hepatitis B virus. FASEB J 13(13):1796–1799 Erratum in: FASEB J 1999 13(15):2339

Katsnelson A, Ransom J, Vermij P, Waltz E (2006) News in brief: USDA approves the first plant-based vaccine. Nat Biotechnol 24(3):233–234 http://www.nature.com/articles/nbt0306-233#t1

Kessans SA, Linhart MD, Meador LR, Kilbourne J, Hogue BG, Fromme P, Matoba N, Mor TS (2016) Immunological characterization of plant-based HIV-1 Gag/Dgp41 virus-like particles. PLoS One 11(3):e0151842. https://doi.org/10.1371/journal.pone.0151842

Khandelwal A, Lakshmi Sita G, Shaila MS (2003) Oral immunization of cattle with hemagglutinin protein of rinderpest virus expressed in transgenic peanut induces specific immune responses. Vaccine 21(23):3282–3289

Kim S, Kim D, Cho SW, Kim J, Kim JS (2014) Highly efficient RNA-guided genome editing in human cells via delivery of purified Cas9 ribonucleoproteins. Genome Res 24(6):1012–1019. https://doi.org/10.1101/gr.171322.113

Kim MY, Reljic R, Kilbourne J, Ceballos-Olvera I, Yang MS, Reyes-Del Valle J, Mason HS (2015) Novel vaccination approach for dengue infection based on recombinant immune complex universal platform. Vaccine 33:1830–1880. https://doi.org/10.1016/j.vaccine.2015.02.036

Klein C, Lammens A, Schäfer W, Georges G, Schwaiger M, Mössner E, Hopfner KP, Umaña P, Niederfellner G (2013) MAbs 5(1):22–33. https://doi.org/10.4161/mabs.22771

Ko K, Steplewski Z, Glogowska M, Koprowski H (2005) Inhibition of tumor growth by plant-derived mAb. Proc Natl Acad Sci U S A 102(19):7026–7030. https://doi.org/10.1073/pnas.0502533102

Kolotilin I, Kaldis A, Devriendt B, Joensuu J, Cox E, Menassa R (2012) Production of a subunit vaccine candidate against porcine postweaning diarrhea in high-biomass transplastomic tobacco. PLoS One 7:e42405. https://doi.org/10.1371/journal.pone.0042405

Komarova TV, Baschieri S, Donini M, Marusic C, Benvenuto E, Dorokhov YL (2010) Transient expression systems for plant-derived biopharmaceuticals. Expert Rev Vaccines 9(8):859–876. https://doi.org/10.1586/erv.10.85

Komarova TV, Kosorukov VS, Frolova OY, Petrunia IV, Skrypnik KA, Gleba YY, Dorokhov YL (2011) Plant-made trastuzumab (herceptin) inhibits HER2/Neu+ cell proliferation and retards tumor growth. PLoS One 6(3):e17541. https://doi.org/10.1371/journal.pone.0017541

Kushnir N, Streatfield SJ, Yusibov V (2012) Virus-like particles as a highly efficient vaccine platform: diversity of targets and production systems and advances in clinical development. Vaccine 31(1):58–83. https://doi.org/10.1016/j.vaccine.2012.10.083

Lai H, Engle M, Fuchs A, Keller T, Johnson S, Gorlatov S, Diamond MS, Chen Q (2010) Monoclonal antibody produced in plants efficiently treats West Nile virus infection in mice. Proc Natl Acad Sci U S A 107(6):2419–2424. https://doi.org/10.1073/pnas.0914503107

Lai H, He J, Engle M, Diamond MS, Chen Q (2012) Robust production of virus-like particles and monoclonal antibodies with geminiviral replicon vectors in lettuce. Plant Biotechnol J 10:95–104. https://doi.org/10.1111/j.1467-7652.2011.00649.x

Lai H, He J, Hurtado J, Stahnke J, Fuchs A, Mehlhop E, Gorlatov S, Loos A, Diamond MS, Chen Q (2014) Structural and functional characterization of an anti-West Nile virus monoclonal antibody and its single-chain variant produced in glycoengineered plants. Plant Biotechnol J 12:1098–1107. https://doi.org/10.1111/pbi.12217

Landry N, Ward BJ, Trépanier S, Montomoli E, Dargis M, Lapini G, Vézina LP (2010) Preclinical and clinical development of plant-made virus-like particle vaccine against avian H5N1 influenza. PLoS One 5(12):e15559. https://doi.org/10.1371/journal.pone.0015559

Landry N, Pillet S, Favre D, Poulin JF, Trépanier S, Yassine-Diab B, Ward BJ (2012) Influenza virus-like particle vaccines made in Nicotiana benthamiana elicit durable, poly-functional and cross-reactive T cell responses to influenza HA antigens. Clin Immunol 154(2):164–177. https://doi.org/10.1016/j.clim.2014.08.003

Lee JE, FuscoML HAJ, Oswald WB, Burton DR, Saphire EO (2008) Structure of the Ebola virus glycoprotein bound to an antibody from a human survivor. Nature 454:177–182. https://doi.org/10.1038/nature07082

Lee JW, Heo W, Lee J, Jin N, Yoon SM, Park KY, Kim EY, Kim WT, Kim JY (2018) The B cell death function of obinutuzumab-HDEL produced in plant (Nicotiana benthamiana L.) is equivalent to obinutuzumab produced in CHO cells. PLoS One 13(1):e0191075. https://doi.org/10.1371/journal.pone.0191075

Li JF, Norville JE, Aach J, McCormack M, Zhang D, Bush J, Church GM, Sheen J (2013) Multiplex and homologous recombination-mediated genome editing in Arabidopsis and Nicotiana benthamiana using guide RNA and Cas9. Nat Biotechnol 31(8):688–691. https://doi.org/10.1038/nbt.2654

Lienard D, Sourrouille C, Gomord V, Faye L (2007) Pharming and transgenic plants. Biotechnol Annu Rev 13:115–147. https://doi.org/10.1016/S1387-2656(07)13006-4

List T, Neri D (2013) Immunocytokines: a review of molecules in clinical development for cancer therapy. Clin Pharmacol 5:29–45. https://doi.org/10.2147/CPAA.S49231

Loh HS, Green BJ, Yusibov V (2017) Using transgenic plants and modified plant viruses for the development of treatments for human diseases. Curr Opin Virol 26:81–89. https://doi.org/10.1016/j.coviro.2017.07.019

Ma JK, Hiatt A, Hein M, Vine ND, Wang F, Stabila P, van Dolleweerd C, Mostov K, Lehner T (1995) Generation and assembly of secretory antibodies in plants. Science 268(5211):716–719

Ma JK, Drake PM, Christou P (2003) The production of recombinant pharmaceutical proteins in plants. Nat Rev Genet 4(10):794–805. https://doi.org/10.1038/nrg1177

Ma JK, Drossard J, Lewis D, Altmann F, Boyle J, Christou P, Cole T, Dale P, van Dolleweerd CJ, Isitt V, Katinger D, Lobedan M, Mertens H, Paul MJ, Rademacher T, Sack M, Hundleby PA, Stiegler G, Stoger E, Twyman RM, Vcelar B, Fischer R (2015) Regulatory approval and a first-in-human phase I clinical trial of a monoclonal antibody produced in transgenic tobacco plants. Plant Biotechnol J 13(8):1106–1120. https://doi.org/10.1111/pbi.12416

Marcus R, Davies A, Ando K, Klapper W, Opat S, Owen C, Phillips E, Sangha R, Schlag R, Seymour JF, Townsend W, Trněný M, Wenger M, Fingerle-Rowson G, Rufibach K, Moore T, Herold M, Hiddemann W (2017) Obinutuzumab for the first-line treatment of follicular lymphoma. N Engl J 377(14):1331–1344. https://doi.org/10.1056/NEJMoa1614598

Marillonnet S, Giritch A, Gils M, Kandzia R, Klimyuk V, Gleba Y (2004) In planta engineering of viral RNA replicons: efficient assembly by recombination of DNA modules delivered by Agrobacterium. PNAS 101:6852–6857

Marusic C, Novelli F, Salzano AM, Scaloni A, Benvenuto E, Pioli C, Donini M (2016) Production of an active anti-CD20-hIL-2 immunocytokine in Nicotiana benthamiana. Plant Biotechnol J 14(1):240–251. https://doi.org/10.1111/pbi.12378

Marusic C, Pioli C, Stelter S, Novelli F, Lonoce C, Morrocchi E, Benvenuto E, Salzano AM, Scaloni A, Donini M (2018) N-glycan engineering of a plant-produced anti-CD20-hIL-2 immunocytokine significantly enhances its effector functions. Biotechnol Bioeng 115(3):565–576. https://doi.org/10.1002/bit.26503

Mason HS, Lam DM, Arntzen CJ (1992) Expression of hepatitis B surface antigen in transgenic plants. Proc Natl Acad Sci U S A 89(24):11745–11749

Massa S, Franconi R, Brandi R, Muller A, Mett V, Yusibov V, Venuti A (2007) Anti-cancer activity of plant-produced HPV16 E7 vaccine. Vaccine 25(16):3018–3021. https://doi.org/10.1016/j.vaccine.2007.01.018

Massa S, Paolini F, Spanò L, Franconi R, Venuti A (2011) Mutants of plant genes for developing cancer vaccines. Hum Vaccin 7(Suppl):147–155

Massa S, Paolini F, Curzio G, Cordeiro MN, Illiano E, Demurtas OC, Franconi R, Venuti A (2017) A plant protein signal sequence improved humoral immune response to HPV prophylactic and therapeutic DNA vaccines. Hum Vaccin Immunother 13(2):271–282. https://doi.org/10.1080/21645515.2017.1264766

Mayfield SP, Franklin SE, Lerner RA (2003) Expression and assembly of a fully active antibody in algae. Proc Natl Acad Sci U S A 100:438–442. https://doi.org/10.1073/pnas.0237108100

McCarthy M (2014) US signs contract with ZMapp maker to accelerate development of the Ebola drug. BMJ 349:g5488. https://doi.org/10.1136/bmj.g5488

McCormick AA, Reddy S, Reinl SJ, Cameron TI, Czerwinkski DK, Vojdani F, Hanley KM, Garger SJ, White EL, Novak J, Barrett J, Holtz RB, Tusé D, Levy R (2008) Plant-produced idiotype vaccines for the treatment of non-Hodgkin's lymphoma: safety and immunogenicity in a phase I clinical study. Proc Natl Acad Sci U S A 105(29):10131–10136. https://doi.org/10.1073/pnas.0803636105

Mett V, Chichester JA, Stewart ML, Musiychuk K, Bi H, Reifsnyder CJ, Hull AK, Albrecht MT, Goldman S, Baillie LW, Yusibov V (2011) A non-glycosylated, plant-produced human monoclonal antibody against anthrax protective antigen protects mice and non-human primates from B. anthracis spore challenge. Hum Vaccin 7:183–190

Meyers B, Zaltsman A, Lacroix B, Kozlovsky SV, Krichevsky A (2010) Nuclear and plastid genetic engineering of plants: Comparison of opportunities and challenges. Biotechnol Adv 28(6):747–756. https://doi.org/10.1016/j.biotechadv.2010.05.022

Monroy-García A, Gómez-Lim MA, Weiss-Steider B, Hernández-Montes J, Huerta-Yepez S, Rangel-Santiago JF, Santiago-Osorio E, Mora García Mde L (2014) Immunization with an HPV-16 L1-based chimeric virus-like particle containing HPV-16 E6 and E7 epitopes elicits long-lasting prophylactic and therapeutic efficacy in an HPV-16 tumor mice model. Arch Virol 159(2):291–305. https://doi.org/10.1007/s00705-013-1819-z

Musiychuk K, Stephenson N, Bi H, Farrance CE, Orozovic G, Brodelius M, Brodelius P, Horsey A, Ugulava N, Shamloul AM, Mett V, Rabindran S, Streatfield SJ, Yusibov V (2007) A launch vector for the production of vaccine antigens in plants. Influenza Other Respir Viruses 1(1):19–25. https://doi.org/10.1111/j.1750-2659.2006.00005.x

Nahampun HN, Bosworth B, Cunnick J, Mogler M, Wang K (2015) Expression of H3N2 nucleoprotein in maize seeds and immunogenicity in mice. Plant Cell Rep 34:969–980. https://doi.org/10.1007/s00299-015-1758-0

Nandi S, Yalda D, Lu S, Nikolov Z, Misaki R, Fujiyama K, Huang N (2005) Process development and economic evaluation of recombinant human lactoferrin expressed in rice grain. Transgenic Res 14(3):237–249

Nekrasov V, Staskawicz B, Weigel D, Jones JD, Kamoun S (2013) Targeted mutagenesis in the model plant Nicotiana benthamiana using Cas9 RNA-guided endonuclease. Nat Biotechnol 31(8):691–693. https://doi.org/10.1038/nbt.2655

Oey M, Lohse M, Kreikemeyer B, Bock R (2009) Exhaustion of the chloroplast protein synthesis capacity by massive expression of a highly stable protein antibiotic. Plant J 57(3):436–445

Olinger GG Jr, Pettitt J, Kim D, Working C, Bohorov O, Bratcher B, Hiatt E, Hume SD, Johnson AK, Morton J, Pauly M, Whaley KJ, Lear CM, Biggins JE, Scully C, Hensley L, Zeitlin L (2012) Delayed treatment of Ebola virus infection with plant-derived monoclonal antibodies provides protection in rhesus macaques. PNAS 109:18030–18035. https://doi.org/10.1073/pnas.1213709109

Oliphant T, Engle M, Nybakken GE, Doane C, Johnson S, Huang L, Gorlatov S, Mehlhop E, Marri A, Chung KM, Ebel GD, Kramer LD, Fremont DH, Diamond MS (2005) Development of a humanized monoclonal antibody with therapeutic potential against West Nile virus. Nat Med 11:522–530. https://doi.org/10.1038/nm1240

Paul M, Reljic R, Klein K, Drake PM, van Dolleweerd C, Pabst M, Windwarder M, Arcalis E, Stoger E, Altmann F, Cosgrove C, Bartolf A, Baden S, Ma JK (2014) Characterization of a plant-produced recombinant human secretory IgA with broad neutralizing activity against HIV. mAbs 6:1585–1597. https://doi.org/10.4161/mabs.36336

Pellegrino E, Bedini S, Nuti M, Ercoli L (2018) Impact of genetically engineered maize on agronomic, environmental and toxicological traits: a meta-analysis of 21 years of field data. Sci Rep 8(1):3113–3124. https://doi.org/10.1038/s41598-018-21284-2

Peréz Aguirreburualde MS, Gómez MC, Ostachuk A, Wolman F, Albanesi G, Pecora A, Odeon A, Ardila F, Escribano JM, Dus Santos MJ, Wigdorovitz A (2013) Efficacy of a BVDV subunit vaccine produced in alfalfa transgenic plants. Vet Immunol Immunopathol 151(3–4):315–324. https://doi.org/10.1016/j.vetimm.2012.12.004

Pérez Filgueira D, Zamorano P, Domínguez M, Taboga O, Del Médico ZM, Puntel M, Romera SA, Morris TJ, Borca MV, Sadir AM (2003) Bovine herpes virus gD protein produced in plants using a recombinant tobacco mosaic virus (TMV) vector possesses authentic antigenicity. Vaccine 21:4201–4209

Peyret H, Lomonossoff GP (2015) When plant virology met Agrobacterium: the rise of the deconstructed clones. Plant Biotechnol J 13(8):1121–1135. https://doi.org/10.1111/pbi.12412

Phoolcharoen W, Bhoo SH, Lai H, Ma J, Arntzen CJ, Chen Q, Mason HS (2011) Expression of an immunogenic Ebola immune complex in Nicotiana benthamiana. Plant Biotechnol J 9(7):807–816. https://doi.org/10.1111/j.1467-7652.2011.00593.x

Pillet S, Aubin É, Trépanier S, Bussière D, Dargis M, Poulin JF, Yassine-Diab B, Ward BJ, Landry N (2016) A plant-derived quadrivalent virus like particle influenza vaccine induces cross-reactive antibody and T cell response in healthy adults. Clin Immunol 168:72–87. https://doi.org/10.1016/j.clim.2016.03.008

Plosker GL, Figgitt DP (2003) Rituximab: a review of its use in non-Hodgkin's lymphoma and chronic lymphocytic leukaemia. Drugs 63:803–843

Pua TL, Chan XY, Loh HS, Omar AR, Yusibov V, Musiychuk K, Hall AC, Coffin MV, Shoji Y, Chichester JA, Bi H, Streatfield SJ (2017) Purification and immunogenicity of hemagglutinin from highly pathogenic avian influenza virus H5N1 expressed in Nicotiana benthamiana. Hum Vaccin Immunother 13:306–313. https://doi.org/10.1080/21645515.2017.1264783

Pujol M, Gavilondo J, Ayala M, Rodríguez M, González EM, Pérez L (2007) Fighting cancer with plant-expressed pharmaceuticals. Trends Biotechnol 25(10):455–459. https://doi.org/10.1016/j.tibtech.2007.09.001

Qiu X, Wong G, Audet J, Bello A, Fernando L, Alimonti JB, Fausther-Bovendo H, Wei H, Aviles J, Hiatt E, Johnson A, Morton J, Swope K, Bohorov O, Bohorova N, Goodman C, Kim D, Pauly MH, Velasco J, Pettitt J, Olinger GG, Whaley K, Xu B, Strong JE, Zeitlin L, Kobinger GP (2014) Reversion of advanced Ebola virus disease in nonhuman primates with ZMapp. Nature 514:47–53. https://doi.org/10.1038/nature13777

Rademacher T (2013) Method for the generation and cultivation of a plant cell pack. Patent WO 2013113504

Reff ME, Carner K, Chambers KS, Chinn PC, Leonard JE, Raab R, Newman RA, Hanna N, Anderson DR (1994) Depletion of B cells in vivo by a chimeric mouse human monoclonal antibody to CD20. Blood 83(2):435–445

Reski R, Parsons J, Decker EL (2015) Moss-made pharmaceuticals: from bench to bedside. Plant Biotechnol J 13(8):1191–1198. https://doi.org/10.1111/pbi.12401

Rios-Huerta R, Monreal-Escalante E, Govea-Alonso DO, Angulo C, Rosales-Mendoza S (2017) Expression of an immunogenic LTB-based chimeric protein targeting Zaire ebolavirus epitopes from GP1 in plant cells. Plant Cell Rep 36:355–365. https://doi.org/10.1007/s00299-016-2088-6

Rodriguez M, Ramirez NI, Ayala M, Freyre F, Pérez L, Triguero A, Mateo C, Selman-Housein G, Gavilondo JV, Pujol M (2005) Transient expression in tobacco leaves of an aglycosylated recombinant antibody against the epidermal growth factor receptor. Biotechnol Bioeng 89(2):188–194. https://doi.org/10.1002/bit.20333

Rubio-Infante N, Govea-Alonso DO, Romero-Maldonado A, García-Hernández AL, Ilhuicatzi-Alvarado D, Salazar-González JA, Korban SS, Rosales-Mendoza S, Moreno-Fierros L (2015) A plant-derived multi-HIV antigen induces broad immune responses in orally immunized mice. Mol Biotechnol 57(7):662–674. https://doi.org/10.1007/s12033-015-9856-3

Rukavtsova EB, Rudenko NV, Puchko EN, Zakharchenko NS, Buryanov YI (2015) Study of the immunogenicity of hepatitis B surface antigen synthesized in transgenic potato plants with increased biosafety. J Biotechnol 203:84–88. https://doi.org/10.1016/j.jbiotec.2015.03.019

Rybicki EP (2010) Plant-made vaccines for humans and animals. Plant Biotechnol J 8(5):620–637. https://doi.org/10.1111/j.1467-7652.2010.00507.x

Rybicki EP (2014) Plant-based vaccines against viruses. Virol J 11:205–225. https://doi.org/10.1186/s12985-014-0205-0

Sabalza M, Vamvaka E, Christou P, Capell T (2013) Seeds as a production system for molecular pharming applications: status and prospects. Curr Pharm Des 19(31):5543–5552

Sainsbury F, Thuenemann EC, Lomonossoff GP (2009) pEAQ: versatile expression vectors for easy and quick transient expression of heterologous proteins in plants. Plant Biotechnol J 7(7):682–693. https://doi.org/10.1111/j.1467-7652.2009.00434.x

Sainsbury F, Sack M, Stadlmann J, Quendler H, Fischer R, Lomonossoff GP (2010) Rapid transient production in plants by replicating and non-replicating vectors yields high quality functional anti-HIV antibody. PLoS One 5(11):e13976. https://doi.org/10.1371/journal.pone.0013976

Salazar-González JA, Angulo C, Rosales-Mendoza S (2015) Chikungunya virus vaccines: current strategies and prospects for developing plant-made vaccines. Vaccine 33(31):3650–3658. https://doi.org/10.1016/j.vaccine.2015.05.104

Santos RB, Abranches R, Fischer R, Sack M, Holland T (2016) Putting the spotlight back on plant suspension cultures. Front Plant Sci 7:60–72. https://doi.org/10.3389/fpls.2016.00297

Sarrion-Perdigones A, Falconi EE, Zandalinas SI, Juárez P, Fernández-del-Carmen A, Granell A, Orzaez D (2011) GoldenBraid: an iterative cloning system for standardized assembly of reusable genetic modules. PLoS One 6(7):e21622. https://doi.org/10.1371/journal.pone.0021622

Schillberg S, Raven N, Fischer R, Twyman RM, Schiermeyer A (2013) Molecular farming of pharmaceutical proteins using plant suspension cell and tissue cultures. Curr Pharm Des 19(31):5531–5542

Scotti N, Rybicki EP (2013) Virus-like particles produced in plants as potential vaccines. Expert Rev Vaccines 12(2):211–224. https://doi.org/10.1586/erv.12.147

Shaaltiel Y, Bartfeld D, Hashmueli S, Baum G, Brill-Almon E, Galili G, Dym O, Boldin-Adamsky SA, Silman I, Sussman JL, Futerman AH, Aviezer D (2007) Production of glucocerebrosidase with terminal mannose glycans for enzyme replacement therapy of Gaucher's disease using a plant cell system. Plant Biotechnol J 5(5):579–590. https://doi.org/10.1111/j.1467-7652.2007.00263.x

Shaaltiel Y, Gingis-Velitski S, Tzaban S, Fiks N, Tekoah Y, Aviezer D (2015) Plant-based oral delivery of β-glucocerebrosidase as an enzyme replacement therapy for Gaucher's disease. Plant Biotechnol J 13(8):1033–1040. https://doi.org/10.1111/pbi.12366

Shan Q, Wang Y, Li J, Zhang Y, Chen K, Liang Z, Zhang K, Liu J, Xi JJ, Qiu JL, Gao C (2013) Targeted genome modification of crop plants using a CRISPR-Cas system. Nat Biotechnol 31(8):686–688. https://doi.org/10.1038/nbt.2650

Shoji Y, Chichester JA, Palmer GA, Farrance CE, Stevens R, Stewart MGoldschmidt L, Deyde V, Gubareva L, Klimov A, Mett V, Yusibov V (2011) An influenza N1 neuraminidase-specific monoclonal antibody with broad neuraminidase inhibition activity against H5N1 HPAI viruses. Hum Vaccin 7(Suppl):199–204

Shoji Y, Jones RM, Mett V, Chichester JA, Musiychuk K, Sun X, Tumpey TM, Green BJ, Shamloul M, Norikane J, Bi H, Hartman CE, Bottone C, Stewart M, Streatfield SJ, Yusibov V (2013) A plant-produced H1N1 trimeric hemagglutinin protects mice from a lethal influenza virus challenge. Hum Vaccin Immunother 9(3):553–560

Shoji Y, Prokhnevsky A, Leffet B, Vetter N, Tottey S, Satinover S, Musiychuk K, Shamloul M, Norikane J, Jones RM, Chichester JA, Green BJ, Streatfield SJ, Yusibov V (2015) Immunogenicity of H1N1 influenza virus-like particles produced in Nicotiana benthamiana. Hum Vaccin Immunother 11(1):118–123. https://doi.org/10.4161/hv.34365

Simmons G, Wool-Lewis RJ, Baribaud F, Netter RC, Bates P (2002) Ebola virus glycoproteins induce global surface protein down-modulation and loss of cell adherence. J Virol 76:2518–2528

Singh A, Srivastava S, Chouksey A, Panwar BS, Verma PC, Roy S, Singh PK, Saxena G, Tuli R (2015) Expression of rabies glycoprotein and ricin toxin B chain (RGP-RTB) fusion protein in tomato hairy roots: a step towards oral vaccination for rabies. Mol Biotechnol 57(4):359–370. https://doi.org/10.1007/s12033-014-9829-y

Skarjinskaia M, Ruby K, Araujo A, Taylor K, Gopalasamy-Raju V, Musiychuk K, Chichester JA, Palmer GA, de la Rosa P, Mett V, Ugulava N, Streatfield SJ, Yusibov V (2013) Hairy roots as a vaccine production and delivery system. Adv Biochem Eng Biotechnol 134:115–134. https://doi.org/10.1007/10_2013_184

Sparrow P, Broer I, Hood EE, Eversole K, Hartung F, Schiemann J (2013) Risk assessment and regulation of molecular farming - a comparison between Europe and US. Curr Pharm Des 19(31):5513–5530

Strasser R, Stadlmann J, Schahs M, Stiegler G, Quendler H, Mach L, Glössl J, Weterings K, Pabst M, Steinkellner H (2008) Generation of glyco-engineered *Nicotiana benthamiana* for the production of monoclonal antibodies with a homogeneous human-like N-glycan structure. Plant Biotechnol J 6(4):392–402. https://doi.org/10.1111/j.1467-7652.2008.00330.x

Streatfield SJ (2006) Mucosal immunization using recombinant plant-based oral vaccines. Methods 38(2):150–157. https://doi.org/10.1016/j.ymeth.2005.09.013

Tacket CO (2007) Plant-based vaccines against diarrheal diseases. Trans Am Clin Climatol Assoc 118:79–87

Tacket CO, Mason HS, Losonsky G, Clements JD, Levine MM, Arntzen CJ (1998) Immunogenicity in humans of a recombinant bacterial antigen delivered in a transgenic potato. Nat Med 4(5):607–609

Tacket CO, Mason HS, Losonsky G, Estes MK, Levine MM, Arntzen CJ (2000) Human immune responses to a novel Norwalk virus vaccine delivered in transgenic potatoes. J Infect Dis 182(1):302–305. https://doi.org/10.1086/315653

Takeyama N, Kiyono H, Yuki Y (2015a) Plant-based vaccines for animals and humans: recent advances in technology and clinical trials. Ther Adv Vaccin 3(5–6):139–154. https://doi.org/10.1177/2051013615613272

Takeyama N, Yuki Y, Tokuhara D, Oroku K, Mejima M, Kurokawa S, Kuroda M, Kodama T, Nagai S, Ueda S, Kiyono H (2015b) Oral rice-based vaccine induces passive and active immunity against enterotoxigenic E. coli-mediated diarrhea in pigs. Vaccine 33(39):5204–5211. https://doi.org/10.1016/j.vaccine.2015.07.074

Thanavala Y, Mahoney M, Pal S, Scott A, Richter L, Natarajan N, Goodwin P, Arntzen CJ, Mason HS (2005) Immunogenicity in humans of an edible vaccine for hepatitis B. Proc Natl Acad Sci U S A 102(9):3378–3382. https://doi.org/10.1073/pnas.0409899102

Tinland B, Hohn B (1995) Recombination between prokaryotic and eukaryotic DNA: integration of Agrobacterium tumefaciens T-DNA into the plant genome. Genet Eng (NY) 17:209–229

Torosantucci A, Chiani P, Bromuro C, De Bernardis F, Palma AS, Liu Y, Mignogna G, Maras B, Colone M, Stringaro A, Zamboni S, Feizi T, Cassone A (2009) Protection by anti-beta-glucan antibodies is associated with restricted beta-1,3 glucan binding specificity and inhibition of fungal growth and adherence. PLoS One 4(4):e5392. https://doi.org/10.1371/journal.pone.0005392

Tremblay R, Wang D, Jevnikar AM, Ma S (2010) Tobacco, a highly efficient green bioreactor for production of therapeutic proteins. Biotechnol Adv 28(2):214–221. https://doi.org/10.1016/j.biotechadv.2009.11.008

Tschofen M, Knopp D, Hood E, Stöger E (2016) Plant molecular farming: much more than medicines. Annu Rev Anal Chem (Palo Alto, Calif) 9(1):271–294. https://doi.org/10.1146/annurev-anchem-071015-041706

Tuboly T, Yu W, Bailey A, Degrandis S, Du S, Erickson L, Nagy E (2000) Immunogenicity of porcine transmissible gastroenteritis virus spike protein expressed in plants. Vaccine 18(19):2023–2028

Ullisch DA, Müller CA, Maibaum S, Kirchhoff J, Schiermeyer A, Schillberg S, Roberts JL, Treffenfeldt W, Büchs J (2012) Comprehensive characterization of two different Nicotiana tabacum cell lines leads to doubled GFP and HA protein production by media optimization. J Biosci Bioeng 113(2):242–248. https://doi.org/10.1016/j.jbiosc.2011.09.022

Vaquero C, Sack M, Schuster F, Finnern R, Drossard J, Schumann D, Reimann A, Fischer R (2002) A carcinoembryonic antigen-specific diabody produced in tobacco. FASEB J 16(3):408–410. https://doi.org/10.1096/fj.01-0363fje

Venuti A, Massa S, Mett V, Vedova LD, Paolini F, Franconi R, Yusibov V (2009) An E7-based therapeutic vaccine protects mice against HPV16 associated cancer. Vaccine 27(25–26):3395–3397. https://doi.org/10.1016/j.vaccine.2009.01.068

von Schomberg R (2013) Prospects for technology assessment in a framework of responsible research and innovation. In: Dusseldorp M, Beecroft R (eds) Estimate technology consequences, educational potential of transdisciplinary methods. Wiley, London, pp 39–62. ISBN 978-3-531-17908-7. ISBN 978-3-531-93468-6 (eBook). https://doi.org/10.1007/978-3-531-93468-6

Ward BJ, Landry N, Trépanier S, Mercier G, Dargis M, Couture M, D'Aoust MA, Vézina LP (2014) Human antibody response to N-glycans present on plant-made influenza virus-like particle (VLP) vaccines. Vaccine 32(46):6098–6106. https://doi.org/10.1016/j.vaccine.2014.08.079

Whaley KJ, Morton J, Hume S, Hiatt E, Bratcher B, Klimyuk V, Hiatt A, Pauly M, Zeitlin L (2014) Emerging antibody-based products. Curr Top Microbiol Immunol 375:107–126. https://doi.org/10.1007/82_2012_240

Wilson JA, Hevey M, Bakken R, Guest S, Bray M, Schmaljohn AL, Hart MK (2000) Epitopes involved in antibody-mediated protection from Ebola virus. Science 287:1664–1666

Wongsamuth R, Doran PM (1997) Production of monoclonal antibodies by tobacco hairy roots. Biotechnol Bioeng 54(5):401–415. https://doi.org/10.1002/(SICI)1097-0290(19970605)54:5<401::AID-BIT1>3.0.CO;2-I

Woo JW, Kim J, Kwon SI, Corvalán C, Cho SW, Kim H, Kim SG, Kim ST, Choe S, Kim JS (2015) DNA-free genome editing in plants with preassembled CRISPR-Cas9 ribonucleoproteins. Nat Biotechnol 33(11):1162–1164. https://doi.org/10.1038/nbt.3389

Woods RR, Geyer BC, Mor TS (2008) Hairy-root organ cultures for the production of human acetylcholinesterase. BMC Biotechnol 8:95102. https://doi.org/10.1186/1472-6750-8-95

Wu J, Yu L, Li L, Hu J, Zhou J, Zhou X (2007) Oral immunization with transgenic rice seeds expressing VP2 protein of infectious bursal disease virus induces protective immune responses in chickens. Plant Biotechnol J 5(5):570–578. https://doi.org/10.1111/j.1467-7652.2007.00270.x

Xu J, Dolan MC, Medrano G, Cramer CL, Weathers PJ (2012) Green factory: plants as bioproduction platforms for recombinant proteins. Biotechnol Adv 30(5):1171–1184. https://doi.org/10.1016/j.biotechadv.2011.08.020

Yang CD, Liao JT, Lai CY, Jong MH, Liang CM, Lin YL, Lin NS, Hsu YH, Liang SM (2007) Induction of protective immunity in swine by recombinant bamboo mosaic virus expressing foot-and-mouth disease virus epitopes. BMC Biotechnol 7:62–73. https://doi.org/10.1186/1472-6750-7-62

Young JA, Collier RJ (2007) Anthrax toxin: receptor binding, internalization, pore formation, and translocation. Annu Rev Biochem 76:243–265. https://doi.org/10.1146/annurev.biochem.75.103004.142728

Yusibov V, Hooper DC, Spitsin SV, Fleysh N, Kean RB, Mikheeva T, Deka D, Karasev A, Cox S, Randall J, Koprowski H (2002) Expression in plants and immunogenicity of plant virus-based experimental rabies vaccine. Vaccine 20(25–26):3155–3164

Zeitlin L, Pettitt J, Scully C, Bohorova N, Kim D, Pauly M, Hiatt A, Ngo L, Steinkellner H, Whaley KJ, Olinger GG (2011) Enhanced potency of a fucose-free monoclonal antibody being developed as an Ebola virus immunoprotectant. PNAS 108:20690–20694. https://doi.org/10.1073/pnas.1108360108

Zhou JY, Cheng LQ, Zheng XJ, Wu JX, Shang SB, Wang JY, Chen JG (2004) Generation of the transgenic potato expressing full-length spike protein of infectious bronchitis virus. J Biotechnol 111(2):121–130. https://doi.org/10.1016/j.jbiotec.2004.03.012

Sequencing Plant Genomes

Daniel G. Peterson and Mark Arick II

Contents

Communicated by Francisco M. Cánovas

D. G. Peterson (✉) and M. Arick II
Institute for Genomics, Biocomputing & Biotechnology (IGBB), Mississippi State University, Mississippi State, MS, USA
e-mail: dp127@msstate.edu

Progress in Botany (2019) 80: 109–194, DOI 10.1007/124_2018_18,
Published online: 10 May 2018

Abstract The first decade of the twenty-first century witnessed the development and commercialization of the so-called "second-generation" sequencing techniques. These short-read sequencing methods produced such large amounts of sequence data at such low prices that the costly and time-consuming physical map-based sequencing techniques, used to generate the human and the *Arabidopsis* genomes, were largely abandoned. The rise of second-generation sequencing at the cost of physical mapping resulted in a tremendous increase in the number and diversity of plant genome projects, a proliferation in the number of individuals and institutions involved in genome sequencing endeavors, and an overall decrease in the quality of resulting genome assemblies. Single-molecule ("third-generation") sequencing techniques, which came onto the scene more recently, provide much longer (and currently lower quality) reads than second-generation techniques, and a combination of second- and third-generation technologies is resulting in higher quality, more complete genome assemblies than second-generation techniques alone. At present, excellent results are being obtained by assembling genomes with third-generation reads, polishing the resulting contigs/scaffolds with second-generation (Illumina) data, and using Hi-C chromatin conformation capture and/or optical mapping techniques to increase contig/scaffold accuracy. Because improvements in sequencing have occurred so quickly, bioinformatics tools for assembling, analyzing, and annotating genomes have not been standardized; indeed, such tools can vary widely with regard to input needs, output quality, and scalability. With hundreds of plant genomes in various states of completion, much is being learned about general trends in plant genome evolution – for example, the predominance of polyploidy and paleopolyploidy events in angiosperms and the enormous contribution of LTR retrotransposons to genome size expansions/contractions in land plants. Plant genome sequencing projects have also made apparent the difficulties associated with identifying and quantifying gene numbers. While genome sequences are powerful tools, they are simply maps; they contain information about genomic landmarks and their possible functions, but they are not equivalent to the complex organisms they represent. Nonetheless, like any maps, they represent a means by which one can explore more expeditiously. Integration of genome sequence maps with chromatin configuration, biochemical, and developmental data is critical in advancing understanding of how genes and genomes function and evolve in vivo.

1 Introduction

In each semester at the beginning of the first lecture of the graduate course "Genomes and Genomics," co-author Peterson asks his students the same question: "Why do we do so much DNA sequencing?" Typically, the students look down at the floor and sink lower in their chairs; they have sadly, and perhaps wisely, learned to avoid trying to answer such open-ended questions, especially when it is clear that the professor has a specific answer in mind. Eventually, he gives them the answer he thinks is most accurate: "We sequence DNA because we can!"

Clearly researchers sequence genomes, transcriptomes, etc., for specific scientific reasons, but generating nucleic acid sequences has become so fast and so inexpensive that scientists often use sequencing even when other approaches would afford more direct answers. As shown in Fig. 1, the cost of sequencing DNA (machine time and sequencing reagents only) has plummeted from $5,300 per megabase (Mb) in 2001 to $0.01 per Mb in 2016. For the cost of a case of pipet tips, a researcher can generate more sequence data than they can manually annotate/examine in a lifetime. However, even if a scientist is only interested in a small part of the sequence data they generate (e.g., a gene family), by making the data available to the scientific community through the US National Center for Biotechnology Information (NCBI), the DNA Data Bank of Japan (DDBJ), or the European Molecular Biology Laboratory-European Bioinformatics Institute (EMBL-EBI), the data producer is

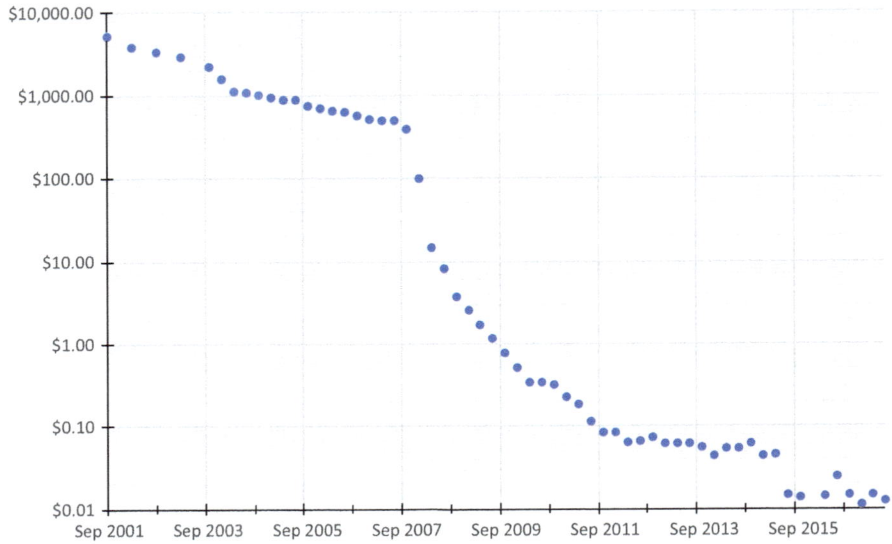

Fig. 1 Cost (USD) per megabase of DNA sequence (raw data) over the last 15 years. These numbers, obtained from the National Human Genome Research Institute website (see Wetterstrand 2017), do not include DNA isolation, library preparation, and instrument purchase costs

enabling other scientists from around the world to utilize that data in exploring other biological questions.[1] Such is the fundamental value of genomics (see Fig. 2).

Genome sequences represent the "genetic blueprints" of an organism. They contain the instructions for building proteins and generating structural and regulatory RNA molecules. While the difference between a genome and a functioning entity (or blueprints and a constructed building) is tremendous, generation of a quality genome sequence is a crucial step in understanding how entities function and evolve. Comparison of genome sequences reveals how changes in DNA can affect phenotype and drive speciation.

Modern phylogenetic research typically classifies plants into an unranked eukaryotic clade called the *Viridiplantae* (Latin for *green plants*) which includes green algae, hornworts, liverworts, ferns, mosses, club mosses, angiosperms, and gymnosperms.[2] Plants possess cell walls containing cellulose and are typically photosynthetic due to the presence of chloroplasts, photosynthetic organelles derived from

[1]NCBI, DDBJ, and EMBL-EBI share and synchronize data on a daily basis (Kulikova et al. 2007). The value of this is inestimable! Without such sharing, one would have to search each database separately.

[2]In some treatments (e.g., Adl et al. 2005), red algae and glaucophytes and the *Viridiplantae* are place together into a group known as the *Archaeplastida*.

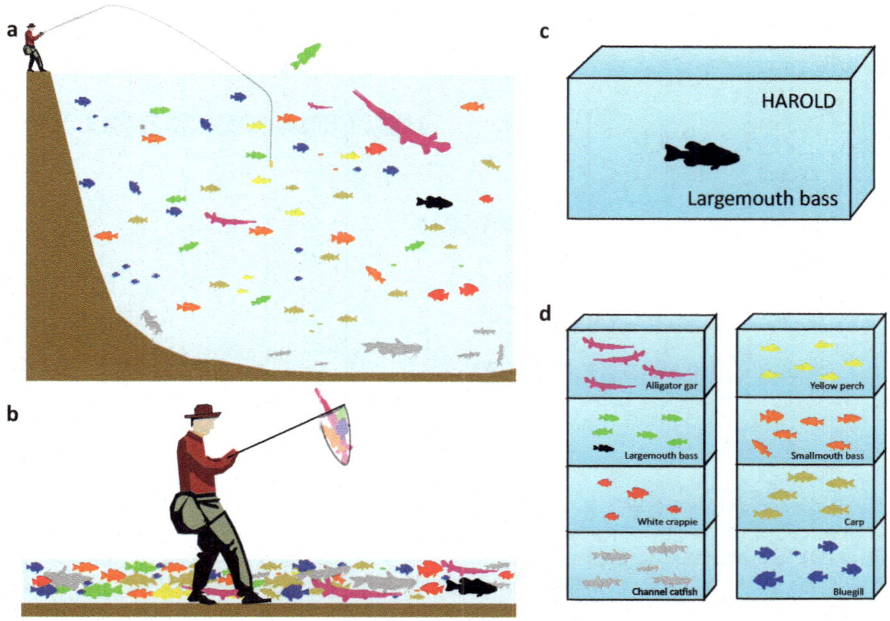

Fig. 2 Pond analogy for genomics. Imagine a genome as a pond full of many different kinds of fish with each fish representing a gene or other genome sequence. Fish of the same species represent similar/related sequences (e.g., gene or repeat families). Now let's say you are looking to catch a fish, not just any fish but a particular largemouth bass known as Harold (shown in black). (**a**) In days of old (i.e., the pre-genomics era), the traditional way of catching Harold would be to walk along the shoreline trying different baits and lures. If you could afford it, you might even rent a boat and try fishing from the middle of the pond. Eventually, through trial and error, you learn which lures/baits are particularly attractive to largemouth bass. You may also discover places where largemouth bass are more prevalent. After 6 years of fishing, you finally catch Harold. Congratulations! You then decide you want to study some of Harold's relatives. (**b**) In the modern (i.e., genomics) era, catching Harold is still your primary goal. However, rather than taking 6 years to catch one fish, you adopt a strategy that benefits and informs a much larger group of fish lovers (scientists). First you and your colleagues drain the pond of most of its water so that it is less than 30 cm deep. You then walk through the shallow water and net every fish until the pond has been emptied. You certainly don't have time to study all the fish, even your buddies are only interested in a studying a handful of the tens of thousands of fish you capture, but you know that there are other fish lovers that will be interested in studying different fish groups. (**c**) After 2 weeks of work, you capture Harold. (**d**) By the end of the month, you and your buddies have transferred all the bluegills to one tank, all the white crappies to another, all the largemouth bass to another, etc. You make the tanks and their contents publicly available for study via an international aquarium (i.e., Fishbank). You inventory all the fish and describe their diversity, relative numbers, etc. and publish your results in *Modern Fish*. You write a special paper on Harold which you publish in *Bass*. The fish you have placed in the aquarium fuel research by scientists from around the world. While you are the world's foremost expert on Harold, you make him available for others to study as well (he can be found in the tank with the other largemouth bass)

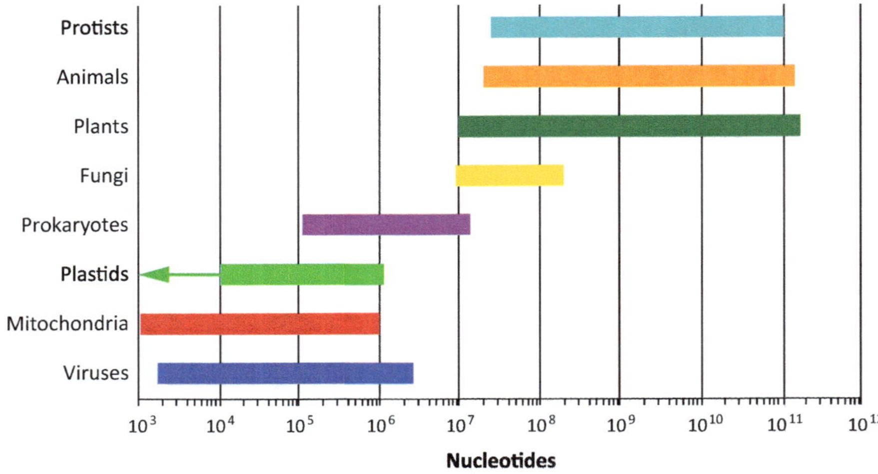

Fig. 3 Genome size ranges for viruses (Mankertz 2008; Philippe et al. 2013), mitochondria (NCBI Genome; Kolesnikov and Gerasimov 2012), chloroplasts (Bellot and Renner 2015; Munoz-Gomez et al. 2017), prokaryotes (Martinez-Cano et al. 2014; Schneiker et al. 2007), fungi (Mohanta and Bae 2015), plants (Derelle et al. 2006; Pellicer et al. 2010), animals (Gregory 2018), and protists (Corradi et al. 2010; Shuter et al. 1983). For plastids, the bar shows the range in plastids containing functioning genes, while the arrow to the left of the bar is used to represent those plastids that still exist as membrane-bound structures but have lost their genomes (Molina et al. 2014; Smith and Lee 2014). Of note, plants show the greatest range (greater than four orders of magnitude) in DNA content

endosymbiotic cyanobacteria.[3] The *Viridiplantae* contains 350,000–500,000 species (Chapman 2009; Smith et al. 2009; Ruhfel et al. 2014). The vast majority of these species are angiosperms.

The importance of plants to the evolution of life on earth is difficult to overstate. Due to their ability to convert light energy into chemical energy, plants are the primary source of food for most of the world's terrestrial heterotrophs (notably animals and fungi). Colonization of land by animals would have been unlikely without prior and concomitant colonization of land by plants. The majority of human calories come directly from plants; the remainder comes from other animals that get their food from consuming plants or plant-eating organisms. Humans use plant materials to build shelters/houses, to make clothing, to heat their homes, as a source of medicinal compounds, and to add beauty to the places in which they live and work. In brief, humans owe everything to plants.

Plants exhibit the greatest variation in genome sizes of any group of genome-bearing entities (Fig. 3). In addition, plants exhibit a level of genome plasticity not tolerated by most other eukaryotes. Whole genome duplications (polyploidy), extra

[3]A few plants have adopted parasitic or heterotrophic lifestyles and have chloroplasts that have lost their photosynthetic ability (Keeling 2010). In a few cases, the entire chloroplast genome has been lost although the plastid has been retained (Molina et al. 2014; Smith and Lee 2014).

or missing chromosomes (aneuploidy), interspecific hybridization, large-scale chromosome rearrangements, and massive epigenetic changes are common in plants (Leitch and Leitch 2008). The plasticity of plant genomes appears to be central to the evolutionary success of plants, most notably the angiosperms which account for roughly 90% of plant species (Chapman 2009). It can also make plant genome research (including genome sequencing) extremely challenging.

In this chapter, we will discuss the history and current state of plant genome sequencing. In addition, we will provide an overview of terminology used in discussing genomes and genome sequencing, a review of technology and techniques utilized in sequencing and assembling genomes, and a brief overview of what has been learned from plant genome sequencing projects.

2 Fundamental Concepts When Considering Genomes and Genome Sequences

2.1 What Is a Genome?

This is actually not an easy question to answer. In broad terms, a genome is *nucleic acid* that contains within its nucleotide sequence the genetic information that defines an *entity* and through which that entity transmits, *in full or part*, its genetic information to future generations. Note that we italicized three particular wording choices, specifically, *entity*, *nucleic acid*, and *in full or part*. Here is why.

- *Entity* – We use the term *entity* in the definition because the more obvious word choice, *organism*, implies life, and not all things that have a genome are alive. While it is true that all living creatures possess genomes, viruses and viroids are nonliving infectious agents with their own genomes. Likewise, although plastids and mitochondria evolved from bacteria (i.e., living cells), they are now "lifeless" eukaryotic organelles with genomes that have been stripped of most of the genes required for independent existence. Lastly, while death is typically associated with degeneration of nucleic acids, genomes do not decay immediately and, in certain environmental situations, may be partially preserved for extremely long periods of time. The relative stability of DNA in particular allows the genomes of deceased cells to be used in forensic identification of victims and/or potential culprits of violent crime (e.g., semen, blood, and/or skin tissues left at crime scenes can be used to narrow down suspects in rape and murder cases) (Butler 2015). Moreover, well-preserved tissues may allow study of the genomes of ancient and/or extinct organisms (see Shapiro and Hofreiter 2014, for review).
- *Nucleic acid* – While all known living creatures have genomes composed of DNA, all viroids and the majority of viruses possess RNA genomes (Gelderblom 1996). Indeed, the genomes of viruses may be composed of double-stranded DNA (dsDNA), single-stranded DNA (ssDNA), single-stranded RNA (ssRNA), or double-stranded RNA (dsRNA) (van Regenmortel and Mahy 2004).

- *In full or part* – Viruses, plastids, mitochondria, and prokaryotes normally reproduce by generating exact replicates of their genomes which they pass to their progeny. Granted, mistakes in replication, recombination events, and improper genome partitioning/packaging can lead to progeny with mutations, but passing a full genome complement to progeny is the norm in the aforementioned entities. In most eukaryotes there are two types of cell division – mitosis and meiosis. The majority of eukaryotes are functionally diploid meaning they have two sets of homologous chromosomes in their somatic cells. Mitosis, the eukaryotic form of asexual reproduction, results in two daughter cells identical to the parental cell; the full parental genome (including each homologous pair) is passed to both daughter cells. Meiosis, which involves a highly modified first cell division followed by a fairly standard mitotic division, results in the production of gametes, each containing half the number of chromosomes from which they were derived. Sex involves the fusion of two gametes, often, but not always, from different individuals of the same species; this restores the chromosome number in the resulting zygote to that of the parents. In asexual reproduction, the genome of an entity is passed to its progeny in full, while in sexual reproduction, only part of a cell's genetic information (ca. one-half) is passed to its offspring.

2.2 *Which Genome Is the Plant Genome?*

Plants have three distinct genomes: a nuclear genome, a chloroplast genome, and a mitochondrial genome. While mitochondrial and chloroplasts are derived from once free-living bacteria, these organellar genomes have evolved within the environment of the plant cell and are essential to plant survival. Hence, they are truly plant genomes and can be used in plant phylogenetic research. However, when one says they are sequencing the genome of *Zea mays*, it is assumed that they are sequencing the *Zea mays* nuclear genome. If they are sequencing a chloroplast or mitochondrial genome, they are expected to indicate this (e.g., *Zea mays* mitochondrial genome). With regard to the remainder of this review, all discussions will focus on nuclear genomes.

2.3 *What Are C Value and* n *Value?*

A complication when discussing plant nuclear genomes is rooted in the fact that the amount of DNA in a cell and the number of chromosomes in the cell nucleus differ based upon the cell's stage in the cell cycle and/or its developmental fate. Cells in G_1 are said to contain the 2C DNA amount (where "C" stands for *constant*) (Bennett and Leitch 2005) and two homologous chromosome sets (i.e., they have the $2n$ chromosome number where "n" stands for *number*). A shorthand way of writing this is $2C = 2n$ or $2n = 2C$. Cells that have completed S phase (DNA replication) are

$4C = 2n$. This does not make intuitive sense to most students – if the DNA molecules have doubled, wouldn't the number of chromosomes double? The word chromosome (literally meaning *colored body*) is a cytological term coined long before DNA was known to be a constituent of chromosomes. Whether a chromosome has 1, 2, 4, or thousands of DNA molecules in it, if it appears under the microscope as a single, cohesive/connected unit, it is considered a single chromosome. The term that is now used for each DNA molecule (and its associated proteins) in a chromosome is *chromatid*. Thus, after S phase each chromosome is composed of two sister chromatids. It is critical that the two sister chromatids remain attached together until anaphase of mitosis to ensure their proper segregation into different daughter cells Once the sister chromatids are no longer physically connected, they are called sister chromosomes.

As mentioned above, a few plant cells do not go through the standard cell cycle but rather enter meiosis. At the beginning of meiosis, a cell is $2n$ and 4C. In prophase I of meiosis, homologous chromosomes pair, synapse, and undergo crossing over. When two homologs synapse, they are collectively called a bivalent and are considered a single chromosome as they are now physically connected. Consequently, when synapsis occurs, the DNA content remains 4C, but the chromosome number is halved (i.e., it becomes $1n$). At prometaphase I, bivalents are moved to the metaphase plate, and at anaphase I, homologs, which were held together by their crossovers, separate and are moved to opposite cell ends. The second meiotic division results in four cells, each which possesses the 1C DNA content and the $1n$ chromosome number. In higher plants, the 1C products of meiosis give rise to a 1C gametophyte and, via mitosis, eventually give rise to 1C gametes. Of particular importance to this conversation is the fact that homologs must pair, synapse, undergo crossing over, and stay closely connected until they are pulled into different daughter cells at anaphase I. If these things do not happen, balanced gametes will not be produced.

2.4 *Genome Size*

In any discussion of genomes and sequencing genomes, someone will invariably ask for the genome size of the species of interest. The larger the genome size, the more DNA sequencing that must be done to ensure adequate coverage of any given genetic locus. The genome size of a $1n = 4C$ cell is twice that of a $2n = 2C$ cell and four times that of a $1n = 1C$ cell. Which value should be used?

The general convention is to provide genome size values in terms of 1C DNA content (i.e., the amount of nuclear DNA in an unreplicated gamete). The units in which 1C content are typically provided are megabases (Mb; millions of bases), gigabases (Gb; billions of bases), or picograms (pg; 10^{-12} g) (Bennett and Leitch 2005). The logic behind using the 1C value is that it is the minimum amount of DNA to contain one copy of every locus. If genome size is presented as 2C or 4C DNA amount, this must be indicated.

Methods used to estimate genome size are reviewed in Peterson (2014). The easiest and most widely accepted method of obtaining a genome size estimate is through the use of flow cytometry. In brief, isolated nuclei in solution are stained with a fluorescent quantitative DNA stain (most commonly, propidium iodide). The stained nuclei are gently pulled into a thin stream of liquid whose diameter is so narrow that the fluid exhibits almost laminar flow. The nuclei move into the stream one at a time; if multiple nuclei try and enter the stream, the flow is disrupted. The liquid stream passes through a laser which excites the fluorescent DNA stain in the nuclei. As it passes through the laser, each nucleus gives off a burst of fluorescence that is proportional to the amount of DNA it carries. Measurements from thousands of nuclei can be obtained in just a few minutes. Stained nuclei from an organism of known genome size are usually included as an internal standard facilitating accurate genome size estimation.

One might think that whole genome sequencing would provide the most accurate genome size determinations. However, plant genome assemblies typically contain gaps of uncertain size. Such gaps are especially prevalent at centromeres, telomeres, and in other highly repetitive genome regions. Based on the *Arabidopsis thaliana* genome assembly (https://www.ncbi.nlm.nih.gov/assembly/GCF_000001735.3; accessed on March 2, 2018), the genome size of *A. thaliana* would be predicted to be 119.9 Mb (119.7 Mb in sequence assemblies plus 0.2 Mb in gaps). However, in a study led by Michael Bennett, the world's foremost expert on plant genome sizes (Bennett et al. 2003), the authors convincingly showed that the *Arabidopsis thaliana* genome is about 157 Mb, a value that agrees with previous Feulgen densitometry and flow cytometry measurements. Thus the first and arguably best plant genome assembly is probably only 76% complete. Until considerable advances are made in sequencing and assembly of genomes, Feulgen densitometry and FCM should be viewed as the most accurate means of determining genome sizes (Peterson 2014).

2.5 C-Value Paradox/Enigma

In 1944, it was shown that DNA is the molecule in which genetic information is stored (Avery et al. 1944). In 1948, it was revealed that species had relatively stable genome sizes that could be measured (Boivin et al. 1948), and in 1950 the term 1C DNA value was coined (Swift 1950). By 1951, scientists already had acquired enough genome size estimates to notice that 1C DNA contents of eukaryotic organisms did not seem to be correlated with organismal structural complexity.[4] Mirsky and Ris (1951) deemed this phenomenon the "C-value paradox." The assumption at the time was that since all genes were encoded by DNA, then all

[4]For example, the lily, *Paris japonica*, does not need or have 955 times as many genes as *Arabidopsis thaliana*, although the two species have a 955-fold difference in 1C DNA content (Pellicer et al. 2010; Bennett et al. 2003).

DNA must be genes. We now know that differences in genome size typically are reflective of differences in repetitive sequence contents and, to a much lesser extent, polyploidy and gene duplication (see Peterson 2014, for review). The reasons why natural selection has resulted in obese and skinny genomes is not well understood, but the nature of the differences in genome sizes has been elucidated. Consequently, it has been recommended that the C-value paradox should now be referred to as the C-value enigma (Gregory 2005).

2.6 Homogeneity

When generating a genome sequence for a species or clade, it might seem logical to include DNA from many different individuals as, perhaps, a means of looking at diversity. This is absolutely incorrect. The assembly of sequence reads into long DNA stretches is most easily performed if there is a complete absence of heterogeneity at any locus (or if this is not an option, as little heterogeneity as possible). Ideally a reference genome is prepared from a source of identical haploid cells (e.g., a megagametophyte), a doubled-haploid cell line, or a highly inbred individual. The goal of producing a genome sequence is to produce the most complete representation of a single (haploid) genome as possible. The sequence itself provides no information on genetic diversity among individuals, but once completed, the genome serves as a way in which diversity can be quickly assessed (see Sect. 3.18 below). One way to circumvent problems with heterozygosity when DNA must be obtained from a diploid, heterozygous individual is through a process known as *phasing* (see Sect. 3.23 below).

2.7 Chromatin

The mixture of DNA and its associated proteins is called chromatin. In eukaryotic cells, DNA is always found within the context of chromatin and chromosomes (Fig. 4). On a particular DNA molecule, two loci may be separated by thousands or millions of base pairs, but due to chromatin looping/folding, these loci may be very near to each other in the 3D environment of the nucleus (Fig. 4f). Since genomes evolved within the context of chromatin, it is perhaps not surprising that regulatory DNA sequences may be close to the gene they regulate in 3D space while being distant in base pairs from that gene.

Chromatin from each chromosome occupies a relative distinct domain within the interphase nucleus. More recently, chromatin conformation capture techniques have been coupled with NGS to look at pairwise associations of sequences in nuclei (see Sect. 3.24 below for more details). This research has led to the discovery of *topologically associating domains* (TADs) which are collections of DNA sequences that tend to be closely associated (and ostensibly in contact) with each other in the

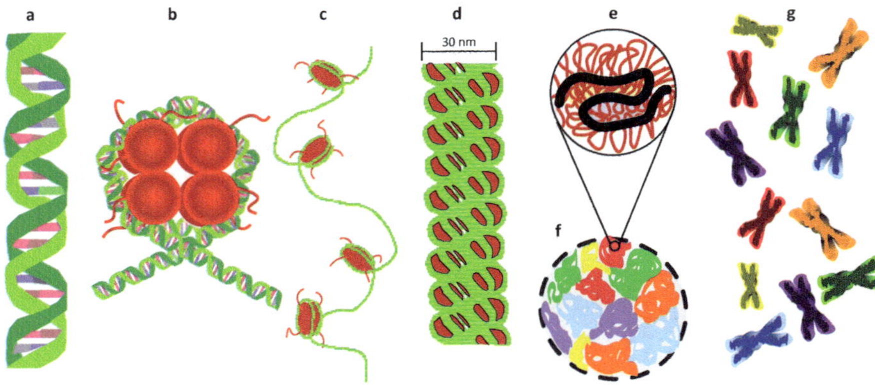

Fig. 4 A simplified drawing of DNA within the context of chromatin and chromosomes. (**a**) A DNA double helix. In vivo DNA is always associated with protein. (**b**) The most fundamental interaction between DNA and protein is the association of the double helix with an octamer of histone proteins; the octamer is composed of two H2A, two H2B, two H3, and two H4 histones (red spheres). Each histone has a "tail" (red lines). The octamer is wrapped 1.75 times (146 bp) by the DNA strand. Each octamer of histones wrapped by 146 bp of DNA is referred to as a *nucleosome core*. (**c**) In the absence of linker histone (LH), which is not a part of the nucleosome core, chromatin has a "beads on a string" appearance, with each nucleosome core separated from flanking nucleosomes by naked *linker DNA*. A nucleosome core and one of its adjacent linker DNA segments are called a *nucleosome*, although many people incorrectly use the term nucleosome when they are referring to a nucleosome core. The mean diameter of a nucleosome fiber is about 11 nm. (**d**) Linker histone interacts with the linker DNA and the nucleosomes to coil the 11 nm fiber into a solenoid with a mean diameter of 30 nm. The exact structure of the 30 nm fiber (and even its existence in vivo) is a subject of considerable debate (Iyer et al. 2011). Regardless, there does appear to be a coiled fiber which is highly dynamic in shape, structure, and condensation. (**e**) In interphase nuclei, the 30 nm fiber is packaged into loops (red lines) which extend from a proteinaceous nuclear matrix (black lines). At the base of chromatin loops are AT-rich matrix attachment regions (MARs; Gluch et al. 2008). DNA replication (Wilson and Coverley 2013) and transcription are preferentially associated with the matrix, and not surprisingly, gene sequences are often near MARs. Due to the extensive looping and folding of chromatin, loci that are separated by many base pairs may be physically close together within the nucleus. DNA sequences that are separated by long stretches of nucleotides but are normally clustered together in the nucleus as determined by Hi-C are referred to as *topologically associating domains* (TADs). One TAD is highlighted in yellow while a second TAD is highlighted in blue. (**f**) Zooming out from the region shown in E provides a view of an entire nucleus. Each chromosome is represented by a different color (homologs are shown in the same color). Note that the chromosomes occupy distinct nuclear regions (in which TADs are the functional unit). (**g**) During mitosis, the 30 nm fibers are coiled and further packaged with proteins to produce chromosomes that are visible by light microscopy. Each mitotic metaphase chromosome is 10,000–20,000 times shorter than its corresponding naked DNA molecule (Woodcock and Ghosh 2010). For example, the tomato nuclear DNA sequence is 1,002 Mb. Using the base pair to meter conversion values from Dickerson et al. (1982), the tomato genome is 0.33 m in length. The average sum length of tomato's 12 metaphase ($1n = 1x = 1C$) chromosomes is 26.76 μm. Thus the level of compaction from naked DNA to mitotic metaphase chromosome is 12,333-fold (i.e., $0.33 \text{ m} \div 0.00002676 \text{ m}$)

3D space of the nucleus (Denker and De Laat 2016). Chromatin contacts within a TAD are greatly preferred over interactions between sequences in different TADs (Liu and Weigel 2015). An understanding of TADs and how they function and evolve will be essential to understanding how plant genomes function and evolve.

2.8 Diploidy and Polyploidy

Most plants are functionally diploid meaning that each chromosome in their $2n$ cells has one (and only one) homolog with which it will pair, synapse, and recombine during meiosis. Being a diploid or behaving like a diploid during meiosis is key to maintaining the ability to reproduce sexually. The majority of plants maintain sexual reproduction even if their seeds/spores are typically outcompeted by asexual propagules; e.g., poplar trees are an example of plants that primarily reproduce via root sprouts but maintain sexual reproduction (DiFazio et al. 2011). The advantages of sexual reproduction in plants are not as well understood as one might think, yet the fact that only about 0.1% of flowering plants have lost their ability to reproduce sexually is indicative of the importance of meiosis and sex to this clade (Charlesworth 2006).

Polyploidy is a phenomenon through which a genome grows in size through the addition of another complete genome. While most plants are functionally diploids, it appears that all plant genomes have undergone polyploidy events in their evolutionary past (Dodsworth et al. 2016). Sometimes these events happened so long ago that they can only be detected through careful whole genome sequence analyses. Species that have had polyploidy events in their ancient past are said to be paleopolyploids. Other species have had polyploidy events happen more recently.

If a species is a true diploid or a paleopolyploid, its diploid cells are said to have the $2x$ ploidy level and its gametes have the $1x$ ploidy level. Such organisms would have $2C = 2n = 2x$ somatic cells and $1C = 1n = 1x$ *gametes*. Plants with 3, 4, 5, 6, 7, 8, 9, 10, 11, or 12 sets of chromosomes in their somatic cells are called triploids, tetraploids, pentaploids, hexaploids, heptaploids, octoploids, nonaploids, decaploids, undecaploids, and dodecaploids, respectively. For higher levels of polyploidy, a number followed by ploid may be used (e.g., 30-ploid). Species that have an odd number ploidy level are infertile. Species with even number ploidy levels may or may not be fertile depending upon how they became polyploids. There are two primary means by which organisms become polyploids: specifically, allopolyploidy and autopolyploidy.

2.9 Allopolyploidy

Allopolyploidy is one mechanism by which polyploidy occurs. It is initiated via hybridization between two related but sexually incompatible species. The species may or may not have different chromosome numbers. The zygote resulting from this hybridization event contains a complete $1n$ set of chromosomes from each parent, and while the organism produced may survive and even thrive, it will be sterile due to the fact that its chromosomes do not have true homologs with which to interact at meiosis (e.g., a mule, the cross between a donkey and a horse, is viable yet sterile). However, on rare occasion, the hybrid cell may replicate its chromosomes during S phase and start a mitotic division only to have the division stall out and fail prior to cytokinesis. In such "restitution nuclei," the sister chromatids come apart to form sister chromosomes (which then function as homologous chromosomes), and the cell chromosome number doubles. If the cell or one of its daughter cells enters meiosis, each chromosome will now have a homolog with which to pair, synapse, and exchange DNA via crossing over, which may give this polyploidy the ability to produce viable gametes. If the two parental species are closely related, there may be issues in meiosis where homologs and their homeologs may form structures other than bivalents, leading to full or partial infertility. Allopolyploidy is common in the evolutionary histories of most plants.

2.10 Autopolyploidy

Sometimes a plant cell that is functionally diploid will have a failure in its mitotic division and produce a restitution nucleus. The resulting cell will have four sets of homologous chromosomes. If the cell or one of its daughter cells enters meiosis, it is likely that a given group of four homologs won't form bivalents but rather quadrivalents (four homologs that synapse together in full or part) or a trivalent (three synapsed homologs) and a univalent (an unpaired chromosome). The result is unbalanced "aneuploid" gametes (aneuploidy is the state of having more or less than a full complement of chromosomes). For this reason, autopolyploids are generally sterile, at least at first.

2.11 Diploidization

Autopolyploids are almost always sterile upon formation, and allopolyploids may have issues that prevent them from producing balanced gametes. However, over time, selective pressure on infertile polyploids may result in evolution of genetic mechanisms that facilitate production of balanced gametes. The phenomenon where a polyploid organism starts behaving like a diploid during meiosis is known as *diploidization*. There are a number of ways in which a polyploid can develop diploid-like behavior and regain fertility including divergence of homoeologous genomes, development of genes that facilitate bivalent formation, and development of mechanisms that limit crossing over to termini of multivalents. Dodsworth et al.

(2016) suggest that diploidization is central to the evolutionary success of angiosperms.

3 Concepts and Terms Associated with DNA Sequencing

3.1 Next-Generation Sequencing (NGS)

In one of the most egregious naming decisions ever made, marketing personnel began referring to 454 and Illumina platforms as "next-generation sequencing" or NGS. Naturally, giving any technological breakthrough, the name "next generation" is illogical as the technology already exists making it current, not next, generation. Moreover, the inevitable development of new techniques will quickly make the so-called next-generation items previous-generation items. Nonetheless, the use of NGS has continued. While originally used to describe 454 and Illumina techniques, NGS is now used to refer to all twenty-first-century sequencing techniques developed after modern Sanger sequencing, the dominant form of sequencing in the 1980s and 1990s. Because the term "next-generation sequencing" is both vague and technically incorrect, there has been a more recent trend to use the terms *first-*, *second-*, and *third-*generation sequencing to describe sequencing technologies. While not exactly accurate, these terms (see below) are preferable to NGS.

3.2 Bioinformatics

Bioinformatics, like most words used by biologists, is a term that is subject to a wide range of interpretations. It was first coined by Paulien Hogeweg and Ben Hesper to describe the "... study of informatics processes in biotic systems" (Hogeweg 2011). However, the term was co-opted to describe the development of computer algorithms, software, pipelines, and databases to aid in the study of biological systems at the macromolecular/genetic level. In this latter, and more widely known, meaning, bioinformatics is engineering informed by biology.

3.3 Computational Biology

In contrast to bioinformatics, computational biology is the use of bioinformatics tools and advanced computing infrastructure to explore biological questions. Consequently, it is scientific exploration that is possible based on advances in computer and software engineering.

3.4 DNA Library

For decades, the term "DNA library" was used to refer to a collection of DNA sequences from an organism of interest that had been ligated into a vector (e.g., a plasmid) and used to transform bacterial cells. The DNA library was the collection of bacterial cells containing the recombinant (vector/insert) molecules. Eventually the term was also used to describe a collection of virions each containing a genome into which a piece of foreign DNA had been ligated. Regardless if the foreign DNA inserts were pieces of nuclear DNA, the library was deemed a *genomic library*. If the inserts were reverse-transcribed mRNA molecules (i.e., cDNAs), the term *cDNA library* was used to describe it. If the vector contained elements that encouraged transcription and translation of the inserts, the term *expression library* was utilized.

Unfortunately, when so-called next-generation sequencing (NGS) came along, the term DNA library was co-opted to simply mean DNA prepared for sequencing on a NGS instrument. Preparation of DNA for NGS is platform dependent, although typically genomic DNA or DNA prepared from reverse-transcribed RNA molecules is sheared into fragments of a particular length (by sonication, nebulization, etc.) and quantified. For many NGS instruments, the remaining preparation process often involves ligation of adapters onto the ends of DNA fragments.

3.5 Physical Mapping

The term "physical mapping" can be used to describe the identification of relatively large pieces of partially overlapping DNA collectively spanning a chromosome or part of a chromosome and/or determination of nucleotide distances between genes and markers. There are numerous physical mapping techniques that can be utilized separately or together. Some of these techniques rely upon the existence of large-insert genomic libraries, most notably bacterial artificial chromosome (BAC) libraries. Each BAC clone can carry a relatively large piece of genomic DNA (>100 kb) (see Peterson et al. 2000 for review). A few of the physical mapping techniques that are utilized include:

1. DNA fingerprinting – BACs can be digested with a series of restriction enzymes and separated by gel electrophoresis. The resulting banding patterns, known as fingerprints, can be compared as a way to identify partially overlapping BACs.
2. BAC-end sequencing (BES) – BAC inserts are cloned into a vector of known sequence, and consequently primers complementary to the vector can be used to sequence the regions at each end of a BAC (these sequences are called BAC-end sequences or BESs, and they are a type of paired-end sequence – see below). BESs mark the termini of a BAC insert whose size can be determined by pulsed-field gel electrophoresis; the distance between the two BESs can be elucidated.

3. Macroarray analysis – BESs and other molecular markers can be hybridized to high-density BAC macroarrays to identify BACs that share molecular loci (i.e., overlapping BACs).
4. Polymerase chain reaction (PCR) – PCR screening of strategically pooled BACs can be used as an alternative to macroarray analysis in identifying clones containing shared loci.
5. Fluorescence in situ hybridization (FISH) – FISH can be used to visualize the locations of markers/genes on chromosome preparations and position BACs with respect to one another along chromosomes.
6. Optical mapping – Optical mapping is a relatively new physical mapping technique. In brief, DNA molecules are placed on a substrate containing microfluidic channels. Single DNA molecules are stretched across the channels. The stretched DNA molecules are treated with enzymes that either cut the DNA or nick the DNA. In the former case, DNA-intercalating dyes can be used to visualize the DNA molecules and the sites of restriction events – the fragment patterns of the stretched DNA molecules can be compared (in a manner similar to DNA fingerprinting) allowing overlapping pieces of DNA to be discovered. In the latter case, the stretched DNA is treated with enzymes that make nicks at specific sequences. The nicks are expanded via an exonuclease, and then a DNA polymerase is used to fill in the gap with fluorochrome-labeled nucleotides. The pattern of fluorescence for a DNA molecule can be compared with other optically mapped molecules to discover shared DNA regions (see, e.g., Lam et al. 2012).

BAC-based physical mapping (including DNA fingerprinting, BAC-end sequencing, BAC macroarray screening or PCR-based BAC library screening, and BAC-based FISH) was traditionally used as a means of positioning large pieces of DNA with respect to one another to obtain a minimum tiling path, i.e., the minimum number of clones collectively representing all, or an extensive part of, a chromosome. Sequencing the clones in a chromosome's minimum tiling path served as a means to produce a high-quality chromosome sequence (see Peterson 2014 for review).

3.6 emPCR and Bead Colonies

The sequencing techniques of 454, SOLiD, and Ion Torrent amplify DNA by using beads containing millions of identical capture sequences. Genomic DNA (target DNA) with adapters ligated to its ends is denatured and mixed with an excess quantity of beads. One of the adapters on each target molecule is complementary to the capture probes on the beads. The low amount of DNA compared to beads ensures that, on average, no bead will have more than one DNA sequence hybridize to one of its capture probe sequences. PCR reagents and oil are added to the beads, and the mixture is agitated to create an emulsion. The beads are of a size that promotes the formation of an aqueous layer (including the PCR reagents) around each bead, while the hydrophobic oil keeps beads separated from one another. Emulsion PCR (emPCR) is performed so that each bead to which a target molecule had hybridized will end up with thousands of identical DNA molecules attached to its surface. The emulsion is then broken, and only those DNA strands synthesized via the elongation of capture probes remain attached to the beads. In 454 and Ion Torrent sequencing, the beads are deposited into the wells of a microfabricated plate. In contrast, in SOLiD, the beads are fixed onto a glass slide (Heather and Chain 2016).

There is no official name for bead-based amplification of DNA molecules. We will refer to a bead attached to clonally amplified molecules as a "bead colony."

3.7 Bridge Amplification and Clusters

Illumina does not use emPCR to create groups of amplified molecules. Instead, target molecules flanked by adapters are hybridized to adapter complementary sequences fixed to lanes of a glass slide. A process known as bridge amplification is used to generate DNA colonies of amplicons (see Fig. 5). The resulting DNA colonies/spots are called clusters.[5]

3.8 First-Generation Sequencing

The first sequencing of nucleic acids was performed in the 1960s using short, readily available tRNA and rRNA molecules as substrates. The general approach, which was later adapted to sequence small viral genomes, involved full and partial restriction digests of a nucleic acid molecule of interest coupled with fractionation of the resulting fragments. These restriction enzyme-based strategies were highly limited in scope and not readily utilized except in a few laboratories (Heather and Chain 2016).

In 1977 two papers describing new DNA sequencing techniques were published back-to-back in the *Proceedings of the National Academy of Sciences of the United States of America* (PNAS). Unlike the methods used before, these two methods could, in theory, be used to sequence any DNA molecule. The first paper described a technique, now known as Maxam-Gilbert sequencing, which involved isolation of a restriction fragment band containing thousands of copies of the same DNA molecule; radiolabeling of fragments at their 5′ ends; separation of the DNA strands by denaturing electrophoresis; selection of one of the strands for sequencing; distribution of the molecule to be sequenced into four separate tubes; chemical treatment of the contents of each tube with an agent(s) that, at low frequency, preferentially

[5]emPCR with beads and bridge amplification are not the only means of producing DNA polymerase colonies (see Fuller et al. 2009 for review).

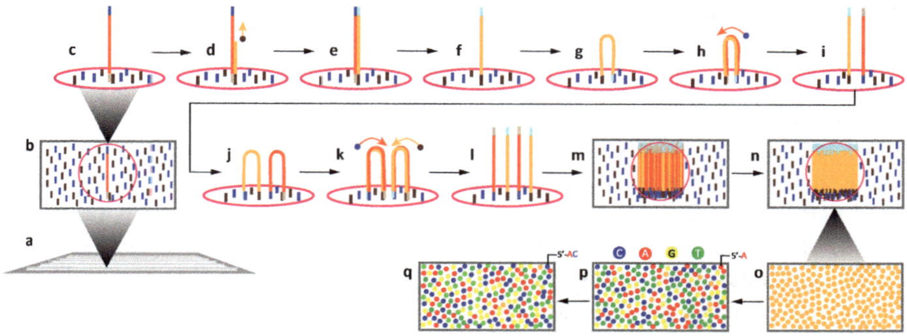

Fig. 5 Creation of clusters via bridge amplification and Illumina sequencing. (**a**) An Illumina flow cell with multiple lanes. Each lane is populated with numerous single-stranded capture probes/primers. (**b**) Magnified region of lane 1 showing capture probes/primers (short dark blue and dark brown lines). A dilute solution of single-stranded DNA molecules with adapter (i.e., end) sequences complementary to the capture probes is hybridized to the flow cell. The red circle encloses one DNA molecule that has hybridized to a capture probe. The forward primer sequence (light brown) on the target molecule (dark orange) is complementary to the dark brown capture probes on the slide. (**c**) The red circle in (**b**) shown at a higher magnification and a different angle. (**d**) The capture probe (dark brown) is extended by DNA polymerase (in the direction of the arrow). The color of the circle at the start of the arrow reflects the primer (color) that is being elongated. (**e**) A dsDNA molecule is the result of the chain elongation process. (**f**) The DNA is denatured and the original DNA strand is washed away. (**g**) The reverse primer sequence (aqua blue) at the terminus of the synthesized DNA strand hybridizes to a complementary capture probe/primer (dark blue). In the process a "bridge" is formed. (**h**) DNA polymerase extends the resulting loop in the direction shown by the arrow. (**i**) Denaturation of the DNA results in two complementary DNA strands. (**j**) When returned to conditions that promote hybridization, the two strands create new bridge loops. (**k**) Primers are extended. (**l**) Denaturation results in four DNA strands. (**m**) The bridge amplification process is continued until the cluster consists of thousands to millions of copies of the same sequence (both forward and reverse strands). (**n**) The reverse strands are cut at their lane-attached primers and washed away leaving only forward strands. The 3′ ends of the forward strands are blocked to prevent further bridge amplification. The cluster is now ready for sequencing. (**o**) Zooming out from the cluster reveals part of a lane in which hundreds of bridge amplification reactions have been completed (i.e., the region contains hundreds of clusters). (**p**) The fluorescently labeled reversible chain terminator nucleotides and DNA polymerase are run through the flow cell lanes. When exposed to light, the clusters omit a fluorescent glow characteristic of the first nucleotide incorporated. Each cluster's color/base is called (e.g., see cluster in upper right corner; the first incorporated nucleotide is A). (**q**) The 3′ hydroxyl groups at the ends of the reversible terminator nucleotides are freed, and a second nucleotide is incorporated into each strand in each cluster. The upper corner cluster has been extended by a C residue. The process in (**p**, **q**) is repeated 50–300 times in total

removes one or two of the four nitrogenous base(s) from target molecules; breakage of strands at sites of base excision; and resolution of fragments (each tube in a different lane) by polyacrylamide gel electrophoresis. The sequence of the target molecule could be deciphered by looking at an autoradiogram of the gel (Maxam and Gilbert 1977). The second PNAS paper (Sanger et al. 1977) described a technique now known as Sanger sequencing. Frederick Sanger and his colleagues demonstrated the value of this technique using pieces of ϕX174, a phage genome that Sanger and Arthur Coulson had sequenced using an early version of their new method (Sanger and Coulson 1975). Restriction fragments from the ϕX174 were isolated and used as primers in chain elongation using intact ϕX174 as the substrate. In brief, a primer and target DNAs were placed in four different tubes, the DNA was denatured, and the primer was allowed to anneal to one of the DNA strands. Each tube contained regular dNTPs (one labeled with a radioisotope), DNA polymerase, and a small amount of one of the four dideoxyribonucleotide triphosphates (ddNTPs – ddATP, ddGTP, ddCTP, and ddTTP). In each reaction tube, the incorporation of a ddNTP into a DNA molecule resulted in chain termination. The net result was four tubes each containing a series of fragments of different lengths. The contents of the tubes were resolved in four adjacent lanes of a polyacrylamide gel. As with Maxam-Gilbert sequencing, the sequence was deciphered by looking at an autoradiogram of the gel. Walter Gilbert and Frederick Sanger went on to share a Nobel Prize for their breakthrough sequencing techniques.

Maxam-Gilbert[6] sequencing was initially the most popular of the two techniques as it was relatively inexpensive. Moreover, it was strictly chemical in nature and did not require primers; using primers required some preknowledge of the sequence of the target DNA or the sequences flanking the target DNA. However, additional advances in technology would soon see Sanger sequencing gaining a wider following and Maxam-Gilbert sequencing disappear (more or less). With regard to Sanger sequencing, molecular cloning, ligation of DNA fragments into plasmids of known sequence, and transformation of bacteria with the recombinant molecules quickly eliminated the primer problem; one could sequence the insert (or its ends) by using primers complementary to the plasmid/vector immediately flanking the insertion site. Moreover, the development of ddNTPs each labeled with a different fluorescent tag permitted the chain termination reaction to be done in a single tube and eliminated the dangers of working with radioisotopes. When the polymerase chain reaction (PCR) emerged on the scene, the amplification of target sequences by PCR was integrated into the chain elongation/termination step (a process called "cycle sequencing"). Lastly, use of microcapillary electrophoresis rather than slab gel electrophoresis improved the throughput of Sanger sequencing and allowed the technique to be automated. Automated Sanger sequencing became the means by which the human genome, the *Arabidopsis* genome, the *Caenorhabditis* genome, and all of the initial eukaryotic genomes were sequenced.

The length of reads produced by Sanger and Maxam-Gilbert sequencing techniques was limited by the resolving power of acrylamide gel electrophoresis. While fragments ranging from 1 to 1,000 nucleotides can be resolved by standard polyacrylamide and/or capillary electrophoresis, polyacrylamide cannot reproducibly resolve fragments above 1,000 nt that differ by only one nucleotide (Karger and Guttman 2009).

[6]While co-author Peterson has been involved in sequencing many genomes, the only time he actually did the "hands-on" portion of sequencing himself (i.e., perform reactions, operate apparatuses, etc.) was when he took a short course on Maxam-Gilbert sequencing in graduate school.

In today's literature, first-generation sequencing refers to automated Sanger sequencing (and possibly the other techniques mentioned above). Though not the first method of sequencing nor the only method of sequencing, automated Sanger sequencing played a key role in the development of genomics as a discipline and allowed whole eukaryote genome sequencing to be a reality. Automated Sanger sequencing is still used today for small projects, filling in gaps in regions refractory to NGS methods, and for BAC-end sequencing, a technique for which there is no NGS equivalent.

3.9 Second-Generation Sequencing

Second-generation sequencing is used to describe sequencing techniques where thousands of bead colonies/clusters are generated on, or loaded onto, a substrate. All sequencing reactions are monitored at once. From their introduction, these massively parallel sequencing approaches produced orders of magnitude more DNA sequence per dollar than modern Sanger sequencing. The read lengths generated by second-generation sequencers were initially quite short compared to Sanger sequence reads, but refinements in technology have increased second-generation read lengths.

The first commercially available second-generation technique was developed by 454 Technologies. Characterized by emPCR followed by massively parallel pyrosequencing of bead colonies (see Table 1 for details), 454's successful *Genome Sequencer 20* instrument was an immediate hit leading to considerable sales and ultimately a $60 M exclusive licensing deal with Roche. Being the first NGS on the market gave 454 a huge advantage and led 454's founder, Jonathan Rothberg, to say, "It's been a 25-year race, we're commercial, and we won" (Bio-IT World Staff 2013). However, homopolymeric runs (e.g., AAAAAAA) greater than six nucleotides in length were a problem for 454 instruments, sample preparation was relatively expensive (Metzker 2010), and Roche/454 were unable to match the tremendous throughput increases achieved by Illumina. After sales of 454 machines dropped precipitously, Roche halted production of 454 instruments in 2013 and stopped sales of 454 instruments in 2015. Despite being first, 454 was outcompeted by Illumina and Ion Torrent instruments (Hollmer 2013).[7] With this said, 454 sequencing is included in this review because it was the first commercial NGS technology and it sparked tremendous innovation by 454 researchers and scientists developing competing technologies. Moreover, 454 played a key role in sequencing several plant genomes.

[7]In 2012 Roche attempted a hostile takeover of Illumina which Illumina successfully thwarted. Also, Jonathan Rothberg, the founder of 454 Technologies, later started Ion Torrent, a company whose success effectively helped to undermine Roche's investment in 454 (Bio-IT World Staff 2013).

Table 1 First-, second-, and third-generation instruments/platforms used in genome sequencing

<table>
<tr><th>Technology (year)</th><th>Description</th></tr>
<tr><td>Automated Sanger (1987)
www.thermofisher.com
Status: used in small-scale projects and in projects involving BAC-end sequencing
Prediction: will cease to be a part of major sequencing projects</td><td>Utilizes chemistry of DNA replication • Prep: target DNA is typically ligated into vector containing reverse and forward primer sequences flanking insertion site • Each clone placed in separate well of microtiter plate(s) • Recombinant DNA isolated • Sequencing: combination of dNTPs and ddNTPs (each ddNTP labeled with a different fluorescent tag) is mixed with primers, recombinant target DNA, and Taq polymerase • During PCR-based primer elongation, insertion of ddNTP into a strand causes chain termination • PCR used to generate DNA molecules with lengths from one to several thousand nucleotides. Each molecule terminates with a labeled ddNTP at its 5′ end • Fragments from each sequencing reaction are separated by size via capillary acrylamide electrophoresis • Each fragment passes through laser which causes its terminal ddNTP to fluoresce
<table>
<tr><th>Generation</th><th>Read length (avg)</th><th>HT model</th><th>Output</th><th>Key reference(s)</th></tr>
<tr><td>First</td><td>700 bp</td><td>ABI3730xl</td><td>269 kb per run</td><td>Sanger et al. (1977)</td></tr>
</table></td></tr>
<tr><td>454/Roche (2005)
Status: production of 454 instruments ceased in 2013, outcompeted by Illumina</td><td>Utilizes chemistry of DNA replication • Prep: sheared target DNA molecules are ligated to A and B primers • ssDNA molecules flanked by an A primer on one end and a B primer on the other end are isolated • DNA hybridized to bead each containing thousands of capture probes complementary to primer A; DNA is diluted so no more than one ssDNA molecule binds to a bead • Bead polonies are generated by emulsion PCR • Beads are allowed to fall into microfabricated wells containing sequencing reagents • Sequencing: B primer allowed to anneal to sequences in polonies • One dNTP (e.g., dATP) is passed over plate containing bead polonies • If nucleotide is added to B primer end, diphosphate (PPi) is released • PPi is used in a reaction involving luciferase to generate light • Image of light production captured • Extra nts washed away or degraded • Second dNTP (e.g., dTTP) is run across plate • Chemiluminescence recorded and plate washed • Third dNTP (e.g., dCTP) is run across plate • Chemiluminescence recorded and plate washed • Fourth dNTP (e.g., dGTP) is run across plate • Chemiluminescence recorded and plate washed • Process repeated tens to hundreds of times • Software generates sequence for each bead colony based on signal pattern and intensity
<table>
<tr><th>Generation</th><th>Read length (max)</th><th>HT model</th><th>Output</th><th>Key reference(s)</th></tr>
<tr><td>Second</td><td>1,000 bp</td><td>FLX+</td><td>450 Mb</td><td>Margulies et al. (2005)</td></tr>
</table></td></tr>
<tr><td>Illumina (2006)[a] www.illumina.com</td><td>Utilizes chemistry of DNA replication • Prep: sheared target DNA molecules are ligated to A and B primers • Target fragments</td></tr>
</table>

(continued)

Table 1 (continued)

<table>
<tr><th>Technology (year)</th><th colspan="5">Description</th></tr>
<tr><td rowspan="3">Status: currently the NGS leader for throughput and accuracy
Prediction: likely to be a part of genome sequencing for some time</td><td colspan="5">with an A primer on one end and a B primer on the other are isolated • Fragments made single-stranded and hybridized to slide containing many copies of sequences complimentary to primer A and primer B • Polonies are generated by bridge amplification (see Fig. 4) • Reverse strand of each cluster sequence is cut and washed away • Sequencing: cluster chain elongation involves DNA polymerase and modified dNTPs where each nucleotide is (A, T, G, C) labeled with its own fluorescent tag and has its 3' OH reversibly blocked (Metzker 2010) • Polymerase adds one base to each primer; extra nts washed away; polonies on slide are illuminated providing color image; each colored spot reveals first nucleotide in target sequences • 3' OH of incorporated base is freed and second nt is incorporated, illuminated, and recorded • This process is repeated 10–250 times</td></tr>
<tr><td>Generation</td><td>Read length (max)</td><td>HT model</td><td>Output</td><td>Key reference(s)</td></tr>
<tr><td>Second</td><td>300 bp</td><td>NovaSeq 6000</td><td>6,000 Gb per run</td><td>Bennett (2004) and Bentley et al. (2008)</td></tr>
<tr><td>SOLiD (2007) www.thermofisher.com
Status: low throughput compared to other second-generation instruments; low cost per base and dedicated users have kept technique afloat
Prediction: will cease to be used in major DNA sequencing endeavors</td><td colspan="5">Utilizes ligation and base complementarity • Prep: sheared target DNA is ligated to P1 and P2 adapters at 5′ and 3′ ends, respectively • Target/adapter molecules denatured, diluted, and hybridized to beads containing thousands of copies of the P1 sequence • Emulsion PCR is used to generate bead polonies • Beads containing amplified DNA molecules are separated from those that did not bind target molecules • Bead polonies are attached to glass slide • Sequencing: a primer complementary to the terminus of the P1 adapter (offset 0 primer) is annealed to molecules on bead polonies • Di-base probes are added; each di-base probe is eight nucleotides long and consists, from 3′ to 5′, of two actual query bases, three universal bases that "complement" any of the four bases, and three additional universal bases attached to a fluorescent tag • There are 16 possible di-base probes based on two nucleotide query combinations; these are grouped into four subgroups of four with each subgroup possessing a different color fluorescent tag • Ligase is added, and the di-base probe complementary to the first two nucleotides next to the free 5′ end of the primer is ligated to the primer • A laser is used to excite the fluorescent tag, and the color is recorded • The three 5′ nts associated and the fluorescent tag are cleaved from the di-base probe • The ligation, color calling, and tag removal processes are repeated y-1 times • The primer is removed, and a second primer with one less nucleotide at its 5′ end (−1 offset primer) is hybridized to the P1 adapter • Di-base ligation, color calling, and tag removal steps are repeated y times • A −2 offset probe is annealed and di-base ligation, color calling, and tag removal steps</td></tr>
</table>

(continued)

Table 1 (continued)

<table>
<tr><th>Technology (year)</th><th colspan="5">Description</th></tr>
<tr><td rowspan="3"></td><td colspan="5">are repeated $y + 1$ times • A −3 offset probe is annealed and di-base ligation, color calling, and tag removal steps are repeated $y + 1$ times • A −4 offset probe is annealed and di-base ligation, color calling, and tag removal steps are repeated $y + 1$ times • Knowledge of the members of each di-base subgroup and the color patterns produced from extension of each of the primer offsets are used to determine the nucleotide sequence</td></tr>
<tr><td>Generation</td><td>Read length (max)</td><td>HT model</td><td>Output</td><td>Reference</td></tr>
<tr><td>Second</td><td>75 bp</td><td>5500xl SOLiD</td><td>300 Gb per run</td><td>Pandey et al. (2008)</td></tr>
<tr><td rowspan="3">Ion Torrent (2011)[a] www.thermofisher.com
Status: popular in small laboratories due to low instrument cost, but read length and per base cost not competitive with Illumina
Prediction: likely to remain a minor contributor to genome sequencing research</td><td colspan="5">Utilizes chemistry of DNA replication • Prep: sheared target DNA is ligated to A and P1 adaptor/primers • Target DNA molecules flanked by an A and a P1 adapter are hybridized to beads, and bead polonies are generated using emulsion PCR • Beads are allowed to fall into wells of a modified semiconductor chip • Sequencing: primer complementary to the B primer is allowed to anneal to sequences in polonies • One dNTP (e.g., dATP) is passed over plate containing bead polonies • If nucleotide added to B primer end, formation of phosphodiester bond results in release of a hydrogen ion • pH in each well is measured; if incorporation occurred, there is a drop in pH where the magnitude of the drop is roughly proportional to the number of bases incorporated (e.g., incorporation of three dATPs will result in a steeper pH drop than would incorporation of one or two dATPs • Extra nts washed away or degraded • Second dNTP (e.g., dTTP) is run across plate • pH changes recorded and plate washed • Third dNTP (e.g., dCTP) is run across plate • pH changes recorded and plate washed • Fourth dNTP (e.g., dGTP) is run across plate • pH changes recorded and plate washed • Process repeated tens to hundreds of times • Software generates sequence for each bead colony based on pH drops and drop intensity for each incorporation</td></tr>
<tr><td>Generation</td><td>Read length (max)</td><td>HT model</td><td>Output</td><td>Key reference(s)</td></tr>
<tr><td>Second</td><td>600 bp</td><td>Ion GeneStudio 5S Prime</td><td>25 Gb per run</td><td>Rothberg et al. (2011)</td></tr>
<tr><td>Pacific Biosciences (2011)[a] www.pacb.com
Status: after rocky start, gaining momentum
Prediction: currently future looks bright, but its continued relevance depends on Oxford Nanopore</td><td colspan="5">Utilizes chemistry of DNA replication • Prep: adapter sequences are typically ligated onto the ends of target DNA molecules; one of the adapters may have a hairpin at their termini • The target DNA molecules are denatured and annealed with a primer complementary to an adapter • Pre-primed target molecules are loaded onto a plate containing tens of thousands of tiny wells known as zero mode waveguides (ZMWs) • Tethered to the bottom of each ZMW is a modified DNA polymerase molecule • Each</td></tr>
</table>

(continued)

Table 1 (continued)

<table>
<tr><th>Technology (year)</th><th colspan="5">Description</th></tr>
<tr><td rowspan="3"></td><td colspan="5">polymerase molecule binds to a target molecule • Each of the four dNTPs added to the ZMWs has a different fluorescent tag attached to its terminal phosphate group • Light shining up through the bottom of the plate illuminates the area immediately around each polymerase molecule and excites the fluorescent tags • The labeled dNTPs move by Brownian motion in and out of the area around the polymerase and produce an overall background fluorescence • However, when a nucleotide is incorporated onto the end of the primer, it is held by the polymerase long enough that its fluorescent signal can be detected above the background noise • During the final part of incorporation, diphosphate (including the fluorescent tag) is cleaved from the molecule • The cleaved tag diffuses away from the polymerase molecule • The polymerase is now ready to incorporate a second complementary nucleotide into the strand • This process is continued with each nucleotide incorporation being recorded for each ZMW • If hairpin adapters are placed on both ends of a target molecule (effectively circularizing it), the molecule may serve as a template for multiple rounds of sequencing; reading the same molecule multiple times allows creation of consensus sequences compensating for the relatively high error rate of the technique</td></tr>
<tr><td>Generation</td><td>Read length (avg)</td><td>HT model</td><td>Output</td><td>Key reference(s)</td></tr>
<tr><td>Third</td><td>20 kb</td><td>Sequel</td><td>128 Gb per run</td><td>Eid et al. (2009)</td></tr>
<tr><td rowspan="3">Oxford Nanopore (2015)[a]
www.nanoporetech.com
Status: gaining strong user base
Prediction: will eventually be primary means of sequencing genomes</td><td colspan="5">Utilizes principles of electrophoresis and membrane conductance • Prep: long dsDNA is ligated to “Y” adaptors (Fig. 5) • A motor enzyme binds to the single-strand overhangs of the Y adaptors • Sequencing: the DNA/adaptor molecules are placed in buffer on one side of a microchamber divided by a membrane containing nanopores • A voltage is applied, and the DNA travels to the nanopore which is the only conduit through which it can reach the positive pole • The motor enzyme binds to the nanopore and ratchets one the “template” strand of the DNA duplex through the nanopore at a rate much slower than the molecule would naturally travel through the pore • Each base that passes through the nanopore produces a characteristic change in the conductance of the membrane in which the nanopore is embedded • Changes in membrane conductance as a DNA molecule is passed through the nanopore can be used to determine the nucleotide sequence • The motor protein bound to the end of the “complementary” strand encourages sequencing of the complementary strand immediately (or soon after) sequencing of the template strand (Fig. 5)</td></tr>
<tr><td>Generation</td><td>Read length (max)</td><td>HT model</td><td>Max output</td><td>Key reference(s)</td></tr>
<tr><td>Third</td><td>882 kb</td><td>GridIonx5</td><td>75 Gb per run</td><td>Jain et al. (2016)</td></tr>
</table>

[a]Note that kit chemistry, flow cells/chips, and protocols are constantly being upgraded and modified

About a year after 454 machines appeared on the market, Solexa (which was quickly purchased by Illumina) began producing NGS instruments that utilized bridge amplification cluster generation and proprietary fluorescently tagged reversible chain terminating nucleotides. Unlike the pyrosequencing technique of 454, all nucleotides could be run across DNA colonies at the same time. Moreover, homopolymer nucleotide runs did not pose a problem for Illumina instruments as no more than one nucleotide could be incorporated into a chain without first unblocking the reversible chain terminator of the incorporated nucleotide. An overview of the steps in Illumina sequencing is provided in Fig. 5.

A third second-generation technique was developed under the name SOLiD (sequencing by oligonucleotide ligation and detection) by Life Technologies. After bead-based emPCR, SOLiD sequencing diverged from 454 and Illumina sequencers in that the SOLiD system utilized extension of primers via ligation reactions using four sets of four different fluorescently labeled di-base probes. Life Technologies was purchased by Thermo Fisher in 2014, and consequently SOLiD sequencers are now sold as Thermo Fisher products.

A fourth technique known as Ion Torrent sequencing was later developed by Jonathan Rothberg, founder of 454 Technologies. Unlike the other second-generation techniques that detect either fluorescence or luminescence using image capture technologies, the Ion Torrent machines detected base incorporations from changes in pH around a bead colony (H^+ is released as a by-product of phosphodiester bond formation). Ion Torrent instruments require no sophisticated optics system as they do not detect fluorescence or light emission. Ion Torrent is now owned by Thermo Fisher.

The basic operation principles of these second-generation machines are detailed in Table 1. At present, Illumina is far and away the most widely utilized of the technologies, a trend that does not appear to be changing anytime soon (Greenleaf and Sidow 2014). Like 454 instruments, Ion Torrent instruments have trouble with homopolymeric runs (Heather and Chain 2016), but their relatively low machine cost has made them moderately popular in smaller labs. SOLiD sequencing does not provide the read length or output of Illumina instruments, but it has a number of unique uses, a devoted user group, and it is competitive on a cost per base basis (Heather and Chain 2016). However, there are currently no efforts to improve SOLiD technology (Shendure et al. 2017).

It should be noted that Complete Genomics (now owned by BGI) developed its own sequencing platform which it has optimized for study of human genome sequencing (Heather and Chain 2016). The Complete Genomics technology utilizes rolling circle amplification to create "DNA nanoballs" which are fixed to a slide and sequenced via ligation using a series of competing, fluorescently labeled probes (Drmanac et al. 2010). Complete Genomics does not sell its instruments and, to my knowledge, has not been involved in sequencing DNA from organisms other than humans (www.completegenomics.com).

All second-generation sequencing instruments utilize target molecule amplification techniques to create DNA colonies/clusters/nanoballs. The amplification step increases sequencing signal intensity which reduces errors, at least at first. However, simultaneous sequencing of clonally amplified DNA molecules is also the major factor limiting the read length of second-generation instruments. Sequencing is based on millions of synchronous elongation or ligation reactions at each DNA cluster. However, the efficiency of any molecular reaction is never 100%. Consequently, during each round of elongation or ligation, some of the amplicons may not be involved in a base incorporation or ligation event. This will cause these particular amplicons to fall out of synch with their sister amplicons. While problems will be minimal at first, those molecules that have become unsynchronized will create noise, while molecular inefficiency in each additional cycle will lead to greater levels of asynchronization. At some point, the noise created by asynchrony will make base calling impossible (Buermans and den Dunnen 2014).

3.10 Third-Generation Sequencing

Third-generation sequencing is used to describe sequencing techniques that can elucidate the sequence of a single (i.e., unamplified) DNA molecule.

Arguably the first third-generation machine was developed by Helicos. The Helicos instrument utilized fluorescently tagged reversible terminator nucleotides in a manner similar to Illumina machines. However, no amplification of the target sequences was performed (Metzker 2010). The Helicos system had some powerful features; e.g., easy sample preparation. However, Helicos went bankrupt in 2012 after selling only a few instruments (GenomeWeb News Staff Writer 2012). Of note, a company called SeqLL has licensed technology from Helicos and appears to be contemplating selling services and/or instruments (see www.seqll.com).

Pacific Biosciences (PacBio) single molecule, real-time (SMRT) sequencing is an approach in which a powerful fluorescence microscope is used to "eavesdrop" on DNA replication. The details of the PacBio approach are discussed in Table 1. The PacBio SMRT technique was so captivating that the company rushed the instruments to market before many of the kinks had been worked out. This ultimately hurt the company's reputation, although PacBio has since worked hard to make improvements without overselling the capabilities of its machines. Greenleaf and Sidow (2014) remarked upon the disappointing initial 2008 release of the first PacBio machines and the critical improvement of the instruments since then by saying, "Perhaps PacBio would have been well advised to have saved the fireworks it sent into the Florida skies in 2008 for this year."

Oxford Nanopore[8] devices do not rely on DNA replication chemistry at all. While adapters are used to help ensure proper orientation of DNA molecules as they interact with the nanopore, polymerase, dNTPs, and sequencing primers are needed for sequencing (Fig. 6). This makes the Oxford Nanopore technique the first technique since Maxam-Gilbert sequencing to not require primers and to not utilize DNA replication or ligation chemistry (at least in the sequencing phase). DNA molecules are currently sequenced using 2D and $1D^2$ strategies (see Fig. 6). The advantage of the 2D strategy is that the template and complementary strand are sequenced in a single read. The advantage of the $1D^2$ strategy is that if the template strand breaks during sequencing, it will not prevent sequencing of the complementary strand. The error rate of Oxford Nanopore reads is currently higher than other NGS methods, but considerable improvements have been made (Jain et al. 2016). Oxford Nanopore read lengths are unparalleled (Table 1).

Both PacBio and Oxford Nanopore technologies have expanded their markets considerably over the last 2 years. PacBio is in the driver's seat at the moment in terms of sales, but there are several indicators that Oxford Nanopore may be close to overthrowing PacBio as the leader in long-read sequencing technology (Greenleaf and Sidow 2014; Heather and Chain 2016).

3.11 Contig

A contig (short for contiguous sequence) is a set of partially overlapping sequence reads ostensibly representing a continuous piece of genomic DNA (Fig. 7). The term contig is also used to describe a set of partially overlapping BACs or other genomic DNA clones.

3.12 Consensus Sequence

A DNA sequence derived from a contig where the most common base found at a given position along aligned sequence reads is used to represent that position (Fig. 7).

---

[8] As their name suggests, nanopores are simply extremely small pores. They can be derived from naturally existing protein-based membrane pores (e.g., α-hemolysin and *Mycobacterium smegmatis* porin A) or simply etched into silicon or other substrates (solid-state nanopores). Numerous laboratories around the world have been working on using biological and solid-state nanopores to characterize a variety of different molecules including DNA, RNA, proteins, enantiomers, drugs, and polymers (Haque et al. 2013). To date, Oxford Nanopore is the only group to produce commercially available nanopore-based DNA sequencing instruments (Heather and Chain 2016; Jain et al. 2016).

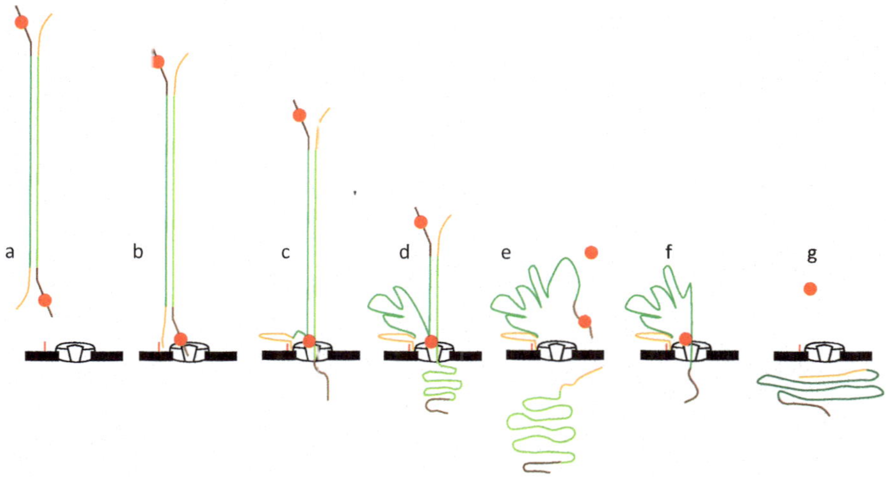

Fig. 6 Oxford Nanopore sequencing strategy for producing "$1D^2$" reads. (**a**) Y-adaptors (brown and gold lines with single-stranded ends) are ligated to both termini of a target DNA molecule. The longer ssDNA strand of the Y adaptor (brown) is bound by a molecular motor protein (orange circle). It does not matter which end of the DNA duplex contacts the nanopore first. The nanopore is flanked by a small docking molecule (orange line). (**b**) The short ssDNA strand of the Y-adaptor binds to the docking molecule. The motor protein binds to the nanopore and begins passing the longer strand of the Y-adaptor through the nanopore. As each nucleotide moves through the nanopore, it causes a base-specific change in conductance across the membrane into which the nanopore is embedded (black). Changes in conductance are recorded and used to call the DNA sequence. (**c**) The template (light green) strand follows the adaptor sequence into the nanopore while the complement strand forms loops as it becomes single-stranded. (**d**, **e**) The process continues until the template strand and its adaptor sequences have been passed through the nanopore. When sequencing of the template strand is complete, its motor molecule drifts away from the nanopore. (**f**, **g**) The complementary strand, bound to its own molecular motor, is already near the nanopore entrance. Consequently, the chances of it attaching to the nanopore and being sequenced are high

3.13 *Scaffold*

A scaffold is composed of a set of contigs that are grouped together because genetic and physical mapping techniques suggest that they are located near each other along the length of a chromosomal DNA molecule (Fig. 7). A scaffold, by definition, contains some gaps.

3.14 *Pseudomolecule*

A pseudomolecule is a collection of contigs and scaffolds that have been grouped into something that, more or less, resembles the DNA sequence of an entire chromosome. Under ideal conditions, a pseudomolecule will approximate the base-by-base sequence of an entire chromosome, and, for a given species, the number of pseudomolecules will be equal to the number of chromosomes.

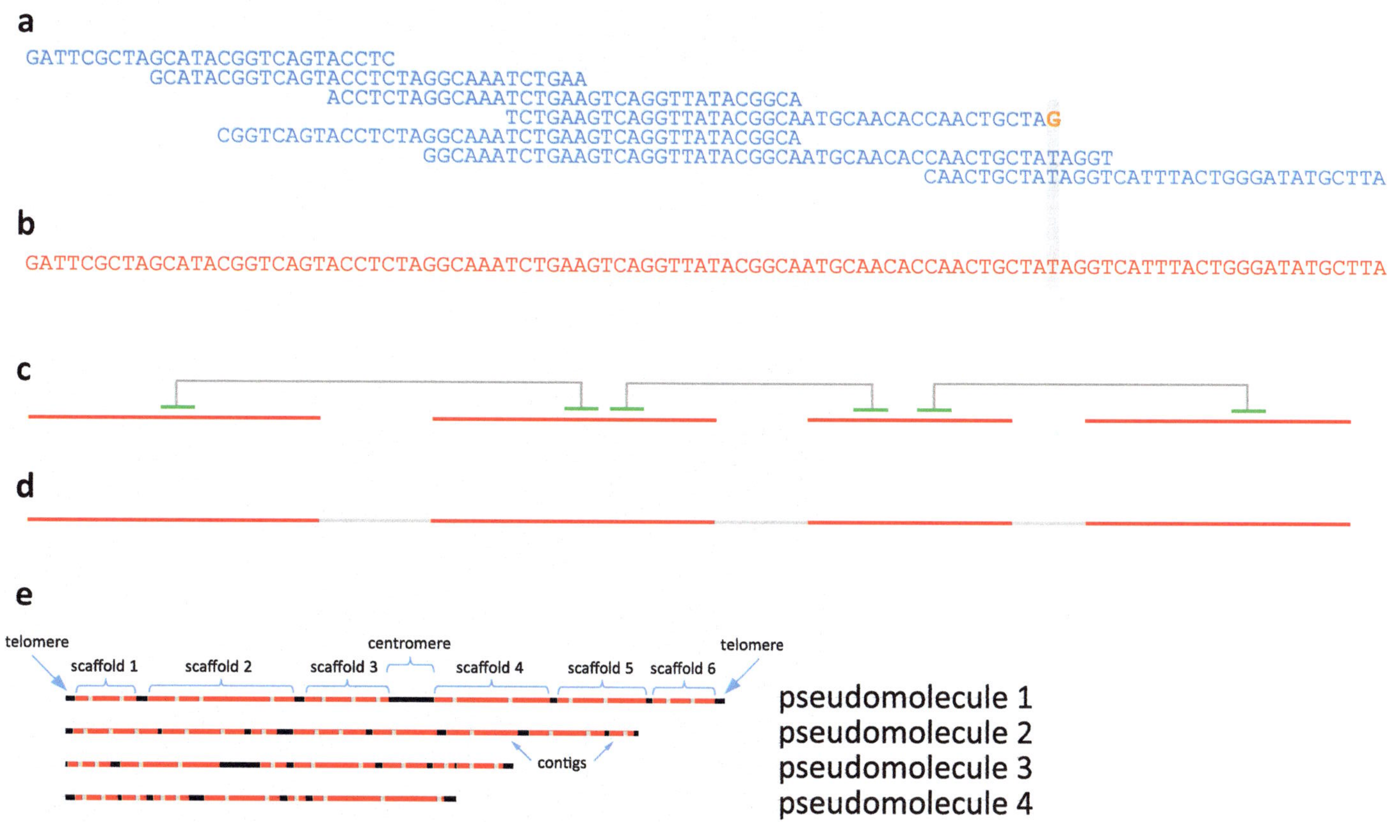

Fig. 7 Illustrating terms associated with genome assembly. (**a**) A contig consisting of partially overlapping DNA reads is assembled by sequence alignment algorithms. Note that the orange G residue at the end of the fourth read from the top differs from the bases at the same position in the two overlapping reads (both

3.15 *Draft Genome*

A draft genome is a collection of sequence reads for a particular species. Draft genomes are often assembled into contigs and may even be grouped into scaffolds.

3.16 *Reference-Quality Genome Sequence*

A reference-quality genome sequence is composed of high-quality pseudomolecules with relatively few gaps between scaffolds. Typically, the DNA used to generate a reference sequence is from a single individual (see Sect. 2.6 above). Even with second- and third-generation sequencing techniques, a reference-quality genome is relatively expensive to produce as efforts must be made to minimize errors in assembly and fill gaps. However, once a reference-quality genome sequence is produced, resequencing (see below) can be used to rapidly and cheaply investigate the molecular nature of genetic diversity.

3.17 *De Novo Genome Sequencing*

De novo (literally "from new" in Latin) is used to describe a genome sequencing project for an organism for which there is not a close reference-quality genome sequence available. Alternatively, it can be used to describe sequencing and assembly of a genome without utilizing an existing reference-quality genome sequence. This latter form of de novo sequencing is used to test new sequencing techniques and assembly algorithms (i.e., will our technique be able to produce something comparable to the reference genome sequence?).

Fig. 7 (continued) have T residues at the position in question). (**b**) The reads in a contig are used to produce a consensus sequence for the contig. As highlighted in the gray rectangle, when there is base discrepancy at a specific position in the contig, the base most commonly found at that position is included in the consensus sequence (i.e., two of the three reads have a T at the position in question, and thus the consensus sequence contains a T). (**c**) Contig consensus sequences, each represented by a red line, are organized into scaffolds. A number of different physical mapping techniques (e.g., end sequencing of long molecules, integration of scaffolds with molecular genetic map data) allow alignment and ordering of contigs in a scaffold. Note that while the scaffold represents a continuous stretch of DNA, there are gaps of unknown sequence (and sometimes unknown length) between contigs. (**d**) A scaffold is shown where red lines represent contig consensus sequences and gray lines represent gaps between contigs. (**e**) Physical mapping techniques are used to arrange scaffolds into pseudomolecules that represent whole chromosomes or linkage groups. Gaps between scaffolds are represented by black lines. Note that centromeric and telomeric regions are common gap regions in pseudomolecules

3.18 Resequencing

Resequencing is another technique that is poorly named as it typically does not refer to sequencing the same DNA over again. Rather, resequencing is the quick sequencing of a genome via second- and/or third-generation techniques and alignment of the resulting sequence reads to a reference-quality genome sequence (Lesk 2012). Resequencing is used to quantify diversity and investigate changes in key genes. When short-read sequences are aligned with a reference-quality genome, single nucleotide polymorphisms (SNPs) and short indels (insertions and deletions) can be readily detected. Alignment of longer reads to a reference genome allows detection of insertions, deletions, and inversions as well. When people say that the human genome can now be sequenced for $1,000, they are referring to resequencing of a human genome, not de novo sequencing and assembly of a human genome.

3.19 N50

N50 has become a common statistic used in describing genome assemblies. In my opinion, most definitions of N50 are vague and rather difficult to comprehend. However, through a diagram the idea of N50 can be relayed relatively easily. In this regard, see Fig. 8.

3.20 Paired-End Sequencing and Unidirectional Sequencing

Paired-end sequencing (also known as bidirectional sequencing) is simply sequencing both ends of a DNA molecule. In some cases, paired-end sequencing provides as much information and is considerably cheaper, than performing two unidirectional (single-end) sequencing runs. In other instances, the paired-end data has value in genome assembly.

The original paired-end sequencing was performed with cloned DNA molecules and Sanger sequencing. The recombinant vector/insert molecules were isolated and divided between two tubes. In the first tube, Sanger sequencing was performed using a primer complementary to a piece of vector sequence immediately adjacent to one side of the insertion site (forward sequencing reaction), while in the second tube, a primer complementary to vector on the other side of the insertion site was used in sequencing (reverse sequencing reaction).[9] Because new DNA strands are only

---

[9]Forward and reverse designations are arbitrary. However, molecular biology companies often use these designations to refer to specific vector-based primers, i.e., for paired-end sequencing or PCR amplification using a particular vector, you will want to purchase both the forward and reverse primers for that vector.

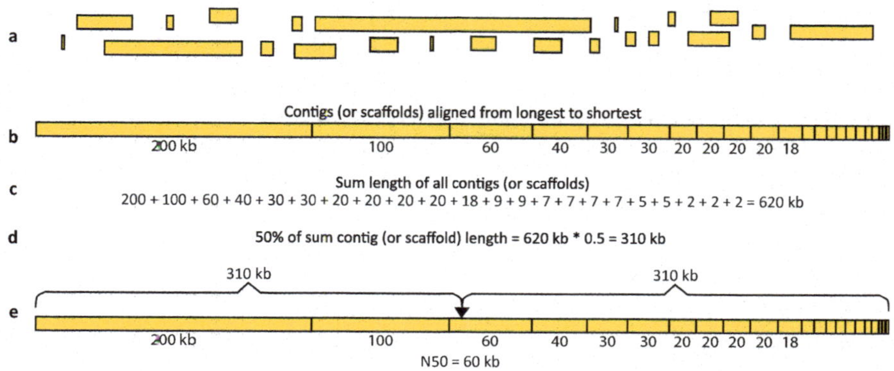

Fig. 8 Understanding N50. (**a**) Unordered contigs representing a genome assembly. Each contig is represented by a rectangle with its width being proportional to the contig's relative length in kb. (**b**) In silico, the contigs are arranged from longest to shortest. (**c**) The sum length of the contigs is determined (620 kb), and (**d**) 50% of the sum contig length is calculated (310 kb). (**d**) Starting from either end of the ordered contigs and moving 50% of the sum contig length (i.e., 310 kb) will bring you to a particular contig (arrow between braces). (**e**) The length of the contig intersected by the arrow (i.e., 60 kb) is the N50. In other words, 50% of the sum contig length is contained in contigs ≥60 kb in length. Note that N50 values can be reported for scaffolds as well as contigs

sequenced from 5′ to 3′, the two primers used were necessarily complementary to different strands of the vector molecule. If the insert size was short enough (e.g., 1,400 bp), the forward and reverse reads might overlap, allowing generation of consensus sequence. If the insert was larger, the paired-end sequences would be separated by a gap of unknown sequence. If the gap size is known, the paired-end sequences can be used as bridges between contigs. For example, let's say you have a BAC with an insert size of 200 kb and you have sequenced 700 nt from each end of the insert. Both the resulting sequence data and positional data can be of considerable use. If alignment of the paired-end sequences (in this case, BAC-end sequences or BESs) with contigs reveals that each BES is found in a different contig, those contigs can be grouped into a scaffold with an estimated space of 200 kb between the BES sites. To our knowledge, BAC-end sequencing is only performed using automated Sanger sequencing.

When NGS technologies were being developed, it was clear that being able to produce paired-end sequences would be desirable, primarily as a means of bridging gaps and increasing output. For Illumina, development of paired-end sequencing was relatively straightforward. As shown in Fig. 5m, n, after bridge amplification, reverse strands are cut from the slide leaving only forward strands to serve as templates for sequencing. However, after sequencing using the forward strands, the 3′ ends of the forward strands can be unblocked, and bridge amplification can be used to regenerate reverse strands. The forward strands can then be cleaved from the slide, the 3′ ends of the reverse strands can be blocked, and sequencing using the reverse strands as templates can be conducted. However, target molecules longer than 2 kb can cause problems with bridge amplification,[10] so paired-end Illumina sequencing is currently limited to target molecules <2 kb in length.

For Illumina platforms, paired-end sequencing is often indicated by the number two followed by a multiplication sign followed by read length in bp or nt. Thus, if you order 2 × 250 nt sequencing from a vendor, you are requesting paired-end sequencing with 250 nt reads (500 nt total sequence per molecule). If sequencing is unidirectional, it is denoted by the read length alone or by the number one times the read length (i.e., 250 bp or 1 × 250 bp).

3.21 Mate-Pair Sequencing

With the exception of Illumina, simple paired-end sequencing is not an option for second-generation techniques. However, there is a technique known as mate-pair sequencing that can be used to sequence the ends of fairly large DNA molecules. Mate-pair approaches involve adding biotin or biotinylated adapters to the ends of fairly long sequences, circularization of DNA molecules, fragmentation of circularized molecules to a size suitable for sequencing on the platform of choice, and enrichment of biotin-containing fragments using streptavidin. Typically, regular sequencing adapters are added to the ends of the mate-pair molecules. Procedures for making mate-pair libraries can be quite sophisticated. For example, the mate-pair library technique of Jarvie and Harkins (2008), developed for 454 sequencing, involves shearing DNA into 3 kb fragments, methylating the DNA fragments (to prevent restriction later on), ligating a biotinylated hairpin adapter to both ends of each fragment, digestion of the hairpin adapters with a restriction enzyme that does not cut methylated DNA (e.g., *Eco*RI) to produce complementary sticky ends, ligation of ends together to create circularized molecules, shearing of the circularized DNA into 250 bp fragments, streptavidin-based isolation of fragments containing the remainder of the two annealed hairpin adapters, and addition of *A* and *B* adapters for emulsion PCR amplification and sequencing.

PacBio and Oxford Nanopore produce long-read lengths from single DNA molecules not fixed to a substrate. Consequently, producing paired-end reads or mate-pair sequences is not a goal of either of these methods.

3.22 Multiplexing, Indexing, and Bar Coding

Second-generation sequencers can produce so many reads that it is often economical to group different DNA libraries together before sequencing them on a single lane/

[10]This is according to Illumina; our own experience suggests a dramatic decrease in quality if target molecule length exceeds 900 bp.

plate/chip, a process known as multiplexing. To avoid the chaos that would result from simply mixing heterogeneous sources of DNA together and trying to assign each read back to its correct source, prior to mixing, each DNA library can be ligated to adapter primers that are identical except for a short region (usually 2–5 nt) deemed an index or bar code. Ligating a different indexed adaptor(s) to each sample allows the resulting sequence reads from the mixture to be correctly deconvoluted and assigned to their appropriate source (via simple in silico sorting). If both the forward and reverse adapters contain indices (called "double indexing"), deconvolution of data is significantly more accurate (Kircher et al. 2012).

3.23 *Phasing*

Phasing is the term used to describe assignment of alleles in a diploid individual(s) into haplotypes (Roach et al. 2011). With regard to genome sequencing, certain types of phasing can be used to circumvent assembly problems associated with heterozygosity. As discussed above, starting with a genetically homogeneous individual or a haploid tissue is an ideal way to start a genome sequencing project. However, homogeneity is not always an option, especially when dealing with obligate outcrossing species. The "gold standard" phasing technique used in genome sequencing is called *trio phasing*. For an individual produced by sexual reproduction, the genomes of that individual's parents are sequenced via a second-generation (i.e., short-read) method (e.g., Illumina sequencing). The offspring's DNA is sequenced using a third-generation (long-read) instrument. *K*-mers from both parents are compared, and those unique to the maternal parent are matched with the reads of the child. Those reads recognized by the maternal *k*-mers are considered part of the maternal haplotype. The same process is repeated using unique *k*-mers from the father. Further assembly of the maternal and paternal haplotypes into contigs/scaffolds is then conducted to produce two haplotype assemblies.

3.24 *Hi-C*

The fact that interphase chromosomes occupy distinct domains within nuclei has been exploited as a means of looking at chromatin organization and determining which DNA sequences are typically near each other in the 3D space of the nucleus. A chromatin conformation capture technique known as Hi-C has become a popular high-throughput means of exploring chromatin relationships (Denker and De Laat 2016). The Hi-C concept/method is illustrated in Fig. 9. In brief, nuclei are treated with formaldehyde which introduces cross-linking between adjacent chromatin regions. Ligation of the DNA molecules from the cross-linked regions produces chimeric DNA molecules that can be sequenced with second- or third-generation sequencing. If two sequences are found associated with each other more than one

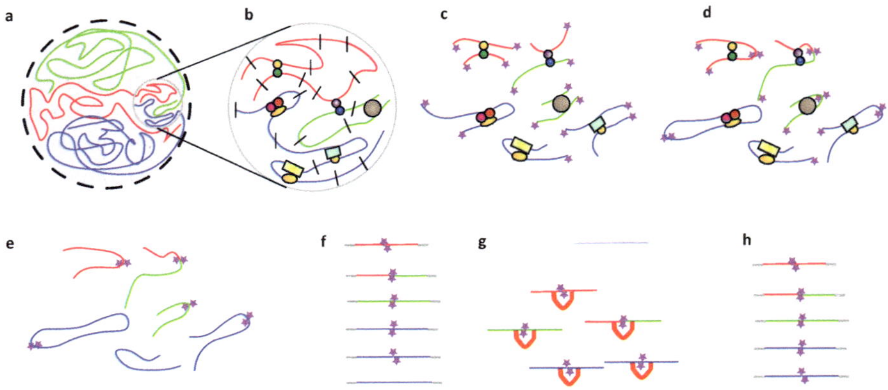

Fig. 9 Hi-C method of studying chromatin interactions. (**a**) Nuclei are embedded in agarose, and treatment with low concentration formaldehyde cross-links DNA to associated/nearby proteins. Moreover, protein-protein cross-linking occurs. DNA-DNA cross-linking is minimal at low formaldehyde concentration (Hoffman et al. 2015). In this simplified drawing, there is a single nucleus containing three chromosomes (represented by blue, red, and green lines, respectively). (**b**) Close-up view of a region from the nucleus. Proteins are represented by shaded circles, ovals, and rectangles, and hash marks bisecting chromatin molecules represent *Hind*III recognition sites. Non-crosslinked proteins are removed by treatment of the chromatin with dilute SDS. (**c**) The sticky ends produced by restriction digestion are filled in by a DNA polymerase with normal dTTP, dGTP, dATP, and biotin-labeled dCTP (biotin = pink stars). The result is blunt-ended DNA molecules. (**d**) Ligase is added under conditions that favor ligation of blunt-ended fragments attached to the same chromatin complex. Blunt-end ligation of two restriction fragments produces a characteristic sequence (i.e., 5-GCTAGC-3′) with biotinylated C residues at the join. (**e**) Proteinase K is used to digest proteins and RNase treatment removes RNA. After purification of the DNA, the fragments are treated with an exonuclease that removes nucleotides from the ends of DNA molecules (i.e., it removes biotinylated nucleotides at the termini of DNA fragments). The reaction is stopped with EDTA. (**f**) Sonication is used to shorten the molecules and produce fragment lengths suitable for second-generation sequencing. The fragments are end-repaired and ligated to sequencing adaptors. (**g**) Streptavidin beads are used to pull down biotinylated DNA fragments. Fragments without internal biotinylated nucleotides are washed away. (**h**) Second-generation sequencing is used to sequence the chimeric DNA molecules. The frequency of each combination of fragments in a chimeric molecule is determined. Any two fragments that are more frequently paired with each other in a chimeric molecule likely are near each other in the interphase nucleus and, because chromosomes are found in distinct domains, are likely on the same chromosome

would expect by chance, it is likely that these sequences are found near each other in the nucleus and are likely parts of the same chromosome. It was through Hi-C research that topographically associating domains (TADs) were discovered. Hi-C can be used to gain insight into sequences involved in long-distance (along the linear DNA molecule) interactions (e.g., promoters and enhancer elements) (Belton et al. 2012; Denker and De Laat 2016).

Of note, Hi-C can be a form of phasing. Homologous chromosomes occupy their own regions within the 3D-space of a nucleus. Thus formaldehyde cross-linking between chromatin from homologous chromosomes is unlikely. If Hi-C sequencing depth is deep enough and reads are long, each homolog/haplotype can be separated prior to assembly.

3.25 mRNA/cDNA/EST/RNA-Seq Data

The discovery and proper annotation of genes is greatly facilitated by utilization of sequence obtained from mRNA molecules. Since RNA is difficult to sequence, all first- and second-generation sequencing techniques have researchers convert mRNA into complementary DNA (cDNA) using reverse transcriptase.[11] Partial cDNA sequences have been referred to as expressed sequence tags (ESTs), while high-throughput sequencing of cDNA using second-generation techniques is called RNA-Seq. Regardless, cDNA sequences (particularly relatively long cDNA sequences) greatly facilitate structural genome annotation (see below). RNA-Seq data can also be useful, although it is perhaps best utilized to study differential gene expression (especially after a quality genome sequence is in place).

3.26 Proteogenomic Mapping

cDNA sequencing can help one learn which genes in a genome are being transcribed to create mRNA. However, not all RNA sequences that are transcribed are translated into protein. Proteogenomic mapping is the use of proteomics (mass spectroscopy) data as a means to facilitate recognition of genes that are actually being translated (Jaffe et al. 2004).

3.27 Structural Annotation

Structural annotation is the process of identifying and demarcating certain sequence types within DNA (or RNA or protein) sequence data. Typically, structural annotation is performed using algorithms that recognize specific motifs (ab initio identification) and/or compare known sequence types to the query sequence (homology-based identification). Unless noted otherwise, the features mentioned below can be identified using both ab initio and homology-based methods:

- Open reading frames (ORFs) – An ORF is any stretch of DNA that has a transcription start codon (5′-ATG) and a transcription stop codon (5′-TAA, TGA, or TAG) separated by a distance that suggests a potential coding region

[11]Oxford Nanopore and PacBio have started advertising direct RNA sequencing.

(of note, all exons are ORFs, but not all ORFs are exons). ORFs are usually detected using ab initio approaches.

- Coding sequences (CDSs) – A CDS is an exon or series of exons that make up the protein coding portions of a mature mRNA. CDSs are usually identified by sequencing of cDNAs and homology-based alignment of cDNAs with a genome sequence. However, ab initio approaches have also proven useful in identifying CDSs.
- Introns – Sequences between the exons within a gene.
- Promoters – Sequence upstream from, and relatively near, an ORF.
- Mobile elements (MEs) – Sequences with motifs and features characteristic of one of the many types of mobile elements. Mobile elements are often initially detected by homology-based comparisons with known MEs followed by ab initio scans.
- Tandem repeats including microsatellites (SSRs).
- miRNAs.

Initial computer-based annotation may be further refined manually.

3.28 Functional Annotation

Functional annotation is assignment of biological data to structurally annotated genome features. Functional annotation may include information about gene function, biochemistry, interactions, expression, and regulation. The assignment of function to a gene is ultimately dependent upon experiments that show how the absence or reduced expression of that gene affects phenotype. Alternatively, biochemical experiments that demonstrate the role of a gene product in a particular metabolic pathway may be used to determine a gene's function (although genes may have multiple functions). The former groups of experiments, often known as "functional genomics" approaches, are relatively easy to perform in *Arabidopsis* but are more difficult in other plants. Consequently, for newly sequenced plant genomes, gene function may be tentatively assigned to a gene based upon the function of its *Arabidopsis* (or other plant species) ortholog (if one exists). Because highly similar gene sequences may have very different functions in vivo, functional annotation not directly based on experiments done in the species whose genome is being annotated should be treated with great caution.

The Gene Ontology (GO) Consortium has worked to develop a controlled, standardized vocabulary for use in assigning functional annotations to genes. Even if the function of a gene is unknown, GO annotation helps place the gene into one of several larger categories based upon its domains and/or the functions of its orthologs. At the highest GO level, genes are categorized as either cellular component, molecular function, or biological process.

4 Bioinformatics and Computer Concerns

In the early days of genome analysis, there were relatively few computer scientists working on issues related to DNA sequencing, and consequently, many of the first sequence analysis scripts were written by biologists. While there are likely some major exceptions, biologists, in general, do not write computer code as efficiently and effectively as those trained specifically to do so. Co-author Peterson once was involved in a discussion with J. Craig Venter, a key (and controversial) figure in the Human Genome Project (HGP). When Dr. Venter was asked what was the biggest scientific breakthrough that allowed the HGP to finish under budget and ahead of schedule (Hilgartner 2017), he said without hesitation, "Hiring computer programmers to rewrite our scripts."

The HGP led to considerable improvement in sequence manipulation/analysis tools. The HGP and the plant and animal genome projects that followed were dependent upon on the use of such tools including the Phred-Phrap-Consed suite (Nickerson et al. 1997), BLAST (Altschul et al. 1990) and its derivatives, RepeatMasker (Smit et al. 2013), and a number of gene/repeat finding scripts. Computer clusters played a large role in sequence manipulation and analysis, although much work was done on individual workstations.

When NGS approaches emerged on the scene, they resulted in such a glut of data that many of the sequencing centers that were beta-testing the machines quickly filled up their computer storage capacities. These researchers were accustomed to saving the electropherogram (raw) data produced by Sanger sequencers in case reanalysis was determined prudent, and they initially tried saving the massive image files produced by 454 and Illumina machines in the same manner. There was not room for these large data files to be stored or archived in NCBI/DDBJ/EMBL as was common practice with Sanger data. Consequently, throwing away raw data files, which was considered very bad form in the pre-NGS days, became a necessity (Richter and Sexton 2009).

NGS has forced genome scientists to start utilizing the power of high-performance computing (HPC). This is an ongoing transition that has not been pain-free. In brief, few early bioinformatics programs were developed for HPC environments. While the programs could often be run on HPC systems, they were not designed to take advantage of the efficiency of HPC systems. Thus running early bioinformatics scripts on a supercomputer was not a huge step up from running the programs on the desktop workstations on which they often were designed. The problem of script inefficiency was partly offset by creation of pipelines where the results of an analysis by one program were automatically fed (by customized scripts) to the next program in a chain of programs. Pipelines are useful even when scripts are written for HPC. However, a major problem in bioinformatics has been a failure to write/adapt scripts to utilize the many advantages of parallel computing, the division of computational labor between multiple (often hundreds or thousands of) processors rather than relying on one high RAM machine or small cluster for analysis. Today, more bioinformaticians are writing new and adapting old tools to utilize parallel computing environments; however, there is still much room for improvement. Whether across campus or in the cloud, HPC/parallel processing is necessary for timely analysis of genomic data (see Ocana and de Oliveira 2015 for review).

A discussion of all the bioinformatics tools, which can be utilized in DNA sequence assembly, annotation, comparison, etc., is beyond the scope of this chapter. However, Table 2 presents a brief description, references, and download links to the major sequence analysis tools currently being utilized by the research team at the Institute for Genomics, Biocomputing & Biotechnology (IGBB) at Mississippi State University.

With regard to the use of bioinformatics tools, we do have a few suggestions based on our experiences. These suggestions are:

- Be wary when using commercial software. Some commercial software is good, but much of it is not. In our experience, many commercial bioinformatics programs aren't capable of handling large volumes of NGS data. One commercial program we tested, which was run using the desktop computer provided with the software, took several months to produce a bacterial DNA assembly; we assembled the same sequence reads using the publicly available ABySS (Jackman et al. 2017) and Velvet (Zerbino and Birney 2008) programs independently on an HPC system. Both the ABySS and Velvet analyses took less than 5 min. The assembly produced by the commercial software was aesthetically pleasing (there were no gaps), but the assembly looked nothing like the highly similar ABySS and Velvet assemblies. Moreover, the commercial software produced an assembly that shared no synteny with published genomes of species from the same bacterial genus. Because of its proprietary nature, we could not figure out why the commercial program did what it did. In short, if you can't look behind the curtains to see what is going on, you better have good reason to trust the commercial software vendor.
- Use Linux and its default command line (terminal) interface. Most bioinformatics tools are written to operate in a Linux environment, and very few have a GUI (graphic user interface) available. Most HPC centers have Linux operating systems.
- Adopt the literate programming model. Literate programming (Knuth 1984) is the practice of placing documentation and source code in the same document. The value of this is that a single document has the *what*, *why*, and *how* information on a particular script. Adding a summary of results to the document increases its usefulness further. Org-mode (https://orgmode.org), an Emacs text processor extension, is an excellent tool for this type of documentation.
- Document and track everything. Experimenting with different tools and parameters is vital to produce the best possible analysis from the data. Even changing the versions of tools can have a significant impact on the results (Salzberg et al. 2012; Conesa et al. 2016). With all this variability, documentation is crucial to the reproducibility of any analysis. To aid in this task, a version control system can be utilized to track the changes through time. Git (https://git-scm.com/) is a popular open-source distributed version control system and the one we use in our lab.

Table 2 A list of publicly available sequence processing, assembly, alignment, comparison, and annotation tools used by the authors at the Institute for Genomics, Biocomputing & Biotechnology (IGBB) at Mississippi State University

Program name hyperlinked to program download site – description (reference)
3D-DNA – 3D de novo assembly (3D DNA) pipeline, i.e., a Hi-C scaffolder (Dudchenko et al. 2017)
ABySS – de novo, parallel, paired-end sequence assembler that is designed for short reads. The single-processor version is useful for assembling genomes up to 100 Mb in size. The parallel version is implemented using MPI and is capable of assembling larger genomes (Simpson et al. 2009; Jackman et al. 2017)
ALLPATHS-LG – short-read genome assembler from the Computational Research and Development group at the Broad Institute (Gnerre et al. 2011)
ArrayIDer – maps EST and probe IDs to their corresponding public database accessions (van den Berg et al. 2009)[a]
Augustus – predicts genes in eukaryotic genomic sequences (Stanke et al. 2004)
BioConductor – tools for the analysis and comprehension of high-throughput genomic data (Zhang et al. 2003)
BioPerl – Perl libraries for biology (Stajich et al. 2002)
Biopython – a set of freely available tools for biological computation (Cock et al. 2009)
BLAST+ – Basic Local Alignment Search Tool (Camacho et al. 2009)
BLAT– sequence alignment tool (Kent 2002)
Bowtie/Bowtie2 – an ultrafast and memory-efficient tool for aligning sequencing reads to long reference sequences (Langmead 2010; Langmead and Salzberg 2012)
BUSCO – assessing genome assembly and annotation completeness with Benchmarking Universal Single-Copy Orthologs (Simao et al. 2015)
BWA – a software package for mapping low-divergent sequences against a large reference genome (Li and Durbin 2009)
Canu – adaptation of the Celera assembler designed for high-noise single-molecule sequence assembly (PacBio & Nanopore) (Koren et al. 2017)
CAP3 – DNA sequence assembly (Huang and Madan 1999)
Circos – a software package for visualizing data and information in a circular format (Krzywinski et al. 2009)
Cufflinks – assembles transcripts, estimates their abundances, and tests for differential expression and regulation in RNA-Seq data (Trapnell et al. 2012)
DISCOVAR de novo – new genome assembly tool made by makers of ALLPATHS-LG (Love et al. 2016)
EA-utils – command-line tools for processing biological sequencing data, barcode demultiplexing, adapter trimming, etc. (Aronesty 2013)
EMBOSS – suite of bioinformatics software (Rice et al. 2000)
Exonerate – a generic tool for pairwise sequence comparison (Slater and Birney 2005)
FastQC – a quality control tool for high-throughput sequence data (Babraham Bioinformatics 2016)
GATK – a wide variety of tools with a primary focus on variant discovery, genotyping, and data quality assurance (McKenna et al. 2010)
GeneMark – a family of gene prediction programs (Isono et al. 1994)
GenomeScope – script that estimates genome heterozygosity, repeat content, and size from sequencing reads using a *k*-mer-based statistical approach (Vurture et al. 2017)

(continued)

Table 2 (continued)

Program name hyperlinked to program download site – description (reference)
GO Annotation Quality (GAQ) Score – calculates the quality of GO annotations associated with a set of gene products (Buza et al. 2008)[a]
GA2GEO – allows users to convert a file between the gene association format and an NCBI GEO (Gene Expression Omnibus) type format[a]
GOanna – allows users to quickly add more GO annotations by transferring GO annotations from similar sequences (McCarthy et al. 2006)[a]
GOanna2ga – used to convert the output from GOanna into a gene association file format (McCarthy et al. 2011)[a]
GOModeler – allows users to conduct a hypothesis-driven interrogation of their functional genomics datasets (Manda et al. 2010)[a]
GORetriever – returns GO annotations for a list of protein accessions or GO IDs. Used to retrieve existing GO annotation (McCarthy et al. 2006)[a]
HISAT2 – a fast and sensitive alignment program for mapping next-generation sequencing reads (Kim et al. 2015)
HMMER – used for searching sequence databases for homologs of protein sequences and for making protein sequence alignments. It implements methods using probabilistic models called profile hidden Markov models (profile HMMs) (Finn et al. 2015)
InterProScan (iprscan) – scans sequences for matches against the InterPro collection of protein signature databases (Zdobnov and Apweiler 2001)
LINKS – long interval nucleotide *k*-mer scaffolder (Warren et al. 2015b)
MAKER2 – a portable and easily configurable genome annotation pipeline (Cantarel et al. 2008; Holt and Yandell 2011)
MGEScan-LTR – identifies long terminal repeats (LTR) (Rho et al. 2007)
Minimap2 – a versatile pairwise aligner for genomic and spliced nucleotide sequences (Li 2016)
Miniasm – ultrafast de novo assembly for long noisy reads (though having no consensus step) (Li 2016)
MCScanX – algorithm for detection of synteny and collinearity (Wang et al. 2012b)
MUMmer – a system for rapidly aligning entire genomes, whether in complete or draft form (Marcais et al. 2018)
MUSCLE – multiple sequence alignment (Edgar 2004)
NextClip – a tool for comprehensive quality analysis and read preparation for Nextera Long Mate Pair (LMP) libraries (Leggett et al. 2014)
Nseg – used to mask nucleic acid sequences, needed by RepeatScout
OMSSA – open mass spectrometry search algorithm (Geer et al. 2004)
OrthoCluster – synteny block identification (Ng et al. 2009)
Picard – a set of tools (in Java) for working with next-generation sequencing data in the BAM (http://www.htslib.org/) format
Pilon – an automated genome assembly improvement and variant detection tool (Walker et al. 2014)
ProQuant – tool for label-free protein quantification of SEQUEST output for the Microsoft Windows environment (Bridges et al. 2007)[a]
Proteogenomic mapping tool – provides experimentally based structural annotations at a complete genome level (Sanders et al. 2011)[a]
ProtIDer – enhances proteomics based on EST or EST assemblies by generating databases of matching highly homologous proteins (McCarthy et al. 2006)[a]

(continued)

Table 2 (continued)

Program name hyperlinked to program download site – description (reference)
Quack – very fast and scalable quality assurance tool for evaluating FASTQ (high-throughput sequence) files (Thrash et al. 2018)[a]
Quast – evaluation of genome assemblies (Gurevich et al. 2013)
Racon – ultrafast consensus module for raw de novo genome assembly of long uncorrected reads (Vaser et al. 2017)
RAxML – tool for maximum likelihood-based phylogenetic inference (Stamatakis 2006)
Recon – a package for automated de novo identification of repeat families from genomic sequences (Bao and Eddy 2002)
RepeatMasker – a program that screens DNA sequences for interspersed repeats and low-complexity DNA sequences (Smit et al. 2013)
RepeatModeler – a de novo repeat family identification and modeling package
RepeatScout – a tool to discover repetitive substrings in DNA (Price et al. 2005)
REVIGO – summarizes and visualizes long lists of Gene Ontology terms (Supek et al. 2011)
Salmon – quantify the expression of transcripts using RNA-Seq data (Patro et al. 2017)
Samtools – a suite of programs for interacting with high-throughput sequencing data (Li et al. 2009)
SNAP – SNAP is a general purpose gene-finding program suitable for both eukaryotic and prokaryotic genomes (Zaharia et al. 2011)
StringTie – a fast and highly efficient assembler of RNA-Seq alignments into potential transcripts (Pertea et al. 2015)
Trimmomatic – performs a variety of useful trimming tasks for Illumina paired-end and single-ended data (Bolger et al. 2014b)
Trinity – efficient and robust de novo reconstruction of transcriptomes from RNA-Seq data (Haas et al. 2013)

[a]Developed at the IGBB or in collaboration with the IGBB

Using Git has the added benefit of services like GitHub and GitLab which facilitate collaboration with others.

5 Sequencing, Assembly, and Annotation of Genomes De Novo

5.1 The Human Genome Project (HGP) Sets the Stage for Plant Genome Sequencing

The HGP spurred many advances in technology and led to the development of a number of protocols that were critical in de novo sequencing and assembly of the first plant genome sequences. Specifically, the HGP spurred:

- The development and use of BACs as cloning vectors (along with the eventual abandonment of yeast artificial chromosomes or YACs) – While YACs could

carry megabase-sized DNA inserts, roughly 50% of YAC inserts were chimeric or were characterized by insert rearrangements. In contrast, BAC inserts were highly stable. Moreover, isolating YACs was relatively difficult, while BACs were as easy to isolate as any single-copy bacterial plasmid (see Peterson et al. 2000 for review). Additionally, BACs could be readily end-sequenced, and BAC-end sequences were a highly efficient way to span gaps (up to 100–300 kb) during genome assembly.
- The design and production of automated Sanger sequencers by Lloyd Smith, Leroy Hood, and Applied Biosystems (Smith et al. 1986; Shendure et al. 2017) – Automation greatly reduced the human hours involved in actual sequencing and post-sequencing data processing.
- The inclusion of random fragment ("shotgun") sequencing in the genome sequencing process – previously considered a complete waste of money, production of shotgun sequence reads became a feasible approach for helping improve assembly of non-repetitive regions of the genome due to the drop in sequencing costs.
- Algorithm development and construction of computational tools to expedite many aspects of the sequencing, assembly, and annotation process, notably Phred, Phrap, Consed, the TIGR assembler, and the Celera assembler (Shendure et al. 2017) – The bioinformatics tools utilized by the HGP were adopted by the plant genome community.

5.2 *BAC-by-BAC Sequencing*

The first plant to have its genome sequenced was *Arabidopsis thaliana*, an angiosperm dicot from the *Brassicaceae*. This mustard plant was selected not for its socioeconomic importance but because it was an established model for plant physiology and genetics and had all the makings of a superlative genomics model. In brief, it (a) has a small genome size, (b) possesses a rapid life cycle, (c) is short and compact and needs little space to grow, (d) can be crossed and selfed, (e) produces numerous seeds, (f) can be transformed, and (g) can be readily used to create mutants using a number of methods, with T-DNA mutagenesis being particularly valuable (see Koornneef and Meinke 2010 for review).

The second plant genome project focused on rice (*Oryza sativa* spp. japonica), an angiosperm monocot with C3 carbon fixation. Rice has many features of a model plant (small genome size), but it also has the added advantage of being the single most important source of human calories (Awika 2011).

Arabidopsis and rice were sequenced and assembled using a BAC-by-BAC strategy (Claros et al. 2012; Jackson 2016) that paralleled the successful strategy employed by the publicly funded HGP. The tenets of the BAC-by-BAC strategy are as follows, although neither the *Arabidopsis* nor rice genomes were sequenced exactly as described below:

1. Several BAC libraries from a single individual or, if possible, a single haploid tissue are generated. The BAC libraries differ with regard to the restriction enzyme used to partially digest the high-molecular-weight DNA. It is assumed that making libraries with different restriction enzymes would help reduce bias and provide better overall coverage of the genome. A common goal is to have BAC libraries whose insert sizes collectively represent >10 genome equivalents of DNA from the organism of interest. Robots are used to pick clones from agar plates and place them into separate wells in 384-well microtiter plates. Each plate has a number and each clone in a plate has its own row and column designation. The plate number and positional coordinates of each BAC are tracked in all subsequent steps.
2. The BACs are fingerprinted.
3. BESs are generated for as many BACs as possible.
4. BAC libraries are gridded onto high-density macroarrays, and the macroarrays are screened with carefully designed pools of molecular markers, probes designed from cDNA molecules, and BESs. The BACs recognized by a particular probe are noted.
5. Fingerprint, BES, and macroarray hybridization data are integrated to develop a BAC-based physical map of each chromosome. BACs are grouped into contigs consisting of BACs with overlapping fingerprints and/or linked by BESs or molecular markers. Molecular linkage maps, fluorescence in situ hybridization, and other techniques are utilized to group contigs into scaffolds.
6. Those specific BACs that are involved in making up a chromosome's minimum tiling path are retrieved from their corresponding plates and wells.
7. Minimum tiling paths of each chromosome are sequenced. This is done by breaking each BAC into pieces, "subcloning" the pieces, and shotgun end sequencing the pieces using automated Sanger sequencing. As was done previously with the whole BACs, the reads for a particular BAC would be assembled into contigs. Gaps in a BAC sequence were filled (when possible) using chromosome-walking techniques.

It should be noted that maize, the third largest supplier of human calories and America's most important crop (Awika 2011), was initially sequenced using a BAC-by-BAC approach. However, the BAC-by-BAC sequence was improved by addition of shotgun sequence data and reduced-representation sequencing[12] data (Schnable et al. 2009).

[12]See Peterson (2005) for review of reduced-representation techniques.

5.3 Shotgun Sequencing with Aspects of BAC-Based Physical Mapping

The next set of pre-NGS plant genomes was assembled using automated Sanger paired-end sequencing of short-insert clones, midsized insert clones (fosmids), and large-insert clones (BACs, PACs, etc.). End sequencing of the clones was primarily shotgun sequencing. Assemblies generated from shotgun sequence reads were verified and corrected using physical maps, physical mapping techniques (e.g., BAC-end sequencing and BAC fingerprinting), and/or genetic recombination map data. Reference-quality plant genomes sequenced using this strategy include grape (French-Italian Public Consortium for Grapevine Genome Characterization 2007), poplar (Tuskan et al. 2006), sorghum (Paterson et al. 2009), *Brachypodium distachyon* (International Brachypodium Initiative 2010), *Medicago truncatula* (Young et al. 2011), and the four chlorophyte algae species (Palenik et al. 2007; Worden et al. 2009; Moreau et al. 2012; Derelle et al. 2006) in Table 3.

5.4 Second Generation + Sanger

When NGS came onto the scene, it was widely, but cautiously, embraced. However, while many groups started to include NGS in their genome sequencing strategies (primarily as shotgun sequence), there was still a fear of abandoning the "gold standard" that was automated Sanger sequencing. A combination of Sanger sequencing, 454 and/or Illumina sequencing, and physical mapping was utilized in sequencing the genomes of bladderwort (Ibarra-Laclette et al. 2013), gray rockcress (Willing et al. 2015), D-genome cotton (Paterson et al. 2012), cacao (Motamayor et al. 2013), tomato (Tomato Genome Consortium 2012), upland cotton (Li et al. 2015a; Zhang et al. 2015a), *Solanum pennellii* (Bolger et al. 2014a), carrot, cucumber (Woycicki et al. 2011), green bean (Vlasova et al. 2016), beet (Dohm et al. 2014), pigeon pea (Varshney et al. 2011), greater duckweed (Wang et al. 2014b), apple (Velasco et al. 2010), Malaysian banana (D'Hont et al. 2012), and African oil palm (Singh et al. 2013). Addition of physical mapping tools could often substantially increase the quality of an assembly. For example, in tomato the initial ordering of scaffolds was based on molecular genetic mapping data (Tomato Genome Consortium 2012). However, Shearer et al. (2014), working as part of the consortium, used FISH and optical mapping to show that 45 of the 91 scaffolds (representing one-third of the tomato genome) were positioned incorrectly. Even modest investment in physical mapping can contribute a great deal to the quality of a genome sequence.

Table 3 Highest-quality publicly available plant genomes[a]

Species • 1C DNA content[b] • common name • chromosome number[b] • angiosperm group (MC = monocot; DC = dicot) if applicable) • family • sequencing strategy[c]	Total assembly length (Kb)	Total assembly gap length (Kb)	Scaffold number	Scaffold N50 (Kb)	Contig number	Contig N50 (Kb)	Gene number	NCBI assembly accession • genome paper reference
Chlorophyte algae								
Ostreococcus lucimarinus • 13.2 Mb[d] • $n = 21$[d] • Bathycoccaceae • S	13,205	0	21	Genome complete			7640[e]	GCF_000092065.1 • Palenik et al. (2007)
Micromonas commoda • 20.9 Mb[f] • $n = 17$[f] • Mamiellaceae • S	21,109	0	17	Genome complete			10,014[e]	GCA_000090985.2 • Worden et al. (2009)
Bathycoccus prasinos • 15 Mb[g] • $n = 19$[f,g] • Bathycoccaceae • S	15,074	36	19	956	41	663	7828[e]	GCF_002220235.1 • Moreau et al. (2012)
Ostreococcus tauri • 13 Mb[h] • $n = 20$[h] • Bathycoccaceae • S	12,917	10	20	770	497	15	7942[e]	GCF_000214015.2 • Derelle et al. (2006)
Angiosperms								
Arabidopsis thaliana • 157 Mb • thale cress • $n = 5$ • DC • *Brassicaceae* • B	119,668	186	7	23,460	102	11,195	37,898[e]	GCF_000001735.3 • Arabidopsis Genome Initiative (2000)
Oryza sativa (japonica) • 489 Mb • Japanese rice • $n = 12$ • MC • Poaceae • B	374,423	117	58	29,958	305	7,711	32,945[e]	GCF_001433935.1 • International Rice Genome Sequencing Project (2005)
Brachypodium distachyon • 355 Mb • stiff brome • $n = 5$ • MC • Poaceae • S	272,133	607	28	59,385	486	6,404	28,924[e]	GCF_000005505.2 • International Brachypodium Initiative (2010)
Schrenkiella parvula • 180 Mb[i] • saltwater cress $n = 7$[j] • DC • *Brassicaceae* • 4I	137,073	3	1,463	9,613	1,490	5,289	30,419[k]	GCA_000218505.1 • Dassanayake et al. (2011)

(continued)

Table 3 (continued)

Species • 1C DNA content[b] • common name • chromosome number[b] • angiosperm group (MC = monocot; DC = dicot) if applicable) • family • sequencing strategy[c]	Total assembly length (Kb)	Total assembly gap length (Kb)	Scaffold number	Scaffold N50 (Kb)	Contig number	Contig N50 (Kb)	Gene number	NCBI assembly accession • genome paper reference
Utricularia gibba • 88 Mb • bladderwort • $n = 14$[l] • DC • Lentibulariaceae • SI4	100,689	Unknown	518	3,446	518	3,446	29,666[m]	GCA_002189035.1 • Ibarra-Laclette et al. (2013)
Citrus maxima • 383 Mb[n] • pummelo • $n = 9$[l] • DC • Rutaceae • PI	345,757	885	1,612	4,211	2,139	2,916	30,123[k]	GCA_002006925.1 • Wang et al. (2017)
***Sorghum bicolor* • 734 Mb • sorghum • $n = 10$ • MC • Poaceae • S**	**709,345**	**33,371**	**869**	**68,658**	**4,775**	**1,310**	**32,716**[e]	**GCF_000003195.3 • Paterson et al.** (2009)
Zea mays • 2,665 Mb • maize • $n = 10$ • MC • Poaceae • BS	2,135,083	30,733	598	10,525	2,789	1,280	47,429[e]	GCF_000005005.2 • Schnable et al. (2009)
Arabis alpina • 372 Mb • gray rockcress • $n = 8$ • DC • *Brassicaceae* • I4S	311,642	10,865	8	36,598	1,637	895	23,265[o]	GCA_900128785.1 • Willing et al. (2015)
Helianthus annuus • 2,372 Mb •common sunflower • $n = 17$ • DC • Compositae • PI	3,027,845	102,549	1,528	178,899	13,954	414	81,134[e]	GCF_002127325.1 • Badouin et al. (2017)
Medicago truncatula • 465 Mb • barrel medic • $n = 8$ • DC • Leguminosae • S	412,924	22,825	2,187	49,172	8,027	339	51,516[e]	GCF_000219495.3 • Young et al. (2011)
Prunus persica • 269 Mb • peach • $n = 8$ • DC • Rosaceae • S	227,569	2,772	192	27,368	2,531	255	26,275[e]	GCF_000346465.2 • International Peach Genome Initiative et al. (2013)
Populus trichocarpa • 484 Mb • black cottonwood • $n = 19$ • DC • Salicaceae • S	417,287	13,387	2,514	18,836	13,561	242	45,555[k]	GCF_000002775.3 • Tuskan et al. (2006)
Glycine max • 1,100 Mb • soybean • $n = 20$ • DC • Leguminosae • S	978,972	23,047	1,191	48,578	17,146	183	58,605[e]	GCF_000004515.4 • Schmutz et al. (2010)

Brassica juncea • 1,491 Mb • brown mustard • $n = 18$ • DC • *Brassicaceae* • I	954,861	194,274	9,746	38,841	18,588	180	80,050[k]	GCA_001687265.1 • Yang et al. (2016)
Setaria italica • 513 Mb • foxtail millet • $n = 9$ • MC • Poaceae • S	405,868	4,824	337	47,253	6,779	126	30,963[e]	GCF_000263155.2 • Bennetzen et al. (2012)
***Gossypium raimondii* • 880 Mb • D5 cotton • $n = 13$ • DC • Malvaceae • S4I**	**761,565**	**13,440**	**1,034**	**62,175**	**20,751**	**120**	**44,550**[e]	**GCF_000327365.1 • Paterson et al.** (2012) **and Wang et al.** (2012a)
Ananas comosus • 538 Mb • pine-apple • $n = 25$[l] • DC • Bromeliaceae • I4P	382,056	6,782	3,129	11,759	9,391	114	25,312[e]	GCF_001540865.1 • Ming et al. (2015)
Vitis vinifera • 416 Mb • wine grape • $n = 19$ • DC • Vitaceae • S	486,197	15,055	2,061	3,426	14,634	103	26,638[e]	GCF_000003745.3 • French-Italian Public Consortium for Grapevine Genome Characterization (2007)
Theobroma cacao • 416 Mb • cacao • $n = 10$ • DC • Malvaceae • S4	324,880	18,500	431	36,364	7,743	87	24,792[e]	GCF_000208745.1 • Motamayor et al. (2013)
Solanum lycopersicum • 1,002 Mb • tomato • $n = 12$ • DC • Solanaceae • S4M	823,786	85,995	3,224	16,468	26,880	87	30,194[e]	GCF_000188115.3 • Tomato Genome Consortium (2012)
Panicum hallii • 550 Mb[p] • Hall's panicgrass • $n = 9$[l] • MC • Poaceae • I	553,845	61,412	8,756	1,484	30,116	83	37,205[e]	GCA_002211085.1
***Gossypium hirsutum* • 2,347 Mb • upland cotton • $n = 26$ • Malvaceae • IS**	**2,189,140**	**62,280**	**9,148**	**70,912**	**57,884**	**78**	**77,846**[e]	**GCF_000987745.1 • Li et al.** (2015a) **and Zhang et al.** (2015a)
Capsicum annuum • 3,090 Mb • pepper • $n = 12$ • DC • Solanaceae • I	2,935,884	94,871	6,478	1,483	108,799	75	41,148[e]	GCF_000710875.1

(continued)

Table 3 (continued)

Species • 1C DNA content[b] • common name • chromosome number[b] • angiosperm group (MC = monocot; DC = dicot) if applicable) • family • sequencing strategy[c]	Total assembly length (Kb)	Total assembly gap length (Kb)	Scaffold number	Scaffold N50 (Kb)	Contig number	Contig N50 (Kb)	Gene number	NCBI assembly accession • genome paper reference
Gossypium arboreum • 1,677 Mb • tree cotton • $n = 13$ • DC • Malvaceae • I	1,694,598	133,606	75,419	121,339	121,185	72	39,964[e]	GCF_000612285.1
Nicotiana attenuata • 2,428 Mb • coyote tobacco • $n = 12$ • DC • Solanaceae • I4P	2,365,683	275,105	37,194	524	103,240	64	39,470[e]	GCF_001879085.1
Solanum pennellii • 1,198 Mb • wild tomato • $n = 12$ • DC • Solanaceae • IS	926,426	71,063	12	77,991	30,753	60	32,519[e]	GCF_001406875.1 • Bolger et al. (2014a)
Triticum turgidum • 12,005 Mb • emmer wheat • $n = 14$ • MC • Poaceae • IH	10,494,955	171,455	149,145	726,428	493,558	57	62,813[k]	GCA_900184675.1 • Avni et al. (2017)
Daucus carota • 978 Mb • carrot • $n = 9$ • DC • Apiaceae • I4S	421,539	34,733	4,826	36,610	21,054	56	36,045[e]	GCF_001625215.1
Lupinus angustiflorus • 905 Mb • narrow-leaved lupin • $n = 20$ • DC • Leguminosae • I	557,909	4,350	20	30,409	41,861	53		GCA_002285895.1 • Yang et al. (2013)
Sesamum indicum • 949 Mb • sesame • $n = 13$ • DC • Pedaliaceae • I	275,059	4,573	16,369	2,060	26,176	52	26,417[e]	GCF_000512975.1 • Wang et al. (2014a)
Leersia perrieri • 323 Mb[q] • cut-grass • $n = 12$[r] • MC • Poaceae • I	266,688	993	12	22,540	9,937	50		GCA_000325765.3
Citrus sinensis • 611 Mb • sweet orange • $n = 9$ • DC • Rutaceae • I	327,830	26,598	4,995	1,779	17,383	50	28,267[e]	GCF_000317415.1 • Xu et al. (2013)

Vigna radiata • 513 Mb • mung bean • $n = 11$ • DC • Leguminosae • I4	463,638	33,470	2,499	25,361	23,501	49	28,917[e]	GCF_000741045.1 • Kang et al. (2014)
Lupinus angustifolius • 905 Mb • narrow-leaved lupin • $n = 20$ • DC • Leguminosae • I	609,203	4,153	14,378	703	55,107	48	38,451[e]	GCF_001865875.1 • Hane et al. (2017)
Asparagus officinalis • 1,320 Mb • asparagus • $n = 10$ • MC • Asparagaceae • I	1,187,538	35,521	11,792	131,340	56,182	47	32,073[e]	GCF_001876935.1
Brassica rapa • 782 Mb • field mustard • $n = 10$ • DC • *Brassicaceae* • I	284,129	10,904	40,432	1,960	51,659	46	46,932[e]	GCF_000309985.1 • Wang et al. (2011)
Oryza punctata • 538 Mb • red rice (BB) • $n = 12$ • MC • Poaceae • I4	393,817	1,659	12	31,245	16,598	43		GCA_000573905.1
Cucumis sativus • 880 Mb • cucumber • $n = 7$ • DC • Cucurbitaceae • S4	195,669	3,493	190	29,076	11,366	42	17,320[e]	GCF_000004075.2 • Woycicki et al. (2011)
Brassica napus • 1,125 Mb • rape-seed • $n = 19$ • DC • *Brassicaceae* • I	930,508	72,627	1,471	41,855	34,110	40	114,227[e]	GCF_000686985.1
Oryza longistaminata • 734 Mb • red rice (AA) • $n = 12$ • MC • Poaceae • I4	362,064	20,310	11,745	364	26,322	39	32,502[k]	GCA_001514335.2 • Zhang et al. (2015b)
Phaseolus vulgaris • 587 Mb • green bean • $n = 11$ • DC • Leguminosae • I4SL	521,077	48,623	708	50,367	41,568	39	32,720[e]	GCF_000499845.1 • Vlasova et al. (2016)
Vigna angularis • 538 Mb • adzuki bean • $n = 11$ • DC • Leguminosae • I	467,301	16,985	37,727	1,292	58,639	38	29,297[e]	GCF_001190045.1 • Yang et al. (2015)

(continued)

Table 3 (continued)

Species • 1C DNA content[b] • common name • chromosome number[b] • angiosperm group (MC = monocot; DC = dicot) if applicable) • family • sequencing strategy[c]	Total assembly length (Kb)	Total assembly gap length (Kb)	Scaffold number	Scaffold N50 (Kb)	Contig number	Contig N50 (Kb)	Gene number	NCBI assembly accession • genome paper reference
Brassica nigra • 758 Mb • black mustard • $n = 8$ • DC • *Brassicaceae* • I	402,145	47,527	2,545	39,062	21,385	38		GCA_001682895.1
Oryza nivara • 455 Mb • Indian wild rice • $n = 12$ • MC • Poaceae • I	337,950	27,953	12	28,646	16,484	38		GCA_000576065.1
Cicer arietinum • 929 Mb • chick-pea • $n = 8$ • DC • Leguminosae • I	530,894	48,145	7,545	698	39,024	34	27,650[e]	GCF_000331145.1 • Varshney et al. (2013)
Ziziphus jujuba • 444 Mb[s] • common jujube • $n = 12^s$ • DC • Rhamnaceae • I	437,754	20,440	5,897	301	28,947	34	33,079[e]	GCF_000826755.1 • Liu et al. (2014)
Camelina sativa • 750 Mb[t] • false flax • $n = 20^t$ • DC • *Brassicaceae* • I4	641,356	45,010	37,212	30,100	72,248	33	97,832[e]	GCF_000633955.1 • Kagale et al. (2014)
Prunus mume • 280 Mb[s] • Japanese apricot • $n = 8^s$ • DC • Rosaceae • I	234,030	16,851	8,626	637	23,761	32	23,553[e]	GCF_000346735.1 • Zhang et al. (2012)
Oryza glumipatula • 489 Mb • common wild rice • $n = 12$ • MC • Poaceae • I	372,860	71,585	12	31,548	17,912	32		GCA_000576495.1
Musa acuminata • 611 Mb • Malaysian banana • $n = 11$ • MC • Musaceae • S4	472,231	81,754	7,512	1,311	29,249	28	34,100[e]	GCF_000313855.2 • D'Hont et al. (2012)

Fragaria vesca • 241 Mb • wild strawberry • $n = 7$ • DC • Rosaceae • I4L	214,373	12,335	3,048	27,880	16,479	28	26,417[e]	GCF_000184155.1 • Shulaev et al. (2011)
Manihot esculenta • 807 Mb • cassava • $n = 36$ • DC • Euphorbiaceae • I	582,279	86,649	2,020	28,119	39,916	27	31,750[e]	GCF_001659605.1
Beta vulgaris • 1,223 Mb • beet • $n = 9$ • DC • Amaranthaceae • I4S	566,550	48,761	40,406	2,013	71,208	27	27,258[e]	GCF_000511025.2 • Dohm et al. (2014)
Arachis duranensis • 1,240 Mb • A-genome peanut • $n = 10$ • DC • Leguminosae • I	1,084,261	144,229	3,189	1,068	130,694	25	45,035[e]	GCF_000817695.2 • Bertioli et al. (2016)
Arachis ipaensis • 2,763 Mb • B-genome peanut • $n = 10$ • DC • Leguminosae • I	1,353,504	96,430	997	6,170	112,021	24	49,769[e]	GCF_000816755.2 • Bertioli et al. (2016)
Oryza sativa (indica) • 489 Mb • long-grained rice • $n = 12$ • MC • Poaceae • S	426,337	15,662	10,627	7,196	50,234	23	46,022–55,615[u]	GCA_000004655.2 • Yu et al. (2002)
Cajanus cajan • 1,839 Mb • pigeon pea • $n = 16$ • DC • Leguminosae • IS	592,971	34,435	36,536	556	72,923	22	31,424[e]	GCF_000340665.1 • Varshney et al. (2011)
Dioscorea rotundata • 611 Mb • Guinea yam • $n = 20$ • MC • Dioscoreaceae • I	456,675	50,569	21	25,273	34,429	22		GCA_002240015.2
Brassica oleracea • 758 Mb • wild cabbage • $n = 9o$ • DC • *Brassicaceae* • I4	488,954	43,000	32,886	48,367	83,956	22	48,927[e]	GCF_000695525.1
Oryza brachyantha • 342 Mb • malo sina • $n = 12$ • MC • Poaceae • I	259,908	17,432	2,491	21,479	23,184	20	24,793[e]	GCF_000231095.1 • Chen et al. (2013a)

(continued)

Table 3 (continued)

Species • 1C DNA content[b] • common name • chromosome number[b] • angiosperm group (MC = monocot; DC = dicot) if applicable) • family • sequencing strategy[c]	Total assembly length (Kb)	Total assembly gap length (Kb)	Scaffold number	Scaffold N50 (Kb)	Contig number	Contig N50 (Kb)	Gene number	NCBI assembly accession • genome paper reference
Oryza barthii • 611 Mb • African wild rice • *n* = 12 • MC • Poaceae • I4	308,272	2,540	12	25,712	25,427	19		GCA_000182155.3
Pennisetum glaucum • 2,616 Mb • pearl millet • *n* = 7 • MC • Poaceae • I	1,794,547	237,740	20,389	240,577	176,257	18	38,579[k]	GCA_002174835.1 • Varshney et al. (2017)
Spirodela polyrhiza • 291 Mb • greater duckweed • *n* = 20[v] • MC • Araceae • S4	136,670	9,920	20	7,641	11,823	18	19,623[k]	GCA_001981405.1 • Wang et al. (2014b)
Malus domestica • 792 Mb • apple • *n* = 17[s] • DC • Rosaceae • S4	1,874,765	1,176,878	1,667	1,894	106,204	12	58,044[e]	GCF_000148765.1 • Velasco et al. (2010)
Elaeis guineensis • 1,839 Mb • African oil palm • *n* = 16 • DC • Arecaceae • S4	1,535,178	478,071	40,349	1,045	202,468	9	30,134[e]	GCF_000442705.1 • Singh et al. (2013)
Oryza meridionalis • 445 Mb • common wild rice • *n* = 12 • MC • Poaceae • I	335,668	80,024	12	30,391	62,778	9		GCA_000338895.2

Trifolium pratense • 418 Mb • red clover • $n = 7$ • DC • Leguminosae • I	345,991	36,978	39,051	22,683	135,023	5	47,398[k]	GCA_900079335.1 • Ištvánek et al. (2014)

Co-author Peterson was involved in sequencing those species in bold text

[a]Genome sequence data was obtained from the National Center for Biotechnology Information (NCBI) Assembly Database (https://www.ncbi.nlm.nih.gov/assembly/) on October 10, 2017. On the Assembly home page, the *Browse by Organism* link was selected. In the search bar at the top of the page, the delimiter for plants – plants (taxid: 3193) – was entered. The plant genome assemblies were sorted using the Assembly Level toggle button. Those plants with an assembly level of "complete genome" or "chromosome" and a RefSeq category of "reference genome" or "representative genome" were included in the table. Note that, at the time of this writing, no charophyte green algae, bryophytes, pteridophytes, lycopods, or gymnosperms have genome assemblies of sufficient quality to be included in the table

[b]All genome size and chromosome number values are from Bennett and Leitch (2012a) unless indicated otherwise

[c]Sequencing strategies are *B* BAC-by-BAC (Sanger), *S* Sanger shotgun, *4* 454, *I* Illumina, *P* PacBio, *L* SOLiD, *H* Hi-C

[d]Palenik et al. (2007)

[e]Calculated from NCBI Genome pages. Gene values include nuclear genes and genes unassigned to chromosomes, plastids, or mitochondria

[f]Worden et al. (2009)

[g]Moreau et al. (2012)

[h]Blanc-Mathieu et al. (2014)

[i]Oh et al. (2010)

[j]Dassanayake et al. (2011)

[k]Gene number obtained from genome reference paper in far right column

[l]Rice et al. (2015)

[m]Gene number from https://genomevolution.org/CoGe/GenomeInfo.pl?gid=29027

[n]Gmitter et al. (2012)

[o]Gene number from http://www.uniprot.org/proteomes/UP000029120

[p]Lowry et al. (2013)

[q]Copetti et al. (2015)

[r]Katayama (1995)

[s]Chagné (2015)

[t]Hutcheon et al. (2010)

[u]Gene number from https://www.ebi.ac.uk/ena/data/view/GCA_000004655.2

[v]Wang et al. (2014b)

5.5 Second Generation Alone

As confidence in NGS techniques increased, more genomes were assembled solely from second-generation reads. Different insert size libraries and mate-pair sequencing were essential in assembly of these genomes. Examples of genome assembled using second-generation sequencing alone include saltwater cress (Dassanayake et al. 2011), brown mustard (Yang et al. 2016), Hall's panicgrass, pepper, A-genome cotton, narrow-leaved lupin (Yang et al. 2013; Hane et al. 2017), sesame (Wang et al. 2014a), cut-grass, sweet orange (Xu et al. 2013), mung bean (Kang et al. 2014), asparagus, field mustard (Wang et al. 2011), *Oryza punctate*, *O. longistaminata* (Zhang et al. 2015b), *O. glumipatula*, *O. brachyantha* (Chen et al. 2013a), *O. meridionalis*, rapeseed, adzuki bean (Yang et al. 2015), black mustard, Indian wild rice, chickpea (Varshney et al. 2013), common jujube (Liu et al. 2014), false flax (Kagale et al. 2014), Japanese apricot (Zhang et al. 2012), wild strawberry (Shulaev et al. 2011), cassava, A- and B-genome peanuts (Bertioli et al. 2016), Guinea yam, African wild rice, pearl millet (Varshney et al. 2017), and red clover (Ištvánek et al. 2014). Although the preceding genomes have high enough quality to make the list in Table 3, the use of only second-generation sequencing, in general, greatly hobbled the quality of genome assemblies. In brief, the genes and unique regions of genomes could often be assembled fairly easily. However, the short-read lengths prevented assembly of many repetitive regions. Of note, Illumina sequencing of wild emmer wheat was greatly enhanced by the addition of Hi-C chromatin configuration sequencing/analysis (Avni et al. 2017).

5.6 Third- + Second-Generation Sequencing

Third-generation sequencing, while still lacking the accuracy and output of Illumina sequencers, provides relatively long-read lengths that can be used to help link contigs into scaffolds. Alternatively, contigs/scaffolds generated using third-generation reads can be "polished" using Illumina reads to correct mistakes and fill in gaps. Genomes that have been sequenced and assembled using a combination of second- and third-generation platforms include pummelo (Wang et al. 2017), sunflower (Badouin et al. 2017), pineapple (Ming et al. 2015), and coyote tobacco. Moreover, several published genomes assembled using first- and second-generation technologies have later been improved by incorporating third-generation long-read sequencing data (e.g., Jiao et al. 2017; Zimin et al. 2017).

5.7 *The Present*

At present, there are many strategies that one could employ to perform de novo genome sequencing of a plant genome. Currently, we would recommend a combination of Illumina and third-generation sequencing along with Hi-C and/or optical mapping support. At the Institute for Genomics, Biocomputing & Biotechnology, we have employed the following steps to assemble a number of plant, vertebrate, and insect genomes:

- Sequence 30-genome equivalent (30X) coverage using a third-generation sequencer. We have utilized both PacBio Sequel and the Oxford Nanopore MinIon/GridIon instruments (the former through contracts and the latter "in-house").
- Assemble the sequence reads into contigs and scaffolds using a Minimap2-Minasm-Racon (see Li 2016; Vaser et al. 2017) pipeline.
- Acquire 100X coverage of the genome using a high-throughput Illumina instrument. At present we are using a NovaSeq 6000, a HiSeq X Ten, or a HiSeq 4000.
- Use Illumina reads to *polish* the third-generation contigs/scaffolds using Pilon (Walker et al. 2014).
- Use Hi-C to improve quality and length of contigs and scaffolds using the 3D-DNA program (Dudchenko et al. 2017).
- Perform structural annotation of the genome with MAKER2 (Holt and Yandell 2011) using evidence from GeneMark, Augustus, BUSCO, Snap, BLAST+, and Exonerate (Isono et al. 1994; Stanke et al. 2004; Simao et al. 2015; Zaharia et al. 2011; Camacho et al. 2009; Slater and Birney 2005).
- Conduct functional annotation of the genome with InterProScan and BLAST+ (Zdobnov and Apweiler 2001; Camacho et al. 2009).

6 The Best Plant Genome Sequences

6.1 *A Different Perspective*

In 2013, Michael and Jackson published a paper entitled "The 50 first plant genomes." As its title suggests, this excellent review focused on the 50 first plant genome publications. The relative quality of the 50 first genomes varies enormously (e.g., the listed contig N50 for *Brachypodium distachyon* is 348 kb, while the contig N50 for *Azadirachta indica* is 1 kb).

To provide a different perspective, we have chosen to focus not on chronological publication/release dates but on the highest-quality plant genome assemblies currently available. Some of these genomes have not been described in publication form, but all are available through the National Center for Biotechnology Information (NCBI) Assembly Database (https://www.ncbi.nlm.nih.gov/assembly/). Table 3 provides a list of those plant genomes that meet the criteria for reference-quality assemblies. As of October 10, 2017, there are 72 plant reference-quality genome assemblies in NCBI.

6.2 The Myth of Plant Genome Sequences

On more or less a biweekly basis, another plant genome paper is published. Many of these papers have the phrase "whole genome sequence" in their titles which has a somewhat misleading effect of suggesting that the entire genome of each of these plants is complete. How many plant genomes have been completely sequenced and assembled without mistakes or holes? The answer, like the majority of biological questions, is probably unknowable. With this said, there are, as of this writing, two plant genomes that appear to be finished – specifically, the chlorophyte green algae species *Ostreococcus lucimarinus* and *Micromonas commode* (Palenik et al. 2007; Worden et al. 2009). The tiny genomes of these algae (13.2 and 29 Mb, respectively) and their relatively low levels of repeat sequences allowed production of ostensibly complete genome assemblies. However, all other plant genomes currently listed in NCBI Genome possess gaps between scaffolds and regions that have been recalcitrant to chromosome/linkage group assignment (Table 3). These gaps are due to problems associated with assembling segmental duplications and highly similar repeat sequences. Moreover, centromeres are missing from all land plant genomes, with the partial exception of rice (Yan and Jiang 2007), due to their high quantities of repeats. This is unfortunate as some centromeric regions have been shown to contain important genes (Wu et al. 2011; Fan et al. 2011; Shahinnia et al. 2012; Chen et al. 2013b; Yan and Jiang 2007). Regardless, even the best land plant genome sequences contain gaps; as mentioned above, even the *Arabidopsis thaliana* genome appears to be only 76% complete.

6.3 The Un- and Underrepresented

Examination of the species listed in Table 3 reveals that all the plant species with reference-quality genome sequences are angiosperms or chlorophyte algae. The angiosperms are the most species-rich plant clade and are the source of most human calories. Consequently, the high representation of angiosperms (especially those species of agronomic value) in Table 3 is not surprising. The chlorophyte algae have relatively small genomes and have been touted as potential tools in bioenergy production which explains their representation in the table. Some of the groups that do not have a reference-quality genome representative include:

- *Charophyte* algae – Land plants (embryophytes) emerged as a subgroup of the *Charophyte* algae. Thus *Charophyte* genomes represent a logical outgroup for exploring the early evolution of land plants (de Vries and Archibald 2018). To

date, the only *Charophyte* draft genome available through NCBI is that of *Klebsormidium nitens* (formerly *K. flaccidum*) (Hori et al. 2014).
- Liverworts – A draft genome for a single liverwort species (*Marchantia polymorpha*) is in NCBI Genome.
- Hornworts – No draft sequences are available.
- Lycophytes (club mosses, firmosses, quillworts, scale trees, and spikemosses) – The lycophytes are represented by two draft genomes, both representing the genus *Selaginella*. The *S. moellendorffii* genome is in fairly good shape (Banks et al. 2011), while the *S. kraussiana* genome is still quite fragmented (Ge et al. 2016).
- Monilophytes (ferns and horsetails) – While fern genomes are being explored, there are no fern draft genomes in NCBI (Wolf et al. 2015). Horsetails have enormous genomes (12.6–29.7 Gb; Bennett and Leitch 2012b), and thus it is unlikely that a reference-quality horsetail genome will be available anytime soon.
- Mosses – The first draft of the *Physcomitrella patens* genome was published 10 years ago (Rensing et al. 2008). Since then, the genome has been greatly improved; it now consists of <2,000 scaffolds (Lang et al. 2018). It is the sole moss draft genome in NCBI.
- Gymnosperms (conifers, gnetophytes, cycads, and ginkgo) – While gnetophytes, cycads, and ginkgo have relictual distributions, some coniferous gymnosperms (most notably, pines, spruces, firs, Douglas firs, larches, cedars, and junipers) dominate many of the world's temperate and boreal forest ecosystems (Farjon 2008). Moreover, in a number of regions, the value of conifers as sources of lumber and pulp/paper is enormous (Farjon 2010).

  All gymnosperms have large genomes (mean 1C DNA content = 17.7 Gb) with the conifers having genomes between 8.9 and 35.3 Gb (Murray et al. 2012). Initially, socioeconomic value was the driving force in determining which gymnosperm genomes were subjects of genome sequencing efforts; in other words, the first gymnosperm species were all large genome conifers in the family Pinaceae. To date, draft genome sequences have been generated for *Picea abies* (Nystedt et al. 2013), *Pinus taeda* (Neale et al. 2014; Zimin et al. 2017), and *Picea glauca* (Birol et al. 2013; Warren et al. 2015a) which possess 1C DNA contents of 19.6, 21.6, and 15.8 Gb, respectively (Murray et al. 2012). None of these genomes approach reference genome quality, although they have provided considerable insight into conifer evolution. A genome sequencing project on Douglas fir (*Pseudotsuga menziesii*) is now underway as well (1C = 18.6 Gb; Murray et al. 2012). More recently, draft genomes have been published for *Ginkgo biloba* (1C = 10.61 Gb; Guan et al. 2016) and the gnetophyte *Gnetum montanum* (1C = 4.2 Gb; Wan et al. 2018).

Although Table 3 is chock full of angiosperms and contains a few chlorophyte algae, there is definitely a skew toward certain clades. Three of the four chlorophytes are in the same family; two are in the same genus. Of the >400 angiosperm families (Haston et al. 2009), members of 23 families are represented in Table 3. The families with the highest levels of representation are the grasses (Poaceae), the legumes (Leguminosae or Fabaceae), the crucifers (*Brassicaceae*), the mallows (Malvaceae), the roses (Rosaceae), and the nightshades (Solanaceae) with 17, 12, 9, 4, 4, and

4 representatives, respectively. The most species-rich angiosperm families (Asteraceae or Compositae and Orchidaceae) have 1 and 0 representatives, respectively.

7 What Genome Sequencing Has Revealed About Plant Genome Structure and Function

Sequencing of plant genomes has resulted in some new discoveries about plant evolution and genetics and reinforced a number of previously observed trends.

7.1 Whole Genome Duplications Are Very Common

It has long been known that whole genome duplications (WGDs) are common in angiosperms. However, it wasn't until whole genome sequencing that the extent to which plant genomes (angiosperms and other lineages) have been shaped by ancient polyploidy events began to become apparent.

Intra- and inter-genome comparisons indicate that paleopolyploidy is a feature of the evolutionary histories of all land plants and probably all eukaryotic organisms[13]; even the yeast *Saccharomyces cerevisiae*, with a genome one-fifth the size of the smallest plant genome, is a paleopolyploid (Wong et al. 2002). In retrospect, this is perhaps not that surprising. New genes rarely arise de novo, they are typically derivations of duplicated existing genes, and whole genome duplication is ostensibly one route that plants have used to increase their gene repertoires time and again.

As discussed by Garsmeur et al. (2014), whole genome duplication events typically are followed by preferential loss of most duplicated genes via intrachromosomal recombination. This process, called *fractionation*, can preferentially target one of the subgenomes (biased fractionation) or may be fairly balanced between subgenomes (unbiased fractionation). Fractionation and divergence of gene duplicates (whether they are homologs, homeologs, orthologs, or paralogs) are critical parts of diploidization (Paterson et al. 2004).

The actual number of times a plant lineage has experienced whole genome duplications is not straightforward. Very ancient paleopolyploid events have likely become undetectable. Likewise, allopolyploidy events may not involve parental species with the same ploidy level, a fact that can complicate paleopolyploidy research.

Figure 10 shows WGDs in plants based on DNA sequence data. The figure was last modified on April 3, 2013, so it is missing information from plants that have had their genomes sequenced since then. Recent sequence analysis suggests that

[13]Even the yeast *Saccharomyces cerevisiae*, with a genome one-fifth the size of the smallest plant genome, is a paleopolyploid (Wong et al. 2002).

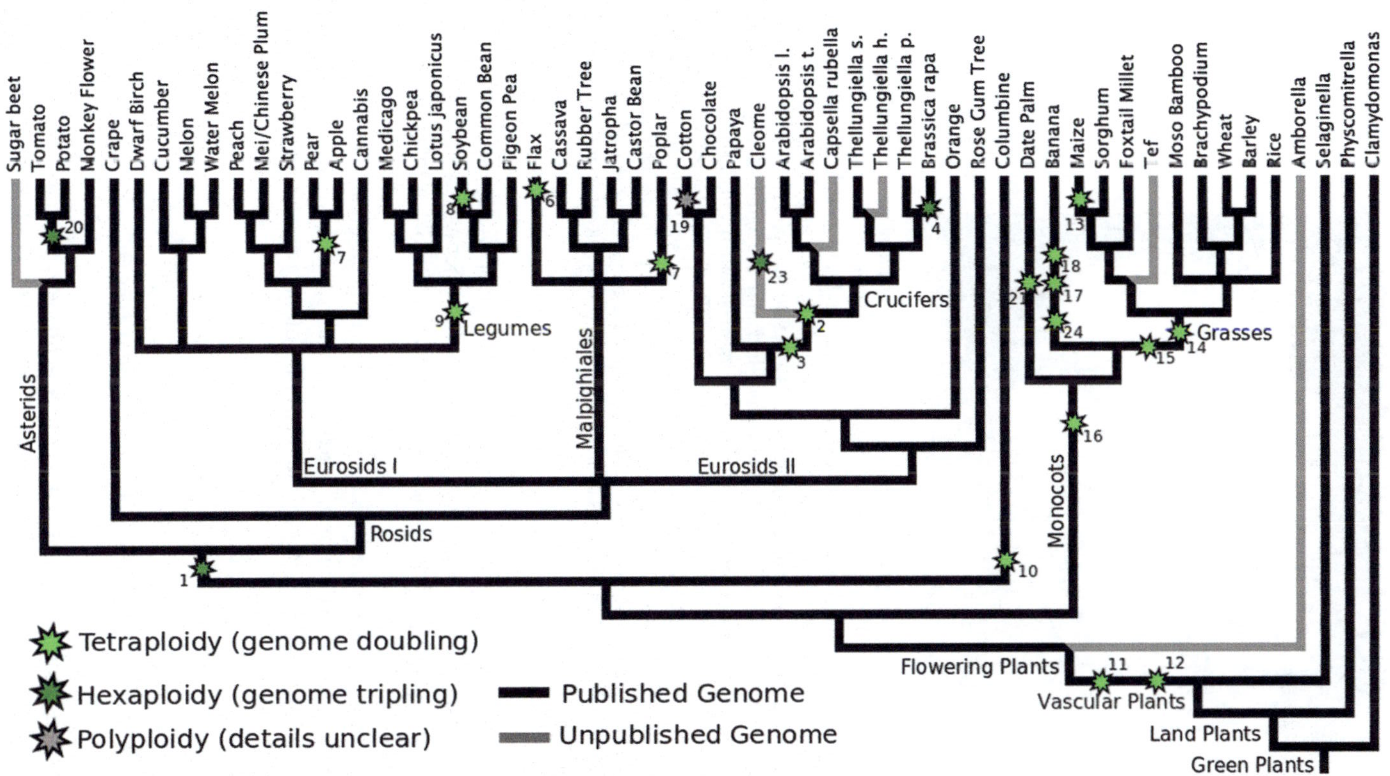

Fig. 10 Ancient whole genome duplications (WGDs) in plants. Polyploidy events are indicated by seven-pointed stars, whereas the colors of the stars indicate the relative increase in ploidy. Note that two WGD events occurred early in the history of the vascular plants (marked by numbers 11 and 12). The first of these WGDs took place before the angiosperm/gymnosperm split (gymnosperms not shown), while the second involves only angiosperms. The ancestor of the eudicots underwent an additional hexaploidy (marked by the number 1) after which many of the eudicot lineages underwent additional WGDs. The monocots share a tetraploidization event (number 16). The figure was generated by James Schnable and Eric Lyons and is freely available for use from https://genomevolution.org/wiki/index.php/Plant_paleopolyploidy through a Creative Commons license (CC-BY-4.0). See the website for detailed notes and references

Fig. 11 Paleopolyploidy in *Gossypium* (cottons). The genus *Gossypium* has a particularly complicated history of WGDs. Production of a reference-quality sequence for *G. raimondii* and comparison of the genome with draft sequences of *G. hirsutum* and several diploid cotton species revealed that all cottons have undergone a fivefold increase in ploidy since their divergence from the paleohexaploid "grape-like" basal eudicot genome (i.e., a net 30-fold ploidy increase from the diploid ancestors of the eudicots). The tetraploid cottons including the agronomically important *G. hirsutum* and *G. barbadense* arose from hybridization between an African diploid *Gossypium* species (similar to, if not, *G. herbaceum*, genotype *AA*) and a New World diploid species (similar to, if not, *G. raimondii*, genotype *DD*) roughly 1–2 million years ago (MYA). The union of the A and D genomes followed by chromosome doubling to produce the tetraploids resulted in a 60-fold increase in ploidy between the tetraploid cottons and the diploid ancestor of the hexaploid eudicots (Paterson et al. 2012). Note that these values do not include the two tetraploidy events early in seed plant and angiosperm history (Fig. 10)

gymnosperms have had several clade-specific WGD events in their histories in addition to the polyploidy event they share with angiosperms (Li et al. 2015b). Figure 11 discusses paleopolyploidy and a relatively recent WGD event in the cotton (*Gossypium*) genus.

There is cytogenetic and some molecular evidence to suggest that polyploidy may be very important in the evolution of green algae, ferns, and bryophytes (Dar and Ruhman 2017). Clearly more genome and transcriptome sequence from these groups would be useful.

7.2 The Trouble with Genes Is

Genes have been traditionally defined as DNA sequences that code for proteins. Whether this definition is appropriate or not is a matter of considerable debate. One complicating factor is that some sequences that appear as if they would be transcribed and translated do not produce detectable transcripts or proteins. Conversely, some sequences that do not have "gene" features are expressed. It would be easy to say that sequences should only be considered genes if there is protein evidence showing that they are translated, but "comprehensive" proteomics data are difficult (perhaps impossible?) and expensive to obtain, and thus many genomes do not have substantive proteomics support. Transcriptome data can also be used to identify which genes are transcribed, but not all transcripts are translated. Additionally, some genes may only be active under very rare circumstances which might make discovery of their corresponding transcripts and/or proteins nearly impossible. Consequently, genes are typically identified using homology-based and ab initio approaches, with most/many reported genes actually being "gene models." Lastly, structural and regulatory RNA molecules perform functions analogous to structural proteins and transcription regulators, so perhaps they should be considered genes.

For the "high-quality" plant genomes shown in Table 3, reported gene numbers range from 7,640 to 114,227, with an average gene number of 37,980.[14] Gymnosperm genome gene estimates, which are not in Table 3 based on the relatively low quality of their genome assemblies, are 16,386 for *Picea glauca*, 27,491 for *Gnetum montanum*, 50,172 for *Pinus taeda*, and 70,968 for *Picea abies* (Warren et al. 2015a; Neale et al. 2014; Wan et al. 2018; Nystedt et al. 2013). These values fit comfortably within the gene number range from Table 3. Of note, the chlorophyte algae have the smallest gene numbers. While the smaller numbers may reflect one of the issues mentioned below, it is likely that the numbers are at least partly reflective of the tremendous amount of evolutionary time separating chlorophytes from the angiosperms.

So what accounts for this variation in gene numbers among the seed plants? Undoubtedly polyploidy and segmental duplication accounts for some of the differences. However, the seemingly wide range in gene counts is most likely due to the fact that identification of genes using computer algorithms is highly dependent upon the quality of the algorithms, the parameters used in gene searching, and the quality and availability of previously annotated "training" sequences (Michael and Jackson 2013). Two highly competent research groups can come up with very different results even with only seemingly modest differences in their approaches. For example, in an analysis of 100 BAC clones from loblolly pine, Wegrzyn et al. (2013) detected only one gene in the BAC clone set using MAKER (Holt and Yandell 2011) informed by Augustus (Stanke et al. 2004) and the eukaryotic orthologous proteins (KOGS) database (Koonin et al. 2004). However, our analysis of the same group of BACs using MAKER informed by Augustus, GeneMark (Isono et al. 1994), FGENESH (Solovyev et al. 2006), and cDNA data resulted in identification of 154 gene models of which 10 had enough ancillary scientific support to be declared "certified genes" (Perera et al. 2018). Clearly, which combinations of algorithms and databases are utilized is not trivial. Moreover, what gene number is

[14]In their review of the 50 first plant genomes, Michael and Jackson (2013) report a range in gene numbers between 20,000 and 94,000 with an average of 32,605.

presented is often based on the relative cautiousness or boldness of the scientists responsible for estimating gene number values. For example, the research group studying *Picea glauca* note in their publications that they have chosen to count only those gene models supported by considerable auxillary evidence (Warren et al. 2015a). Other groups are more liberal in their acceptance of gene models.

7.3 Long Terminal Repeat (LTR) Retrotransposons Account for Much Genome Size Variation

Eukaryotes exhibit a 17-fold difference in transcriptome sizes (Cavalier-Smith 2005). However, variation in 1C DNA content is as great as 64,000-fold (Pellicer et al. 2018). Plant genomes show the greatest variation in genome size among the eukaryotic lineages (Fig. 3) with 2,400-fold variation in 1C DNA content (Pellicer et al. 2018). However, even when taking into account difficulties in estimating plant genes (see preceding paragraph), the predicted gene numbers for the plants in Table 3 show only a 15-fold variation. What accounts for this discrepancy?

While polyploidy and paleopolyploidy have undoubtedly played a key role in the C-value enigma, their role is dwarfed by that of mobile elements (MEs), arguably the greatest driver of plant genome size divergence. Genome sequencing data indicates that, in particular, long terminal repeat (LTR) retrotransposons are the principal mobile elements involved in plant genome size expansions and contractions (El Baidouri and Panaud 2013). Using a *copy-and-paste* transposition mechanism, it is easy to see how LTR retrotransposons could accumulate in genomes. Deletion of LTR retroelements by mechanisms such as illegitimate recombination, transposon-mediated unequal homologous recombination, and deletion-biased double-strand break (DSB) repair (see Schubert and Vu 2016 for review of the latter) represents a key means by which plant genomes become smaller (Pellicer et al. 2018). Based on a study of the genome sequences of eight angiosperm species, El Baidouri and Panaud (2013) note that massive LTR retrotransposon and other ME expansions are not the result of a slow, continuous process. Rather, major ME amplifications occur in amplification bursts. A particular burst is typically followed by the elimination of repeat copies through deletion and recombination. While the plant genomes, large and small, studied by El Baidouri and Panaud have had retrotransposon activity in their recent pasts, only a few have had recent mobile element bursts. The extent to which a particular burst is eliminated from a plant is likely due to evolutionary pressures (or lack thereof) as well as the relative activity of repeat-eliminating mechanisms in that species. The conifers, *Ginkgo biloba*, and the lily genus *Fritillaria* (1C = 30–100 Gb; Kelly et al. 2015) appear to have experienced a number of ME (primarily LTR retrotransposon) bursts followed by long periods in which the mechanisms of repeat removal were not active/efficient. As a result, the repeats have slowly diverged to the point that they are either not recognized as repeats or, when classified by ab initio approaches, are grouped as many different repeat families composed of only or two representatives (Wegrzyn et al. 2013; Perera et al. 2018; Pellicer et al. 2018; Guan et al. 2016). Interestingly, while possessing relatively small genomes, the gnetophyte gymnosperm, *Gnetum montanum* (1C = 4.2 Gb), and *Amborella trichopoda* (1C = 0.87 Gb), the sole surviving member of a sister group to all extant angiosperms (see Fig. 10), also are characterized by ancient, highly diverged repeats (Wan et al. 2018; Amborella Genome Project 2013).

8 Genomes as Maps

What is a map? In general, a map is a simplified representation of something that is much more complex. Its most common purpose is to serve as a tractable guide by which its user can find and/or explore previously studied places while avoiding obstacles/dangers. Maps typically start off quite crude and simple. However, with increased research and improvements in technology, maps become more accurate and informative (Fig. 12). Nonetheless, maps must not be confused with what they represent. A map of Europe is not equivalent to Europe itself; the map provides only a little bit of information about Europe, most notably the relative locations of towns and geographical landmarks. Certainly, more and more information can be added to a map, but a map that contains too much detail as to be difficult to read is as frustrating as a map that is not detailed enough.

Geneticists and molecular biologists have long utilized the word "map." Genetic, molecular, recombination, physical, QTL (quantitative trait loci), cytomolecular, proteogenomic, and optical are just a few of the examples of maps utilized in studying plants and other eukaryotes. Oftentimes, the aforementioned map types are utilized to position DNA sequences along chromosomes.

So is a genome sequence a map or has it moved out of the realm of *mapdom*?

A genome sequence is still very much a map. It provides information on the linear order of nucleotides along chromosomal strands of DNA and, when annotated, shows the locations of landmarks and their potential functions. However, just as a map of Europe represents something many, many orders of magnitude more complex than the map itself, a plant genome sequence also is but a crude glimpse into a plant, one of the amazing products of billions of years of evolution.

There is a very important function of maps that has not yet been mentioned. Maps are used by those seeking adventures, those looking to learn more about features on the map, and those seeking to find novel, exciting things that others have overlooked or not seen. Genome sequence maps are crude sketches of largely unexplored territories. All of these territories are beautiful and wondrous. Some are ancient; some are relatively new. Methods like Hi-C that provide a glimpse into how chromatin/chromosome interacts in living cells will help give a more accurate assessment of how genomes function and evolve. Functional genomics and biochemistry will be needed to understand the roles of coding and noncoding sequences and study their interactions. Integrating the numerous environmental, developmental, and epigenetic signals

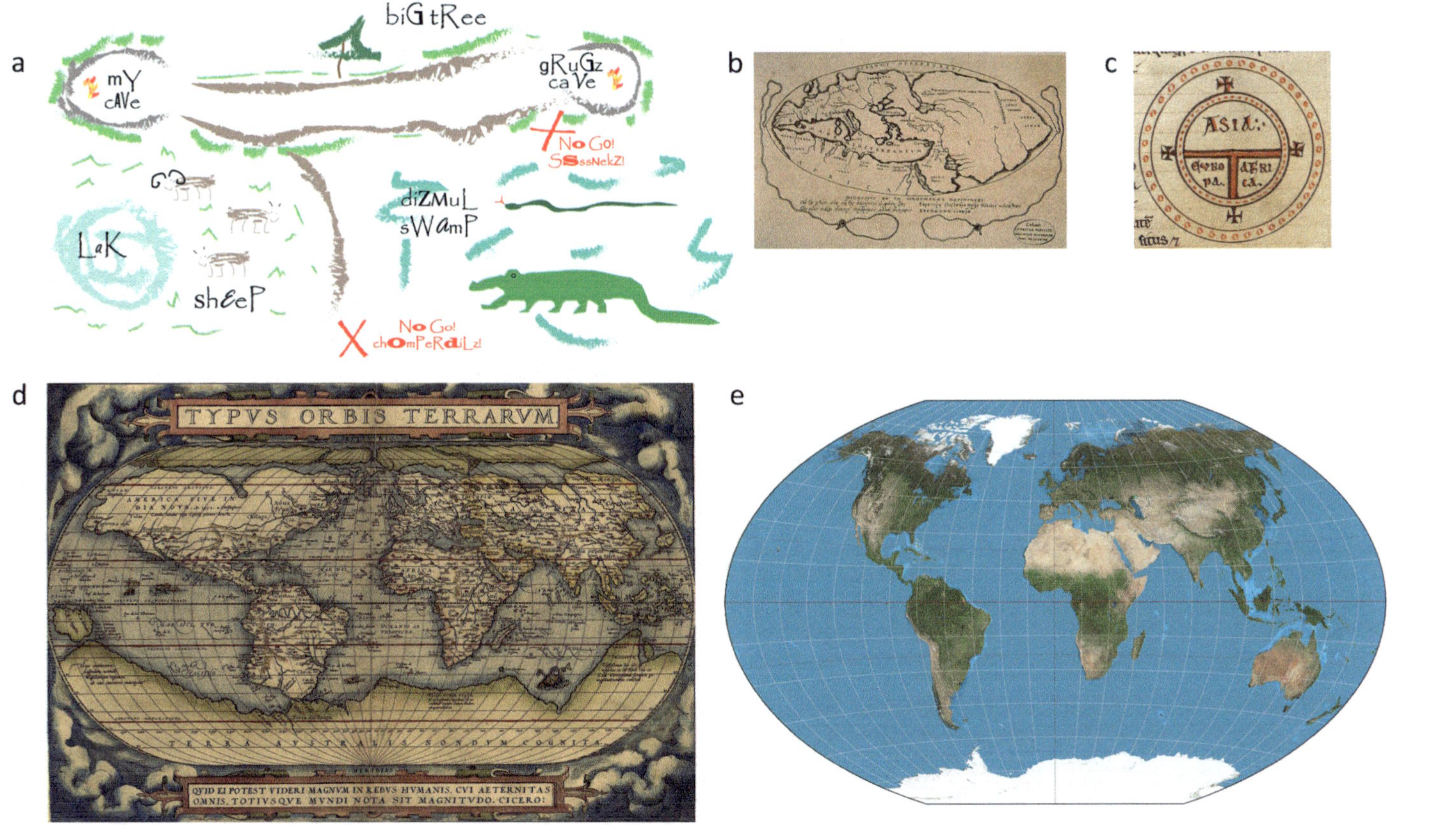

Fig. 12 Maps. (**a**) Ostensibly, the earliest maps helped people keep their bearing and avoid dangerous obstacles. (**b**) World map drawn in 1628 based on Posidonius (150–130 BC; public domain). (**c**) The twelfth-century copy of *Etymologiae* by Isidore of Sevilla (died 636; public domain). (**d**) The world map *Typus Orbis Terrarum* by Ortelius, first published in 1564. Note the significant improvements in coverage. (**e**) A world map on the Winkel tripel projection, a low-error map projection (by Strebe, used according to CC BY-SA 3.0)

that are involved in the production of a cell/organism, especially a multicellular organism with many different tissue types, is also a tall order, but it is part of the frontier made more readily explorable by genome sequence maps.

References

Adl SM, Simpson AG, Farmer MA, Andersen RA, Anderson OR, Barta JR, Bowser SS, Brugerolle G, Fensome RA, Fredericq S, James TY, Karpov S, Kugrens P, Krug J, Lane CE, Lewis LA, Lodge J, Lynn DH, Mann DG, McCourt RM, Mendoza L, Moestrup O, Mozley-Standridge SE, Nerad TA, Shearer CA, Smirnov AV, Spiegel FW, Taylor MF (2005) The new higher level classification of eukaryotes with emphasis on the taxonomy of protists. J Eukaryot Microbiol 52(5):399–451

Altschul SF, Gish W, Miller W, Myers EW, Lipman DJ (1990) Basic local alignment search tool. J Mol Biol 215(3):403–410

Amborella Genome Project (2013) The Amborella genome and the evolution of flowering plants. Science 342(6165):1241089

Arabidopsis Genome Initiative (2000) Analysis of the genome sequence of the flowering plant *Arabidopsis thaliana*. Nature 408(6814):796–815

Aronesty E (2013) Comparison of sequence utility programs. Open Bioinformatics J 7:1–8

Avery OT, Macleod CM, McCarty M (1944) Studies on the chemical nature of the substance inducing transformation of pneumococcal types: induction of transformation by a desoxyribonucleic acid fraction isolated from *Pneumococcus* type III. J Exp Med 79 (2):137–158

Avni R, Nave M, Barad O, Baruch K, Twardziok SO, Gundlach H, Hale I, Mascher M, Spannagl M, Wiebe K, Jordan KW, Golan G, Deek J, Ben-Zvi B, Ben-Zvi G, Himmelbach A, MacLachlan RP, Sharpe AG, Fritz A, Ben-David R, Budak H, Fahima T, Korol A, Faris JD, Hernandez A, Mikel MA, Levy AA, Steffenson B, Maccaferri M, Tuberosa R, Cattivelli L, Faccioli P, Ceriotti A, Kashkush K, Pourkheirandish M, Komatsuda T, Eilam T, Sela H, Sharon A, Ohad N, Chamovitz DA, Mayer KFX, Stein N, Ronen G, Peleg Z, Pozniak CJ, Akhunov ED, Distelfeld A (2017) Wild emmer genome architecture and diversity elucidate wheat evolution and domestication. Science 357 (6346):93–97

Awika JM (2011) Major cereal grains production and use around the world. Am Chem Soc 1089:1–13

Babraham Bioinformatics (2016) Fastqc a quality control tool for high throughput sequence data. v0.11.5. https://www.bioinformatics.babraham.ac.uk/projects/fastqc/. Accessed 8 June 2017

Badouin H, Gouzy J, Grassa CJ, Murat F, Staton SE, Cottret L, Lelandais-Briere C, Owens GL, Carrere S, Mayjonade B, Legrand L, Gill N, Kane NC, Bowers JE, Hubner S, Bellec A, Berard A, Berges H, Blanchet N, Boniface MC, Brunel D, Catrice O, Chaidir N, Claudel C, Donnadieu C, Faraut T, Fievet G, Helmstetter N, King M, Knapp SJ, Lai Z, Le Paslier MC, Lippi Y, Lorenzon L, Mandel JR, Marage G, Marchand G, Marquand E, Bret-Mestries E, Morien E, Nambeesan S, Nguyen T, Pegot-Espagnet P, Pouilly N, Raftis F, Sallet E, Schiex T, Thomas J, Vandecasteele C, Vares D, Vear F, Vautrin S, Crespi M, Mangin B, Burke JM, Salse J, Munos S, Vincourt P, Rieseberg LH, Langlade NB (2017) The sunflower genome provides insights into oil metabolism, flowering and Asterid evolution. Nature 546 (7656):148–152

Banks JA, Nishiyama T, Hasebe M, Bowman JL, Gribskov M, dePamphilis C, Albert VA, Aono N, Aoyama T, Ambrose BA, Ashton NW, Axtell MJ, Barker E, Barker MS, Bennetzen JL, Bonawitz ND, Chapple C, Cheng C, Correa LG, Dacre M, DeBarry J, Dreyer I, Elias M, Engstrom EM, Estelle M, Feng L, Finet C, Floyd SK, Frommer WB, Fujita T, Gramzow L, Gutensohn M, Harholt J, Hattori M, Heyl A, Hirai T, Hiwatashi Y, Ishikawa M, Iwata M, Karol

KG, Koehler B, Kolukisaoglu U, Kubo M, Kurata T, Lalonde S, Li K, Li Y, Litt A, Lyons E, Manning G, Maruyama T, Michael TP, Mikami K, Miyazaki S, Morinaga S, Murata T, Mueller-Roeber B, Nelson DR, Obara M, Oguri Y, Olmstead RG, Onodera N, Petersen BL, Pils B, Prigge M, Rensing SA, Riano-Pachon DM, Roberts AW, Sato Y, Scheller HV, Schulz B, Schulz C, Shakirov EV, Shibagaki N, Shinohara N, Shippen DE, Sorensen I, Sotooka R, Sugimoto N, Sugita M, Sumikawa N, Tanurdzic M, Theissen G, Ulvskov P, Wakazuki S, Weng JK, Willats WW, Wipf D, Wolf PG, Yang L, Zimmer AD, Zhu Q, Mitros T, Hellsten U, Loque D, Otillar R, Salamov A, Schmutz J, Shapiro H, Lindquist E, Lucas S, Rokhsar D, Grigoriev IV (2011) The Selaginella genome identifies genetic changes associated with the evolution of vascular plants. Science 332(6032):960–963

Bao Z, Eddy SR (2002) Automated *de novo* identification of repeat sequence families in sequenced genomes. Genome Res 12(8):1269–1276

Bellot S, Renner SS (2015) The plastomes of two species in the endoparasite genus pilostyles (Apodanthaceae) each retain just five or six possibly functional genes. Genome Biol Evol 8(1):189–201

Belton JM, McCord RP, Gibcus JH, Naumova N, Zhan Y, Dekker J (2012) Hi-C: a comprehensive technique to capture the conformation of genomes. Methods 58(3):268–276

Bennett S (2004) Solexa Ltd. Pharmacogenomics 5(4):433–438

Bennett MD, Leitch IJ (2005) Genome size evolution in plants. In: Gregory TR (ed) The evolution of the genome. Elsevier Academic Press, Boston, pp 89–162

Bennett MD, Leitch IJ (2012a) Plant DNA C-values database (release 6.0, Dec. 2012). http://data.kew.org/cvalues/. Accessed 29 Dec 2017

Bennett MD, Leitch IJ (2012b) Pteridophyte DNA C-values database (release 5.0, Dec. 2012). http://www.kew.org/cvalues/. Accessed 20 Feb 2018

Bennett MD, Leitch IJ, Price HJ, Johnston JS (2003) Comparisons with *Caenorhabditis* (approximately 100 Mb) and *Drosophila* (approximately 175 Mb) using flow cytometry show genome size in *Arabidopsis* to be approximately 157 Mb and thus approximately 25% larger than the Arabidopsis Genome Initiative estimate of approximately 125 Mb. Ann Bot (Lond) 91(5):547–557

Bennetzen JL, Schmutz J, Wang H, Percifield R, Hawkins J, Pontaroli AC, Estep M, Feng L, Vaughn JN, Grimwood J, Jenkins J, Barry K, Lindquist E, Hellsten U, Deshpande S, Wang X, Wu X, Mitros T, Triplett J, Yang X, Ye CY, Mauro-Herrera M, Wang L, Li P, Sharma M, Sharma R, Ronald PC, Panaud O, Kellogg EA, Brutnell TP, Doust AN, Tuskan GA, Rokhsar D, Devos KM (2012) Reference genome sequence of the model plant *Setaria*. Nat Biotechnol 30(6):555–561

Bentley DR, Balasubramanian S, Swerdlow HP, Smith GP, Milton J, Brown CG, Hall KP, Evers DJ, Barnes CL, Bignell HR, Boutell JM, Bryant J, Carter RJ, Keira Cheetham R, Cox AJ, Ellis DJ, Flatbush MR, Gormley NA, Humphray SJ, Irving LJ, Karbelashvili MS, Kirk SM, Li H, Liu X, Maisinger KS, Murray LJ, Obradovic B, Ost T, Parkinson ML, Pratt MR, Rasolonjatovo IM, Reed MT, Rigatti R, Rodighiero C, Ross MT, Sabot A, Sankar SV, Scally A, Schroth GP, Smith ME, Smith VP, Spiridou A, Torrance PE, Tzonev SS, Vermaas EH, Walter K, Wu X, Zhang L, Alam MD, Anastasi C, Aniebo IC, Bailey DM, Bancarz IR, Banerjee S, Barbour SG, Baybayan PA, Benoit VA, Benson KF, Bevis C, Black PJ, Boodhun A, Brennan JS, Bridgham JA, Brown RC, Brown AA, Buermann DH, Bundu AA, Burrows JC, Carter NP, Castillo N, Chiara ECM, Chang S, Neil Cooley R, Crake NR, Dada OO, Diakoumakos KD, Dominguez-Fernandez B, Earnshaw DJ, Egbujor UC, Elmore DW, Etchin SS, Ewan MR, Fedurco M, Fraser LJ, Fuentes Fajardo KV, Scott Furey W, George D, Gietzen KJ, Goddard CP, Golda GS, Granieri PA, Green DE, Gustafson DL, Hansen NF, Harnish K, Haudenschild CD, Heyer NI, Hims MM, Ho JT, Horgan AM, Hoschler K, Hurwitz S, Ivanov DV, Johnson MQ, James T, Huw Jones TA, Kang GD, Kerelska TH, Kersey AD, Khrebtukova I, Kindwall AP, Kingsbury Z, Kokko-Gonzales PI, Kumar A, Laurent MA, Lawley CT, Lee SE, Lee X, Liao AK, Loch JA, Lok M, Luo S, Mammen RM, Martin JW, McCauley PG, McNitt P, Mehta P, Moon KW, Mullens JW, Newington T, Ning Z, Ling Ng B, Novo SM, O'Neill MJ, Osborne MA, Osnowski A, Ostadan O, Paraschos LL, Pickering L, Pike AC, Pike AC, Chris Pinkard D, Pliskin DP, Podhasky J, Quijano VJ, Raczy C, Rae VH,

Rawlings SR, Chiva Rodriguez A, Roe PM, Rogers J, Rogert Bacigalupo MC, Romanov N, Romieu A, Roth RK, Rourke NJ, Ruediger ST, Rusman E, Sanches-Kuiper RM, Schenker MR, Seoane JM, Shaw RJ, Shiver MK, Short SW, Sizto NL, Sluis JP, Smith MA, Ernest Sohna Sohna J, Spence EJ, Stevens K, Sutton N, Szajkowski L, Tregidgo CL, Turcatti G, Vandevondele S, Verhovsky Y, Virk SM, Wakelin S, Walcott GC, Wang J, Worsley GJ, Yan J, Yau L, Zuerlein M, Rogers J, Mullikin JC, Hurles ME, McCooke NJ, West JS, Oaks FL, Lundberg PL, Klenerman D, Durbin R, Smith AJ (2008) Accurate whole human genome sequencing using reversible terminator chemistry. Nature 456(7218):53–59

Bertioli DJ, Cannon SB, Froenicke L, Huang G, Farmer AD, Cannon EK, Liu X, Gao D, Clevenger J, Dash S, Ren L, Moretzsohn MC, Shirasawa K, Huang W, Vidigal B, Abernathy B, Chu Y, Niederhuth CE, Umale P, Araujo AC, Kozik A, Kim KD, Burow MD, Varshney RK, Wang X, Zhang X, Barkley N, Guimaraes PM, Isobe S, Guo B, Liao B, Stalker HT, Schmitz RJ, Scheffler BE, Leal-Bertioli SC, Xun X, Jackson SA, Michelmore R, Ozias-Akins P (2016) The genome sequences of *Arachis duranensis* and *Arachis ipaensis*, the diploid ancestors of cultivated peanut. Nat Genet 48(4):438–446

Bio-IT World Staff (2013) Six years after acquisition, Roche quietly shutters 454. Bio-IT World. http://www.bio-itworld.com/2013/10/16/six-years-after-acquisition-roche-quietly-shutters-454.html

Birol I, Raymond A, Jackman SD, Pleasance S, Coope R, Taylor GA, Yuen MM, Keeling CI, Brand D, Vandervalk BP, Kirk H, Pandoh P, Moore RA, Zhao Y, Mungall AJ, Jaquish B, Yanchuk A, Ritland C, Boyle B, Bousquet J, Ritland K, Mackay J, Bohlmann J, Jones SJ (2013) Assembling the 20 Gb white spruce (Picea glauca) genome from whole-genome shotgun sequencing data. Bioinformatics 29(12):1492–1497

Blanc-Mathieu R, Verhelst B, Derelle E, Rombauts S, Bouget FY, Carre I, Chateau A, Eyre-Walker A, Grimsley N, Moreau H, Piegu B, Rivals E, Schackwitz W, Van de Peer Y, Piganeau G (2014) An improved genome of the model marine alga *Ostreococcus tauri* unfolds by assessing Illumina de novo assemblies. BMC Genomics 15:1103

Boivin A, Vendrely R, Vendrely C (1948) L'acide désoxyribonuclèique du noyau cellulaire dépositaire des caractères héréditaires; arguments d'ordre analytique. C R Acad Sci 226:1061–1063

Bolger A, Scossa F, Bolger ME, Lanz C, Maumus F, Tohge T, Quesneville H, Alseekh S, Sorensen I, Lichtenstein G, Fich EA, Conte M, Keller H, Schneeberger K, Schwacke R, Ofner I, Vrebalov J, Xu Y, Osorio S, Aflitos SA, Schijlen E, Jimenez-Gomez JM, Ryngajllo M, Kimura S, Kumar R, Koenig D, Headland LR, Maloof JN, Sinha N, van Ham RC, Lankhorst RK, Mao L, Vogel A, Arsova B, Panstruga R, Fei Z, Rose JK, Zamir D, Carrari F, Giovannoni JJ, Weigel D, Usadel B, Fernie AR (2014a) The genome of the stress-tolerant wild tomato species *Solanum pennellii*. Nat Genet 46(9):1034–1038

Bolger AM, Lohse M, Usadel B (2014b) Trimmomatic: a flexible trimmer for Illumina sequence data. Bioinformatics 30(15):2114–2120

Bridges SM, Magee GB, Wang N, Williams WP, Burgess SC, Nanduri B (2007) ProtQuant: a tool for the label-free quantification of MudPIT proteomics data. BMC Bioinformatics 8(Suppl 7): S24

Buermans HP, den Dunnen JT (2014) Next generation sequencing technology: advances and applications. Biochim Biophys Acta 1842(10):1932–1941

Butler JM (2015) The future of forensic DNA analysis. Philos Trans R Soc Lond Ser B Biol Sci 370 (1674):20140252

Buza TJ, McCarthy FM, Wang N, Bridges SM, Burgess SC (2008) Gene Ontology annotation quality analysis in model eukaryotes. Nucleic Acids Res 36(2):e12

Camacho C, Coulouris G, Avagyan V, Ma N, Papadopoulos J, Bealer K, Madden TL (2009) BLAST+: architecture and applications. BMC Bioinformatics 10:421

Cantarel BL, Korf I, Robb SM, Parra G, Ross E, Moore B, Holt C, Sanchez Alvarado A, Yandell M (2008) MAKER: an easy-to-use annotation pipeline designed for emerging model organism genomes. Genome Res 18(1):188–196

Cavalier-Smith T (2005) Economy, speed and size matter: evolutionary forces driving nuclear genome miniaturization and expansion. Ann Bot 95(1):147–175

Chagné D (2015) Whole genome sequencing of fruit tree species. Adv Bot Res 74:1–37

Chapman AD (2009) Numbers of living species in Australia and the World, 2nd edn. Australian Government Department of the Environment, Water, Heritage and the Arts, Canberra

Charlesworth B (2006) The evolutionary biology of sex. Curr Biol 16(17):R693–R695

Chen J, Huang Q, Gao D, Wang J, Lang Y, Liu T, Li B, Bai Z, Luis Goicoechea J, Liang C, Chen C, Zhang W, Sun S, Liao Y, Zhang X, Yang L, Song C, Wang M, Shi J, Liu G, Liu J, Zhou H, Zhou W, Yu Q, An N, Chen Y, Cai Q, Wang B, Liu B, Min J, Huang Y, Wu H, Li Z, Zhang Y, Yin Y, Song W, Jiang J, Jackson SA, Wing RA, Wang J, Chen M (2013a) Whole-genome sequencing of *Oryza brachyantha* reveals mechanisms underlying *Oryza* genome evolution. Nat Commun 4:1595

Chen M, Zhao Z, Chen L, Zhou F, Zhong Z, Jiang L, Wan J (2013b) Genetic analysis and fine mapping of a semi-dwarf gene in a centromeric region in rice (Oryza sativa L.) Breed Sci 63(2):164–168

Claros MG, Bautista R, Guerrero-Fernandez D, Benzerki H, Seoane P, Fernandez-Pozo N (2012) Why assembling plant genome sequences is so challenging. Biology 1(2):439–459

Cock PJ, Antao T, Chang JT, Chapman BA, Cox CJ, Dalke A, Friedberg I, Hamelryck T, Kauff F, Wilczynski B, de Hoon MJ (2009) Biopython: freely available Python tools for computational molecular biology and bioinformatics. Bioinformatics 25(11):1422–1423

Conesa A, Madrigal P, Tarazona S, Gomez-Cabrero D, Cervera A, McPherson A, Szczesniak MW, Gaffney DJ, Elo LL, Zhang X, Mortazavi A (2016) A survey of best practices for RNA-seq data analysis. Genome Biol 17:13

Copetti D, Zhang J, El Baidouri M, Gao D, Wang J, Barghini E, Cossu RM, Angelova A, Maldonado LC, Roffler S, Ohyanagi H, Wicker T, Fan C, Zuccolo A, Chen M, Costa de Oliveira A, Han B, Henry R, Hsing YI, Kurata N, Wang W, Jackson SA, Panaud O, Wing RA (2015) RiTE database: a resource database for genus-wide rice genomics and evolutionary biology. BMC Genomics 16:538

Corradi N, Pombert JF, Farinelli L, Didier ES, Keeling PJ (2010) The complete sequence of the smallest known nuclear genome from the microsporidian Encephalitozoon intestinalis. Nat Commun 1:77

D'Hont A, Denoeud F, Aury JM, Baurens FC, Carreel F, Garsmeur O, Noel B, Bocs S, Droc G, Rouard M, Da Silva C, Jabbari K, Cardi C, Poulain J, Souquet M, Labadie K, Jourda C, Lengelle J, Rodier-Goud M, Alberti A, Bernard M, Correa M, Ayyampalayam S, McKain MR, Leebens-Mack J, Burgess D, Freeling M, Mbeguie AMD, Chabannes M, Wicker T, Panaud O, Barbosa J, Hribova E, Heslop-Harrison P, Habas R, Rivallan R, Francois P, Poiron C, Kilian A, Burthia D, Jenny C, Bakry F, Brown S, Guignon V, Kema G, Dita M, Waalwijk C, Joseph S, Dievart A, Jaillon O, Leclercq J, Argout X, Lyons E, Almeida A, Jeridi M, Dolezel J, Roux N, Risterucci AM, Weissenbach J, Ruiz M, Glaszmann JC, Quetier F, Yahiaoui N, Wincker P (2012) The banana (*Musa acuminata*) genome and the evolution of monocotyledonous plants. Nature 488(7410):213–217

Dar T-U-H, Ruhman R-U (2017) Occurrence of polyploidy. In: Polyploidy: recent trends and future perspectives. Springer, New Dehli, pp 31–42

Dassanayake M, Oh DH, Haas JS, Hernandez A, Hong H, Ali S, Yun DJ, Bressan RA, Zhu JK, Bohnert HJ, Cheeseman JM (2011) The genome of the extremophile crucifer *Thellungiella parvula*. Nat Genet 43(9):913–918

de Vries J, Archibald JM (2018) Plant evolution: landmarks on the path to terrestrial life. New Phytol 217(4):1428–1434

Denker A, De Laat W (2016) The second decade of 3C technologies: detailed insights into nuclear organization. Genes Dev 30(12):1357–1382

Derelle E, Ferraz C, Rombauts S, Rouze P, Worden AZ, Robbens S, Partensky F, Degroeve S, Echeynie S, Cooke R, Saeys Y, Wuyts J, Jabbari K, Bowler C, Panaud O, Piegu B, Ball SG, Ral JP, Bouget FY, Piganeau G, De Baets B, Picard A, Delseny M, Demaille J, Van de Peer Y,

Moreau H (2006) Genome analysis of the smallest free-living eukaryote *Ostreococcus tauri* unveils many unique features. Proc Natl Acad Sci U S A 103(31):11647–11652

Dickerson RE, Drew HR, Conner BN, Wing RM, Fratini AV, Kopka ML (1982) The anatomy of A-, B-, and Z-DNA. Science 216(4545):475–485

DiFazio SP, Slavov GT, Joshi CP (2011) Populus: a premier pioneer system for plant genomics. In: Joshi CP, Di Fazio SP, Kole C (eds) Genetics, genomics and breeding of poplar. CRC Press, Boca Raton, pp 1–28

Dodsworth S, Chase MW, Leitch AR (2016) Is post-polyploidization diploidization the key to the evolutionary success of angiosperms? Bot J Linn Soc 180(1):1–5

Dohm JC, Minoche AE, Holtgrawe D, Capella-Gutierrez S, Zakrzewski F, Tafer H, Rupp O, Sorensen TR, Stracke R, Reinhardt R, Goesmann A, Kraft T, Schulz B, Stadler PF, Schmidt T, Gabaldon T, Lehrach H, Weisshaar B, Himmelbauer H (2014) The genome of the recently domesticated crop plant sugar beet (*Beta vulgaris*). Nature 505(7484):546–549

Drmanac R, Sparks AB, Callow MJ, Halpern AL, Burns NL, Kermani BG, Carnevali P, Nazarenko I, Nilsen GB, Yeung G, Dahl F, Fernandez A, Staker B, Pant KP, Baccash J, Borcherding AP, Brownley A, Cedeno R, Chen L, Chernikoff D, Cheung A, Chirita R, Curson B, Ebert JC, Hacker CR, Hartlage R, Hauser B, Huang S, Jiang Y, Karpinchyk V, Koenig M, Kong C, Landers T, Le C, Liu J, McBride CE, Morenzoni M, Morey RE, Mutch K, Perazich H, Perry K, Peters BA, Peterson J, Pethiyagoda CL, Pothuraju K, Richter C, Rosenbaum AM, Roy S, Shafto J, Sharanhovich U, Shannon KW, Sheppy CG, Sun M, Thakuria JV, Tran A, Vu D, Zaranek AW, Wu X, Drmanac S, Oliphant AR, Banyai WC, Martin B, Ballinger DG, Church GM, Reid CA (2010) Human genome sequencing using unchained base reads on self-assembling DNA nanoarrays. Science 327(5961):78–81

Dudchenko O, Batra SS, Omer AD, Nyquist SK, Hoeger M, Durand NC, Shamim MS, Machol I, Lander ES, Aiden AP, Aiden EL (2017) De novo assembly of the Aedes aegypti genome using Hi-C yields chromosome-length scaffolds. Science 356(6333):92–95

Edgar RC (2004) MUSCLE: a multiple sequence alignment method with reduced time and space complexity. BMC Bioinformatics 5:113

Eid J, Fehr A, Gray J, Luong K, Lyle J, Otto G, Peluso P, Rank D, Baybayan P, Bettman B, Bibillo A, Bjornson K, Chaudhuri B, Christians F, Cicero R, Clark S, Dalal R, Dewinter A, Dixon J, Foquet M, Gaertner A, Hardenbol P, Heiner C, Hester K, Holden D, Kearns G, Kong X, Kuse R, Lacroix Y, Lin S, Lundquist P, Ma C, Marks P, Maxham M, Murphy D, Park I, Pham T, Phillips M, Roy J, Sebra R, Shen G, Sorenson J, Tomaney A, Travers K, Trulson M, Vieceli J, Wegener J, Wu D, Yang A, Zaccarin D, Zhao P, Zhong F, Korlach J, Turner S (2009) Real-time DNA sequencing from single polymerase molecules. Science 323(5910):133–138

El Baidouri M, Panaud O (2013) Comparative genomic paleontology across plant kingdom reveals the dynamics of TE-driven genome evolution. Genome Biol Evol 5(5):954–965

Fan C, Walling JG, Zhang J, Hirsch CD, Jiang J, Wing RA (2011) Conservation and purifying selection of transcribed genes located in a rice centromere. Plant Cell 23(8):2821–2830

Farjon A (2008) A natural history of conifers. Timber Press, Portland

Farjon A (2010) The economic importance of conifers. In: A handbook of the world's conifers. Brill Academic Publishers, Leiden, pp 25–28

Finn RD, Clements J, Arndt W, Miller BL, Wheeler TJ, Schreiber F, Bateman A, Eddy SR (2015) HMMER web server: 2015 update. Nucleic Acids Res 43(W1):W30–W38

French-Italian Public Consortium for Grapevine Genome Characterization (2007) The grapevine genome sequence suggests ancestral hexaploidization in major angiosperm phyla. Nature 449 (7161):463–467

Fuller CW, Middendorf LR, Benner SA, Church GM, Harris T, Huang X, Jovanovich SB, Nelson JR, Schloss JA, Schwartz DC, Vezenov DV (2009) The challenges of sequencing by synthesis. Nat Biotechnol 27(11):1013–1023

Garsmeur O, Schnable JC, Almeida A, Jourda C, D'Hont A, Freeling M (2014) Two evolutionarily distinct classes of paleopolyploidy. Mol Biol Evol 31(2):448–454

Ge Y, Liu J, Zeng M, He J, Qin P, Huang H, Xu L (2016) Identification of WOX family genes in Selaginella kraussiana for studies on stem cells and regeneration in lycophytes. Front Plant Sci 7:93
Geer LY, Markey SP, Kowalak JA, Wagner L, Xu M, Maynard DM, Yang X, Shi W, Bryant SH (2004) Open mass spectrometry search algorithm. J Proteome Res 3(5):958–964
Gelderblom HR (1996) Structure and classification of viruses. In: Barron S (ed) Medical microbiology, 4th edn. University of Texas Medical Branch at Galveston, Galveston
GenomeWeb News Staff Writer (2012) Helicos biosciences files for chapter 11 bankruptcy protection. GenomeWeb News. https://www.genomeweb.com/sequencing/helicos-biosciences-files-chapter-11-bankruptcy-protection
Gluch A, Vidakovic M, Bode J (2008) Scaffold/matrix attachment regions (S/MARs): relevance for disease and therapy. In: Klussmann E, Scott J (eds) Handbook of experimental pharmacology. Springer, Berlin, pp 67–103
Gmitter FG Jr, Chen C, Machado MA, Alves de Souza A, Ollitrault P, Froehlicher Y, Shimizu T (2012) Citrus genomics. Tree Genet Genomes 8:611–626
Gnerre S, Maccallum I, Przybylski D, Ribeiro FJ, Burton JN, Walker BJ, Sharpe T, Hall G, Shea TP, Sykes S, Berlin AM, Aird D, Costello M, Daza R, Williams L, Nicol R, Gnirke A, Nusbaum C, Lander ES, Jaffe DB (2011) High-quality draft assemblies of mammalian genomes from massively parallel sequence data. Proc Natl Acad Sci U S A 108(4):1513–1518
Greenleaf WJ, Sidow A (2014) The future of sequencing: convergence of intelligent design and market Darwinism. Genome Biol 15(3):303
Gregory TR (2005) Genome size evolution in animals. In: Gregory TR (ed) The evolution of the genome. Elsevier Academic Press, Boston, pp 3–87
Gregory TR (2018) Animal genome size database. http://www.genomesize.com. Accessed 1 Feb 2018
Guan R, Zhao Y, Zhang H, Fan G, Liu X, Zhou W, Shi C, Wang J, Liu W, Liang X, Fu Y, Ma K, Zhao L, Zhang F, Lu Z, Lee SM, Xu X, Wang J, Yang H, Fu C, Ge S, Chen W (2016) Draft genome of the living fossil Ginkgo biloba. GigaScience 5(1):49
Gurevich A, Saveliev V, Vyahhi N, Tesler G (2013) QUAST: quality assessment tool for genome assemblies. Bioinformatics 29(8):1072–1075
Haas BJ, Papanicolaou A, Yassour M, Grabherr M, Blood PD, Bowden J, Couger MB, Eccles D, Li B, Lieber M, MacManes MD, Ott M, Orvis J, Pochet N, Strozzi F, Weeks N, Westerman R, William T, Dewey CN, Henschel R, LeDuc RD, Friedman N, Regev A (2013) De novo transcript sequence reconstruction from RNA-seq using the Trinity platform for reference generation and analysis. Nat Protoc 8(8):1494–1512
Hane JK, Ming Y, Kamphuis LG, Nelson MN, Garg G, Atkins CA, Bayer PE, Bravo A, Bringans S, Cannon S, Edwards D, Foley R, Gao LL, Harrison MJ, Huang W, Hurgobin B, Li S, Liu CW, McGrath A, Morahan G, Murray J, Weller J, Jian J, Singh KB (2017) A comprehensive draft genome sequence for lupin (*Lupinus angustifolius*), an emerging health food: insights into plant-microbe interactions and legume evolution. Plant Biotechnol J 15(3):318–330
Haque F, Li J, Wu HC, Liang XJ, Guo P (2013) Solid-state and biological nanopore for real-time sensing of single chemical and sequencing of DNA. Nano Today 8(1):56–74
Haston E, Richardson JE, Stevens PF, Chase MW, Harris DJ (2009) The Linear Angiosperm Phylogeny Group (LAPG) III: a linear sequence of the families in APG III. Bot J Linn Soc 161(2):128–131
Heather JM, Chain B (2016) The sequence of sequencers: the history of sequencing DNA. Genomics 107(1):1–8
Hilgartner S (2017) Shaping news and making history. In: Reordering life: knowledge and control in the genomics revolution. The MIT Press, Cambridge, pp 185–222
Hoffman EA, Frey BL, Smith LM, Auble DT (2015) Formaldehyde crosslinking: a tool for the study of chromatin complexes. J Biol Chem 290(44):26404–26411
Hogeweg P (2011) The roots of bioinformatics in theoretical biology. PLoS Comput Biol 7(3): e1002021

Hollmer M (2013) Roche to close 454 life sciences as it reduces gene sequencing focus. FierceBiotech: https://www.fiercebiotech.com/medical-devices/roche-to-close-454-life-sciences-as-it-reduces-gene-sequencing-focus

Holt C, Yandell M (2011) MAKER2: an annotation pipeline and genome-database management tool for second-generation genome projects. BMC Bioinformatics 12:491

Hori K, Maruyama F, Fujisawa T, Togashi T, Yamamoto N, Seo M, Sato S, Yamada T, Mori H, Tajima N, Moriyama T, Ikeuchi M, Watanabe M, Wada H, Kobayashi K, Saito M, Masuda T, Sasaki-Sekimoto Y, Mashiguchi K, Awai K, Shimojima M, Masuda S, Iwai M, Nobusawa T, Narise T, Kondo S, Saito H, Sato R, Murakawa M, Ihara Y, Oshima-Yamada Y, Ohtaka K, Satoh M, Sonobe K, Ishii M, Ohtani R, Kanamori-Sato M, Honoki R, Miyazaki D, Mochizuki H, Umetsu J, Higashi K, Shibata D, Kamiya Y, Sato N, Nakamura Y, Tabata S, Ida S, Kurokawa K, Ohta H (2014) *Klebsormidium flaccidum* genome reveals primary factors for plant terrestrial adaptation. Nat Commun 5:3978

Huang X, Madan A (1999) CAP3: a DNA sequence assembly program. Genome Res 9(9):868–877

Hutcheon C, Ditt RF, Beilstein M, Comai L, Schroeder J, Goldstein E, Shewmaker CK, Nguyen T, De Rocher J, Kiser J (2010) Polyploid genome of *Camelina sativa* revealed by isolation of fatty acid synthesis genes. BMC Plant Biol 10:233

Ibarra-Laclette E, Lyons E, Hernandez-Guzman G, Perez-Torres CA, Carretero-Paulet L, Chang TH, Lan T, Welch AJ, Juarez MJ, Simpson J, Fernandez-Cortes A, Arteaga-Vazquez M, Gongora-Castillo E, Acevedo-Hernandez G, Schuster SC, Himmelbauer H, Minoche AE, Xu S, Lynch M, Oropeza-Aburto A, Cervantes-Perez SA, de Jesus Ortega-Estrada M, Cervantes-Luevano JI, Michael TP, Mockler T, Bryant D, Herrera-Estrella A, Albert VA, Herrera-Estrella L (2013) Architecture and evolution of a minute plant genome. Nature 498(7452):94–98

International Brachypodium Initiative (2010) Genome sequencing and analysis of the model grass *Brachypodium distachyon*. Nature 463(7282):763–768

International Peach Genome Initiative, Verde I, Abbott AG, Scalabrin S, Jung S, Shu S, Marroni F, Zhebentyayeva T, Dettori MT, Grimwood J, Cattonaro F, Zuccolo A, Rossini L, Jenkins J, Vendramin E, Meisel LA, Decroocq V, Sosinski B, Prochnik S, Mitros T, Policriti A, Cipriani G, Dondini L, Ficklin S, Goodstein DM, Xuan P, Del Fabbro C, Aramini V, Copetti D, Gonzalez S, Horner DS, Falchi R, Lucas S, Mica E, Maldonado J, Lazzari B, Bielenberg D, Pirona R, Miculan M, Barakat A, Testolin R, Stella A, Tartarini S, Tonutti P, Arus P, Orellana A, Wells C, Main D, Vizzotto G, Silva H, Salamini F, Schmutz J, Morgante M, Rokhsar DS (2013) The high-quality draft genome of peach (*Prunus persica*) identifies unique patterns of genetic diversity, domestication and genome evolution. Nat Genet 45(5):487–494

International Rice Genome Sequencing Project (2005) The map-based sequence of the rice genome. Nature 436(7052):793–800

Isono K, McIninch JD, Borodovsky M (1994) Characteristic features of the nucleotide sequences of yeast mitochondrial ribosomal protein genes as analyzed by computer program GeneMark. DNA Res 1(6):263–269

Ištvánek J, Jaroš M, Krenek A, Řepková J (2014) Genome assembly and annotation for red clover (*Trifolium pratense*; Fabaceae). Am J Bot 101(2):327–337

Iyer BV, Kenward M, Arya G (2011) Hierarchies in eukaryotic genome organization: insights from polymer theory and simulations. BMC Biophys 4:8

Jackman SD, Vandervalk BP, Mohamadi H, Chu J, Yeo S, Hammond SA, Jahesh G, Khan H, Coombe L, Warren RL, Birol I (2017) ABySS 2.0: resource-efficient assembly of large genomes using a Bloom filter. Genome Res 27(5):768–777

Jackson SA (2016) Rice: the first crop genome. Rice 9(1):14

Jaffe JD, Berg HC, Church GM (2004) Proteogenomic mapping as a complementary method to perform genome annotation. Proteomics 4(1):59–77

Jain M, Olsen HE, Paten B, Akeson M (2016) The Oxford nanopore MinION: delivery of nanopore sequencing to the genomics community. Genome Biol 17(1):239

Jarvie T, Harkins T (2008) 3K long-tag paired end sequencing with the genome sequencer FLX system. Nat Methods 5. https://doi.org/10.1038/nmeth.f.1212

Jiao Y, Peluso P, Shi J, Liang T, Stitzer MC, Wang B, Campbell MS, Stein JC, Wei X, Chin CS, Guill K, Regulski M, Kumari S, Olson A, Gent J, Schneider KL, Wolfgruber TK, May MR, Springer NM, Antoniou E, McCombie WR, Presting GG, McMullen M, Ross-Ibarra J, Dawe RK, Hastie A, Rank DR, Ware D (2017) Improved maize reference genome with single-molecule technologies. Nature 546(7659):524–527

Kagale S, Koh C, Nixon J, Bollina V, Clarke WE, Tuteja R, Spillane C, Robinson SJ, Links MG, Clarke C, Higgins EE, Huebert T, Sharpe AG, Parkin IA (2014) The emerging biofuel crop *Camelina sativa* retains a highly undifferentiated hexaploid genome structure. Nat Commun 5:3706

Kang YJ, Kim SK, Kim MY, Lestari P, Kim KH, Ha BK, Jun TH, Hwang WJ, Lee T, Lee J, Shim S, Yoon MY, Jang YE, Han KS, Taeprayoon P, Yoon N, Somta P, Tanya P, Kim KS, Gwag JG, Moon JK, Lee YH, Park BS, Bombarely A, Doyle JJ, Jackson SA, Schafleitner R, Srinives P, Varshney RK, Lee SH (2014) Genome sequence of mungbean and insights into evolution within *Vigna* species. Nat Commun 5:5443

Karger BL, Guttman A (2009) DNA sequencing by capillary electrophoresis. Electrophoresis 30(Suppl 1):S196–S202

Katayama T (1995) Cytogenetical studies on the genus *Oryza*. XIV. Intergeneric hybridizations between tetraploid *Oryza* species and diploid *Leersia* species. Jpn J Genet 70:47–55

Keeling PJ (2010) The endosymbiotic origin, diversification and fate of plastids. Philos Trans R Soc Lond Ser B Biol Sci 365(1541):729–748

Kelly LJ, Renny-Byfield S, Pellicer J, Macas J, Novak P, Neumann P, Lysak MA, Day PD, Berger M, Fay MF, Nichols RA, Leitch AR, Leitch IJ (2015) Analysis of the giant genomes of Fritillaria (Liliaceae) indicates that a lack of DNA removal characterizes extreme expansions in genome size. New Phytol 208(2):596–607

Kent WJ (2002) BLAT – the BLAST-like alignment tool. Genome Res 12(4):656–664

Kim D, Langmead B, Salzberg SL (2015) HISAT: a fast spliced aligner with low memory requirements. Nat Methods 12(4):357–360

Kircher M, Sawyer S, Meyer M (2012) Double indexing overcomes inaccuracies in multiplex sequencing on the Illumina platform. Nucleic Acids Res 40(1):e3

Knuth DE (1984) Literate programming. Comput J 27(2):97–111

Kolesnikov AA, Gerasimov ES (2012) Diversity of mitochondrial genome organization. Biochemistry (Mosc) 77(13):1424–1435

Koonin EV, Fedorova ND, Jackson JD, Jacobs AR, Krylov DM, Makarova KS, Mazumder R, Mekhedov SL, Nikolskaya AN, Rao BS, Rogozin IB, Smirnov S, Sorokin AV, Sverdlov AV, Vasudevan S, Wolf YI, Yin JJ, Natale DA (2004) A comprehensive evolutionary classification of proteins encoded in complete eukaryotic genomes. Genome Biol 5(2):R7

Koornneef M, Meinke D (2010) The development of *Arabidopsis* as a model plant. Plant J 61(6):909–921

Koren S, Walenz BP, Berlin K, Miller JR, Bergman NH, Phillippy AM (2017) Canu: scalable and accurate long-read assembly via adaptive k-mer weighting and repeat separation. Genome Res 27(5):722–736

Krzywinski M, Schein J, Birol I, Connors J, Gascoyne R, Horsman D, Jones SJ, Marra MA (2009) Circos: an information aesthetic for comparative genomics. Genome Res 19(9):1639–1645

Kulikova T, Akhtar R, Aldebert P, Althorpe N, Andersson M, Baldwin A, Bates K, Bhattacharyya S, Bower L, Browne P, Castro M, Cochrane G, Duggan K, Eberhardt R, Faruque N, Hoad G, Kanz C, Lee C, Leinonen R, Lin Q, Lombard V, Lopez R, Lorenc D, McWilliam H, Mukherjee G, Nardone F, Pastor MP, Plaister S, Sobhany S, Stoehr P, Vaughan R, Wu D, Zhu W, Apweiler R (2007) EMBL Nucleotide Sequence Database in 2006. Nucleic Acids Res 35(Database issue): D16–D20

Lam ET, Hastie A, Lin C, Ehrlich D, Das SK, Austin MD, Deshpande P, Cao H, Nagarajan N, Xiao M, Kwok PY (2012) Genome mapping on nanochannel arrays for structural variation analysis and sequence assembly. Nat Biotechnol 30(8):771–776

Lang D, Ullrich KK, Murat F, Fuchs J, Jenkins J, Haas FB, Piednoel M, Gundlach H, Van Bel M, Meyberg R, Vives C, Morata J, Symeonidi A, Hiss M, Muchero W, Kamisugi Y, Saleh O, Blanc G, Decker EL, van Gessel N, Grimwood J, Hayes RD, Graham SW, Gunter LE, McDaniel SF, Hoernstein SNW, Larsson A, Li FW, Perroud PF, Phillips J, Ranjan P, Rokshar DS, Rothfels CJ, Schneider L, Shu S, Stevenson DW, Thummler F, Tillich M, Villarreal Aguilar JC, Widiez T, Wong GK, Wymore A, Zhang Y, Zimmer AD, Quatrano RS, Mayer KFX, Goodstein D, Casacuberta JM, Vandepoele K, Reski R, Cuming AC, Tuskan GA, Maumus F, Salse J, Schmutz J, Rensing SA (2018) The Physcomitrella patens chromosome-scale assembly reveals moss genome structure and evolution. Plant J 93(3):515–533

Langmead B (2010) Aligning short sequencing reads with Bowtie. Curr Protoc Bioinformatics 32:11.17.11–11.17.14

Langmead B, Salzberg SL (2012) Fast gapped-read alignment with Bowtie 2. Nat Methods 9(4):357–359

Leggett RM, Clavijo BJ, Clissold L, Clark MD, Caccamo M (2014) NextClip: an analysis and read preparation tool for Nextera Long Mate Pair libraries. Bioinformatics 30(4):566–568

Leitch AR, Leitch IJ (2008) Genomic plasticity and the diversity of polyploid plants. Science 320 (5875):481–483

Lesk AM (2012) Introduction to genomics, 2nd edn. Oxford University Press, New York

Li H (2016) Minimap and miniasm: fast mapping and de novo assembly for noisy long sequences. Bioinformatics 32(14):2103–2110

Li H, Durbin R (2009) Fast and accurate short read alignment with Burrows-Wheeler transform. Bioinformatics 25(14):1754–1760

Li H, Handsaker B, Wysoker A, Fennell T, Ruan J, Homer N, Marth G, Abecasis G, Durbin R, 1000 Genome Project Data Processing Subgroup (2009) The Sequence Alignment/Map format and SAMtools. Bioinformatics 25(16):2078–2079

Li F, Fan G, Lu C, Xiao G, Zou C, Kohel RJ, Ma Z, Shang H, Ma X, Wu J, Liang X, Huang G, Percy RG, Liu K, Yang W, Chen W, Du X, Shi C, Yuan Y, Ye W, Liu X, Zhang X, Liu W, Wei H, Wei S, Huang G, Zhang X, Zhu S, Zhang H, Sun F, Wang X, Liang J, Wang J, He Q, Huang L, Wang J, Cui J, Song G, Wang K, Xu X, Yu JZ, Zhu Y, Yu S (2015a) Genome sequence of cultivated Upland cotton (*Gossypium hirsutum* TM-1) provides insights into genome evolution. Nat Biotechnol 33(5):524–530

Li Z, Baniaga AE, Sessa EB, Scascitelli M, Graham SW, Rieseberg LH, Barker MS (2015b) Early genome duplications in conifers and other seed plants. Sci Adv 1(10):e1501084

Liu C, Weigel D (2015) Chromatin in 3D: progress and prospects for plants. Genome Biol 16:170

Liu MJ, Zhao J, Cai QL, Liu GC, Wang JR, Zhao ZH, Liu P, Dai L, Yan G, Wang WJ, Li XS, Chen Y, Sun YD, Liu ZG, Lin MJ, Xiao J, Chen YY, Li XF, Wu B, Ma Y, Jian JB, Yang W, Yuan Z, Sun XC, Wei YL, Yu LL, Zhang C, Liao SG, He RJ, Guang XM, Wang Z, Zhang YY, Luo LH (2014) The complex jujube genome provides insights into fruit tree biology. Nat Commun 5:5315

Love RR, Weisenfeld NI, Jaffe DB, Besansky NJ, Neafsey DE (2016) Evaluation of DISCOVAR de novo using a mosquito sample for cost-effective short-read genome assembly. BMC Genomics 17:187

Lowry DB, Purmal CT, Juenger TE (2013) A population genetic transect of *Panicum hallii* (Poaceae). Am J Bot 100(3):592–601

Manda P, Freeman MG, Bridges SM, Jankun-Kelly TJ, Nanduri B, McCarthy FM, Burgess SC (2010) GOModeler – a tool for hypothesis-testing of functional genomics datasets. BMC Bioinformatics 11(Suppl 6):S29

Mankertz P (2008) Molecular biology of porcine circoviruses. In: Animal viruses: molecular biology. Academic Press, Caister, pp 355–374

Marcais G, Delcher AL, Phillippy AM, Coston R, Salzberg SL, Zimin A (2018) MUMmer4: a fast and versatile genome alignment system. PLoS Comput Biol 14(1):e1005944

Margulies M, Egholm M, Altman WE, Attiya S, Bader JS, Bemben LA, Berka J, Braverman MS, Chen YJ, Chen Z, Dewell SB, Du L, Fierro JM, Gomes XV, Godwin BC, He W, Helgesen S,

Ho CH, Irzyk GP, Jando SC, Alenquer ML, Jarvie TP, Jirage KB, Kim JB, Knight JR, Lanza JR, Leamon JH, Lefkowitz SM, Lei M, Li J, Lohman KL, Lu H, Makhijani VB, McDade KE, McKenna MP, Myers EW, Nickerson E, Nobile JR, Plant R, Puc BP, Ronan MT, Roth GT, Sarkis GJ, Simons JF, Simpson JW, Srinivasan M, Tartaro KR, Tomasz A, Vogt KA, Volkmer GA, Wang SH, Wang Y, Weiner MP, Yu P, Begley RF, Rothberg JM (2005) Genome sequencing in microfabricated high-density picolitre reactors. Nature 437:376–380

Martinez-Cano DJ, Reyes-Prieto M, Martinez-Romero E, Partida-Martinez LP, Latorre A, Moya A, Delaye L (2014) Evolution of small prokaryotic genomes. Front Microbiol 5:742

Maxam AM, Gilbert W (1977) A new method for sequencing DNA. Proc Natl Acad Sci U S A 74(2):560–564

McCarthy FM, Wang N, Magee GB, Nanduri B, Lawrence ML, Camon EB, Barrell DG, Hill DP, Dolan ME, Williams WP, Luthe DS, Bridges SM, Burgess SC (2006) AgBase: a functional genomics resource for agriculture. BMC Genomics 7:229

McCarthy FM, Gresham CR, Buza TJ, Chouvarine P, Pillai LR, Kumar R, Ozkan S, Wang H, Manda P, Arick T, Bridges SM, Burgess SC (2011) AgBase: supporting functional modeling in agricultural organisms. Nucleic Acids Res 39(Database issue):D497–D506

McKenna A, Hanna M, Banks E, Sivachenko A, Cibulskis K, Kernytsky A, Garimella K, Altshuler D, Gabriel S, Daly M, DePristo MA (2010) The genome analysis toolkit: a MapReduce framework for analyzing next-generation DNA sequencing data. Genome Res 20(9):1297–1303

Metzker ML (2010) Sequencing technologies – the next generation. Nat Rev Genet 11(1):31–46

Michael TP, Jackson S (2013) The first 50 plant genomes. Plant Genome 6(2). https://doi.org/10.3835/plantgenome2013.3803.0001in

Ming R, VanBuren R, Wai CM, Tang H, Schatz MC, Bowers JE, Lyons E, Wang ML, Chen J, Biggers E, Zhang J, Huang L, Zhang L, Miao W, Zhang J, Ye Z, Miao C, Lin Z, Wang H, Zhou H, Yim WC, Priest HD, Zheng C, Woodhouse M, Edger PP, Guyot R, Guo HB, Guo H, Zheng G, Singh R, Sharma A, Min X, Zheng Y, Lee H, Gurtowski J, Sedlazeck FJ, Harkess A, McKain MR, Liao Z, Fang J, Liu J, Zhang X, Zhang Q, Hu W, Qin Y, Wang K, Chen LY, Shirley N, Lin YR, Liu LY, Hernandez AG, Wright CL, Bulone V, Tuskan GA, Heath K, Zee F, Moore PH, Sunkar R, Leebens-Mack JH, Mockler T, Bennetzen JL, Freeling M, Sankoff D, Paterson AH, Zhu X, Yang X, Smith JA, Cushman JC, Paull RE, Yu Q (2015) The pineapple genome and the evolution of CAM photosynthesis. Nat Genet 47(12):1435–1442

Mirsky AE, Ris H (1951) The desoxyribonucleic acid content of animal cells and its evolutionary significance. J Gen Physiol 34:451–462

Mohanta TK, Bae H (2015) The diversity of fungal genome. Biol Proced Online 17:8

Molina J, Hazzouri KM, Nickrent D, Geisler M, Meyer RS, Pentony MM, Flowers JM, Pelser P, Barcelona J, Inovejas SA, Uy I, Yuan W, Wilkins O, Michel CI, Locklear S, Concepcion GP, Purugganan MD (2014) Possible loss of the chloroplast genome in the parasitic flowering plant *Rafflesia lagascae* (Rafflesiaceae). Mol Biol Evol 31(4):793–803

Moreau H, Verhelst B, Couloux A, Derelle E, Rombauts S, Grimsley N, Van Bel M, Poulain J, Katinka M, Hohmann-Marriott MF, Piganeau G, Rouze P, Da Silva C, Wincker P, Van de Peer Y, Vandepoele K (2012) Gene functionalities and genome structure in *Bathycoccus prasinos* reflect cellular specializations at the base of the green lineage. Genome Biol 13(8):R74

Motamayor JC, Mockaitis K, Schmutz J, Haiminen N, Livingstone D 3rd, Cornejo O, Findley SD, Zheng P, Utro F, Royaert S, Saski C, Jenkins J, Podicheti R, Zhao M, Scheffler BE, Stack JC, Feltus FA, Mustiga GM, Amores F, Phillips W, Marelli JP, May GD, Shapiro H, Ma J, Bustamante CD, Schnell RJ, Main D, Gilbert D, Parida L, Kuhn DN (2013) The genome sequence of the most widely cultivated cacao type and its use to identify candidate genes regulating pod color. Genome Biol 14(6):r53

Munoz-Gomez SA, Mejia-Franco FG, Durnin K, Colp M, Grisdale CJ, Archibald JM, Slamovits CH (2017) The new red algal subphylum Proteorhodophytina comprises the largest and most divergent plastid genomes known. Curr Biol 27(11):1677–1684.e1674

Murray BG, Leitch IJ, Bennett MD (2012) Gymnosperm DNA C-values database (release 5.0, Dec. 2012). http://www.kew.org/cvalues/. Accessed 28 Jan 2017

Neale DB, Wegrzyn JL, Stevens KA, Zimin AV, Puiu D, Crepeau MW, Cardeno C, Koriabine M, Holtz-Morris AE, Liechty JD, Martínez-García PJ, Vasquez-Gross HA, Lin BY, Zieve JJ, Dougherty WM, Fuentes-Soriano S, Wu LS, Gilbert D, Marçais G, Roberts M, Holt C, Yandell M, Davis JM, Smith KE, Dean JFD, Lorenz WW, Whetten RW, Sederoff R, Wheeler N, McGuire PE, Main D, Loopstra CA, Mockaitis K, deJong PJ, Yorke JA, Salzberg SL, Langley CH (2014) Decoding the massive genome of loblolly pine using haploid DNA and novel assembly strategies. Genome Biol 15(3):R59

Ng MP, Vergara IA, Frech C, Chen Q, Zeng X, Pei J, Chen N (2009) OrthoClusterDB: an online platform for synteny blocks. BMC Bioinformatics 10:192

Nickerson DA, Tobe VO, Taylor SL (1997) PolyPhred: automating the detection and genotyping of single nucleotide substitutions using fluorescence-based resequencing. Nucleic Acids Res 25(14):2745–2751

Nystedt B, Street NR, Wetterbom A, Zuccolo A, Lin YC, Scofield DG, Vezzi F, Delhomme N, Giacomello S, Alexeyenko A, Vicedomini R, Sahlin K, Sherwood E, Elfstrand M, Gramzow L, Holmberg K, Hallman J, Keech O, Klasson L, Koriabine M, Kucukoglu M, Kaller M, Luthman J, Lysholm F, Niittyla T, Olson A, Rilakovic N, Ritland C, Rossello JA, Sena J, Svensson T, Talavera-Lopez C, Theissen G, Tuominen H, Vanneste K, Wu ZQ, Zhang B, Zerbe P, Arvestad L, Bhalerao R, Bohlmann J, Bousquet J, Garcia GR, Hvidsten TR, de Jong P, MacKay J, Morgante M, Ritland K, Sundberg B, Thompson SL, Van de Peer Y, Andersson B, Nilsson O, Ingvarsson PK, Lundeberg J, Jansson S (2013) The Norway spruce genome sequence and conifer genome evolution. Nature 497(7451):579–584

Ocana K, de Oliveira D (2015) Parallel computing in genomic research: advances and applications. Adv Appl Bioinforma Chem 8:23–35

Oh DH, Dassanayake M, Haas JS, Kropornika A, Wright C, d'Urzo MP, Hong H, Ali S, Hernandez A, Lambert GM, Inan G, Galbraith DW, Bressan RA, Yun DJ, Zhu JK, Cheeseman JM, Bohnert HJ (2010) Genome structures and halophyte-specific gene expression of the extremophile *Thellungiella parvula* in comparison with *Thellungiella salsuginea* (*Thellungiella halophila*) and *Arabidopsis*. Plant Physiol 154(3):1040–1052

Palenik B, Grimwood J, Aerts A, Rouze P, Salamov A, Putnam N, Dupont C, Jorgensen R, Derelle E, Rombauts S, Zhou K, Otillar R, Merchant SS, Podell S, Gaasterland T, Napoli C, Gendler K, Manuell A, Tai V, Vallon O, Piganeau G, Jancek S, Heijde M, Jabbari K, Bowler C, Lohr M, Robbens S, Werner G, Dubchak I, Pazour GJ, Ren Q, Paulsen I, Delwiche C, Schmutz J, Rokhsar D, Van de Peer Y, Moreau H, Grigoriev IV (2007) The tiny eukaryote *Ostreococcus* provides genomic insights into the paradox of plankton speciation. Proc Natl Acad Sci U S A 104(18):7705–7710

Pandey V, Nutter RC, Prediger E (2008) Applied biosystems SOLiD™ system: ligation-based sequencing. In: Janitz M (ed) Next generation genome sequencing: towards personalized medicine. Wiley-VCH, Weinheim, pp 29–41

Paterson AH, Bowers JE, Chapman BA (2004) Ancient polyploidization predating divergence of the cereals, and its consequences for comparative genomics. Proc Natl Acad Sci U S A 101(26):9903–9908

Paterson AH, Bowers JE, Bruggmann R, Dubchak I, Grimwood J, Gundlach H, Haberer G, Hellsten U, Mitros T, Poliakov A, Schmutz J, Spannagl M, Tang H, Wang X, Wicker T, Bharti AK, Chapman J, Feltus FA, Gowik U, Grigoriev IV, Lyons E, Maher CA, Martis M, Narechania A, Otillar RP, Penning BW, Salamov AA, Wang Y, Zhang L, Carpita NC, Freeling M, Gingle AR, Hash CT, Keller B, Klein P, Kresovich S, McCann MC, Ming R, Peterson DG, Mehboob-ur-Rahman, Ware D, Westhoff P, Mayer KF, Messing J, Rokhsar DS (2009) The *Sorghum bicolor* genome and the diversification of grasses. Nature 457 (7229):551–556

Paterson AH, Wendel JF, Gundlach H, Guo H, Jenkins J, Jin D, Llewellyn D, Showmaker KC, Shu S, Udall J, Yoo MJ, Byers R, Chen W, Doron-Faigenboim A, Duke MV, Gong L,

Grimwood J, Grover C, Grupp K, Hu G, Lee TH, Li J, Lin L, Liu T, Marler BS, Page JT, Roberts AW, Romanel E, Sanders WS, Szadkowski E, Tan X, Tang H, Xu C, Wang J, Wang Z, Zhang D, Zhang L, Ashrafi H, Bedon F, Bowers JE, Brubaker CL, Chee PW, Das S, Gingle AR, Haigler CH, Harker D, Hoffmann LV, Hovav R, Jones DC, Lemke C, Mansoor S, Rahman M, Rainville LN, Rambani A, Reddy UK, Rong JK, Saranga Y, Scheffler BE, Scheffler JA, Stelly DM, Triplett BA, Van Deynze A, Vaslin MF, Waghmare VN, Walford SA, Wright RJ, Zaki EA, Zhang T, Dennis ES, Mayer KF, Peterson DG, Rokhsar DS, Wang X, Schmutz J (2012) Repeated polyploidization of *Gossypium* genomes and the evolution of spinnable cotton fibres. Nature 492(7429):423–427

Patro R, Duggal G, Love MI, Irizarry RA, Kingsford C (2017) Salmon provides fast and bias-aware quantification of transcript expression. Nat Methods 14(4):417–419

Pellicer J, Fay MF, Leitch IJ (2010) The largest eukaryotic genome of them all? Bot J Linn Soc 164:10–15

Pellicer J, Hidalgo O, Dodsworth S, Leitch IJ (2018) Genome size diversity and its impact on the evolution of land plants. Genes (Basel) 9(2):88

Perera D, Magbanua ZV, Thummasuwan S, Mukherjee D, Arick M II, Chouvarine P, Nairn CJ, Schmutz J, Grimwood J, Dean JFD, Peterson DG (2018) Exploring the loblolly pine (*Pinus taeda*L.) genome by BAC sequencing and Cot analysis. Gene. https://doi.org/10.1016/j.gene.2018.04.024

Pertea M, Pertea GM, Antonescu CM, Chang TC, Mendell JT, Salzberg SL (2015) StringTie enables improved reconstruction of a transcriptome from RNA-seq reads. Nat Biotechnol 33(3):290–295

Peterson DG (2005) Reduced representation strategies and their application to plant genomes. In: Meksem K, Kahl G (eds) The handbook of genome mapping: genetic and physical mapping. Wiley-VCH Verlag GmbH, KGaA, Weinheim, pp 307–335

Peterson DG (2014) Evolution of plant genome analysis. In: Paterson AH (ed) Advances in botanical research, Genomes of herbaceous land plants, vol 69. Academic Press, Amsterdam, pp 13–46

Peterson DG, Tomkins JP, Frisch DA, Wing RA, Paterson AH (2000) Construction of plant bacterial artificial chromosome (BAC) libraries: an illustrated Guide. J Agric Genomics 5:1–100. http://wheat.pw.usda.gov/jag/

Philippe N, Legendre M, Doutre G, Coute Y, Poirot O, Lescot M, Arslan D, Seltzer V, Bertaux L, Bruley C, Garin J, Claverie JM, Abergel C (2013) Pandoraviruses: amoeba viruses with genomes up to 2.5 Mb reaching that of parasitic eukaryotes. Science 341(6143):281–286

Price AL, Jones NC, Pevzner PA (2005) De novo identification of repeat families in large genomes. Bioinformatics 21(Suppl 1):i351–i358

Rensing SA, Lang D, Zimmer AD, Terry A, Salamov A, Shapiro H, Nishiyama T, Perroud PF, Lindquist EA, Kamisugi Y, Tanahashi T, Sakakibara K, Fujita T, Oishi K, Shin IT, Kuroki Y, Toyoda A, Suzuki Y, Hashimoto S, Yamaguchi K, Sugano S, Kohara Y, Fujiyama A, Anterola A, Aoki S, Ashton N, Barbazuk WB, Barker E, Bennetzen JL, Blankenship R, Cho SH, Dutcher SK, Estelle M, Fawcett JA, Gundlach H, Hanada K, Heyl A, Hicks KA, Hughes J, Lohr M, Mayer K, Melkozernov A, Murata T, Nelson DR, Pils B, Prigge M, Reiss B, Renner T, Rombauts S, Rushton PJ, Sanderfoot A, Schween G, Shiu SH, Stueber K, Theodoulou FL, Tu H, Van de Peer Y, Verrier PJ, Waters E, Wood A, Yang L, Cove D, Cuming AC, Hasebe M, Lucas S, Mishler BD, Reski R, Grigoriev IV, Quatrano RS, Boore JL (2008) The Physcomitrella genome reveals evolutionary insights into the conquest of land by plants. Science 319 (5859):64–69

Rho M, Choi JH, Kim S, Lynch M, Tang H (2007) De novo identification of LTR retrotransposons in eukaryotic genomes. BMC Genomics 8:90

Rice P, Longden I, Bleasby A (2000) EMBOSS: the European molecular biology open software suite. Trends Genet 16(6):276–277

Rice A, Glick L, Abadi S, Einhorn M, Kopelman NM, Salman-Minkov A, Mayzel J, Chay O, Mayrose I (2015) The Chromosome Counts Database (CCDB) – a community resource of plant chromosome numbers. New Phytol 206(1):19–26

Richter BG, Sexton DP (2009) Managing and analyzing next-generation sequence data. PLoS Comput Biol 5(6):e1000369

Roach JC, Glusman G, Hubley R, Montsaroff SZ, Holloway AK, Mauldin DE, Srivastava D, Garg V, Pollard KS, Galas DJ, Hood L, Smit AF (2011) Chromosomal haplotypes by genetic phasing of human families. Am J Hum Genet 89(3):382–397

Rothberg JM, Hinz W, Rearick TM, Schultz J, Mileski W, Davey M, Leamon JH, Johnson K, Milgrew MJ, Edwards M, Hoon J, Simons JF, Marran D, Myers JW, Davidson JF, Branting A, Nobile JR, Puc BP, Light D, Clark TA, Huber M, Branciforte JT, Stoner IB, Cawley SE, Lyons M, Fu Y, Homer N, Sedova M, Miao X, Reed B, Sabina J, Feierstein E, Schorn M, Alanjary M, Dimalanta E, Dressman D, Kasinskas R, Sokolsky T, Fidanza JA, Namsaraev E, McKernan KJ, Williams A, Roth GT, Bustillo J (2011) An integrated semiconductor device enabling non-optical genome sequencing. Nature 475(7356):348–352

Ruhfel BR, Gitzendanner MA, Soltis PS, Soltis DE, Burleigh JG (2014) From algae to angiosperms-inferring the phylogeny of green plants (Viridiplantae) from 360 plastid genomes. BMC Evol Biol 14:23

Salzberg SL, Phillippy AM, Zimin A, Puiu D, Magoc T, Koren S, Treangen TJ, Schatz MC, Delcher AL, Roberts M, Marcais G, Pop M, Yorke JA (2012) GAGE: a critical evaluation of genome assemblies and assembly algorithms. Genome Res 22(3):557–567

Sanders WS, Wang N, Bridges SM, Malone BM, Dandass YS, McCarthy FM, Nanduri B, Lawrence ML, Burgess SC (2011) The proteogenomic mapping tool. BMC Bioinformatics 12:115

Sanger F, Coulson AR (1975) A rapid method for determining sequences in DNA by primed synthesis with DNA polymerase. J Mol Biol 94(3):441–448

Sanger F, Nicklen S, Coulson AR (1977) DNA sequencing with chain-terminating inhibitors. Proc Natl Acad Sci U S A 74(12):5463–5467

Schmutz J, Cannon SB, Schlueter J, Ma J, Mitros T, Nelson W, Hyten DL, Song Q, Thelen JJ, Cheng J, Xu D, Hellsten U, May GD, Yu Y, Sakurai T, Umezawa T, Bhattacharyya MK, Sandhu D, Valliyodan B, Lindquist E, Peto M, Grant D, Shu S, Goodstein D, Barry K, Futrell-Griggs M, Abernathy B, Du J, Tian Z, Zhu L, Gill N, Joshi T, Libault M, Sethuraman A, Zhang XC, Shinozaki K, Nguyen HT, Wing RA, Cregan P, Specht J, Grimwood J, Rokhsar D, Stacey G, Shoemaker RC, Jackson SA (2010) Genome sequence of the palaeopolyploid soybean. Nature 463(7278):178–183

Schnable PS, Ware D, Fulton RS, Stein JC, Wei F, Pasternak S, Liang C, Zhang J, Fulton L, Graves TA, Minx P, Reily AD, Courtney L, Kruchowski SS, Tomlinson C, Strong C, Delehaunty K, Fronick C, Courtney B, Rock SM, Belter E, Du F, Kim K, Abbott RM, Cotton M, Levy A, Marchetto P, Ochoa K, Jackson SM, Gillam B, Chen W, Yan L, Higginbotham J, Cardenas M, Waligorski J, Applebaum E, Phelps L, Falcone J, Kanchi K, Thane T, Scimone A, Thane N, Henke J, Wang T, Ruppert J, Shah N, Rotter K, Hodges J, Ingenthron E, Cordes M, Kohlberg S, Sgro J, Delgado B, Mead K, Chinwalla A, Leonard S, Crouse K, Collura K, Kudrna D, Currie J, He R, Angelova A, Rajasekar S, Mueller T, Lomeli R, Scara G, Ko A, Delaney K, Wissotski M, Lopez G, Campos D, Braidotti M, Ashley E, Golser W, Kim H, Lee S, Lin J, Dujmic Z, Kim W, Talag J, Zuccolo A, Fan C, Sebastian A, Kramer M, Spiegel L, Nascimento L, Zutavern T, Miller B, Ambroise C, Muller S, Spooner W, Narechania A, Ren L, Wei S, Kumari S, Faga B, Levy MJ, McMahan L, Van Buren P, Vaughn MW, Ying K, Yeh CT, Emrich SJ, Jia Y, Kalyanaraman A, Hsia AP, Barbazuk WB, Baucom RS, Brutnell TP, Carpita NC, Chaparro C, Chia JM, Deragon JM, Estill JC, Fu Y, Jeddeloh JA, Han Y, Lee H, Li P, Lisch DR, Liu S, Liu Z, Nagel DH, McCann MC, SanMiguel P, Myers AM, Nettleton D, Nguyen J, Penning BW, Ponnala L, Schneider KL, Schwartz DC, Sharma A, Soderlund C, Springer NM, Sun Q, Wang H, Waterman M, Westerman R, Wolfgruber TK, Yang L, Yu Y, Zhang L, Zhou S, Zhu Q, Bennetzen JL, Dawe RK, Jiang J, Jiang N, Presting GG, Wessler SR, Aluru S,

Martienssen RA, Clifton SW, McCombie WR, Wing RA, Wilson RK (2009) The B73 maize genome: complexity, diversity, and dynamics. Science 326(5956):1112–1115

Schneiker S, Perlova O, Kaiser O, Gerth K, Alici A, Altmeyer MO, Bartels D, Bekel T, Beyer S, Bode E, Bode HB, Bolten CJ, Choudhuri JV, Doss S, Elnakady YA, Frank B, Gaigalat L, Goesmann A, Groeger C, Gross F, Jelsbak L, Jelsbak L, Kalinowski J, Kegler C, Knauber T, Konietzny S, Kopp M, Krause L, Krug D, Linke B, Mahmud T, Martinez-Arias R, McHardy AC, Merai M, Meyer F, Mormann S, Munoz-Dorado J, Perez J, Pradella S, Rachid S, Raddatz G, Rosenau F, Ruckert C, Sasse F, Scharfe M, Schuster SC, Suen G, Treuner-Lange A, Velicer GJ, Vorholter FJ, Weissman KJ, Welch RD, Wenzel SC, Whitworth DE, Wilhelm S, Wittmann C, Blocker H, Puhler A, Muller R (2007) Complete genome sequence of the myxobacterium Sorangium cellulosum. Nat Biotechnol 25(11):1281–1289

Schubert I, Vu GT (2016) Genome stability and evolution: attempting a holistic view. Trends Plant Sci 21(9):749–757

Shahinnia F, Druka A, Franckowiak J, Morgante M, Waugh R, Stein N (2012) High resolution mapping of Dense spike-ar (dsp.ar) to the genetic centromere of barley chromosome 7H. Theor Appl Genet 124(2):373–384

Shapiro B, Hofreiter M (2014) A paleogenomic perspective on evolution and gene function: new insights from ancient DNA. Science 343(6169):1236573

Shearer LA, Anderson LK, de Jong H, Smit S, Goicoechea JL, Roe BA, Hua A, Giovannoni JJ, Stack SM (2014) Fluorescence in situ hybridization and optical mapping to correct scaffold arrangement in the tomato genome. G3 (Bethesda) 4(8):1395–1405

Shendure J, Balasubramanian S, Church GM, Gilbert W, Rogers J, Schloss JA, Waterston RH (2017) DNA sequencing at 40: past, present and future. Nature 550:345–353

Shulaev V, Sargent DJ, Crowhurst RN, Mockler TC, Folkerts O, Delcher AL, Jaiswal P, Mockaitis K, Liston A, Mane SP, Burns P, Davis TM, Slovin JP, Bassil N, Hellens RP, Evans C, Harkins T, Kodira C, Desany B, Crasta OR, Jensen RV, Allan AC, Michael TP, Setubal JC, Celton JM, Rees DJ, Williams KP, Holt SH, Ruiz Rojas JJ, Chatterjee M, Liu B, Silva H, Meisel L, Adato A, Filichkin SA, Troggio M, Viola R, Ashman TL, Wang H, Dharmawardhana P, Elser J, Raja R, Priest HD, Bryant DW Jr, Fox SE, Givan SA, Wilhelm LJ, Naithani S, Christoffels A, Salama DY, Carter J, Lopez Girona E, Zdepski A, Wang W, Kerstetter RA, Schwab W, Korban SS, Davik J, Monfort A, Denoyes-Rothan B, Arus P, Mittler R, Flinn B, Aharoni A, Bennetzen JL, Salzberg SL, Dickerman AW, Velasco R, Borodovsky M, Veilleux RE, Folta KM (2011) The genome of woodland strawberry (*Fragaria vesca*). Nat Genet 43(2):109–116

Shuter BJ, Thomas JE, Taylor WD, Zimmerman AM (1983) Phenotypic correlates of genomic DNA content in unicellular eukaryotes and other cells. Am Nat 122(1):26–44

Simao FA, Waterhouse RM, Ioannidis P, Kriventseva EV, Zdobnov EM (2015) BUSCO: assessing genome assembly and annotation completeness with single-copy orthologs. Bioinformatics 31(19):3210–3212

Simpson JT, Wong K, Jackman SD, Schein JE, Jones SJ, Birol I (2009) ABySS: a parallel assembler for short read sequence data. Genome Res 19(6):1117–1123

Singh R, Ong-Abdullah M, Low ET, Manaf MA, Rosli R, Nookiah R, Ooi LC, Ooi SE, Chan KL, Halim MA, Azizi N, Nagappan J, Bacher B, Lakey N, Smith SW, He D, Hogan M, Budiman MA, Lee EK, DeSalle R, Kudrna D, Goicoechea JL, Wing RA, Wilson RK, Fulton RS, Ordway JM, Martienssen RA, Sambanthamurthi R (2013) Oil palm genome sequence reveals divergence of interfertile species in old and new worlds. Nature 500(7462):335–339

Slater GS, Birney E (2005) Automated generation of heuristics for biological sequence comparison. BMC Bioinformatics 6:31

Smit AFA, Hubley R, Green P (2013) RepeatMasker open-4.0. 2013–2015. http://www.repeatmasker.org. Accessed 18 Feb 2018

Smith DR, Lee RW (2014) A plastid without a genome: evidence from the nonphotosynthetic green algal genus *Polytomella*. Plant Physiol 164(4):1812–1819

Smith LM, Sanders JZ, Kaiser RJ, Hughes P, Dodd C, Connell CR, Heiner C, Kent SBH, Hood LE (1986) Fluorescence detection in automated DNA sequence analysis. Nature 321:674

Smith SA, Beaulieu JM, Donoghue MJ (2009) Mega-phylogeny approach for comparative biology: an alternative to supertree and supermatrix approaches. BMC Evol Biol 9:37

Solovyev V, Kosarev P, Seledsov I, Vorobyev D (2006) Automatic annotation of eukaryotic genes, pseudogenes and promoters. Genome Biol 7(Suppl 1):S10.11–S10.12

Stajich JE, Block D, Boulez K, Brenner SE, Chervitz SA, Dagdigian C, Fuellen G, Gilbert JG, Korf I, Lapp H, Lehvaslaiho H, Matsalla C, Mungall CJ, Osborne BI, Pocock MR, Schattner P, Senger M, Stein LD, Stupka E, Wilkinson MD, Birney E (2002) The Bioperl toolkit: perl modules for the life sciences. Genome Res 12(10):1611–1618

Stamatakis A (2006) RAxML-VI-HPC: maximum likelihood-based phylogenetic analyses with thousands of taxa and mixed models. Bioinformatics 22(21):2688–2690

Stanke M, Steinkamp R, Waack S, Morgenstern B (2004) AUGUSTUS: a web server for gene finding in eukaryotes. Nucleic Acids Res 32(Web Server issue):W309–W312

Supek F, Bosnjak M, Skunca N, Smuc T (2011) REVIGO summarizes and visualizes long lists of gene ontology terms. PLoS One 6(7):e21800

Swift H (1950) The constancy of desoxyribose nucleic acid in plant nuclei. Proc Natl Acad Sci U S A 36(11):643–654

Thrash A, Arick M 2nd, Peterson DG (2018) Quack: a quality assurance tool for high throughput sequence data. Anal Biochem 548:38–43

Tomato Genome Consortium (2012) The tomato genome sequence provides insights into fleshy fruit evolution. Nature 485(7400):635–641

Trapnell C, Roberts A, Goff L, Pertea G, Kim D, Kelley DR, Pimentel H, Salzberg SL, Rinn JL, Pachter L (2012) Differential gene and transcript expression analysis of RNA-seq experiments with TopHat and Cufflinks. Nat Protoc 7(3):562–578

Tuskan GA, Difazio S, Jansson S, Bohlmann J, Grigoriev I, Hellsten U, Putnam N, Ralph S, Rombauts S, Salamov A, Schein J, Sterck L, Aerts A, Bhalerao RR, Bhalerao RP, Blaudez D, Boerjan W, Brun A, Brunner A, Busov V, Campbell M, Carlson J, Chalot M, Chapman J, Chen GL, Cooper D, Coutinho PM, Couturier J, Covert S, Cronk Q, Cunningham R, Davis J, Degroeve S, Dejardin A, Depamphilis C, Detter J, Dirks B, Dubchak I, Duplessis S, Ehlting J, Ellis B, Gendler K, Goodstein D, Gribskov M, Grimwood J, Groover A, Gunter L, Hamberger B, Heinze B, Helariutta Y, Henrissat B, Holligan D, Holt R, Huang W, Islam-Faridi N, Jones S, Jones-Rhoades M, Jorgensen R, Joshi C, Kangasjarvi J, Karlsson J, Kelleher C, Kirkpatrick R, Kirst M, Kohler A, Kalluri U, Larimer F, Leebens-Mack J, Leple JC, Locascio P, Lou Y, Lucas S, Martin F, Montanini B, Napoli C, Nelson DR, Nelson C, Nieminen K, Nilsson O, Pereda V, Peter G, Philippe R, Pilate G, Poliakov A, Razumovskaya J, Richardson P, Rinaldi C, Ritland K, Rouze P, Ryaboy D, Schmutz J, Schrader J, Segerman B, Shin H, Siddiqui A, Sterky F, Terry A, Tsai CJ, Uberbacher E, Unneberg P, Vahala J, Wall K, Wessler S, Yang G, Yin T, Douglas C, Marra M, Sandberg G, Van de Peer Y, Rokhsar D (2006) The genome of black cottonwood, *Populus trichocarpa* (Torr. & Gray). Science 313 (5793):1596–1604

van den Berg BH, Konieczka JH, McCarthy FM, Burgess SC (2009) ArrayIDer: automated structural re-annotation pipeline for DNA microarrays. BMC Bioinformatics 10:30

van Regenmortel MH, Mahy BW (2004) Emerging issues in virus taxonomy. Emerg Infect Dis 10(1):8–13

Varshney RK, Chen W, Li Y, Bharti AK, Saxena RK, Schlueter JA, Donoghue MT, Azam S, Fan G, Whaley AM, Farmer AD, Sheridan J, Iwata A, Tuteja R, Penmetsa RV, Wu W, Upadhyaya HD, Yang SP, Shah T, Saxena KB, Michael T, McCombie WR, Yang B, Zhang G, Yang H, Wang J, Spillane C, Cook DR, May GD, Xu X, Jackson SA (2011) Draft genome sequence of pigeonpea (*Cajanus cajan*), an orphan legume crop of resource-poor farmers. Nat Biotechnol 30(1):83–89

Varshney RK, Song C, Saxena RK, Azam S, Yu S, Sharpe AG, Cannon S, Baek J, Rosen BD, Tar'an B, Millan T, Zhang X, Ramsay LD, Iwata A, Wang Y, Nelson W, Farmer AD, Gaur PM,

Soderlund C, Penmetsa RV, Xu C, Bharti AK, He W, Winter P, Zhao S, Hane JK, Carrasquilla-Garcia N, Condie JA, Upadhyaya HD, Luo MC, Thudi M, Gowda CL, Singh NP, Lichtenzveig J, Gali KK, Rubio J, Nadarajan N, Dolezel J, Bansal KC, Xu X, Edwards D, Zhang G, Kahl G, Gil J, Singh KB, Datta SK, Jackson SA, Wang J, Cook DR (2013) Draft genome sequence of chickpea (*Cicer arietinum*) provides a resource for trait improvement. Nat Biotechnol 31(3):240–246

Varshney RK, Shi C, Thudi M, Mariac C, Wallace J, Qi P, Zhang H, Zhao Y, Wang X, Rathore A, Srivastava RK, Chitikineni A, Fan G, Bajaj P, Punnuri S, Gupta SK, Wang H, Jiang Y, Couderc M, Katta M, Paudel DR, Mungra KD, Chen W, Harris-Shultz KR, Garg V, Desai N, Doddamani D, Kane NA, Conner JA, Ghatak A, Chaturvedi P, Subramaniam S, Yadav OP, Berthouly-Salazar C, Hamidou F, Wang J, Liang X, Clotault J, Upadhyaya HD, Cubry P, Rhone B, Gueye MC, Sunkar R, Dupuy C, Sparvoli F, Cheng S, Mahala RS, Singh B, Yadav RS, Lyons E, Datta SK, Hash CT, Devos KM, Buckler E, Bennetzen JL, Paterson AH, Ozias-Akins P, Grando S, Wang J, Mohapatra T, Weckwerth W, Reif JC, Liu X, Vigouroux Y, Xu X (2017) Pearl millet genome sequence provides a resource to improve agronomic traits in arid environments. Nat Biotechnol 35(10):969–976

Vaser R, Sovic I, Nagarajan N, Sikic M (2017) Fast and accurate de novo genome assembly from long uncorrected reads. Genome Res 27(5):737–746

Velasco R, Zharkikh A, Affourtit J, Dhingra A, Cestaro A, Kalyanaraman A, Fontana P, Bhatnagar SK, Troggio M, Pruss D, Salvi S, Pindo M, Baldi P, Castelletti S, Cavaiuolo M, Coppola G, Costa F, Cova V, Dal Ri A, Goremykin V, Komjanc M, Longhi S, Magnago P, Malacarne G, Malnoy M, Micheletti D, Moretto M, Perazzolli M, Si-Ammour A, Vezzulli S, Zini E, Eldredge G, Fitzgerald LM, Gutin N, Lanchbury J, Macalma T, Mitchell JT, Reid J, Wardell B, Kodira C, Chen Z, Desany B, Niazi F, Palmer M, Koepke T, Jiwan D, Schaeffer S, Krishnan V, Wu C, Chu VT, King ST, Vick J, Tao Q, Mraz A, Stormo A, Stormo K, Bogden R, Ederle D, Stella A, Vecchietti A, Kater MM, Masiero S, Lasserre P, Lespinasse Y, Allan AC, Bus V, Chagne D, Crowhurst RN, Gleave AP, Lavezzo E, Fawcett JA, Proost S, Rouze P, Sterck L, Toppo S, Lazzari B, Hellens RP, Durel CE, Gutin A, Bumgarner RE, Gardiner SE, Skolnick M, Egholm M, Van de Peer Y, Salamini F, Viola R (2010) The genome of the domesticated apple (*Malus* x *domestica* Borkh.) Nat Genet 42(10):833–839

Vlasova A, Capella-Gutierrez S, Rendon-Anaya M, Hernandez-Onate M, Minoche AE, Erb I, Camara F, Prieto-Barja P, Corvelo A, Sanseverino W, Westergaard G, Dohm JC, Pappas GJ Jr, Saburido-Alvarez S, Kedra D, Gonzalez I, Cozzuto L, Gomez-Garrido J, Aguilar-Moron MA, Andreu N, Aguilar OM, Garcia-Mas J, Zehnsdorf M, Vazquez MP, Delgado-Salinas A, Delaye L, Lowy E, Mentaberry A, Vianello-Brondani RP, Garcia JL, Alioto T, Sanchez F, Himmelbauer H, Santalla M, Notredame C, Gabaldon T, Herrera-Estrella A, Guigo R (2016) Genome and transcriptome analysis of the Mesoamerican common bean and the role of gene duplications in establishing tissue and temporal specialization of genes. Genome Biol 17:32

Vurture GW, Sedlazeck FJ, Nattestad M, Underwood CJ, Fang H, Gurtowski J, Schatz MC (2017) GenomeScope: fast reference-free genome profiling from short reads. Bioinformatics 33 (14):2202–2204

Walker BJ, Abeel T, Shea T, Priest M, Abouelliel A, Sakthikumar S, Cuomo CA, Zeng Q, Wortman J, Young SK, Earl AM (2014) Pilon: an integrated tool for comprehensive microbial variant detection and genome assembly improvement. PLoS One 9(11):e112963

Wan T, Liu ZM, Li LF, Leitch AR, Leitch IJ, Lohaus R, Liu ZJ, Xin HP, Gong YB, Liu Y, Wang WC, Chen LY, Yang Y, Kelly LJ, Yang J, Huang JL, Li Z, Liu P, Zhang L, Liu HM, Wang H, Deng SH, Liu M, Li J, Ma L, Liu Y, Lei Y, Xu W, Wu LQ, Liu F, Ma Q, Yu XR, Jiang Z, Zhang GQ, Li SH, Li RQ, Zhang SZ, Wang QF, Van de Peer Y, Zhang JB, Wang XM (2018) A genome for gnetophytes and early evolution of seed plants. Nat Plants 4(2):82–89

Wang X, Wang H, Wang J, Sun R, Wu J, Liu S, Bai Y, Mun JH, Bancroft I, Cheng F, Huang S, Li X, Hua W, Wang J, Wang X, Freeling M, Pires JC, Paterson AH, Chalhoub B, Wang B, Hayward A, Sharpe AG, Park BS, Weisshaar B, Liu B, Li B, Liu B, Tong C, Song C, Duran C, Peng C, Geng C, Koh C, Lin C, Edwards D, Mu D, Shen D, Soumpourou E, Li F, Fraser F,

Conant G, Lassalle G, King GJ, Bonnema G, Tang H, Wang H, Belcram H, Zhou H, Hirakawa H, Abe H, Guo H, Wang H, Jin H, Parkin IA, Batley J, Kim JS, Just J, Li J, Xu J, Deng J, Kim JA, Li J, Yu J, Meng J, Wang J, Min J, Poulain J, Wang J, Hatakeyama K, Wu K, Wang L, Fang L, Trick M, Links MG, Zhao M, Jin M, Ramchiary N, Drou N, Berkman PJ, Cai Q, Huang Q, Li R, Tabata S, Cheng S, Zhang S, Zhang S, Huang S, Sato S, Sun S, Kwon SJ, Choi SR, Lee TH, Fan W, Zhao X, Tan X, Xu X, Wang Y, Qiu Y, Yin Y, Li Y, Du Y, Liao Y, Lim Y, Narusaka Y, Wang Y, Wang Z, Li Z, Wang Z, Xiong Z, Zhang Z, Brassica rapa Genome Sequencing Project Consortium (2011) The genome of the mesopolyploid crop species *Brassica rapa*. Nat Genet 43(10):1035–1039

Wang K, Wang Z, Li F, Ye W, Wang J, Song G, Yue Z, Cong L, Shang H, Zhu S, Zou C, Li Q, Yuan Y, Lu C, Wei H, Gou C, Zheng Z, Yin Y, Zhang X, Liu K, Wang B, Song C, Shi N, Kohel RJ, Percy RG, Yu JZ, Zhu YX, Wang J, Yu S (2012a) The draft genome of a diploid cotton *Gossypium raimondii*. Nat Genet 44(10):1098–1103

Wang Y, Tang H, Debarry JD, Tan X, Li J, Wang X, Lee TH, Jin H, Marler B, Guo H, Kissinger JC, Paterson AH (2012b) MCScanX: a toolkit for detection and evolutionary analysis of gene synteny and collinearity. Nucleic Acids Res 40(7):e49

Wang L, Yu S, Tong C, Zhao Y, Liu Y, Song C, Zhang Y, Zhang X, Wang Y, Hua W, Li D, Li D, Li F, Yu J, Xu C, Han X, Huang S, Tai S, Wang J, Xu X, Li Y, Liu S, Varshney RK, Wang J, Zhang X (2014a) Genome sequencing of the high oil crop sesame provides insight into oil biosynthesis. Genome Biol 15(2):R39

Wang W, Haberer G, Gundlach H, Glasser C, Nussbaumer T, Luo MC, Lomsadze A, Borodovsky M, Kerstetter RA, Shanklin J, Byrant DW, Mockler TC, Appenroth KJ, Grimwood J, Jenkins J, Chow J, Choi C, Adam C, Cao XH, Fuchs J, Schubert I, Rokhsar D, Schmutz J, Michael TP, Mayer KF, Messing J (2014b) The *Spirodela polyrhiza* genome reveals insights into its neotenous reduction fast growth and aquatic lifestyle. Nat Commun 5:3311

Wang X, Xu Y, Zhang S, Cao L, Huang Y, Cheng J, Wu G, Tian S, Chen C, Liu Y, Yu H, Yang X, Lan H, Wang N, Wang L, Xu J, Jiang X, Xie Z, Tan M, Larkin RM, Chen LL, Ma BG, Ruan Y, Deng X, Xu Q (2017) Genomic analyses of primitive, wild and cultivated citrus provide insights into asexual reproduction. Nat Genet 49(5):765–772

Warren RL, Keeling CI, Yuen MM, Raymond A, Taylor GA, Vandervalk BP, Mohamadi H, Paulino D, Chiu R, Jackman SD, Robertson G, Yang C, Boyle B, Hoffmann M, Weigel D, Nelson DR, Ritland C, Isabel N, Jaquish B, Yanchuk A, Bousquet J, Jones SJ, MacKay J, Birol I, Bohlmann J (2015a) Improved white spruce (Picea glauca) genome assemblies and annotation of large gene families of conifer terpenoid and phenolic defense metabolism. Plant J 83(2):189–212

Warren RL, Yang C, Vandervalk BP, Behsaz B, Lagman A, Jones SJ, Birol I (2015b) LINKS: scalable, alignment-free scaffolding of draft genomes with long reads. GigaScience 4:35

Wegrzyn JL, Lin BY, Zieve JJ, Dougherty WM, Martínez-García PJ, Koriabine M, Holtz-Morris A, deJong P, Crepeau M, Langley CH, Puiu D, Salzberg SL, Neale DB, Stevens KA (2013) Insights into the loblolly pine genome: characterization of BAC and fosmid sequences. PLoS One 8(9):e72439

Wetterstrand KA (2017) DNA sequencing costs: data from the NHGRI genome sequencing program (GSP). www.genome.gov/sequencingcostsdata. Accessed 20 Dec 2017

Willing EM, Rawat V, Mandakova T, Maumus F, James GV, Nordstrom KJ, Becker C, Warthmann N, Chica C, Szarzynska B, Zytnicki M, Albani MC, Kiefer C, Bergonzi S, Castaings L, Mateos JL, Berns MC, Bujdoso N, Piofczyk T, de Lorenzo L, Barrero-Sicilia C, Mateos I, Piednoel M, Hagmann J, Chen-Min-Tao R, Iglesias-Fernandez R, Schuster SC, Alonso-Blanco C, Roudier F, Carbonero P, Paz-Ares J, Davis SJ, Pecinka A, Quesneville H, Colot V, Lysak MA, Weigel D, Coupland G, Schneeberger K (2015) Genome expansion of *Arabis alpina* linked with retrotransposition and reduced symmetric DNA methylation. Nat Plants 1:14023

Wilson RH, Coverley D (2013) Relationship between DNA replication and the nuclear matrix. Genes Cells 18(1):17–31

Wolf PG, Sessa EB, Marchant DB, Li FW, Rothfels CJ, Sigel EM, Gitzendanner MA, Visger CJ, Banks JA, Soltis DE, Soltis PS, Pryer KM, Der JP (2015) An exploration into fern genome space. Genome Biol Evol 7(9):2533–2544

Wong S, Butler G, Wolfe KH (2002) Gene order evolution and paleopolyploidy in hemiascomycete yeasts. Proc Natl Acad Sci U S A 99(14):9272–9277

Woodcock CL, Ghosh RP (2010) Chromatin higher-order structure and dynamics. Cold Spring Harb Perspect Biol 2(5):a000596

Worden AZ, Lee JH, Mock T, Rouze P, Simmons MP, Aerts AL, Allen AE, Cuvelier ML, Derelle E, Everett MV, Foulon E, Grimwood J, Gundlach H, Henrissat B, Napoli C, McDonald SM, Parker MS, Rombauts S, Salamov A, Von Dassow P, Badger JH, Coutinho PM, Demir E, Dubchak I, Gentemann C, Eikrem W, Gready JE, John U, Lanier W, Lindquist EA, Lucas S, Mayer KF, Moreau H, Not F, Otillar R, Panaud O, Pangilinan J, Paulsen I, Piegu B, Poliakov A, Robbens S, Schmutz J, Toulza E, Wyss T, Zelensky A, Zhou K, Armbrust EV, Bhattacharya D, Goodenough UW, Van de Peer Y, Grigoriev IV (2009) Green evolution and dynamic adaptations revealed by genomes of the marine picoeukaryotes *Micromonas*. Science 324 (5924):268–272

Woycicki R, Witkowicz J, Gawronski P, Dabrowska J, Lomsadze A, Pawelkowicz M, Siedlecka E, Yagi K, Plader W, Seroczynska A, Smiech M, Gutman W, Niemirowicz-Szczytt K, Bartoszewski G, Tagashira N, Hoshi Y, Borodovsky M, Karpinski S, Malepszy S, Przybecki Z (2011) The genome sequence of the North-European cucumber (*Cucumis sativus* L.) unravels evolutionary adaptation mechanisms in plants. PLoS One 6(7):e22728

Wu Y, Kikuchi S, Yan H, Zhang W, Rosenbaum H, Iniguez AL, Jiang J (2011) Euchromatic subdomains in rice centromeres are associated with genes and transcription. Plant Cell 23(11):4054–4064

Xu Q, Chen LL, Ruan X, Chen D, Zhu A, Chen C, Bertrand D, Jiao WB, Hao BH, Lyon MP, Chen J, Gao S, Xing F, Lan H, Chang JW, Ge X, Lei Y, Hu Q, Miao Y, Wang L, Xiao S, Biswas MK, Zeng W, Guo F, Cao H, Yang X, Xu XW, Cheng YJ, Xu J, Liu JH, Luo OJ, Tang Z, Guo WW, Kuang H, Zhang HY, Roose ML, Nagarajan N, Deng XX, Ruan Y (2013) The draft genome of sweet orange (*Citrus sinensis*). Nat Genet 45(1):59–66

Yan H, Jiang J (2007) Rice as a model for centromere and heterochromatin research. Chromosom Res 15(1):77–84

Yang H, Tao Y, Zheng Z, Zhang Q, Zhou G, Sweetingham MW, Howieson JG, Li C (2013) Draft genome sequence, and a sequence-defined genetic linkage map of the legume crop species *Lupinus angustifolius* L. PLoS One 8(5):e64799

Yang K, Tian Z, Chen C, Luo L, Zhao B, Wang Z, Yu L, Li Y, Sun Y, Li W, Chen Y, Li Y, Zhang Y, Ai D, Zhao J, Shang C, Ma Y, Wu B, Wang M, Gao L, Sun D, Zhang P, Guo F, Wang W, Li Y, Wang J, Varshney RK, Wang J, Ling HQ, Wan P (2015) Genome sequencing of adzuki bean (*Vigna angularis*) provides insight into high starch and low fat accumulation and domestication. Proc Natl Acad Sci U S A 112(43):13213–13218

Yang J, Liu D, Wang X, Ji C, Cheng F, Liu B, Hu Z, Chen S, Pental D, Ju Y, Yao P, Li X, Xie K, Zhang J, Wang J, Liu F, Ma W, Shopan J, Zheng H, Mackenzie SA, Zhang M (2016) The genome sequence of allopolyploid *Brassica juncea* and analysis of differential homoeolog gene expression influencing selection. Nat Genet 48(10):1225–1232

Young ND, Debelle F, Oldroyd GE, Geurts R, Cannon SB, Udvardi MK, Benedito VA, Mayer KF, Gouzy J, Schoof H, Van de Peer Y, Proost S, Cook DR, Meyers BC, Spannagl M, Cheung F, De Mita S, Krishnakumar V, Gundlach H, Zhou S, Mudge J, Bharti AK, Murray JD, Naoumkina MA, Rosen B, Silverstein KA, Tang H, Rombauts S, Zhao PX, Zhou P, Barbe V, Bardou P, Bechner M, Bellec A, Berger A, Berges H, Bidwell S, Bisseling T, Choisne N, Couloux A, Denny R, Deshpande S, Dai X, Doyle JJ, Dudez AM, Farmer AD, Fouteau S, Franken C, Gibelin C, Gish J, Goldstein S, Gonzalez AJ, Green PJ, Hallab A, Hartog M, Hua A, Humphray SJ, Jeong DH, Jing Y, Jocker A, Kenton SM, Kim DJ, Klee K, Lai H, Lang C, Lin S, Macmil SL, Magdelenat G, Matthews L, McCorrison J, Monaghan EL, Mun JH, Najar FZ, Nicholson C, Noirot C, O'Bleness M, Paule CR, Poulain J, Prion F, Qin B, Qu C, Retzel EF, Riddle C,

Sallet E, Samain S, Samson N, Sanders I, Saurat O, Scarpelli C, Schiex T, Segurens B, Severin AJ, Sherrier DJ, Shi R, Sims S, Singer SR, Sinharoy S, Sterck L, Viollet A, Wang BB, Wang K, Wang M, Wang X, Warfsmann J, Weissenbach J, White DD, White JD, Wiley GB, Wincker P, Xing Y, Yang L, Yao Z, Ying F, Zhai J, Zhou L, Zuber A, Denarie J, Dixon RA, May GD, Schwartz DC, Rogers J, Quetier F, Town CD, Roe BA (2011) The *Medicago* genome provides insight into the evolution of rhizobial symbioses. Nature 480(7378):520–524

Yu J, Hu S, Wang J, Wong GK, Li S, Liu B, Deng Y, Dai L, Zhou Y, Zhang X, Cao M, Liu J, Sun J, Tang J, Chen Y, Huang X, Lin W, Ye C, Tong W, Cong L, Geng J, Han Y, Li L, Li W, Hu G, Huang X, Li W, Li J, Liu Z, Li L, Liu J, Qi Q, Liu J, Li L, Li T, Wang X, Lu H, Wu T, Zhu M, Ni P, Han H, Dong W, Ren X, Feng X, Cui P, Li X, Wang H, Xu X, Zhai W, Xu Z, Zhang J, He S, Zhang J, Xu J, Zhang K, Zheng X, Dong J, Zeng W, Tao L, Ye J, Tan J, Ren X, Chen X, He J, Liu D, Tian W, Tian C, Xia H, Bao Q, Li G, Gao H, Cao T, Wang J, Zhao W, Li P, Chen W, Wang X, Zhang Y, Hu J, Wang J, Liu S, Yang J, Zhang G, Xiong Y, Li Z, Mao L, Zhou C, Zhu Z, Chen R, Hao B, Zheng W, Chen S, Guo W, Li G, Liu S, Tao M, Wang J, Zhu L, Yuan L, Yang H (2002) A draft sequence of the rice genome (*Oryza sativa* L. ssp. *indica*). Science 296(5565):79–92

Zaharia M, Bolosky WJ, Curtis K, Fox A, Patterson D, Shenker S, Stoica I, Karp RM, Sittler T (2011) Faster and more accurate sequence alignment with SNAP. arXiv 1111.5572v1: November 2011

Zdobnov EM, Apweiler R (2001) InterProScan – an integration platform for the signature-recognition methods in InterPro. Bioinformatics 17(9):847–848

Zerbino DR, Birney E (2008) Velvet: algorithms for de novo short read assembly using de Bruijn graphs. Genome Res 18(5):821–829

Zhang J, Carey V, Gentleman R (2003) An extensible application for assembling annotation for genomic data. Bioinformatics 19(1):155–156

Zhang Q, Chen W, Sun L, Zhao F, Huang B, Yang W, Tao Y, Wang J, Yuan Z, Fan G, Xing Z, Han C, Pan H, Zhong X, Shi W, Liang X, Du D, Sun F, Xu Z, Hao R, Lv T, Lv Y, Zheng Z, Sun M, Luo L, Cai M, Gao Y, Wang J, Yin Y, Xu X, Cheng T, Wang J (2012) The genome of *Prunus mume*. Nat Commun 3:1318

Zhang T, Hu Y, Jiang W, Fang L, Guan X, Chen J, Zhang J, Saski CA, Scheffler BE, Stelly DM, Hulse-Kemp AM, Wan Q, Liu B, Liu C, Wang S, Pan M, Wang Y, Wang D, Ye W, Chang L, Zhang W, Song Q, Kirkbride RC, Chen X, Dennis E, Llewellyn DJ, Peterson DG, Thaxton P, Jones DC, Wang Q, Xu X, Zhang H, Wu H, Zhou L, Mei G, Chen S, Tian Y, Xiang D, Li X, Ding J, Zuo Q, Tao L, Liu Y, Li J, Lin Y, Hui Y, Cao Z, Cai C, Zhu X, Jiang Z, Zhou B, Guo W, Li R, Chen ZJ (2015a) Sequencing of allotetraploid cotton (*Gossypium hirsutum* L. acc. TM-1) provides a resource for fiber improvement. Nat Biotechnol 33(5):531–537

Zhang Y, Zhang S, Liu H, Fu B, Li L, Xie M, Song Y, Li X, Cai J, Wan W, Kui L, Huang H, Lyu J, Dong Y, Wang W, Huang L, Zhang J, Yang Q, Shan Q, Li Q, Huang W, Tao D, Wang M, Chen M, Yu Y, Wing RA, Wang W, Hu F (2015b) Genome and comparative transcriptomics of African wild rice *Oryza longistaminata* provide insights into molecular mechanism of rhizomatousness and self-incompatibility. Mol Plant 8(11):1683–1686

Zimin AV, Stevens KA, Crepeau MW, Puiu D, Wegrzyn JL, Yorke JA, Langley CH, Neale DB, Salzberg SL (2017) An improved assembly of the loblolly pine mega-genome using long-read single-molecule sequencing. GigaScience 6(1):1–4

NAC Transcription Factors in Woody Plants

M[a] Belén Pascual, Fernando de la Torre, Rafael A. Cañas, Francisco M. Cánovas, and Concepción Ávila

Contents

Abstract NAC proteins are a family of plant-specific transcription factors (TFs) that are involved in the regulation of diverse biological processes, including stress tolerance, the formation and differentiation of vascular tissues and wood formation, the last of which is particularly important in trees. Evolutionary studies conducted on NAC proteins suggest a prominent role of these regulatory proteins in the adaptation and survival of plants to land habitats. The characterization of this family of TFs in trees is scarce, and the function of most of the family members remains unknown. However, current advances in sequencing technology, genomic analysis and expression studies from different woody plants have provided novel insights into the role of some of these TFs in trees. Recent research activity has mostly focused on the NAC proteins involved in processes

---

Communicated by Francisco M. Cánovas

M. B. Pascual (✉), F. de la Torre, R. A. Cañas, F. M. Cánovas, and C. Ávila
Departamento de Biología Molecular y Bioquímica, Facultad de Ciencias, Universidad de Málaga, Campus Universitario de Teatinos, Málaga, Spain
e-mail: bpascual@uma.es; fdelatorre@uma.es; rcanas@uma.es; canovas@uma.es; cavila@uma.es

Progress in Botany (2019) 80: 195–222, DOI 10.1007/124_2018_19,
Published online: 11 May 2018

related to improving plant biomass production, which is strongly affected by environmental conditions and has important ecological and economic implications. Studies performed in angiosperm and gymnosperm woody plants suggest that the regulatory network involved in stress responses and wood formation could be conserved.

1 Introduction

Trees represent nearly 80% of the plant biomass on Earth and have an important role as carbon sinks. Plant biomass production is strongly affected by environmental conditions, and its regulation is under complex and dynamic transcriptional control. In this sense, secondary vascular growth represents a key adaptation and a major carbon sink in woody plants. The regulation of gene expression plays a fundamental role in numerous biological processes, and transcription factors (TFs) are key regulatory players that act mainly at the transcriptional level. TFs activate or repress the expression of their target genes by binding to cis-regulatory elements to regulate different signalling pathways. The difficulty to elucidate the transcriptional regulatory network that controls a particular biological process is correlated to the complex network of TFs and/or regulatory molecules, such as miRNAs, controlling the expression of most genes (Hobert 2008; Hernández and Sanan-Mishra 2017).

The identification and organization of both TF families and their specific roles in transcriptional regulation are necessary steps to understand the regulatory networks associated with key processes in trees. TFs can be defined as proteins that generally present a DNA-binding domain (DBD) that recognizes specific DNA sequences in the promoters of target genes to regulate their expression. The classification of TFs into gene families is generally based on their characteristic DBDs (Riechmann et al. 2000), and their size varies considerably among organisms (Wray et al. 2003). Most of the existing knowledge about plant TFs was obtained from studies in *Arabidopsis thaliana*, which contains 2,296 TF genes classified into 58 families (Jin et al. 2013). This number is significantly larger than that found in *Drosophila melanogaster* (708 TFs, Hammonds et al. 2013) or *Caenorhabditis elegans* (934 putative TFs, Reece-Hoyes et al. 2005), which have genomes of a similar size (Riechmann et al. 2000). In addition, *Arabidopsis* shows a greater diversity of TF compared to these two species and to other animals and fungi, suggesting that transcriptional regulation may be more complex and diversified in plants as a consequence of an adaptive response as a result of their lifestyles. Approximately 75% of the TF families in common between plants and animals are more diversified in plants (Shiu et al. 2005). To date, 320,370 TFs from 165 species of green plants, classified into 58 families, have been identified. The average percentage of TFs in different taxonomic lineages of green plants ranges from 1.2% in Chlorophyta to 5.1% in Eudicots of the genomes encoding for putative TFs (Jin et al. 2017b), which often belong to large gene families, such as WRKY, bZIP, MYB, AP2/EREBP and NAC (Shao et al. 2015). Despite the differences in the genome size of green plants and in the total number of proteins

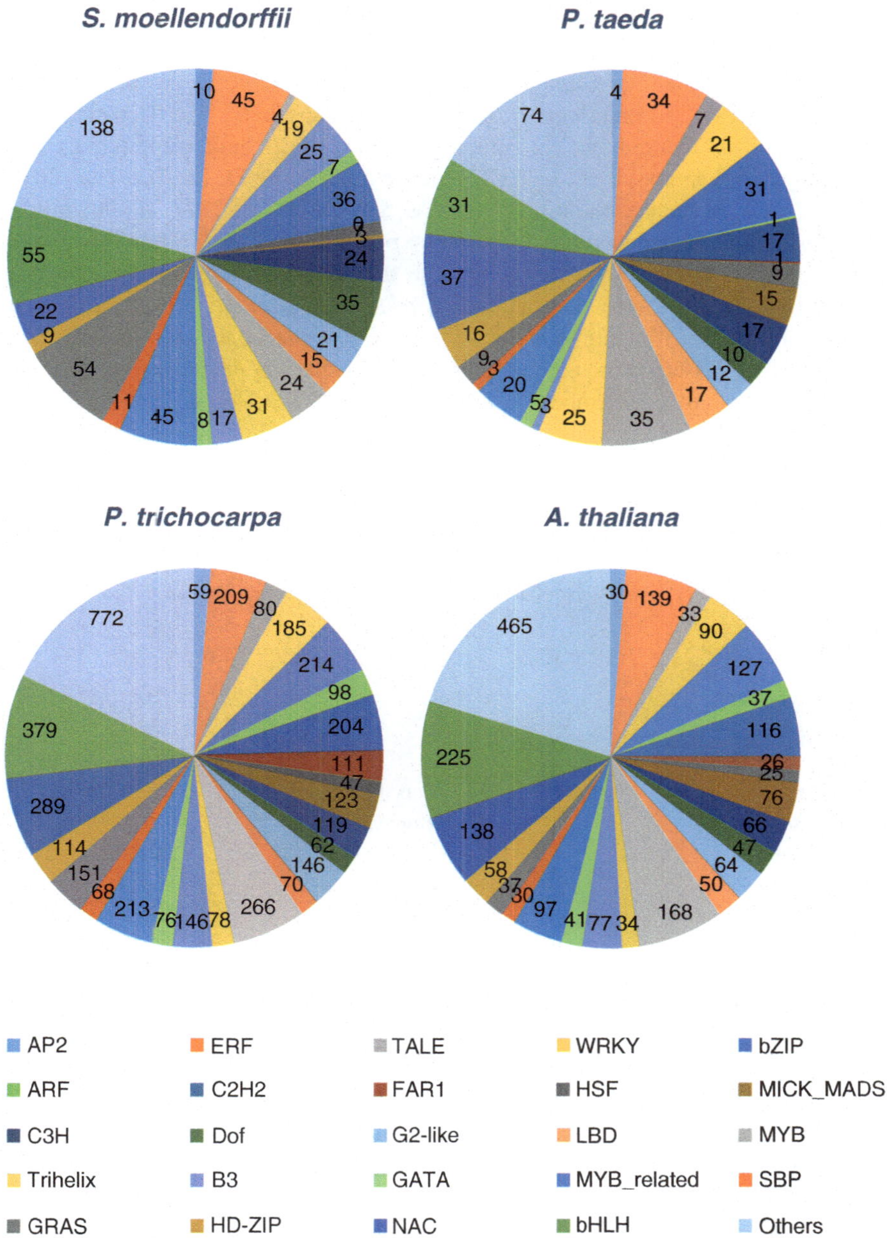

Fig. 1 Representation of the different families of transcription factors, as well as the number of members in each of them, in four plants: *Selaginella moellendorffii*, *Populus trichocarpa*, *Pinus taeda* and *Arabidopsis thaliana*. Those families with a number of members less than ten were not considered for representation. The distribution in each of the TF families was performed using data from the plant TF database PlantTFDB v4. 0 (Jin et al. 2017b)

encoding TFs, the percentage of TFs in each family is usually conserved (Fig. 1). In woody plants, although their genome size and TF number are different, the patterns of TF genes are similar. For example, *Vitis vinifera* contains 1,276 TF genes, *Eucalyptus grandis* contains 1,731 and *Populus trichocarpa* contains 2,466 TFs. This number of TF is higher than that reported for conifers (442–892), even though conifers have an extremely large genome size (20–30 Gb) (Canales et al. 2014; De La Torre et al. 2014; Jin et al. 2017b). In the maritime pine transcriptome, at least 877 TFs have been identified and are distributed into 30 families, but some of these TF families contain fewer members than do their counterparts in the genomes of several angiosperms. The Dof family has 47 members in *Arabidopsis*, 62 in poplar and over 30 in eucalyptus but only ten in maritime and loblolly pines (Rueda-López et al. 2013). Similarly, the NAC TF family includes more than 100 members in *Arabidopsis*, poplar and *Eucalyptus* but only 37 members in *Pinus pinaster* (Pascual et al. 2015) and 32 in *Picea glauca* (Raherison et al. 2015) (Table 1). These findings suggest that, as has occurred in other gene families, an expansion of the NAC family took place after the divergence of tracheophytes through gene duplication or whole-genome duplication, resulting in the appearance of paralogous genes with a high sequence similarity

Table 1 Number of NAC proteins found in different species of the green plant linage

Species	Green Linage	No. of NAC TFs
Arabidopsis thaliana	Eudicots	113
Eucalyptus grandis		202
Vitis vinifera		71
Populus trichocarpa		167
Oryza sativa	Monocots	123
Picea glauca	Coniferophyta	32
Pinus pinaster		37
Psilotum nudum	Monilophytes	21
Selaginella moellendorffii	Lycophytes	22
Leucodon brachypus	Mosses	16
Physcomitrella patens		34
Ricciocarpos natans	Liverworts	9
Nothoceros aenigmaticus	Hornworts	4
Roya obtusa	Zygnematophyceae	4
Cylindrocystis brebissonii		6
Coleochaete irregularis	Coleochaetales	4
Nitella mirabilis	Charales	1
Klebsormidium subtile		1
Klebsormidium flaccidum	Klebsormidiales	3
Chlorokybus atmophyticus	Chlorokybales	0
Mesostigma viride	Mesostigmatales	0
Uronema sp.	Chlorophyta	0
Monomastix opisthostigma		0

The number of NAC proteins has been taken from Maugarny-Calès et al. (2016), Pascual et al. (2015) and PlantTFDB database v4.0 (Jin et al. 2017b)

and functional redundancy. These duplications also contributed to the evolution of novel functions, such as adaptation to stress, disease resistance and improvements in traits of agronomic interest (grain quality and flowering time) (Panchy et al. 2016). The remarkable amplification of the NAC family across the plant kingdom reflects their numerous functions, including plant development, hormone signalling, grain nutrient remobilization, senescence, cell wall biosynthesis and biotic and abiotic stress responses. It is difficult to elucidate the regulatory role of these TFs since their response is mediated by interactions with other TFs or other regulatory proteins and because they show a complex regulation mediated by alternative splicing, miRNA and post-translational modifications such as phosphorylation and ubiquitination.

NAC proteins have a modular arrangement composed of a conserved N-terminal DNA-binding domain (DBD), known as the NAC domain, which is responsible for their oligomerization into dimeric proteins, and of a C-terminal domain (CTD) that is intrinsically disordered and possesses transcriptional regulatory activity. However, some variations from this common structure have been described (Puranik et al. 2012).

Several NAC proteins from *Arabidopsis* have been reported to be involved in different processes of plant biology; however, the studies performed in forest trees are much more limited, and the functions of most of these TF remain unknown. Recent advances in the acquisition of data from different woody plants allow us to have more information about the putative role of these TFs in trees. This chapter reviews the structural features, evolution and regulatory role of NAC proteins in key biological processes in trees, such as the response to biotic and abiotic stresses, xylogenesis and wood formation.

2 Common Structure of NAC Proteins

2.1 *The Conserved N-Terminal NAC Domain*

NAC TFs are modular proteins that are mainly formed by one highly conserved DNA-binding domain (named the NAC domain) in the N-terminal of the protein, which is responsible for binding to *cis*-elements of their target genes, and by a highly divergent transactivation domain placed in the C-terminal region. The NAC domain consists of approximately 160 amino acid residues and is divided into five subdomains (A–E), which have been implicated in nuclear localization and the formation of homodimers or heterodimers with other NAC domain proteins (Olsen et al. 2005) (Fig. 2). Subdomains A, C and D are usually highly conserved in most N-terminal NAC domains, while subdomains B and E are highly divergent and might confer functional diversity to NAC proteins (Ernst et al. 2004; Jensen et al. 2008). However, some NAC proteins have recently been reported that do not share this conserved feature. In *Brachypodium distachyon*, one NAC protein lacks the conserved A and B subdomains, and three members of the family do not contain the conserved C, D and E subdomains (You et al. 2015). *P. pinaster* NAC proteins belonging to Group II contain only subdomains A and D and lack subdomains B, C and E; these are replaced by

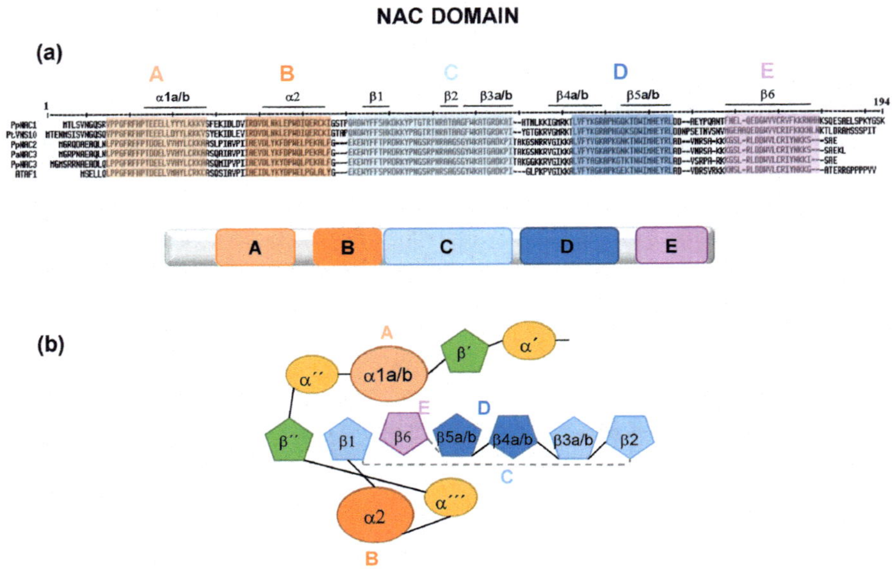

Fig. 2 Structure of NAC domain. (**a**) Sequence alignment of the N-terminal NAC domain from *A. thaliana* (ATAF1), *P. trichocarpa* (PtVNS10), *P. pinaster* (PpNAC1, PpNAC2 and PpNAC3) and *P. abies* (PaNAC3). Amino acid subdomains (A–E) in the NAC domain are represented by coloured boxes, and secondary structure elements (α, β helices) are indicated above the alignment. (**b**) Proposed topology diagram of the ANAC NAC domain according to Ernst et al. (2004). Two short helices (orange) surrounding a twisted β-sheet (blue and purple pentagons). For more details, see Ernst et al. (2004)

a motif homologous to the NAC domain (PF02365) (Pascual et al. 2015), which is present in some NAC proteins (BpNAC4 and BpNAC12) from *Betula platyphylla* (Guo et al. 2017) and in a small number of NAC proteins from other plants (Hussey et al. 2015). In other species, a premature stop codon after the NAC domain has been identified, while other NAC proteins present NAC domain repeats in tandem (Singh et al. 2013). Although some studies suggest that the typical structure of the five subdomains (A–E) or at least subdomains C and D are essential for the accurate functioning of the NAC TFs, additional research is needed to elucidate the role of these structural modifications in some family members.

Despite the numerous studies and scientific interest in the NAC family, only two crystal structures of NAC domains have been reported. One structure corresponds to the NAC domain from the *Arabidopsis* ANAC019 protein (Ernst et al. 2004), and the other corresponds to the rice SNAC1 protein (Chen et al. 2011). The NAC domain of ANAC019 shares structural similarity with AtNAC1 (Zhu et al. 2014) and StNAC from *Solanum tuberosum* (Singh et al. 2013) and BdNAC from *Brachypodium distachyon* (You et al. 2015) proteins. The NAC domain consists of two asymmetric short α-helix monomers surrounding a twisted β-sheet. Both monomers differ in the N-terminal tail region, which is in an extended conformation and is poorly defined in all structures (Ernst et al. 2004) (Fig. 2). The structure of the ANAC019/DNA complex based on X-ray crystallography has also been reported (Welner et al. 2012), showing that the β3-strand and, specifically, the motif WKATGTDK (Fig. 2) contact the major groove of the DNA. This region, which is localized in the C subdomain, is highly conserved, except for the second threonine, and an important structural and functional role has been proposed for the conserved glycine residue (Yoshiyama et al. 2009; Welner et al. 2016). To gain further insights into the structural feature of the NAC domain, additional X-ray crystallography and nuclear magnetic resonance spectroscopy studies would be required.

2.2 *The Regulatory Region Is Placed in the C-Terminal Domain*

The C-terminal domain plays an essential role in the regulation of transcription (Olsen et al. 2005), but it has also been shown to be involved in protein–protein interactions (Kim et al. 2007a; Jeong et al. 2009; Kjaersgaard et al. 2011). Despite being a highly variable and mostly disordered region with a high content of low-complexity amino acid repeats (Jensen et al. 2010), specific motifs have been identified in several NAC subgroups that have been evolutionarily maintained (Nuruzzaman et al. 2012). The W-motif (WNY) and L-motif (SLPPL) have been identified in NAC proteins from *Arabidopsis*, *Picea abies* and *P. pinaster* and seem to be associated with organ initiation and differentiation (Takada et al. 2001; Larsson et al. 2012; Pascual et al. 2015). The W-motif proved to be essential for transcriptional activation, but the function of the L-motif remains unknown. In the C-terminal region of stress-related NAC genes from *Arabidopsis* and *P. abies*, specifically those of subgroup III (ANAC019, ANAC055, ANAC072, ATAF1, ATAF2 and PaNAC03), a motif with a high negative charge density, SEKEE (V/I) QSSFRLE, that forms the transactivation domain has been identified (Jensen et al. 2010; Dalman et al. 2017). In silico analyses indicated that several NAC proteins, such as NAP, a positive regulator of senescence (Guo and Gan 2006), and SHYG (*S*peedy *Hy*ponastic *G*rowth), which regulates flooding-induced leaf movement in *Arabidopsis* (Rauf et al. 2013), have highly conserved intrinsic disorder profiles immediately behind the NAC domain (Stender et al. 2015). Bioinformatic analyses have, in some NAC proteins, identified a sequence motif that is substrate-specific for hydrolases and proteases. However, other members of the NAC family contain transmembrane motifs in the C-terminal region that are mostly formed by polar and negatively charged residues with conserved hydrophobic residues embedded in the polar matrix (Jensen et al. 2010). These NAC TFs are bound to the plasma membrane or endoplasmic reticulum and activated by proteases during stress responses (Seo et al. 2008, 2010; Chi et al. 2017). In summary, our knowledge of the function of these protein motifs is very limited, and further research is required to elucidate their biological importance. The combination of computational methods with experimental data could help identify regulatory motifs with high confidence for a particular TF that can be used to recognize interacting proteins (Stender et al. 2015). This approach could be used to reconstruct regulatory networks that are involved in the control of important processes in trees.

3 Classification and Evolution of the NAC Family

3.1 The Family of NAC Proteins

Phylogenetic analyses with NAC proteins from different plant species have shown that the TFs belonging to this family are involved in different aspects of plant growth, development and stress responses. They can be classified into different clades, suggesting that evolutionarily related NAC genes could play similar roles (Ooka et al. 2003; Shen et al. 2009; Jensen et al. 2010; Pascual et al. 2015). Most studies of the NAC TF family have been conducted using only the highly conserved N-terminal NAC subdomains A to E. A systematic analysis of NAC proteins in *Arabidopsis*, rice and soybean classified them into different subgroups (Ooka et al. 2003; Pinheiro et al. 2009), showing a clear relationship between structure and function. Shen et al. (2009) performed a genome-wide bioinformatics survey of 1,232 NAC proteins from 11 plant species, including woody plants such as poplar and grapevine, and showed that NAC proteins can be divided into eight subfamilies, termed NAC-a to NAC-h (Shen et al. 2009). More recently, a phylogenetic analysis of NAC proteins from maritime pine classified the members of this conifer family into six subfamilies according to those previously defined by Shen et al. (2009) (Pascual et al. 2015). The clade NAC-a includes proteins involved in responses to biotic and abiotic stress, and it is the largest subfamily in all species analysed, accounting for 16–34% of the total NAC proteins. This group included ATAF1, ATAF2, ANAC019 and ANAC55, which have been shown to be induced by drought, salinity and ABA and to enhance tolerance to drought when ectopically overexpressed (Tran et al. 2004). The clade NAC-b includes proteins with transmembrane motifs in their C-terminal domains that play an important role in cytokinin signalling during cell division or endoplasmic reticulum stress responses (Kim et al. 2007b; Silva et al. 2015). NAC proteins associated with secondary cell wall biosynthesis and wood formation are grouped into the clade NAC-c, and the proteins implicated in organ initiation and differentiation (CUC1/2/3 of *Arabidopsis*) belong to the NAC-d subfamily. The NAC-e subfamily includes proteins such as FEZ, which are involved in the orientation of cell division in root meristem (Willemsen et al. 2008), although very few proteins of this group have been characterized to date. NAC proteins involved in the secondary cell wall transcriptional network, such as SND2 and SND3 from *Arabidopsis* and PgNAC8 from *P. glauca*, are grouped in the NAC-g subfamily. OsIDEF2 (IRON DEFICIENCY-BINDING FACTOR 2) belongs to the NAC-f subfamily, and it regulates iron homeostasis-related genes in rice (Ogo et al. 2008; Shen et al. 2009), and SENU5 members are included in the NAC-h subfamily, which have been demonstrated to be involved in leaf senescence (Guo and Gan 2006; Shen et al. 2009). No NAC protein belonging to NAC-f was found in moss (*P. patens*, bryophyte), spike moss (*Selaginella moellendorffii*, lycophyte) or gymnosperms (*P. pinaster*), and the clade NAC-h was also absent in *P. patens* or *P. pinaster*, suggesting that these two subfamilies appeared later during evolution and possibly with more specific and specialized functions.

In the analysis of the NAC TF family in *P. trichocarpa*, *Arabidopsis* and rice (Hu et al. 2010), an additional subfamily (NAC-j) was identified that contained only proteins from *Populus*, suggesting that they may have been acquired in this species and they could have specialized roles related to the poplar perennial habitat. This analysis also revealed a subfamily containing only sequences from poplar and *Arabidopsis*, suggesting that these NAC genes were likely acquired or differentially retained in the post-divergence of dicots from monocots. When *E. grandis* and *Carica papaya* sequences were included in the phylogenetic analysis, a subfamily of NAC genes that contained only *P. trichocarpa*, *E. grandis* and *C. papaya* members was observed (Zhu et al. 2012; Hussey et al. 2015). This suggests that this subfamily may represent tree-specific NAC proteins involved in regulatory mechanisms controlling specialized aspects of wood formation (Zhu et al. 2012; Hussey et al. 2015).

In summary, the number of subfamilies proposed for the NAC TF family in different plants is highly variable (Ooka et al. 2003; Shen et al. 2009; Jensen et al. 2010; Zhu et al. 2012; Hussey et al. 2015), and additional subfamilies will likely be described in the future with increased number of plant genomes sequenced and analysed. However, most NAC subfamilies have representatives in the genomes of angiosperms, with some subfamilies restricted to tracheophytes, monocots, dicots or, exceptionally, the TNACS subfamily that appears restricted to the Solanaceae (Rushton et al. 2008; Singh et al. 2013).

3.2 Evolutionary Aspects of the NAC Family

All of the NAC TFs characterized so far belong to land plants (Zhu et al. 2012; Nagata et al. 2015; Pascual et al. 2015). These TFs are involved in many regulatory processes associated with vegetative and reproductive plant developmental programs (Larsson et al. 2012; Maugarny et al. 2015) or with senescence and biotic and abiotic stress responses (Puranik et al. 2012; Pascual et al. 2015). NAC proteins also regulate the formation and differentiation of the xylem and phloem, specialized tissues for transporting water and organic substances that are essential for land plant growth. For this reason and the lack of identification of NAC proteins within the Chlorophyta, the appearance of NAC TFs has always been associated with the emergence of land plants. However, through an analysis of transcriptomic data generated by the 1,000 Plants projects (www.onekp.com, Wickett et al. 2014) using data from species ranging from monilophytes to chlorophytes, putative homologous NAC TFs have been identified in some groups of streptophyte green algae, such as Zygnematophyceae and Coleochaetales. This analysis also identified NAC-like proteins in organisms belonging to Klebsormidiales and proteins containing a partial predicted NAC domain in *Nitella mirabilis*, a green algae of the Characeae family. However, no sequences corresponding to NAC proteins were identified in Chlorophyta (Maugarny-Calès et al. 2016).

Zygnematophyceae is the largest group within the streptophyte green algae, and it includes some species that can grow in terrestrial environments (Ettl and Gärtner 2014). The NAC proteins identified in Zygnematophyceae are proteins with a NAC domain having high homology with *Arabidopsis* NAC proteins associated with stress responses such as SOG1 (SUPPRESSOR OF GAMMA RESPONSE 1) (Yoshiyama 2016) and with vascular development such as SND2 (SECONDARY WALL-ASSOCIATED NAC DOMAIN PROTEIN 2) and SND3 involved in secondary cell wall formation. The identification of NAC TFs in this group of green algae could suggest an essential role for NAC proteins in protecting the cell against the stresses associated with the adaptation of plants to a terrestrial environment (Maugarny-Calès et al. 2016).

If the number of NAC genes present in different species is compared throughout evolution, an expansion of the NAC TF family is observed after the divergence of tracheophytes. This expansion appears to have arisen through gene duplication or whole-genome duplication, creating paralogous genes with a high degree of sequence similarity and functional redundancy (Singh et al. 2013; Wang et al. 2013b). Recent studies suggest that both tandem and segmental duplication might have played a key role in the expansion of the NAC family (Puranik et al. 2013; Singh et al. 2013; Wang et al. 2013b; You et al. 2015). In *E. grandis*, more than 60% of the NAC genes are distributed in tandem in their genome, while in *S. tuberosum*, *V. vinifera* and *P. trichocarpa*, such expansion has occurred mainly by segmental duplication (Hu et al. 2010; Singh et al. 2013; Wang et al. 2013b). In addition, although some duplicated genes lose their function throughout evolution, in *P. trichocarpa* and other species, it appears that a majority of the NAC genes generated by gene duplication have been retained by substantial subfunctionalization (Shang et al. 2013). With advances in genome sequencing and assembly techniques, an increase in the availability of full genome sequences from gymnosperm species and primitive vascular plants is enabling a further understanding of genome evolution across all plant linages. Recently, the draft genome sequence of *Gnetum montanum*, a gymnosperm species of gnetophytes, has been reported, and five NAC genes corresponding to the VNS group were identified. The authors proposed that these genes are orthologs of *VND4/5/6* from *Arabidopsis* and *VND* genes from gymnosperms, suggesting that the vessel-like water-conducting cells present in *Gnetum* are morphologically closer to conifer tracheids than to angiosperm vessels (Wan et al. 2018).

4 Regulation and Mechanism of Action of NAC Proteins

NAC proteins interact with many different types of proteins, including viral proteins, E3 ubiquitin ligases, phosphatases and kinases, using both their NAC and C-terminal domains (Welner et al. 2016). In addition, they activate or repress the expression of their target genes by binding to cis-regulatory elements named NAC recognition sequences (NACRS), whose consensus sequence varies depending on the NAC protein. NAC TFs that are involved in stress responses and senescence bind to a promoter region containing the CACG core DNA-binding motif. Other sequences, with GCTT or CA(A/C)G(T/C) (T/C/A)(T/C/A) as a core-binding motif, have been reported as NAC-binding sites (Kim et al. 2007a; Wang et al. 2017). One of the best characterized sequences to which NAC TFs are attached is a common cis-acting element of 19 bp, namely, the secondary wall NAC-binding element (SNBE) with (T/A)NN (C/T)(T/C/G)TNNNNNNNA(A/C)GN(A/C/T)(A/T) as a consensus sequence. This *cis*-element is present in many promoter genes involved in secondary wall biosynthesis, cell wall modification and programmed cell death (Zhong et al. 2010b; Wang et al. 2011), and the element is conserved in different species, including poplar, rice, eucalyptus and maritime pine (Ohtani et al. 2011; Zhong et al. 2011b; Pascual et al. 2017). It has also been shown that the DNA sequences flanking the core motif are in contact with the NAC domain (Welner et al. 2012) and contribute to both the affinity and the specificity of the target promoter binding.

For NAC proteins, in addition to regulating the expression of several target genes involved in multiple cellular processes, their expression can be regulated at three levels:

a. *Transcriptional regulation at the specific TF binding level.* The promoter region of *PpNAC2* and *PpNAC3*, two stress-responsive NAC genes of *Pinus pinaster*, contain several *cis*-elements involved in stress response: DPBF1 (ABA-responsive element); MYB binding sites; dehydration and low-temperature responsive elements, hormonal-responsive elements (jasmonic acid and salicylic acid) and ethylene-responsive GCC-boxes; and binding site of ERF1 transcription factors. Additionally, W-boxes were identified as being present in the promoter of defence genes (Rushton et al. 1996; Eulgem et al. 1999; Maleck et al. 2000). These cis-elements are the recognition sites for WRKY transcription factors (Eulgem et al. 1999).
b. *miRNA-mediated cleavage or alternative splicing at the post-transcriptional level.* NAC proteins are also potential targets of miRNAs. It has been observed that the miR164-targeted NAC domain TF module is conserved across species. For example, *CUC1* and *CUC2* of *Arabidopsis* are post-transcriptionally downregulated by miR164 cleavage, while *PaNAC01* of *P. abies*, a putative orthologue of *CUC* NAC genes of *Arabidopsis*, also harbours the miR164 recognition site. However, no miR164 and no NAC protein within the binding site for this microRNA has been identified in primitive vascular plants, such as *S. moellendorffii* and *P. patens* (Axtell et al. 2007). This finding suggests that this type of post-transcriptional regulation appeared before the separation between gymnosperm and angiosperms but after the origin of seed plants.

   In contrast, in *Populus*, it has been shown that *PtrSND1*, a gene involved in secondary cell wall biosynthesis, has alternative splicing variants that lack DNA-binding and transactivation abilities but retain dimerization capability. This splice variant protein (PtrSND1-A2) is a dominant negative of *PtrSND1* and can antagonize the autoregulation of its *PtrSND1* member genes, thus contributing to transcriptional homeostasis associated with wood formation and plant growth (Li et al. 2012).
c. *NAC post-translational modifications.* The NAC proteins can also undergo post-translational modifications such as ubiquitination, phosphorylation, proteolysis, dimerization and/or interaction with other proteins (Xie et al. 2002; Puranik et al. 2012). NTL6 is a membrane-associated NAC TF that has been implicated in cold-induced pathogen resistance, and its activity is regulated with

chemical modification. First, NTL6 is proteolytically activated to a mature nuclear-associated form (Seo et al. 2010), and it is then phosphorylated into a threonine residue (Thr142) of the NAC domain to facilitate the translocation of NTL6 into the nucleus to induce a drought-resistant response. The mutation of this threonine to alanine produced NTL6 proteins that were poorly phosphorylated and failed to move into the nucleus (Kim et al. 2012). The recent results of Guan et al. (2014) showed that RCF2/CPL1, a positive regulator of heat-stress responsive gene expression and thermotolerance in *Arabidopsis*, dephosphorylates ANAC019 and that both proteins, through protein–protein interactions, induce the synthesis of chaperones, which are important components of the response to heat stress (Guan et al. 2014).

5 Outstanding Roles for NAC Proteins

5.1 NAC Proteins Involved in Plant Stress Responses

Increased environmental stresses associated with climate change, particularly high temperature and prolonged drought, are critical factors that reduce tree growth and development and, therefore, forest productivity. Accordingly, it is important to understand how forest trees adapt to hostile environmental conditions, so it is necessary to increase our knowledge of the plant signalling pathways involved in the response to stress and to identify key regulatory proteins (Harfouche et al. 2014).

Many families of TFs, including DREB, bZIP, MYB, WRKY and NAC, regulate plant defences and stress responses in herbaceous plants (Seki et al. 2002; Mao et al. 2012; Shen et al. 2012). However, much less information is available about the regulatory proteins involved in the responses of woody plants to environmental stress. Even less information is available for conifers, which are economically and ecologically important trees in the Northern Hemisphere.

NAC proteins are among the most important families of TFs involved in adaptive stress response in plants (Wu et al. 2009; Puranik et al. 2012; Pascual et al. 2015). In addition, one of the putative NAC proteins identified in the group of green algae Zygnematophyceae presented high similarity with SOG1 of *Arabidopsis*, a NAC protein that has been implicated in the DNA damage response in plants (Yoshiyama 2016), suggesting that the evolutionary emergence of these TFs was directly associated with the appearance of novel defence mechanisms against stress.

Most stress-related NAC proteins characterized in *Arabidopsis*, rice, barley and woody plants are positive effectors that increase biotic and abiotic stress tolerance. For instance, the overexpression of ANAC019, ANAC055, ANAC072 (RD26), ATAF1 and ATAF2 in *Arabidopsis* improved drought and salt tolerance (Tran et al. 2004), although these transgenic plants also showed increased susceptibility to necrotrophic pathogens such as *Botrytis cinerea* and *Fusarium oxysporum* (Delessert et al. 2005; Wang et al. 2009; Wu et al. 2009). In addition, these genes showed co-expression in response to stress hormones (Bu et al. 2008; Wu et al. 2009; Jensen et al. 2010).

A similar response was observed for JcNAC1 from *Jatropha curcas*, a woody plant used for biodiesel production, whose overexpression produced plants with enhanced tolerance to drought and increased susceptibility to pathogens (Qin et al. 2014). OsNAC6 and OsNAC5 increased high-salinity and drought tolerance, and the overexpression of OsNAC10 produced rice plants with enhanced drought tolerance at the reproductive stage and increased the grain yield by 25–42% under drought conditions (Jin and Kim 2010). Several NAC proteins from *Brassica napus* were upregulated in response to biotic and abiotic stresses, including wounding, insect feeding, cold shock, pathogen infection and drought stress (Hegedus et al. 2003; Niu et al. 2014; Liu et al. 2015), and, interestingly, the overexpression of *BnNAC2* and *BnNAC5* genes in yeast enhanced the cells' sensitivity to high-salinity and osmotic stresses (Zhong et al. 2012).

In Norway spruce, orthologs of these stress-characterized *Arabidopsis* genes have been identified in transcriptome studies in response to biotic stress (Danielsson et al. 2011; Lunden et al. 2015), indicating that a regulatory network involved in plant responses to stress could be conserved between angiosperms and gymnosperms, despite their evolutionary divergence. Recently, it has been shown that *PaNAC03* is responsive to biotic stress and may be implicated in the control of defence-associated secondary metabolite production. Its overexpression produced aberrant embryos and several genes involved in flavonoid pathways such as *chalcone synthase* (*CHS*), *flavonoid-3′-hydroxylase* (*F3′H*) and *leucoanthocyanidin reductase 3* (*PaLAR3*) were misregulated. These lines also showed reduced levels of specific flavonoids (Dalman et al. 2017).

The characterization of CsNAN1 in *Citrus reshni* also showed an extensive similarity in its action mode with ANAC019, ANAC055 and ANAC072 from *Arabidopsis* and was strongly induced by drought stress, salt, cold and ABA (de Oliveira et al. 2011).

In addition, these TFs have been demonstrated to have a key function in the tolerance of some woody plants to unfavourable environmental conditions. In *Populus euphratica*, a woody plant that grows in dry deserts and maintains high growth and photosynthetic rates at high salinity, the expression of *PeNAC1* is increased by salt and drought stress, and its overexpression in *Arabidopsis* enhances tolerance to salt stress (Wang et al. 2013a). Similar results were observed in the woody plants *Paulownia tomentosa* for several NAC TFs (Zhao et al. 2017) and *Tamarix hispida* for the ThNAC13 TF (Wang et al. 2014). Transgenic plants overexpressing this TF had a higher activity of antioxidant enzymes, such as superoxide dismutase (SOD) and peroxidase (POD), and reduced reactive oxygen species levels compared to control plants under salt or osmotic stress. The presence of NACRS elements in the promoters of SOD and POD suggests that ThNAC13 might improve tolerance by regulating the transcription levels of these enzymes (Wang et al. 2017).

In a recent study, Chen et al. (2017) showed a significantly high expression level of NAC genes after abscisic acid (ABA) treatment in *Populus hopeiensis*, a poplar with high drought and cold tolerance (Wang et al. 2014). This plant hormone plays an essential role in mediating the adaptation to stress and regulates root growth (McAdam et al. 2016). In another woody plant species that is extremely cold-hardy and drought resistant, *Pyrus betulifolia*, the overexpression of a NAC gene, *PbeNAC1*, produced plants with enhanced cold and drought tolerance. This study shows that PbeNAC1 is involved in the cold and drought signalling network, in which NAC TFs regulate the expression of stress-responsive genes, probably by interacting with C-repeat binding factor (CBF) genes (Jin et al. 2017a).

Several *NAC* genes are also upregulated by phytohormones, such as jasmonic acid (JA), salicylic acid (SA) and ethylene (ET), which play key roles in plant defence responses to different pathogens, insect feeding and wounding (Nuruzzaman et al. 2013). ATAF1 induces JA-/ET-associated defence signalling genes (Jensen et al. 2008), while ATAF2 controls responses to JA and SA (Delessert et al. 2005). In the conifer *P. pinaster*, *PpNAC2* and *PpNAC3* encode stress-responsive NAC TFs that are involved in the jasmonate response. Both genes were strongly induced by mechanical wounding, high salinity and methyl jasmonate (MeJA), and the promoter region of these genes contains, among others, several E-boxes, which are *cis*-elements commonly found in MeJA-responsive promoters, and binding sites for bHLH TFs. This study suggests that the jasmonate signalling pathway might be conserved between angiosperms and gymnosperms (Pascual et al. 2015).

Although a single NAC gene may be mainly associated with a type of stress response, these NAC TFs do not act alone. Recent research has shown that NAC proteins interact with other TFs and proteins to exert their biological functions through complex regulatory networks in plants (Jensen and Skriver 2014; Barah et al. 2016). For this reason, it is essential to identify the key components of the signalling networks in which these factors interact with the objective of generating plants with improved yield and stress tolerance.

5.2 NAC Proteins Involved in Developmental Processes

The first NAC genes identified in a plant belong to the *NAM* class; the plant was a petunia mutant that failed to develop the shoot apical meristem (SAM) (Souer et al. 1996). The *CUC* genes in double mutants of *A. thaliana* showed a similar phenotype with no apical meristem and fused cotyledons (Aida et al. 1997). Three *CUC* genes have been identified and characterized in *Arabidopsis*, and they act together with the class-1 *KNOX* gene *SHOOT MERISTEMLESS (STM)* to establish the embryonal SAM and promote the formation of two separated cotyledons, which is critical for proper leaf and flower patterning (Aida et al. 1999; Scofield et al. 2013).

Phylogenetic analyses from different species have found the NAM/CUC proteins in the same subfamily, suggesting a functional similarity. Orthologs of these NAC class genes have been identified in *Zea mays* (Zimmermann and Werr 2005), *Solanum lycopersicum* (Berger et al. 2009), *Medicago truncatula* (Cheng et al. 2012), *Antirrhinum majus* (Weir et al. 2004), *P. pinaster* (Pascual et al. 2015) and *P. abies* (Larsson et al. 2012). Based on the sequence of the NAC domain, the proteins that belong to this group can be divided into two clades: the NAM clade, which includes the petunia NAM and *Arabidopsis* CUC1 and CUC2 proteins, and the CUC3 clade. In angiosperms, the NAM/CUC1/CUC2 genes contain a binding site for the microRNA164, while the CUC3 genes do not, but this classification is not maintained in gymnosperms (Larsson et al. 2012). In addition, the miR164 recognition site has not been found in any NAC gene in *P. patens* or *S. moellendorffii* (Axtell et al. 2007), suggesting that the miR164-mediated post-transcriptional regulation of the *CUC* genes was present in a common ancestor of angiosperms and that a duplication event generated two clades in the angiosperm lineage after its divergence from gymnosperms.

The characterization of *P. abies PaNAC01* and *PaNAC02* showed that *PaNAC01* is a functional ortholog of *CUC1* and *CUC2* in *Arabidopsis*, thus suggesting an evolutionary conservation of a regulatory pathway during somatic embryo development (Larsson et al. 2012). Another *Arabidopsis* NAC gene of this family, *NAC1*, and its orthologs in the woody plant *Callerya speciosa* play an essential role in the auxin-induced development of lateral roots (Xie et al. 2000; Xu et al. 2016).

5.3 NAC Proteins Involved in Wood Formation

Although the number of NAC TFs characterized in trees is very low, most research has focused on the NAC proteins involved in secondary cell wall biosynthesis, wood formation and processes related to improve plant biomass production (Zhong and Ye 2010; Ohtani et al. 2011; Pascual et al. 2017). Woody plant species irreversibly immobilize large quantities of carbon skeletons in wood during their long life cycles. Wood is the secondary xylem of vascular plants and is composed of lignin, cellulose and hemicellulose (Sjostrom 1993), and its properties and composition vary widely between angiosperms and gymnosperms (Pascual et al. 2016). The secondary xylem of gymnosperms is composed of single cells known as tracheids that serve as both water transport and mechanical support, while angiosperm wood is composed mainly of vessels that are conduits formed of many dead cells to conduct water and xylem fibres that provide structural support. Few studies on secondary cell wall biosynthesis and tracheid development have been carried out in conifers (Jokipii-Lukkari et al. 2017), which limits our current knowledge on the evolutionary origin and the structural and functional diversification of the angiosperm vessels with respect to the gymnosperm tracheids or vessel-like water-conducting cells present in primitive vascular plants (Wan et al. 2018).

The subfamily of NAC genes involved in vascular development is formed by three types of proteins: (1) VASCULAR-RELATED NAC DOMAIN (VND) proteins, whose expression is associated with developing vascular tissues and might regulate vessels formation. (2) NAC SECONDARY WALL THICKENING PROMOTING FACTOR 1 (NST1) and NST2/SECONDARY WALL-ASSOCIATED NAC DOMAIN PROTEIN1 (SND1) working as regulators for fibre cell differentiation (Zhong et al. 2006; Mitsuda et al. 2007). (3) SOMBRERO (SMB), BEARSKIN1 (BRN1) and BRN2 proteins that are expressed in the root cap region, and their overexpression causes ectopic secondary cell wall deposition (Willemsen et al. 2008; Bennett et al. 2010).

All these NAC proteins involved in the biosynthesis of secondary cell wall (SCW) have been grouped as VNS (*V*ND, *N*ST/SND, *S*BM) subfamily (Ohtani et al. 2011; Xu et al. 2014; Pascual et al. 2017). Interestingly, the studies published so far show that there is no correlation between the number of VNS genes present in each species and the size of its genome or the abundance of lignified tissues (Zhu et al. 2012; Nakano et al. 2015). Table 2 shows the number of VNS genes identified in several woody plants. *P. trichocarpa* have 16 VNS genes in its genome (Zhong et al. 2010b; Ohtani et al. 2011; Li et al. 2012), while *V. vinifera* and *E. grandis* have eight and seven VNS genes, respectively (Zhu et al. 2012; Hussey et al. 2015), a similar number that the moss *P. patens* (Xu et al. 2014), a non-vascular land plant, have. In *P. abies* (Nystedt et al. 2013), *P. glauca* (Duval et al. 2014) and *P. pinaster* (Pascual et al. 2015, 2017), conifer species with long life cycles and in which the biosynthesis of lignin is a quantitatively very important process (Pascual et al. 2016), between two and four VNS genes have been identified. Five VNS genes have also been recognized in the genome of *G. montanum*, of smaller size (2–4 Gb) than the genomes of other gymnosperms (20–30 Gb).

The VND subgroup is usually the most diversified, representing about 50% of the VNS proteins in most species, while the NST and SMB groups consist of one to four genes. In *Arabidopsis* seven VND genes have been characterized, termed *VND1–7*, involved in the transcriptional regulation of secondary wall biosynthesis in vessels, three NST genes (*AtNST1*, *AtNST2* and *AtNST3/SND1*) and three SMB genes (*AtBRN1*, *AtBRN2* and *AtSMB*) (Nakano et al. 2015). It has been shown that some poplar wood-associated NAC TFs (*PtrWNDs*) and a *Eucalyptus* wood-associated NAC (*EgWND1*) (Zhong and Ye 2010; Zhong et al. 2010a, b) are functional orthologs of the *Arabidopsis* SND1, NST1/2 and VND6/7 (Zhong and Ye 2007) and their overexpression in *Arabidopsis* produce plants with increased expression of secondary wall biosynthetic genes and ectopic deposition of secondary walls.

In *P. glauca* and *P. abies*, no members of the NST group have been identified so far. Duval et al. (2014) showed that *PgNAC7* is a VND-type gene functionally similar to *AtVND6/7* of Arabidopsis and its expression is preferentially associated with vascular tissue in the stem, while *PgNAC4* is grouped with the SMB proteins and its expression was related to root tips (Duval et al. 2014). In a recent study, three

Table 2 VNS genes in woody plants

	VNS GENES			
	VND	NST/SND	SMB	Reference
Populus trichocarpa	8	4	4	Zhong et al. (2012), Ohtani et al. (2011) and Li et al. (2012)
Carica papaya	3	1	1	Zhu et al. (2012)
Eucalyptus grandis	4	2	1	Hussey et al. (2015)
Vitis vinifera	4	2	2	Zhu et al. (2012)
Pinus pinaster	1	1	1	Pascual et al. (2015, 2017)
Picea abies	2	–	2	Nystedt et al. (2013)
Picea glauca	1	–	1	Duval et al. (2014)

VND genes were identified in *P. pinaster*, each of which is classified in a different subgroup. *PpNAC30* clustered into the VND group with *P. glauca PgNAC7* and *PtVNS01* and *PtVNS04* from *P. trichocarpa*, putative orthologs of *VND6/7* of *Arabidopsis*. The expression of *PpNAC30* is preferentially associated with vascular tissue of cotyledons and roots in pine seedlings (Fig. 3). *PpNAC31* belongs to the SMB group, and its expression is particularly high in developing root tissues (Fig. 3), similarly to the *PgNAC4* of white spruce (Duval et al. 2014). *PpNAC1* is the unique NST gene that has been identified in gymnosperms and whose expression is clearly predominant in tissues undergoing secondary cell wall biosynthesis such as secondary xylem and compression wood of adult trees (Pascual et al. 2017). It exhibits an expression pattern consistent with the previously reported for *PtrWNDS* in *Populus* and *SND1* and *NST1* in *Arabidopsis* (Zhong et al. 2007, 2010b). Its silencing in maritime pine causes several phenotypes associated with vascular development and a downregulation of the expression of several genes involved in cell wall biosynthesis and secondary metabolism (Pascual et al. 2017). The decrease of its expression was particularly important affecting the level of expression of certain Myb TF that have been associated with lignin biosynthesis in other conifers, such as Myb1, Myb4 and Myb8 (Patzlaff et al. 2003a, b; Bomal et al. 2008; Bedon et al. 2010; Craven-Bartle et al. 2013). Furthermore, the over-expression of this TF in *Arabidopsis* led to plants with small rosette size and curled leaves, phenotype characteristic of ectopic lignin deposition and increased secondary wall thickening (Pascual et al. 2017). These plants also showed the upregulation of secondary wall biosynthesis genes and secondary cell wall-associated TFs, mainly Myb TFs directly regulated by SND1, such as Myb46 and Myb83 that have been described as master switches of secondary wall biosynthesis in *Arabidopsis* (Zhong et al. 2007).

These results are consistent with previous studies carried out in poplar (Lin et al. 2013; Wang et al. 2014) and eucalyptus and strongly suggest a functional role of *PpNAC1* as an ortholog of the *Arabidopsis SND1* (Zhong et al. 2006) or *Populus PtrWND2B* and *PtrWND6B* (Zhong et al. 2010b, 2011a; Ohtani et al. 2011), supporting the idea that a transcriptional network involved in the regulation of xylem development could be conserved between angiosperms and gymnosperms.

Several members of the Myb and NAC TF families have been described as strong candidates participating in the transcriptional regulatory network that governs the biosynthesis of the main components of the secondary cell wall. More than 6,200 genes, including NAC and MYB transcription factors and genes involved in lignin and cellulose biosynthesis, exhibited differential expression during xylogenesis in *Neolamarckia cadamba*, a fast-growing tropical woody tree. The expression of these genes, which corresponds to most of the known transcription factors involved in regulatory networks in *A. thaliana*, was significantly higher in both the middle and basal stem compared to the apical stem. The characterization of genes and transcription factors involved in wood formation in trees has a great potential for increasing wood biomass production or modifying wood composition for biofuel production (Ouyang et al. 2016).

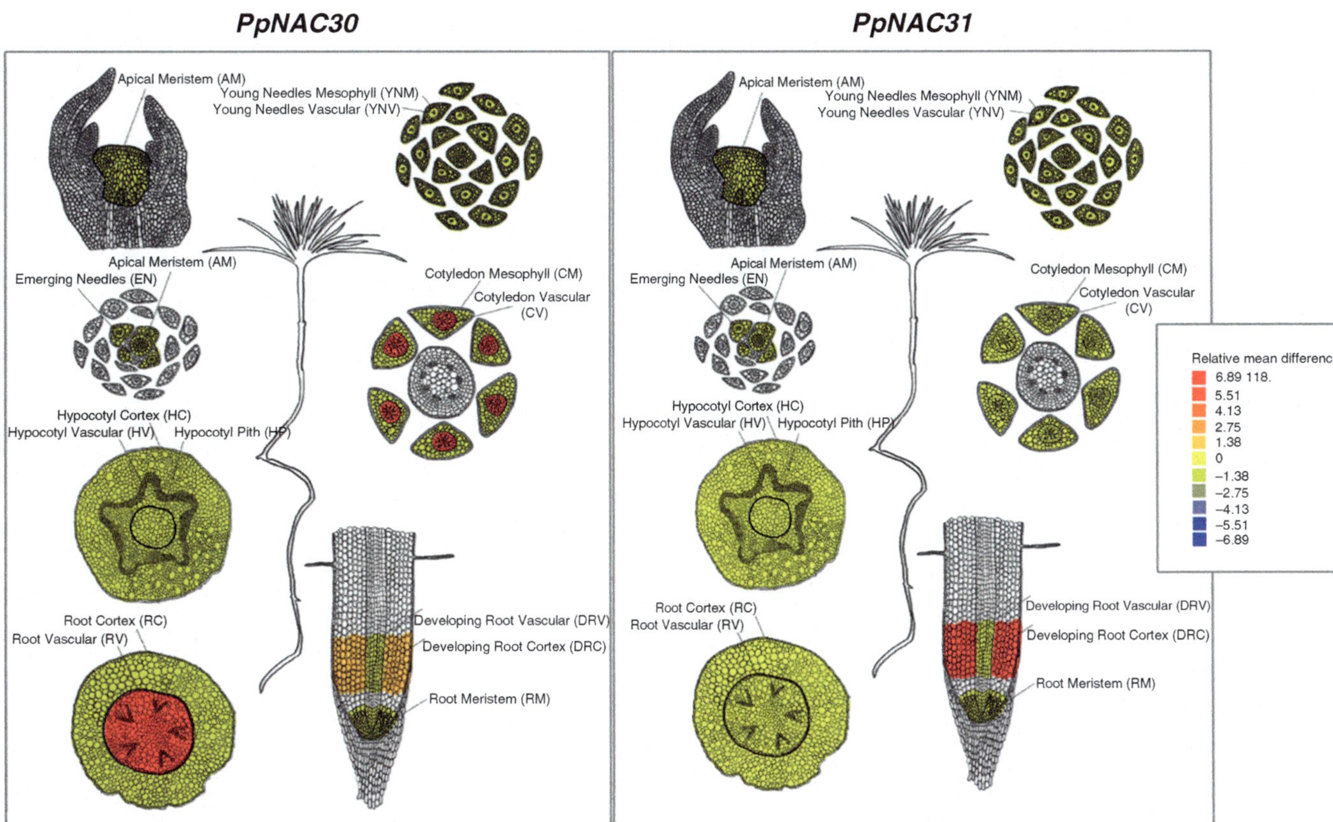

Fig. 3 Image capture using the exImage tool at ConGenIE.org (http://v22.popgenie.org/microdisection/) (Cañas et al. 2017) that shows the expression patterns of *PpNAC30* and *PpNAC31* in maritime pine seedlings

NAC TFs, such as PpNAC1 and PpNAC30 from *P. pinaster*, PgNAC7 from white spruce or PtrWND2B and PtrWND6B from *P. trichocarpa*, could act at the first level of the transcriptional regulatory network involved in wood formation, while Myb TFs could function at the second level, along with other MYB and NAC TFs (Fig. 4). For example, PtrMyb20 and PtrMyb3 from poplar and EgMyb2 and Pt/PpMyb4 from eucalyptus and pine, respectively, are functional orthologs of

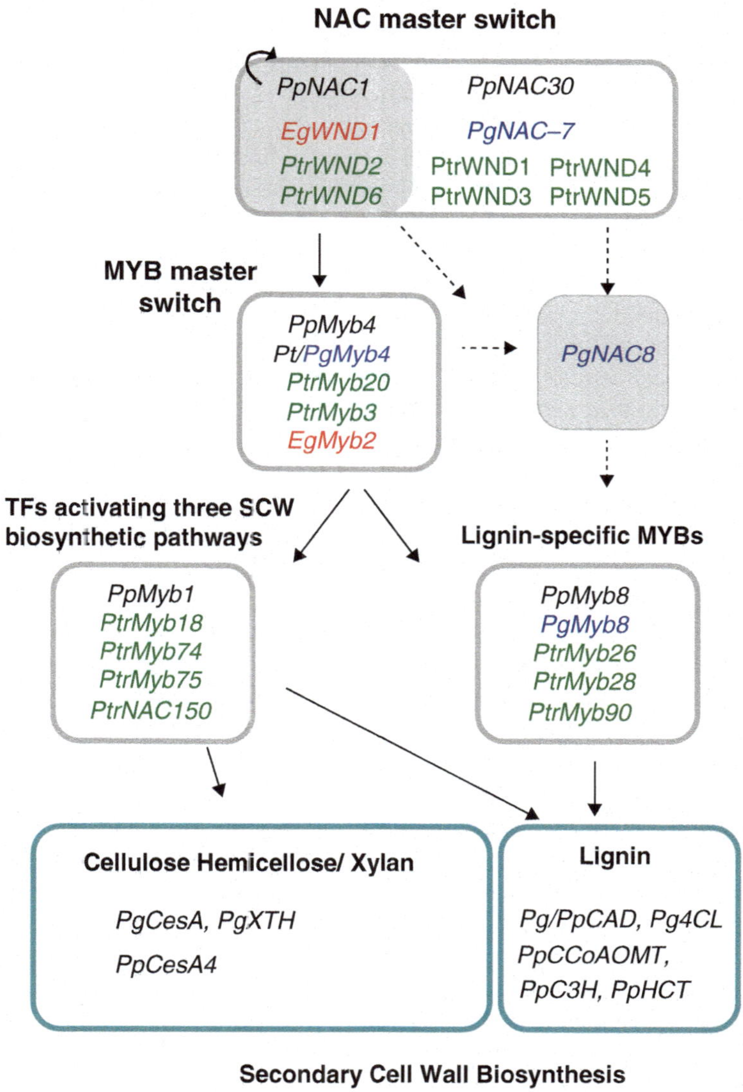

Fig. 4 Proposed transcriptional network controlling secondary cell wall biosynthesis in trees. NAC and Myb TFs from *P. pinaster*, *P. glauca and P. trichocarpa* are shown. Continuous lines indicate that the interactions have been functionally demonstrated

Arabidopsis Myb46 and Myb83 and regulate secondary wall biosynthesis during wood formation (Zhong et al. 2013). In *P. pinaster*, it has been shown that PpNAC1 can activate its own expression and regulate PpMyb4 by binding to the SNBE elements present in their promoter regions, while PpMyb4 can activate PpMyb8, a transcriptional regulator of lignin biosynthesis in maritime pine (Craven-Bartle et al. 2013; Pascual et al. 2017). In addition, Zhong et al. (2013) reported that PpMyb4 and its orthologs, PtMyb4 and EgMyb2, could be involved in the regulation of cellulose and xylan biosynthesis (Zhong et al. 2013). The cellulose and xylan biosynthetic pathways have been well described in plants (Mizrachi et al. 2012), but little information about their regulation in trees has been reported. The overexpression of SND2 in *Arabidopsis* showed an upregulation of biosynthetic genes associated with cellulose, xylan, mannan and lignin polymerization but with minor alteration of the cell wall chemistry. In addition, the effect of this gene on secondary cell wall thickness in *Arabidopsis* is unclear and may be dose-dependent (Hussey et al. 2011; Zhong et al. 2012). However, overexpression of this gene in eucalyptus stems significantly increased fibre cell area, and the overexpression of *Populus NAC154*, a putative ortholog of SND2, resulted also in an increase of bark production compared with xylem in poplar trees (Grant et al. 2010). These results show phenotypic differences of SND2 overexpression between woody and herbaceous plants that could be attributed to a greater channelling of carbon skeletons toward wood biosynthesis in woody plants (Hussey et al. 2011).

Duval et al. (2014) reported that the white spruce PgNAC8, a potential ortholog of SND2/SND3 of *Arabidopsis*, could regulate cellulose biosynthesis in coordination with PgNAC7 (Lamara et al. 2016). In a study of *B. platyphylla*, it was observed that *BplNAC1* and *BplNAC2* genes, which are closely related to *SND2* and *NST1* of *Arabidopsis*, were upregulated by exogenous gibberellin (GA3) application, a key signal molecule inducing xylem development and the expression of secondary wall biosynthesis-related genes such as *CESA*, *PAL* and *GA* oxidase (Guo et al. 2015).

In summary, several studies suggest that, despite the evolutionary distance and different properties and composition of wood between angiosperms and gymnosperms, many transcriptional regulators involved in the transcriptional network associated with secondary cell wall biosynthesis could be conserved. In angiosperms, secondary wall biosynthesis in vessels is regulated by VNS genes, with VND7 being a master regulator of vessel formation in *Arabidopsis* and VND1–6 being a downstream regulator of VND7, while SND1 is a master regulator of xylem fibre cells. Some authors have related vessel formation in angiosperms to the presence of *VND7* or *VND1–VND3* genes because no orthologs of these genes have been identified in the genome of plants lacking vessels, such as gymnosperms. This suggests that the ability to form vessels may have occurred after angiosperms diverged and may be associated with their distinctive growth shapes (Wan et al. 2018).

6 Conclusion

NAC proteins are among the families of plant-specific TFs with the most number of members. This is reflected in the variety of processes that they regulate, including development, stress tolerance and wood formation. Most NAC proteins have a modular organization formed by a conserved N-terminal DNA-binding domain and by a C-terminal domain that is highly variable and plays a role in transcriptional regulation and in protein–protein interaction. These TFs are able to be regulated by alternative splicing, miRNA and post-translational modifications such as phosphorylation or ubiquitination; they can also interact with many different types of proteins. Several NAC proteins associated with vascular development and stress responses have been conserved along plant evolution, suggesting a prominent role of these proteins in their survival and colonization of land habitats. Further, the bulk of the tree biomass that is potentially useful for diverse purposes is based on the trees' trunk volume. Because NAC transcription factors are key regulators of the regulatory network controlling wood formation, this raises the possibility of their use in biotechnological processes to modify the quantity and potential quality of the arboreal biomass.

Acknowledgements Research work in the authors's laboratory was supported by grants from "Ministerio de Economía y Competitividad" (BIO2015–69285-R) and the European Union (FP7-KBBE-289841).

References

Aida M, Tetsuya I, Fukaki H, Fujisawa H, Tasaka M (1997) Genes involved in organ separation in *Arabidopsis*: an analysis of the cup-shaped cotyledon mutant. Plant Cell 9:841–857

Aida M, Ishida T, Tasaka M (1999) Shoot apical meristem and cotyledon formation during *Arabidopsis* embryogenesis: interaction among the *CUP-SHAPED COTYLEDON* and *SHOOT MERISTEMLESS* genes. Development 126:1563–1570

Axtell MJ, Snyder JA, Bartel DP (2007) Common functions for diverse small RNAs of land plants. Plant Cell 19:1750–1769

Barah P, Naika MBN, Jayavelu ND, Sowdhamini R, Shameer K, Bones AM (2016) Transcriptional regulatory networks in *Arabidopsis thaliana* during single and combined stresses. Nucleic Acids Res 44:3147–3164

Bedon F, Bomal C, Caron S, Levasseur C, Boyle B, Mansfield SD, Schmidt A et al (2010) Subgroup 4 R2R3-MYBs in conifer trees: gene family expansion and contribution to the isoprenoid- and flavonoid-oriented responses. J Exp Bot 61:3847–3864

Bennett T, van den Toorn A, Sanchez-Perez GF, Campilho A, Willemsen V, Snel B, Scheres B (2010) SOMBRERO, BEARSKIN1, and BEARSKIN2 regulate root cap maturation in *Arabidopsis*. Plant Cell 22:640–654

Berger Y, Harpaz-Saad S, Brand A, Melnik H, Sirding N, Alvarez JP, Zinder M, Samach A, Eshed Y, Ori N (2009) The NAC-domain transcription factor *GOBLET* specifies leaflet boundaries in compound tomato leaves. Development 136:823–832

Bomal C, Bedon F, Caron S, Mansfield SD, Levasseur C, Cooke JEK, Blais S et al (2008) Involvement of *Pinus taeda* MYB1 and MYB8 in phenylpropanoid metabolism and secondary cell wall biogenesis: a comparative in planta analysis. J Exp Bot 59:3925–3939

Bu Q, Jiang H, Li CB, Zhai Q, Zhang J, Wu X, Sun J, Xie Q, Li C (2008) Role of the *Arabidopsis thaliana* NAC transcription factors ANAC019 and ANAC055 in regulating jasmonic acid-signaled defense responses. Cell Res 18:756–767

Canales J, Bautista R, Label P, Gómez-Maldonado J, Lesur I, Fernández-Pozo N et al (2014) De novo assembly of maritime pine transcriptome: implications for forest breeding and biotechnology. Plant Biotechnol J 12:286–299

Cañas RA, Li Z, Pascual MB, Castro-Rodríguez V, Ávila C, Sterck L, Van de Peer Y, Cánovas FM (2017) The gene expression landscape of pine seedling tissues. Plant J 91:1064–1087

Chen Q, Wang Q, Xiong L, Lou Z (2011) A structural view of the conserved domain of rice stress-responsive NAC1. Protein Cell 2:55–63

Chen Z, Ji L, Wang J, Jin J, Yang X, Rao P, Gao K, Liao W, Ye M, An X (2017) Dynamic changes in the transcriptome of *Populus hopeiensis* in response to abscisic acid. Sci Rep 7:42708

Cheng X, Peng J, Ma J, Tang Y, Chen R, Mysore KS, Wen J (2012) No apical meristem (MtNAM) regulates floral organ identity and lateral organ separation in *Medicago truncatula*. New Phytol 195:71–84

Chi YH, Melencion SMB, Alinapon CV, Kim MJ, Lee ES, Paeng SK, Park JH, Nawkar GM, Jung YJ, Chae HB et al (2017) The membrane-tethered NAC transcription factor, AtNTL7, contributes to ER-stress resistance in *Arabidopsis*. Biochem Biophys Res Commun 488:641–647

Craven-Bartle B, Pascual MB, Cánovas FM, Ávila C (2013) A Myb transcription factor regulates genes of the phenylalanine pathway in maritime pine. Plant J 74:755–766

Dalman K, Wind JJ, Nemesio-Gorriz M, Hammerbacher A, Lundén K, Ezcurra I, Elfstrand M (2017) Overexpression of PaNAC03, a stress induced NAC gene family transcription factor in Norway spruce leads to reduced flavonol biosynthesis and aberrant embryo development. BMC Plant Biol 17:6

Danielsson M, Lunden K, Elfstrand M, Hu J, Zhao T, Arnerup J, Ihrmark K, Swedjemark G, Borg-Karlson AK, Stenlid J (2011) Chemical and transcriptional responses of *Norway spruce* genotypes with different susceptibility to *Heterobasidion* spp. infection. BMC Plant Biol 11:154

De La Torre A, Birol I, Bousquet J, Ingvarsson PK, Jansson S, Jones SJM, Keeling CI, Mackay J, Nilsson O, Ritland K et al (2014) Insights into conifer giga-genomes. Plant Physiol 166:1724–1732

de Oliveira TM, Cidade LC, Gesteira AS, Coelho Filho MA, Soares Filho WS, Costa MGC (2011) Analysis of the NAC transcription factor gene family in citrus reveals a novel member involved in multiple abiotic stress responses. Tree Genet Genomes 7:1123–1134

Delessert C, Kazan K, Wilson IW, Van Der Straeten D, Manners J, Dennis ES et al (2005) The transcription factor ATAF2 represses the expression of pathogenesis-related genes in *Arabidopsis*. Plant J 43:745–757

Duval I, Lachance D, Giguère I, Bomal C, Morency MJ, Pelletier G, Boyle B et al (2014) Large-scale screening of transcription factor-promoter interactions in spruce reveals a transcriptional network involved in vascular development. J Exp Bot 65:2319–2333

Ernst HA, Olsen AN, Larsen S, Lo Leggio L (2004) Structure of the conserved domain of ANAC, a member of the NAC family of transcription factors. EMBO Rep 5:297–303

Ettl H, Gärtner G (2014) Syllabous der Boden-, Luft- und Flechtenalgen, 2nd edn. Berlin, Springer Spektrum, p 773

Eulgem T, Rushton PJ, Schmelzer E, Hahlbrock K, Somssich IE (1999) Early nuclear events in plant defence signalling: rapid gene activation by WRKY transcription factors. EMBO J 8:4689–4699

Grant EH, Fujino T, Beers EP, Brunner AM (2010) Characterization of NAC domain transcription factors implicated in control of vascular cell differentiation in *Arabidopsis* and *Populus*. Planta 232:337–352

Guan Q, Yue X, Zeng H, Zhu J (2014) The protein phosphatase RCF2 and its interacting partner NAC019 are critical for heat stress-responsive gene regulation and thermotolerance in *Arabidopsis*. Plant Cell 26:438–453

Guo Y, Gan S (2006) AtNAP, a NAC family transcription factor, has an important role in leaf senescence. Plant J 46:601–612

Guo H, Wang Y, Liu H, Hu P, Jia Y, Zhang C, Wang Y, Gu S, Yang C, Wang C (2015) Exogenous GA3 application enhances xylem development and induces the expression of secondary wall biosynthesis related genes in *Betula platyphylla*. Int J Mol Sci 16:22960–22975

Guo H, Cui Z, Zhang Y, Wang C (2017) Sequence characterization and expression analysis of *NAC* genes from *Betula platyphylla*. Trees 31:1919–1931

Hammonds AS, Bristow CA, Fisher WW, Weiszmann R, Wu S, Hartenstein V, Kellis M, Yu B, Frise E, Celniker SE (2013) Spatial expression of transcription factors in *Drosophila* embryonic organ development. Genome Biol 14:R140

Harfouche A, Meilan R, Altman A (2014) Molecular and physiological responses to abiotic stress in forest trees and their relevance to tree improvement. Tree Physiol 34:1181–1198

Hegedus D, Yu M, Baldwin D, Gruber M, Sharpe A, Parkin I, Whitwill S, Lydiate D (2003) Molecular characterization of *Brassica napus* NAC domain transcriptional activators induced in response to biotic and abiotic stress. Plant Mol Biol 53:383–397

Hernández Y, Sanan-Mishra N (2017) miRNA mediated regulation of NAC transcription factors in plant development and environment stress response. Plant Gene 11:190–198

Hobert O (2008) Gene regulation by transcription factors and microRNAs. Science 319:1785

Hu R, Qi G, Kong Y, Kong D, Gao Q, Zhou G (2010) Comprehensive analysis of NAC domain transcription factor gene family in *Populus trichocarpa*. Plant Biol 10:145

Hussey SG, Mizrachi E, Spokevicius AV, Bossinger G, Berger DK, Myburg AA (2011) *SND2*, a NAC transcription factor gene, regulates genes involved in secondary cell wall development in *Arabidopsis* fibres and increases fibre cell area in *Eucalyptus*. BMC Plant Biol 11:173

Hussey SG, Saïdi MN, Hefer CA, Myburg AA, Grima-Pettenati J (2015) Structural, evolutionary and functional analysis of the NAC domain protein family in Eucalyptus. New Phytol 206:1337–1350

Jensen MK, Skriver K (2014) NAC transcription factor gene regulatory and protein-protein interaction networks in plant stress responses and senescence. IUBMB Life 66:156–166

Jensen MK, Hagedorn PH, de Torres-Zabala M, Grant MR, Rung JH, Collinge DB et al (2008) Transcriptional regulation by an NAC (NAM–ATAF1, 2–CUC2) transcription factor attenuates ABA signalling for efficient basal defence towards *Blumeria graminis* f. sp. *hordei* in *Arabidopsis*. Plant J 56:867–880

Jensen MK, Kjaersgaard T, Nielsen MM, Galberg P, Petersen K, O'Shea C, Skriver K (2010) The *Arabidopsis* thaliana NAC transcription factor family: structure-function relationships and determinants of ANAC019 stress signalling. Biochem J 426:183–196

Jeong JS, Park YT, Jung H, Park SH, Kim JK (2009) Rice NAC proteins act as homodimers and heterodimers. Plant Biotechnol Rep 3:127–134

Jin SJ, Kim JK (2010) Root-specific expression of OsNAC10 improves drought tolerance and grain yield in rice under field drought conditions. Plant Physiol 153:185–197

Jin J, Zhang H, Kong L, Gao G, Luo J (2013) PlantTFDB 3.0: a portal for the functional and evolutionary study of plant transcription factors. Nucleic Acids Res 42:D1182–D1187

Jin C, Li KQ, Xu XY, Zhang HP, Chen HX, Chen YH, Hao J, Wang Y, Huang XS, Zhang SL (2017a) A novel NAC transcription factor, PbeNAC1, of *Pyrus betulifolia* confers cold and drought tolerance via interacting with PbeDREBs and activating the expression of stress-responsive genes. Front Plant Sci 8:1049

Jin J, Tian F, Yang DC, Meng YQ, Kong L, Luo J et al (2017b) PlantTFDB 4.0: toward a central hub for transcription factors and regulatory interactions in plants. Nucleic Acids Res 45:D1040–D1045

Jokipii-Lukkari S, Sundell D, Nilsson O, Hvidsten TR, Street NR, Tuominen H (2017) NorWood: a gene expression resource for evo-devo studies of conifer wood development. New Phytol 216:482–494

Kim HS, Park BO, Yoo JH, Jung MS, Lee SM, Han HJ et al (2007a) Identification of a calmodulin-binding NAC protein as a transcriptional repressor in *Arabidopsis*. J Biol Chem 282:36292–36302

Kim SY, Kim SG, Kim YS, Seo PJ, Bae M, Yoon HK, Park CM (2007b) Exploring membrane-associated NAC transcription factors in *Arabidopsis*: implications for membrane biology in genome regulation. Nucleic Acids Res 35:203–213

Kim MJ, Park MJ, Seo PJ, Song JS, Kim HJ, Park CM (2012) Controlled nuclear import of the transcription factor NTL6 reveals a cytoplasmic role of SnRK2.8 in the drought-stress response. Biochem J 448:353–363

Kjaersgaard T, Jensen MK, Christiansen MW, Gregersen P, Kragelund BB, Skriver K (2011) Senescence-associated barley NAC (NAM, ATAF1,2, CUC) transcription factor interacts with radical-induced cell death 1 through a disordered regulatory domain. J Biol Chem 14:35418–35429

Lamara M, Raherison E, Lenz P, Beaulieu J, Bousquet J, Mackay J (2016) Genetic architecture of wood properties based on association analysis and co-expression networks in white spruce. New Phytol 210:240–255

Larsson E, Sundström JF, Sitbon F, von Arnold S (2012) Expression of PaNAC01, a *Picea abies* CUP-SHAPED COTYLEDON orthologue, is regulated by polar auxin transport and associated with differentiation of the shoot apical meristem and formation of separated cotyledons. Ann Bot 110:923–934

Li Q, Lin Y-C, Sun Y-H, Song J, Chen H, Zhang XH, Sederoff RR, Chiang VL (2012) Splice variant of the SND1 transcription factor is a dominant negative of SND1 members and their regulation in *Populus trichocarpa*. Proc Natl Acad Sci U S A 109:14699–14704

Lin YC, Li W, Sun YH, Kumari S, Wei H, Li Q, Tunlaya-Anukit S, Sederoff RR, Chiang VL (2013) SND1 transcription factor-directed quantitative functional hierarchical genetic regulatory network in wood formation in *Populus trichocarpa*. Plant Cell 25:4324–4441

Liu C, Zhang X, Zhang K, An H, Hu K, Wen J, Shen J, Ma C, Yi B, Tu J, Fu T (2015) Comparative analysis of the *Brassica napus* root and leaf transcript profiling in response to drought stress. Int J Mol Sci 16:18752–18777

Lunden K, Danielsson M, Durling MB, Ihrmark K, Nemesio-Gorriz M, Stenlid J, Asiegbu FO, Elfstrand M (2015) Transcriptional responses associated with virulence and defence in the interaction between *Heterobasidion annosum* s. s. and Norway spruce. PLoS One 10:e0131182

Maleck K, Levine A, Eulgem T, Morgan A, Schmid J, Lawton KA, Dangl JL, Dietrich RA (2000) The transcriptome of *Arabidopsis thaliana* during systemic acquired resistance. Nat Genet 26:403–410

Mao X, Zhang H, Qian X, Li A, Zhao G, Jing R (2012) TaNAC2, a NAC-type wheat transcription factor conferring enhanced multiple abiotic stress tolerances in *Arabidopsis*. J Exp Bot 63:2933–2946

Maugarny A, Goncalves B, Arnaud N, Laufs P (2015) CUC transcription factors: to the meristem and beyond. In: Gonzalez DH (ed) Plant transcription factors. Academic Press, Boston, MA, pp 229–247

Maugarny-Calès A, Gonçalves B, Jouannic S, Melkonian M, Wong GKS, Laufs P (2016) Apparition of the NAC transcription factors predates de emergence of land plants. Mol Plant 9:1345–1348

McAdam SA, Brodribb TJ, Ross JJ (2016) Shoot-derived abscisic acid promotes root growth. Plant Cell Environ 39:652–659

Mitsuda N, Iwase A, Yamamoto H, Yoshida M, Seki M, Shinozaki K, Ohme-Takagi M (2007) NAC transcription factors, NST1 and NST3, are key regulators of the formation of secondary walls in woody tissues of *Arabidopsis*. Plant Cell 19:270–280

Mizrachi E, Mansfield SD, Myburg AA (2012) Cellulose factories: advancing bioenergy production from forest trees. New Phytol 194:54–62

Nagata T, Hosaka-Sasaki A, Kikuchi S (2015) The evolutionary diversification of genes that encode transcription factor proteins in plants. In: Gonzalez DH (ed) Plant transcription factors. Academic Press, Boston, MA, pp 73–97

Nakano Y, Yamaguchi M, Endo H, Rejab NA, Ohtani M (2015) NAC MYB-based transcriptional regulation of secondary cell wall biosynthesis in land plants. Front Plant Sci 6:288

Niu F, Wang B, Wu F, Yan J, Li L, Wang C, Wang Y, Yang B, Jiang YQ (2014) Canola (*Brassica napus* L.) NAC103 transcription factor gene is a novel player inducing reactive oxygen species accumulation and cell death in plants. Biochem Biophys Res Commun 454:30–35

Nuruzzaman M, Sharoni AM, Satoh K, Kondoh H, Hosaka A, Kikuchi S (2012) A genome-wide survey of the NAC transcription factor family in monocots and eudicots. In: Introduction to genetics – DNA methylation, histone modification and gene regulation. iConcept Press, Hong Kong. ISBN 978-14775549-4-4

Nuruzzaman M, Sharoni AM, Kikuchi S (2013) Roles of NAC transcription factors in the regulation of biotic and abiotic stress responses in plants. Front Microbiol 4:248

Nystedt B, Street NR, Wetterbom A, Zuccolo A, Lin YC, Scofield DG et al (2013) The Norway spruce genome sequence and conifer genome evolution. Nature 497:579–584

Ogo Y, Kobayashi T, Nakanishi Itai R, Nakanishi H, Kakei Y, Takahashi M et al (2008) A novel NAC transcription factor, IDEF2, that recognizes the iron deficiency-responsive element 2 regulates the genes involved in iron homeostasis in plants. J Biol Chem 283:13407–13417

Ohtani M, Nishikubo N, Xu B, Yamaguchi M, Mitsuda N, Goue N, Shi F, Ohme-Takagi M, Demura T (2011) A NAC domain protein family contributing to the regulation of wood formation in poplar. Plant J 67:499–512

Olsen AN, Ernst HA, Leggio LL, Skriver K (2005) NAC transcription factors: structurally distinct, functionally diverse. Trends Plant Sci 10:79–87

Ooka H, Satoh K, Doi K, Nagata T, Otomo Y, Murakami K et al (2003) Comprehensive analysis of NAC family genes in *Oryza sativa* and *Arabidopsis thaliana*. DNA Res 10:239–247

Ouyang K, Li J, Zhao X, Que Q, Li P, Juang H, Deng X, Singh SK, Wu AM, Chen X (2016) Transcriptomic analysis of multipurpose timber yielding tree *Neolamarckia cadamba* during xylogenesis using RNA-Seq. PLoS One 11:e0159407

Panchy N, Lehti-Shiu M, Shiu SH (2016) Evolution of gene duplication in plants. Plant Physiol 171:2294–2316

Pascual MB, Cánovas FM, Ávila C (2015) The NAC transcription factor family in maritime pine (*Pinus pinaster*): molecular regulation of two genes involved in stress responses. BMC Plant Biol 15:254

Pascual MB, El-Azaz J, de la Torre F, Cañas RA, Ávila C, Cánovas FM (2016) Biosynthesis and metabolic fate of phenylalanine in conifers. Front Plant Sci 7:1030

Pascual MB, Llebrés-Ávila MT, Craven-Bartle B, Cañas AR, Cánovas FM, Ávila C (2017) PpNAC1, a main regulator of phenylalanine biosynthesis and utilization in maritime pine. Plant Biotechnol J. https://doi.org/10.1111/pbi.12854

Patzlaff A, McInnis S, Courtenay A, Surman S, Newman LJ, Smith C, Bevan MW et al (2003a) Characterisation of a pine MYB that regulates lignification. Plant J 36:743–754

Patzlaff A, Newman LJ, Dubos C, Whetten RW, Smith C, McInnis S, Bevan MW et al (2003b) Characterization of PtMYB1, an R2R3-MYB from pine xylem. Plant Mol Biol 53:597–608

Pinheiro GL, Marques CS, Costa MDBL, Reis PAB, Alves MS, Carvalho CM, Fietto LG, Fontes EPB (2009) Complete inventory of soybean NAC transcription factors: sequence conservation and expression analysis uncover their distinct roles in stress response. Gene 444:10–23

Puranik S, Sahu PP, Srivastava PS, Prasad M (2012) NAC proteins: regulation and role in stress tolerance. Trends Plant Sci 17:369–381

Puranik S, Sahu PP, Mandal SN, Suresh VB, Parida SK, Prasad M (2013) Comprehensive genome-wide survey, genomic constitution and expression profiling of the NAC transcription factor family in foxtail millet (*Setaria italica* L.). PLoS One 8:e64594

Qin X, Zheng X, Huang X, Lii Y, Shao C, Xu Y, Chen F (2014) A novel transcription factor JcNAC1 response to stress in new model woody plant *Jatropha curcas*. Planta 239:511–520

Raherison ES, Giguère I, Caron S, Lamara M, MacKay JJ (2015) Modular organization of the white spruce (*Picea glauca*) transcriptome reveals functional organization and evolutionary signatures. New Phytol 207:172–187

Rauf M, Arif M, Fisahn J, Xue GP, Balazadeh S, Mueller-Roeber B (2013) NAC transcription factor speedy hyponastic growth regulates flooding-induced leaf movement in *Arabidopsis*. Plant Cell 25:4941–4955

Reece-Hoyes JS, Deplancke B, Shingles J, Grove CA, Hope IA, Walhout AJ (2005) A compendium of *Caenorhabditis elegans* regulatory transcription factors: a resource for mapping transcription regulatory networks. Genome Biol 6:R110

Riechmann JL, Heard J, Martin G, Reuber L, Jiang C, Keddie J, Adam K, Pineda O et al (2000) Arabidopsis transcription factors: genome-wide comparative analysis among eukaryotes. Science 290:2105–2110

Rueda-López M, García-Gutiérrez A, Cánovas FM, Ávila C (2013) The family of Dof transcription factors in pine. Trees 27:1547–1557

Rushton PJ, Torres JT, Parniske M, Wernert P, Hahlbrock K, Somssich IE (1996) Interaction of elicitor-induced DNA-binding proteins with elicitor response elements in the promoters of parsley PR1 genes. EMBO J 15:5690–5700

Rushton PJ, Bokowiec MT, Han S, Zhang H, Brannock JF, Chen X, Laudeman TW, Timko MP (2008) Tobacco transcription factors: novel insights into transcriptional regulation in the Solanaceae. Plant Physiol 147:280–295

Scofield S, Dewitte W, Nieuwland J, Murray JAH (2013) The *Arabidopsis* homeobox gene *SHOOT MERISTEMLESS* has cellular and meristem-organisational roles with differential requirements for cytokinin and CYCD3 activity. Plant J 75:53–66

Seki M, Narusaka M, Ishida J, Nanjo T, Fujita M, Oono Y et al (2002) Monitoring the expression profiles of 7000 *Arabidopsis* genes under drought, cold and high salinity stresses using a full-length cDNA microarray. Plant J 31:279–292

Seo PJ, Kim SG, Park CM (2008) Membrane-bound transcription factors in plants. Trends Plant Sci 13:550–556

Seo PJ, Kim MJ, Park JY, Kim SY, Jeon J, Lee YH, Kim J, Park CM (2010) Cold activation of a plasma membrane-tethered NAC transcription factor induces a pathogen resistance response in *Arabidopsis*. Plant J 61:661–671

Shang H, Li W, Zou C, Yuan Y (2013) Analyses of the NAC transcription factor gene family in Gossypium raimondii Ulbr.: chromosomal location, structure, phylogeny, and expression patterns. J Integr Plant Biol 55:663–676

Shao H, Wang H, Tang X (2015) NAC transcription factors in plant multiple abiotic stress responses: progress and prospects. Front Plant Sci 6:902

Shen H, Yin Y, Chen F, Xu Y, Dixon RA (2009) A bioinformatic analysis of NAC genes for plant cell wall development in relation to lignocellulosic bioenergy production. Bioenerg Res 2:217–232

Shen H, Liu C, Zhang Y, Meng X, Zhou X, Chu C et al (2012) OsWRKY30 is activated by MAP kinases to confer drought tolerance in rice. Plant Mol Biol 80:241–253

Shiu SH, Shih MC, Li WH (2005) Transcription factor families have much higher expansion rates in plants than in animals. Plant Physiol 139:18–26

Silva PA, Silva JCF, Caetan HDN, Machado JPB, Mendes GC, Reis PAB et al (2015) Comprehensive analysis of the endoplasmic reticulum stress response in the soybean genome: conserved and plant-specific features. BMC Genomics 16:783

Singh AK, Sharma V, Pal AK, Acharya V, Ahuja PS (2013) Genome-wide organization and expression profiling of the NAC transcription factor family in potato (*Solanum tuberosum L.*). DNA Res 20:403–423

Sjostrom E (1993) Wood chemistry. Academic Press, San Diego

Souer E, van Houwelingen A, Kloos D, Mol J, Koes R (1996) The *no apical meristem gene* of *Petunia* is required for pattern formation in embryos and flowers and is expressed at meristem and primordia boundaries. Cell 85:159–170

Stender EG, O'Shea C, Skriver K (2015) Subgroup-specific intrinsic disorder profiles of *Arabidopsis* NAC transcription factors: identification of functional hotspots. Plant Signal Behav 10:e1010967

Takada S, Hibara K-i, Ishida T, Tasaka M (2001) The CUC-SHAPED COTYLEDON1 gene of *Arabidopsis* regulates shoot apical meristem formation. Development 128:1127–1135

Tran LS, Nakashima K, Sakuma Y, Simpson SD, Fujita Y, Maruyama K et al (2004) Isolation and functional analysis of *Arabidopsis* stress-inducible NAC transcription factors that bind to a drought-responsive cis-element in the early responsive to dehydration stress 1 promoter. Plant Cell 16:2481–2498

Wan T, Liu ZM, Li LF, Leitch AR, Leitch IJ, Lohaus R, Liu ZJ, Xin HP et al (2018) A genome for gnetophytes and early evolution of seed plants. Nat Plants 4:82–89

Wang X, Goregaoker SP, Culver JN (2009) Interaction of the tobacco mosaic virus replicase protein with a NAC domain transcription factor is associated with the suppression of systemic host defenses. J Virol 83:9720–9730

Wang H, Zhao Q, Chen F, Wang M, Dixon RA (2011) NAC domain function and transcriptional control of a secondary cell wall master switch. Plant J 68:1104–1114

Wang JY, Wang JP, He-Yuan (2013a) A *Populus euphratica* NAC protein regulating Na+/K+ homeostasis improves salt tolerance in *Arabidopsis thaliana*. Gene 521:265–273

Wang N, Zheng Y, Xin H, Fang L, Li S (2013b) Comprehensive analysis of NAC domain transcription factor gene family in *Vitis vinifera*. Plant Cell Rep 32:61–75

Wang L, Wang C, Wang D, Wang Y (2014) Molecular characterization and transcript profiling of NAC genes in response to abiotic stress in *Tamarix hispida*. Tree Genet Genomes 10:157–171

Wang L, Li Z, Lu M, Wang Y (2017) ThNAC13, a NAC transcription factor from *Tamarix hispida*, confers salt and osmotic stress tolerance to transgenic *Tamarix* and *Arabidopsis*. Front Plant Sci 8:635

Weir I, Lu JP, Cook H, Causier B, Schwarz-Sommer Z, Davies B (2004) *CUPULIFORMIS* establishes lateral organ boundaries in *Antirrhinum*. Development 131:915–922

Welner DH, Lindemose S, Grossman GJ, Mollegaard NE, Olsen AN, Helgstrand C, Skriver K, Lo Leggio L (2012) DNA binding by plant NAC proteins: a firm link to WRKY and mammalian GCM transcription factors. Biochem J 444:395–404

Welner DH, Deeba F, Lo Leggio L, Skriver K (2016) NAC transcription factors: from structure to function in stress-associated networks. In: Gonzalez DH (ed) Plant transcription factors: evolutionary, structural and functional aspects. Academic Press, San Diego, CA, pp 199–212

Wickett NJ, Mirarab S, Nguyen N, Warnow T, Carpenter E, Matasci N, Ayyampalayam S, Barker MS, Burleigh JG, Gitzendanner MA et al (2014) Phylotranscriptomic analysis of the origin and early diversification of land plants. Proc Natl Acad Sci U S A 111:E4859–E4868

Willemsen V, Bauch M, Bennett T, Campilho A, Wolkenfelt H, Xu J et al (2008) The NAC domain transcription factors FEZ and SOMBRERO control the orientation of cell division plane in *Arabidopsis* root stem cells. Dev Cell 15:913–922

Wray GA, Hahn MW, Abouheif E, Balhoff JP, Margaret P, Rockman MV, Romano LA (2003) The evolution of transcriptional regulation in eukaryotes. Mol Biol Evol 20:1377–1419

Wu Y, Deng Z, Lai J, Zhang Y, Yang C, Yin B, Zhao Q, Zhang L, Li Y, Yang C et al (2009) Dual function of *Arabidopsis* ATAF1 in abiotic and biotic stress responses. Cell Res 19:1279–1290

Xie Q, Frugis G, Colgan D, Chua NH (2000) *Arabidopsis* NAC1 transduces auxin signal downstream of TIR1 to promote lateral root development. Genes Dev 14:3024–3036

Xie Q, Guo HS, Dallman G, Fang S, Weissman AM, Chua NH (2002) SINAT5 promotes ubiquitin-related degradation of NAC1 to attenuate auxin signals. Nature 419:167–170

Xu B, Ohtani M, Yamaguchi M, Toyooka K, Wakazaki M, Sato M et al (2014) Contribution of NAC transcription factors of plant adaptation to land. Science 343:1505–1508

Xu L, Wang J, Lei M, Li L, Fu Y, Wang Z, Ao M, Li Z (2016) Transcriptome analysis of storage roots and fibrous roots of the traditional medicinal herb *Callerya speciosa* (Champ.) Schot. PLoS One 11:e0160338

Yoshiyama KO (2016) SOG1: a master regulator of the DNA damage response in plants. Genes Genet Syst 90:209–216

Yoshiyama K, Conklin PA, Huefner ND, Britt AB (2009) Suppressor of gamma response 1 (SOG1) encodes a putative transcription factor governing multiple responses to DNA damage. Proc Natl Acad Sci U S A 106:12843–12848

You J, Zhang L, Song B, Qi X, Chan Z (2015) Systematic analysis and identification of stress-responsive genes of the NAC gene family in *Brachypodium distachyon*. PLoS One 10:e0122027

Zhao Z, Li Y, Liu H, Zhai X, Deng M, Dong Y, Fan G (2017) Genome-wide expression analysis of salt-stressed diploid and autotetraploid *Paulownia tomentosa*. PLoS One 12:e0185455

Zhong R, Ye ZH (2007) Regulation of cell wall biosynthesis. Curr Opin Plant Biol 10:564–572

Zhong R, Ye ZH (2010) The poplar PtrWNDs are transcriptional activators of secondary cell wall biosynthesis. Plant Signal Behav 5:469–472

Zhong R, Demura T, Ye ZH (2006) SND1, a NAC domain transcription factor, is a key regulator of secondary wall synthesis in fibers of Arabidopsis. Plant Cell 18:3158–3170

Zhong R, Richardson EA, Ye ZH (2007) The MYB46 transcription factor is a direct target of SND1 and regulates secondary cell wall biosynthesis in *Arabidopsis*. Plant Cell 19:2776–2791

Zhong R, Lee C, Ye ZH (2010a) Evolutionary conservation of the transcriptional network regulating secondary cell wall biosynthesis. Trends Plant Sci 15:625–632

Zhong R, Lee C, Ye ZH (2010b) Functional characterization of poplar wood associated NAC domain transcription factors. Plant Physiol 152:1044–1055

Zhong R, McCarthy RL, Lee C, Ye ZH (2011a) Dissection of the transcriptional program regulating secondary wall biosynthesis during wood formation in poplar. Plant Physiol 157:1452–1468

Zhong R, Lee C, McCarthy RL, Reeves CK, Jones EG, Ye ZH (2011b) Transcriptional activation of secondary wall biosynthesis by rice and maize NAC and MYB transcription factors. Plant Cell Physiol 52:1856–1871

Zhong H, Guo QQ, Chen L, Ren F, Wang QQ, Zheng Y, Li XB (2012) Two *Brassica napus* genes encoding NAC transcription factors are involved in response to high-salinity stress. Plant Cell Rep 31:1991–2003

Zhong R, McCarthy RL, Haghighat M, Ye ZH (2013) The poplar MYB master switches bind to the SMRE site and activate the secondary wall biosynthetic program during wood formation. PLoS One 8:e69219

Zhu T, Nevo E, Sun D, Peng J (2012) Phylogenic analyses unravel the evolutionary of NAC proteins in plants. Evolution 66:1833–1848

Zhu Q, Zou J, Zhu M, Liu Z, Feng P, Fan G, Wang W, Liao H (2014) In silico analysis on structure and DNA binding mode of AtNAC1, a NAC transcription factor from *Arabidopsis thaliana*. J Mol Model 20:2117

Zimmermann R, Werr W (2005) Pattern formation in the monocot embryo as revealed by NAM and CUC3 orthologues from *Zea mays* L. Plant Mol Biol 58:669–685

On the Elaborate Network of Thioredoxins in Higher Plants

Ina Thormählen, Belén Naranjo, José Abraham Trujillo-Hernandez, Jean-Philippe Reichheld, Francisco Javier Cejudo, and Peter Geigenberger

Contents

Communicated by Ulrich Lüttge

I. Thormählen (✉) and P. Geigenberger
Fakultät für Biologie – Pflanzenmetabolismus, Ludwig-Maximilians-Universität München, Munich, Germany
e-mail: thormaehlen@bio.lmu.de

B. Naranjo
Fakultät für Biologie – Molekularbiologie der Pflanzen, Ludwig-Maximilians-Universität München, Munich, Germany

J. A. Trujillo-Hernandez and J.-P. Reichheld
Laboratoire Génome et Développement des Plantes, Université de Perpignan Via Domitia, Perpignan, France

Laboratoire Génome et Développement des Plantes, Centre National de la Recherche Scientifique, Paris, France

F. J. Cejudo
Instituto de Bioquímica Vegetal y Fotosíntesis, Universidad de Sevilla and CSIC, Sevilla, Spain

Progress in Botany (2019) 80: 223–252, DOI 10.1007/124_2018_16,
Published online: 13 March 2018

Abstract Thioredoxins represent ubiquitous small proteins acting as redox regulators of diverse metabolic and developmental processes in almost all organisms. These proteins contain highly conserved cysteines in their redox-active sites, which enable the modification of target enzyme conformation and activity by reversible thiol-disulfide exchanges. Since their discovery in plants around 40 years ago, the number of thioredoxin family members as well as the knowledge about their distinct functions are still increasing and under investigation. Originally, the first plant thioredoxin was found in chloroplasts, while further analyses demonstrated additional cytosolic, nuclear, mitochondrial, endomembrane, and non-photosynthetic plastid locations. This chapter provides an overview on the complexity of the thioredoxin family in higher plants and discusses its role in integrating metabolism, stress responses, development, and gene expression. This will help to understand why plants harbor the most versatile thioredoxin system among all organisms.

Abbreviations

ACHT	Atypical cysteine/histidine-rich thioredoxin
ADP	Adenosine diphosphate
AGPase	ADP-glucose pyrophosphorylase
AMP	Adenosine monophosphate
AOX	Alternative oxidase
APS	AGPase small subunit
ATP	Adenosine triphosphate
ATPase	ATP synthase
CBC	Calvin-Benson cycle
cDNA	Complementary DNA
CDSP	Chloroplastic drought-induced stress protein
CHLM	Mg-protoporphyrin methyl transferase
CxxS	Atypical thioredoxin h with cysteine-*x*-*x*-serine active site
Cys	Cysteine
Cyt b_6f	Cytochrome b_6f complex
DNA	Deoxyribonucleic acid
ER	Endoplasmic reticulum
FAD	Flavin adenine dinucleotide
FBPase	Fructose 1,6-bisphosphatase
FDX	Ferredoxin
FNR	Ferredoxin $NADP^+$ reductase
FTR	Ferredoxin thioredoxin reductase
FUM	Fumarase
Gb	*Gossypium barbadense*
GFP	Green fluorescent protein
GGLC	Glycine-glycine-leucine-cysteine motif
Gly	Glycine
GRX	Glutaredoxin

GSH	Glutathione
HCF	High-chlorophyll-fluorescence-mutant protein
HCGPC	Histidine-cysteine-glycine-proline-cysteine motif
His	Histidine
LOV	Locus orchestrating victorin effects protein
Met	Methionine
MSR	Methionine sulfoxide reductase
NADP	Nicotinamide adenine dinucleotide phosphate
$NADP^+$-MDH	$NADP^+$-dependent malate dehydrogenase
NPQ	Non-photochemical quenching
NPR	Non-pathogenesis-related protein expressor
NRX	Nucleoredoxin
NTR	NADPH-dependent thioredoxin reductase
OAA	Oxaloacetate
ox	Oxidized
P	Phosphate
PC	Plastocyanin
PCNA	Proliferating cell nuclear antigen
PGR	Proton gradient regulation complex
P_i	Inorganic phosphate
PP_i	Inorganic pyrophosphate
PQ	Plastoquinon
PRK	Phosphoribulokinase
protoMME	Protomonomethylester
PRX	Peroxiredoxin
PS	Photosystem
Ps	*Pisum sativum*
PsbS	Photosystem II subunit S
qE	Energy- or ΔpH-dependent quenching
RbcS	Gene of the Rubisco small subunit
red	Reduced
RNA	Ribonucleic acid
ROS	Reactive oxygen species
Rubisco	Ribulose 1,5-bisphosphate carboxylase/oxygenase
SAR	Systemic acquired resistance
SBPase	Sedoheptulose 1,7-bisphosphatase
SDH	Succinate dehydrogenase
S_P	Peroxidatic cysteine residue
S_R	Resolving cysteine residue
SRX	Sulfiredoxin
Ssb	Single-stranded DNA binding protein
TCA	Tricarboxylic acid
TDX	Tetratricoredoxin
THL	h-type thioredoxins in *Brassica napus*
TRX	Thioredoxin

WCEVC	Tryptophan-cysteine-glutamic acid-valine-cysteine motif
WCGPC	Tryptophan-cysteine-glycine-proline-cysteine motif
WCRKC	Atypical thioredoxin with tryptophan-cysteine-arginine-lysine-cysteine active site
YF	Tyrosine-phenylalanine motif

1 Introduction

Redox regulation plays a crucial role in a large number of plant cellular processes (Geigenberger and Fernie 2014). Thioredoxins (TRX), small polypeptides of 12–13 kDa with protein disulfide reductase activity, catalyze Cys-based posttranslational modifications, which affect the conformation and function of a large number of proteins. Plants harbor a multiplicity of TRX isoforms and reduction pathways (Lemaire et al. 2007; Meyer et al. 2012; Geigenberger et al. 2017). In *Arabidopsis thaliana*, the gene family of typical TRX encodes for up to 20 different isoforms located in different subcellular compartments. While the plastid uses ferredoxin (FDX) to reduce TRX via the FDX TRX reductase (FTR), extraplastidial TRX are reduced by NADPH-dependent TRX reductases (NTR). Plastids also contain an unusual NADPH-dependent TRX system termed NTRC, which is characterized by a tethered TRX domain, providing a complete NTR-TRX system in a single polypeptide. In the last years, progress has been made to improve our understanding of the organization and biological roles of this complex thiol-based redox network. The aim of this review is to provide a synthesis of these recent evolvements focusing on the emerging roles of TRX in regulating metabolism, stress responses, development, and gene expression of higher plants.

2 The Plastidial Thioredoxin Systems

2.1 Ferredoxin Thioredoxin Reductase-Dependent Thioredoxin System

2.1.1 Plastids Contain the Most Diverse TRX System of Higher Plant Cells

About half of the total TRX isoforms in *Arabidopsis* cells are localized in the stroma of plastids (Meyer et al. 2012; Geigenberger et al. 2017). Typical TRX consist of a small single TRX domain carrying a highly conserved motif (WCGPC) in the redox-active site, which modulates the redox state of target enzymes by the reversible exchange of an oxidized disulfide bridge to two reduced thiols. The ten typical TRX isoforms localized in plastids are divided into five types, namely f (1–2), m (1–4), x, y (1–2), and z. Interestingly, the two TRX isoforms f1 and f2 contain an additional

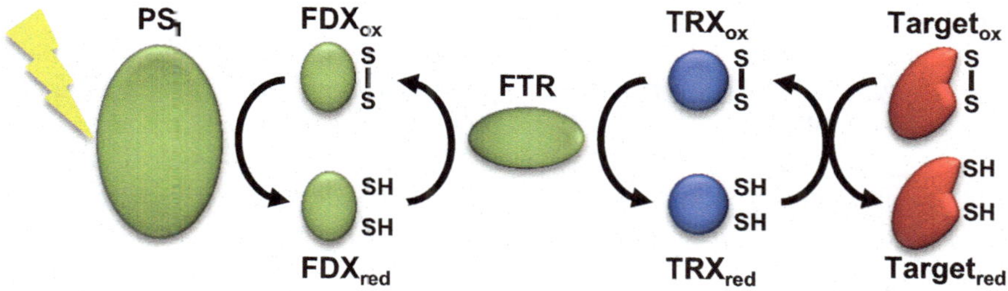

Fig. 1 The light-dependent reduction pathway of TRX and their targets in chloroplasts of higher plants. Absorbed light enables the photosynthetic electron transport chain to provide reducing power from the PS_I to FDX. FDX transmits electrons to the FTR, which reduces oxidized TRX (blue). Reduced TRX modify the redox activation states of distinct target enzymes (red) cleaving the disulfide bridge between Cys residues to two thiols (see above for Abbreviations)

conserved Cys and are suggested to be of eukaryotic origin based on their high homology with cytosolic TRX, while all other plastidial TRX are considered to have prokaryotic ancestors (Sahrawy et al. 1996; Schürmann and Buchanan 2008). Tissue-specific gene expression analyses revealed different patterns of expression for each of the single TRX isoforms in *Arabidopsis* (Bohrer et al. 2012; Belin et al. 2015). TRXm3, y1, and z were almost not detectable in green tissues, while TRXf1, m1, m2, m4, and x were expressed at high levels. In comparison to leaves, roots contained overall low plastidial TRX transcript abundance; however, TRXm2 displayed the highest expression values.

The source of reducing power for TRX reduction differs between plastids and other compartments (Meyer et al. 2012; Geigenberger et al. 2017). While in plastids the delivery of reducing power to TRX depends on reduced FDX, TRX in other compartments rely on NADPH and are reduced via NTR of which *Arabidopsis* contains two isoforms termed NTRA and NTRB (see below). In chloroplasts, photosynthetically reduced FDX provides electrons to the stromal FTR system, which in turn reduces TRX and is thus a strictly light-dependent process (see Fig. 1). In non-photosynthetic plastids, FDX is reduced by the FDX $NADP^+$ reductase (FNR) with NADPH as electron donor. However, the reduction concept of the plastidial TRXz isoform is not clear so far. Chibani et al. (2011) observed a reduction of TRXz only by the FTR system, while Bohrer et al. (2012) stated that it is exclusively reduced by its plastidial TRX family members. More recently, Yoshida and Hisabori (2016b) demonstrated an additional reduction pathway of TRXz via the alternative NADPH-dependent TRX system NTRC (see below) present in plastids.

2.1.2 The Redox Regulation of Target Enzymes Involves Multiple Functions of TRX in Plastidial Metabolism

The first discovery of TRX in plants led to a deeper understanding of the light-dependent regulation of the photosynthetic carbon assimilation reactions, namely the

Calvin-Benson cycle (CBC) enzymes (Buchanan 1980). It was observed that chloroplast TRX act as electron carrier able to modulate enzyme activity by changing the redox state of the target proteins dependent on light. Thus, the previously suggested idea that the increased CO_2 fixation under light is due to the provision of the energy and redox carriers ATP and NADPH during photosynthesis, which might facilitate a promoted protein synthesis of the CBC enzymes, could be ruled out. Following the initial finding about TRX around 40 years ago, extensive in vitro studies with isolated or recombinant TRX proteins were performed (reviewed by Meyer et al. 2012; Geigenberger et al. 2017). These studies extended the list of potential TRX targets and demonstrated distinct functions for the single isoforms. TRX types f and m were found to be predominantly important for the activation of anabolic pathways, like the CBC (Wolosiuk and Buchanan 1977; Breazale et al. 1978; Collin et al. 2003; Marri et al. 2009; Yoshida and Hisabori 2016b), the biosynthesis of starch (Ballicora et al. 2000; Skryhan et al. 2015; Thormählen et al. 2013) and lipids (Sasaki et al. 1997; Yamaryo et al. 2006), and for balancing the energy and redox homeostasis in chloroplasts by participating in ATP synthesis (McKinney et al. 1978) and the malate valve (Wolosiuk et al. 1977; Collin et al. 2003; Yoshida and Hisabori 2016b). Additionally, many other enzymes involved in diverse chloroplast processes were observed to be redox regulated by TRXf and m in vitro, like the oxidative pentose phosphate (Scheibe and Anderson 1981; Née et al. 2009) and shikimate pathway (Entus et al. 2002), the nitrogen assimilation (Lichter and Häberlein 1998) or protein import processes (Bartsch et al. 2008). In vitro studies investigating the regulation of the cyclic electron transport in the photosynthetic light reaction showed a preference for TRXm isoforms, but described so far contrasting effects leading to activation (Hertle et al. 2013) or inhibition (Courteille et al. 2013) of this process. TRXx, y, and z showed higher affinities for enzymes involved in oxidative stress responses (Collin et al. 2003, 2004; Navrot et al. 2006; Vieira Dos Santos et al. 2007; Chibani et al. 2011; Bohrer et al. 2012). The TRX-dependent redox activation of starch degradation (Mikkelsen et al. 2005; Sparla et al. 2006; Valerio et al. 2011; Seung et al. 2013; Silver et al. 2013) and chlorophyll synthesis (Ikegami et al. 2007; Luo et al. 2012; Yoshida and Hisabori 2016b) presented no clear distinction in the preference to different TRX isoforms until now. Additionally, it remains unclear how the starch degradation process in leaves could be activated by TRX reduction, since mobilization of transient starch occurs mainly during the night, when the photosynthetic reduction pathway of TRX is inactive.

2.1.3 The Plastidial TRX System Shows Surprisingly Redundant Activities In Vivo

Within the last decade, reverse genetic approaches were used to investigate *Arabidopsis* mutant lines lacking or overexpressing single or multiple TRX, so that in vivo evidences for the functional roles of the individual TRX *in planta* could be addressed. Surprisingly, under normal light conditions most single mutant lines deficient in plastidial TRX showed no or only minor visible phenotypes as compared to the wild type (Pulido et al. 2010; Thormählen et al. 2013, 2017; Laugier et al. 2013), despite the diverse roles previously proposed for these TRX based on analyses in vitro. For example, TRXf was known to be the only TRX type, which activates the redox regulated CBC enzymes fructose 1,6-bisphosphatase (FBPase) or glyceraldehyde 3-phosphate dehydrogenase (Collin et al. 2003; Marri et al. 2009). However, growth under moderate light conditions and CO_2 fixation rates were not impaired in the *Arabidopsis trxf1* single mutant, showing a strong deficiency in TRXf protein level in leaves (Thormählen et al. 2013; Naranjo et al. 2016a). Even the double mutant *trxf1 trxf2* with complete lack of TRXf presented only slightly decreased rosette fresh weights under short day conditions (Naranjo et al. 2016a). The low impact of single TRX deficiencies on growth is most probably due to underestimated redundant functions of the other TRX isoforms or of additional enzymes in the plastidial redox regulatory network, like NTRC (see below). There is, however, one exception. The recently discovered *Arabidopsis trxz* single mutant presented an albino phenotype with strongly impaired growth compared to wild type plants (Arsova et al. 2010). Here, obviously, other TRX isoforms were not able to compensate for the deficiency in TRXz.

Some of the observed metabolic phenotypes in *Arabidopsis* mutant lines devoid of TRXf or m isoforms were confirmatory of the in vitro results mentioned above. The redox activation state of ADP-glucose pyrophosphorylase (AGPase), which catalyzes the first committed step in starch synthesis, was attenuated in leaves of *trxf1* mutants during the day, accompanied by a reduced accumulation of starch (Thormählen et al. 2013) in agreement with previous in vitro results (Ballicora et al. 2000; Geigenberger et al. 2005). However, leaves of *Nicotiana tabacum* overexpression lines with elevated levels of TRXf accumulated higher levels of starch, while no change in AGPase redox activation was observed (Sanz-Barrio et al. 2013). Plants lacking TRXf isoforms additionally showed an impaired light-dependent activation of the CBC enzyme FBPase (Thormählen et al. 2015; Yoshida et al. 2015; Naranjo et al. 2016a), and the ribulose 1,5-bisphosphate carboxylase/oxigenase (Rubisco) activase (Naranjo et al. 2016a). Okegawa and Motohashi (2015) stated that the TRXm isoforms might be the predominant redox regulators of the CBC enzymes, since *Arabidopsis* triple mutant lines simultaneously deficient in TRXm1, m2, and m4 showed attenuated CO_2 fixation rates, smaller rosettes, and impaired redox activation states of FBPase and sedoheptulose 1,7-bisphosphatase (SBPase) as well as the malate valve enzyme $NADP^+$-dependent malate dehydrogenase ($NADP^+$-MDH). Since the plastidial FBPase is not activated by TRXm isoforms in vitro (Collin et al. 2003; Yoshida and Hisabori 2016b), the in vivo observation on the impaired FBPase redox activation (Okegawa and Motohashi 2015) might not be due to direct TRX effects, but most probably due to secondary reasons, like an inhibited electron flow to downstream metabolic processes. The latter suggestion would be in agreement with additional in vivo studies, which found TRXm impacts on processes involved in the photosynthetic light reaction, like an impaired biogenesis of photosystem (PS) II in *Arabidopsis* mutant lines simultaneously lacking TRXm1, m2, and m4 (Wang et al. 2013) or the inhibition of cyclic electron transport by TRXm4 orthologues in *N. tabacum* (Courteille et al. 2013). Very recently *Arabidopsis trxm1* and *trxm2* single and double mutants were compared to wild type plants with respect to the transient activation of the $NADP^+$-MDH after the onset of light (Thormählen et al. 2017). It was demonstrated that the light-dependent redox activation of the $NADP^+$-MDH decreased upon combined deficiencies of TRXm1 and m2, which corresponded to an impaired acclimation to fluctuating light intensities during the day. In vitro experiments revealed the ability of TRXf and at least partially TRXm to reductively activate the $NADP^+$-MDH (Collin et al. 2003; Yoshida and Hisabori 2016b). However, this contrasts with illuminated leaves of *trxf Arabidopsis* mutant lines exhibiting even higher $NADP^+$-MDH activation states (Thormählen et al. 2015). Further studies are necessary to investigate the distinct involvements of TRXf and m isoforms in redox regulated processes playing a role in the fluctuating light response.

More recently, the findings about TRX-dependent redox regulation of the chlorophyll biosynthesis pathway were extended in vivo. Transgenic pea plants with simultaneously silenced *TRXf* and *m* genes were impaired in redox activation of Mg-chelatase subunits (Luo et al. 2012). Da et al. (2017) demonstrated in *Arabidopsis* that the combined silencing of TRXm1, m2, and m4 impairs the chlorophyll biosynthesis pathway by attenuating the redox control of the Mg-protoporphyrin methyl transferase (CHLM). Furthermore, the in vivo insights about TRXm3 revealed functions in meristem development by modulating symplastic protein transport (Benitez-Alfonso et al. 2009). In response to oxidative stress, mutant lines lacking TRXy2 showed attenuated protein repair mechanisms in leaves due to impaired redox activation of methionine sulfoxide reductases (MSR) (Laugier et al. 2013). TRXz was demonstrated to be essential for autotrophic growth by influencing plastidial gene expression as structural component of the plastid-encoded RNA polymerase (Arsova et al. 2010), although this effect seems to be independent of TRXz redox activity (Wimmelbacher and Börnke 2014). Taken together, the diverse functions of typical TRX in plastidial processes investigated within the last decades illustrate a complex redox regulatory system, which is still not fully elucidated.

2.1.4 Plastids Contain Additional Atypical TRX with Mainly Unresolved Functions

Contrary to the TRX described above, the members of the atypical TRX are not yet well analyzed in their functional roles of plastidial metabolism (Meyer et al. 2012; Belin et al. 2015). These unusual TRX have different protein properties such as atypical active site motifs, multiple TRX domains, or combinations of TRX and other domains. The proteins of the atypical Cys/His-rich TRX (ACHT) family consist of six isoforms carrying diverse non-canonical active site motifs (Dangoor et al. 2009; Meyer et al. 2012; Belin et al. 2015). The chloroplastic drought-induced stress protein CDSP32 contains two TRX domains and several additional conserved Cys (Rey et al. 1998). One of the TRX domains is functional with an unusual redox-active motif (HCGPC), while the second domain is degenerated. The transmembrane high-chlorophyll-fluorescence-mutant protein HCF164 is localized in the thylakoid membranes with atypical TRX motifs (WCEVC) at the stromal as well as at the lumen side (Motohashi and Hisabori 2006; Meyer et al. 2012). About the two isoforms WCRKC1 and WCRKC2 not much is known; however, these proteins contain non-canonical redox sites as their names suggest (Meyer et al. 2012). ACHT2, ACHT4, ACHT6, CDSP32, HCF164, and WCRKC1 are mostly expressed in green tissues (Belin et al. 2015). The genes encoding ACHT1, ACHT3, and ACHT5 have an overall low expression in tested tissues, except for mature and germinating pollen, in which the *ACHT3* gene showed very high transcript level. How the reduction pathways for the atypical TRX are organized is not yet sufficiently investigated. One tested member was HCF164, which was reduced by TRXm (Motohashi and Hisabori 2006). TRXm enables a transmembrane electron transfer to the luminal HCF164 part, so that HCF164 acts as a transducer of reducing equivalents to the inner parts of the thylakoids. Whether the FTR-dependent reduction pathway in chloroplasts is directly interacting with atypical TRX needs to be tested.

The functional insights about atypical TRX are weak compared to the other plastidial TRX. Members of the ACHT family have a less reducing redox midpoint potential than typical TRX and prefer 2-Cys peroxiredoxins (PRX; see below) in comparison to $NADP^+$-MDH as redox targets in vitro (Dangoor et al. 2009). Eliyahu et al. (2015) observed in *Arabidopsis* mutant studies that ACHT4 is deactivating the key enzyme of starch synthesis AGPase through oxidation of the regulatory Cys in the AGPase small subunit APS1 under low light conditions. CDSP32 was shown to be involved in photooxidative stress responses by interacting with plastidial PRX (PRXQ and 2-Cys PRX) and MSRB1 (Rey et al. 2005; Tarrago et al. 2010). HCF164 acts as reductant for luminal target proteins of the photosynthetic electron transport chain and is essential for assembling the chloroplast cytochrome b6f complex (Lennartz et al. 2001; Motohashi and Hisabori 2006). Future studies are needed to identify additional biological roles of atypical TRX in plastidial processes and the connection to their typical relatives.

2.2 NADPH-Dependent Thioredoxin Reductase C System

2.2.1 The *NTRC* Gene Is Exclusive of Aerobic Photosynthetic Organisms

In heterotrophic organisms, and non-photosynthetic plant cells and compartments, the reduction of TRX relies on NADPH with the participation of an NTR. NTR are flavoenzymes able to transfer reducing equivalents from NADPH via FAD to a double Cys at their active site, which reduce the disulfide at the active site of TRX. Like TRX, NTR are widely distributed in prokaryotic and eukaryotic organisms; however, two forms of the enzyme have evolved with different molecular and kinetic properties (Williams et al. 2000). While the enzyme from prokaryotes is, in its native form, a homodimer formed by subunits of approximately 35 kDa with NADPH, FAD, and central domains, the enzyme from mammals has a larger molecular mass (approximately 55 kDa) and contains selenocysteine, which is active in catalysis (Arscott et al. 1997). In plants, the first NTR cloned and characterized, from *Arabidopsis* (Jacquot et al. 1994) and wheat (Serrato et al. 2002), revealed that the enzyme is similar to the one in prokaryotes.

When the genome sequences from *Arabidopsis* and rice were available, a search of the genes encoding NTR in these plants revealed the presence of three genes termed *NTRA*, *NTRB*, and *NTRC* (Serrato et al. 2002). The *NTRA* and *NTRB* genes encode polypeptides corresponding to those previously reported from *Arabidopsis* and wheat. Further analyses in *Arabidopsis* revealed that NTRB is the predominant isoform in mitochondria whereas NTRA is predominant in the cytosol (Reichheld et al. 2005). The polypeptide deduced of the *NTRC* gene contained the expected active site double Cys as well as the NADPH and FAD domains, hence being very similar to NTRA and NTRB. However, in contrast to these proteins, the deduced NTRC polypeptide contained a putative transit peptide at the N-terminus and, most intriguingly, a putative TRX domain at the C-terminus; thus, it contained a complete NTR-TRX system in a single polypeptide (Serrato et al. 2004). Both the NTR and TRX domains were expressed in the bacterium *Escherichia coli* as N-terminal His-tagged proteins, and the biochemical analysis confirmed that these truncated polypeptides displayed NTR and TRX activity, respectively, in vitro, leading to the proposal that NTRC is a bifunctional enzyme (Serrato et al. 2004). Moreover, the search in genomes available at the moment, and recently confirmed, showed that the *NTRC* gene is exclusively found in aerobic photosynthetic organisms including some, but not all, cyanobacteria, algae, and plants (Serrato et al. 2004; Najera et al. 2017). Furthermore, it was shown that NTRC is localized in the chloroplast stroma both in *Arabidopsis* and rice (Serrato et al. 2004; Moon et al. 2006), hence implying that the putative transit peptide deduced of the *NTRC* gene sequence actually serves to target the enzyme to this organelle. More recently, the localization of NTRC was analyzed by expressing NTRC::GFP fusion proteins in *Arabidopsis*. This study confirmed the localization of NTRC in chloroplasts and showed the presence of the enzyme in plastids from non-photosynthetic tissues such as root amyloplasts, though NTRC is much less abundant in these tissues as compared with green tissues (Kirchsteiger et al. 2012). The phylogenetic analysis suggests that the *NTRC* gene originated in cyanobacteria most probably by mutational joining of adjacent but independent *NTR* and *TRX* genes. The combination of NTR and TRX activities in a single enzyme might provide selective advantage, which explains the presence of the *NTRC* gene in all eukaryotic photosynthetic organisms (Najera et al. 2017).

2.2.2 NTRC Is an Efficient Reductant of 2-Cys PRX

An important breakthrough for the biochemical characterization of NTRC was the identification of 2-Cys PRX as a target of the enzyme. 2-Cys PRX are thiol-dependent PRX able to reduce hydrogen or organic peroxides to water or the corresponding alcohol (Perkins et al. 2015). There are different types of 2-Cys PRX, the one found to be a target of NTRC belongs to the typical form, which is a homodimer each of the monomers containing two catalytically active Cys residues, peroxidatic (S_P) and resolving (S_R). A catalytic cycle is initiated by the reaction of hydrogen peroxide with the thiolate form of S_P, which is oxidized to sulfenic (–SOH). This intermediate is then attacked by S_R thus forming a disulfide with the liberation of water (see Fig. 2). For a new catalytic cycle, the S_P-S_R disulfide has to be reduced. This reaction is in chloroplasts efficiently performed by NTRC. Indeed, it is possible to reconstitute in vitro a system formed by NTRC and 2-Cys PRX which is able to use NADPH to reduce hydrogen peroxide (Perez-Ruiz et al. 2006; Moon et al. 2006).

The finding that NTRC is an efficient reductant of 2-Cys PRX was important because it allowed an in-depth characterization of the biochemical properties of this novel enzyme. It was first shown that NTRC conjugates its NTR and TRX activities to efficiently support the hydrogen peroxide scavenging activity of 2-Cys PRX. However, equal amount of the truncated polypeptides containing the NTR and TRX domains of NTRC showed a very poor activity, indicating that the conformation of the enzyme is important to display its biochemical activity. Moreover, it was established that NTRC interacts with 2-Cys PRX by the TRX domain and that the

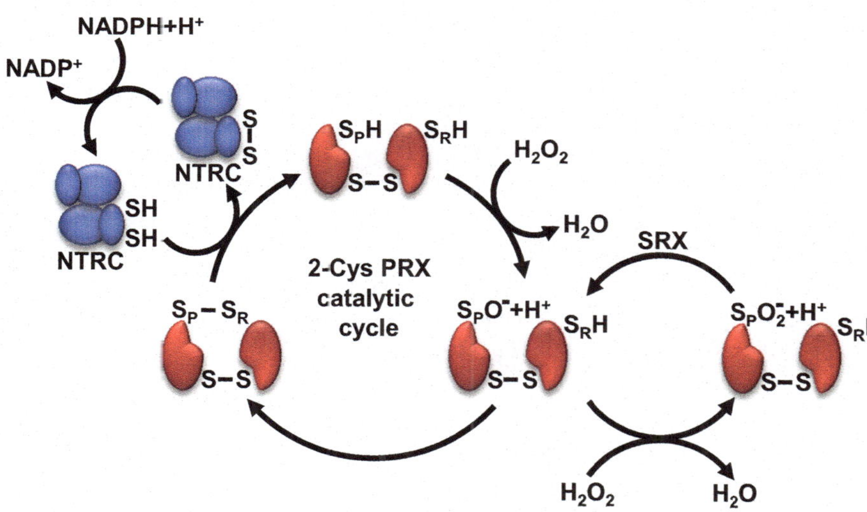

Fig. 2 NTRC controls the redox state of 2-Cys PRX. The catalytically active form of NTRC (blue) is a dimer arranged in a head-to-tail conformation. NTRC interacts with 2-Cys PRX (red) through its TRX domain and efficiently reduces the disulfide bond linking the peroxidatic (S_P) and resolving (S_R) Cys residues of oxidized 2-Cys PRX using NADPH as source of reducing equivalents. The thiolic form of S_P reacts with H_2O_2 releasing H_2O and rendering the peroxidatic Cys oxidized as a sulfenic intermediate (S_PO^-), which is then attacked by the resolving Cys (S_RH) to form the disulfide form. Under oxidizing conditions, the sulfenic intermediate (S_PO^-) can be overoxidized to sulfinic ($S_PO_2^-$), hence yielding an inactive form of the enzyme. This state is reversed by the reductive action of SRX (see above for Abbreviations)

redox-sensitive Cys residues at the active sites of the NTR and the TRX domains are essential for activity (Perez-Ruiz and Cejudo 2009). Based on these data it was proposed that the catalytically active form of NTRC is a homodimer arranged in a head-to-tail conformation. The enzyme shows much higher affinity for NADPH than for NADH. Thus, NADPH is the source of reducing power, which is transferred to FAD and then to the double Cys at the active site of the NTR domain of one of the subunits. There is then an inter-subunit transfer of electrons to the active site double Cys of the TRX domain of the other subunit of the enzyme, which interacts with the disulfide of the 2-Cys PRX (Perez-Ruiz and Cejudo 2009). According to this model, stating that NTRC interacts with its targets by the TRX domain, it could be considered that NTRC is a TRX with its own reductase, which might be the reason of the high catalytic efficiency of this enzyme (Cejudo et al. 2012). The high affinity of NTRC for NADPH and the proposed electron transfer pathway was further confirmed by stopped-flow spectroscopy (Bernal-Bayard et al. 2012), and the interaction between NTRC and 2-Cys PRX was also demonstrated in vivo by bimolecular fluorescence complementation analyses (Bernal-Bayard et al. 2014). In line with this notion it was found that NTRC is unable to interact with other plastidial TRX, hence, having no activity as NTR (Bohrer et al. 2012). However, the *Arabidopsis ntrc* mutant, which is devoid of NTRC, was partially complemented by the expression of a mutated version of the *NTRC* gene encoding an enzyme with intact NTR and inactive TRX domains, indicating that the NTR domain of NTRC might interact with other plastidial TRX among which TRXf was proposed as the most likely partner (Toivola et al. 2013). Therefore, whether NTRC acts exclusively as a TRX with its reductase incorporated or is able to interact with other plastidial TRX, hence showing NTR activity, still requires further analysis.

Finally, it should be mentioned that NTRC shows ability to oligomerize in vitro. While for the enzyme from rice this ability is redox sensitive, the enzyme being detected as dimer in the presence of NADPH and as oligomer in its absence (Perez-Ruiz et al. 2006), the oligomerization capacity of the enzyme from barley is redox insensitive (Wulff et al. 2011). The oligomeric form of NTRC shows chaperone and holdase activity, which might be the reason of the tolerance to elevated temperature of plants overexpressing NTRC (Chae et al. 2013).

2.2.3 NTRC Controls the Redox State of 2-Cys PRX

Based on the fact that NTRC is an efficient reductant of 2-Cys PRX, hence supporting the hydrogen peroxide scavenging activity of these enzymes, it was proposed that NTRC has antioxidant activity (Perez-Ruiz et al. 2006). However, it was also shown that different chloroplast TRX are able to reduce 2-Cys PRX, the most efficient one being TRXx (Collin et al. 2003). Thus, a relevant issue was to establish the hierarchy of the different chloroplast redox systems for the reduction of 2-Cys PRX. The redox state of the 2-Cys PRX in vivo, determined as the ratio of dimeric (oxidized) to monomeric (reduced) forms of the enzyme, was severely impaired in an *Arabidopsis* mutant devoid of NTRC, which contained almost no monomeric form of 2-Cys PRX (Kirchsteiger et al. 2009). In contrast, mutant plants devoid of TRXx showed levels of oxidized and reduced 2-Cys PRX indistinguishable of the wild type, indicating that NTRC is more relevant reductant of 2-Cys PRX than TRXx (Pulido et al. 2010).

In the presence of excess hydrogen peroxide, 2-Cys PRX may be inactivated by over-oxidation of the sulfenic acid intermediate of S_P (–SOH) to sulfinic acid (–SO_2H), which can be reverted by sulfiredoxin (SRX) (Biteau et al. 2003), or even to sulfonic acid (–SO_3H), which is irreversible. Interestingly, 2-Cys PRX from eukaryotes are more sensitive to over-oxidation than those from prokaryotes due to structural constraints imposed by the presence of two motifs, GGLC and YF, in the enzyme from eukaryotes (Wood et al. 2003). Since over-oxidation causes the inactivation of 2-Cys PRX, it was proposed that this feature is a gain-of-function of the enzymes from eukaryotes that allows the accumulation of hydrogen peroxide and, thus, its signaling function, the so-called floodgate hypothesis (Wood et al. 2003). Chloroplast 2-Cys PRX are sensitive to over-oxidation, hence behaving as eukaryotic type enzymes (Kirchsteiger et al. 2009). However, since plant chloroplasts have prokaryotic origin similar to modern cyanobacteria (Gould et al. 2008) the question arising was whether or not 2-Cys PRX from cyanobacteria are sensitive to over-oxidation. The finding that the enzyme from *Anabaena* is sensitive to over-oxidation while the one from *Synechocystis* is not (Pascual et al. 2010) showed the existence of sensitive 2-Cys PRX in prokaryotes and revealed that cyanobacteria have developed two strategies to cope with hydrogen peroxide, the strategy of *Anabaena* being remarkably similar to the one in chloroplasts.

The level of over-oxidation of chloroplast 2-Cys PRX and, thus, the balance of antioxidant and signaling activities of these enzymes depends on the reductants (NTRC and TRX of this organelle), but also on SRX, which catalyzes the reversion of the over-oxidized to the reduced form of the enzyme (Puerto-Galan et al. 2013). However, a double mutant of *Arabidopsis* devoid of NTRC and SRX showed the phenotype of the *ntrc* mutant (Puerto-Galan et al. 2015). Altogether, the results of the biochemical analyses of NTRC in conjunction with the genetic interaction of NTRC with TRXx and SRX indicate that a central function of NTRC is the control of the redox state of the 2-Cys PRX, which is important for the balance of the harmful and signaling activities of hydrogen peroxide.

2.2.4 NTRC Participates in the Redox Regulation of Light Energy Utilization and Several Chloroplast Biosynthetic Pathways

To address the issue of the function of NTRC the use of *Arabidopsis* mutants has been a useful tool. The *Arabidopsis* knock-out mutant *ntrc* shows pale green leaves with a lower content of chlorophylls and retarded growth, as compared to wild type plants. Moreover, the retarded growth phenotype of the *ntrc* mutant is dependent on the photoperiod, short-day conditions aggravating the phenotype (Serrato et al. 2004; Lepistö et al. 2009). More in-depth analyses revealed that the *ntrc* mutant displays lower CO_2 fixation rates, especially at low light intensities, and abnormal chloroplast structure (Perez-Ruiz et al. 2006; Lepistö et al. 2009). In addition, plants devoid of NTRC have been reported to be hypersensitive to different abiotic stresses, such as salinity and drought (Serrato et al. 2004), prolonged darkness (Perez-Ruiz et al. 2006), and heat (Chae et al. 2013), and also to biotic stress (Ishiga et al. 2012, 2016). These phenotypic characteristics of the *ntrc* mutant might be explained by the antioxidant function of NTRC as an efficient reductant of 2-Cys PRX. In this regard, it is worth mentioning that the NTRC-2-Cys PRX system has also been reported to exist in other photosynthetic organisms including green algae, such as *Chlorella* (Machida et al. 2012), and cyanobacteria, such as *Anabaena* (Pascual et al. 2011). Indeed, an *Anabaena* mutant strain devoid of NTRC (*ΔntrC*) is more sensitive to oxidative stress, especially to treatments with hydrogen peroxide, and the redox balance of 2-Cys PRX is impaired, more oxidized, in the mutant (Sanchez-Riego et al. 2016).

Beside the phenotypic features related with the response to stress, which might be due to the antioxidant activity of the NTRC-2-Cys PRX system, it should be noted that the *Arabidopsis ntrc* mutant has a rather pleiotropic phenotype, suggesting additional functions for the enzyme. This notion was further supported by the phenotype of an *Arabidopsis* mutant (*Δ2cp*) with severely decreased levels of 2-Cys PRX (Pulido et al. 2010), which is almost like the wild type.

It is well known that AGPase, a key regulatory enzyme of the starch biosynthesis pathway, is subject to allosteric regulation but also to sugar and light-dependent redox regulation (Geigenberger et al. 2005). On the other hand, as stated above, NTRC uses as source of reducing power NADPH, which can be generated from sugars during the night by the oxidative pentose phosphate pathway. These results suggested the possibility that NTRC participates in the redox regulation of starch biosynthesis, a notion further supported by the lower content of starch in roots and leaves of the *ntrc* mutant. In vitro and in vivo analyses confirmed the participation of NTRC in the redox regulation of AGPase (Michalska et al. 2009; Lepistö et al. 2013) and, thus, in starch biosynthesis. Interestingly, the redox state of the AGPase was found to be altered not only in leaf chloroplasts but also in roots, suggesting a role of NTRC in non-photosynthetic tissues (Michalska et al. 2009), in line with the localization of the enzyme in any type of plastids. However, *Arabidopsis* transgenic plants expressing NTRC exclusively in green tissues, by expression of the NTRC cDNA under the control of the *RbcS* gene promoter in the *ntrc* background, was sufficient to rescue the wild type phenotype with respect to growth and pigmentation, while expression exclusively in roots had no effect (Kirchsteiger et al. 2012). These results show the essential function of chloroplasts for the growth of photosynthetic and non-photosynthetic organs.

The characteristic pale green leaf phenotype of the *ntrc* mutant, which is due to lower chlorophyll content than in the leaves of the wild type (Serrato et al. 2004; Lepistö et al. 2009), suggested the participation of NTRC in chlorophyll biosynthesis. Several reports have shown the NTRC-dependent redox regulation of different enzymes involved in the tetrapyrrole biosynthesis pathway including the glutamyl-transfer RNA reductase GluTR1, CHLM (Richter et al. 2013) and the Mg-chelatase subunit CHLI (Perez-Ruiz et al. 2014).

More recently, it has been reported that NTRC is required for efficient light energy utilization. Under growth light conditions, the *ntrc* mutant displays high levels of energy dissipation by non-photochemical quenching (NPQ) with the concomitant decrease of the photosynthetic performance (Thormählen et al. 2015; Naranjo et al. 2016b; Carrillo et al. 2016). Plants lacking NTRC accumulate more protons in the thylakoid lumen at low and moderate light intensities leading to the synthesis of zeaxanthin and the activation of the rapid inducible and reversible component of NPQ, namely the energy- or ΔpH-dependent quenching (qE) (Naranjo et al. 2016b; Carrillo et al. 2016). The higher accumulation of protons in the absence of NTRC has been attributed to a lower reduction of the ATP synthase (ATPase) γ-subunit (Naranjo et al. 2016b; Carrillo et al. 2016), though other mechanisms may be also involved. The lower photosynthetic performance of the *ntrc* mutant leads to a state of permanent starvation for light energy, which may explain its retarded growth phenotype, particularly under conditions of light limitation like short-day photoperiod. Blocking NPQ in the *ntrc* mutant background by the lack of the PS II subunit S (PsbS), in the double mutant *ntrc psbs*, causes the partial recovery of the photosynthetic performance and growth (Naranjo et al. 2016b). Furthermore, since the induction and recovery of qE are essential to cope with rapid changes in light intensities, the growth of the *ntrc* mutant is even more compromised under fluctuating light conditions (Thormählen et al. 2017).

2.3 NADPH-Dependent Thioredoxin Reductase C Acts Concertedly with Other Chloroplast Thioredoxins

The finding of the participation of NTRC in the redox regulation of biosynthetic pathways previously known to be regulated by TRX raised the issue of the relationship of NTRC with the different TRX in chloroplasts (see Fig. 3). This issue has been addressed in recent years by the analysis of *Arabidopsis* mutants combining the deficiency of NTRC and TRX. The different mutants analyzed lacking NTRC and f-type TRX (Thormählen et al. 2015; Ojeda et al. 2017) or NTRC and TRXx (Ojeda et al. 2017) have in common a very severe growth inhibition phenotype, in agreement with the severe impairment of the efficiency of use of light energy and redox regulation of CBC enzymes in these mutants. In line with these results, *Arabidopsis* mutants combining the lack of NTRC and the deficiency of m-type TRX show as well a severe growth retard and impairment of the redox regulation of enzymes of the tetrapyrrole biosynthesis pathway (Da et al. 2017). Altogether, these results support the notion of a concerted action of NTRC with other plastidial TRX, such as those of types f, x, and m in chloroplast redox regulation. The finding that the double mutant combining the deficiency of NTRC and the catalytic subunit of FTR is not viable (Yoshida and Hisabori 2016b) lends further support to this notion.

An initial possibility to explain the concerted action of NTRC and TRX in chloroplast redox regulation is that both systems share common targets, so that the

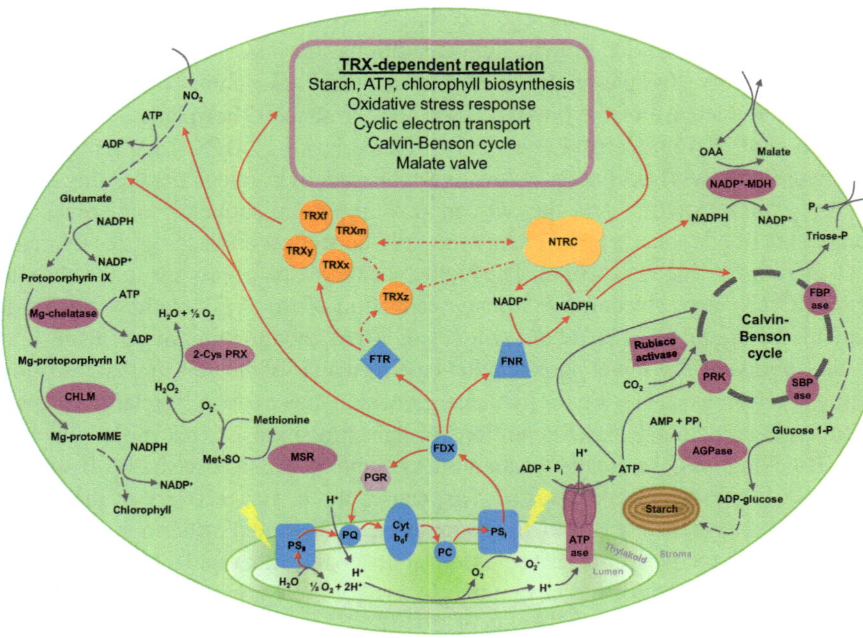

Fig. 3 TRX-dependent redox regulation of metabolic pathways in chloroplasts of higher plants. Light enables the energy-driven electron flow (red arrows) through the thylakoid membranes to FDX, which distributes the reducing power to different systems. The FNR system produces NADPH consumed by the Calvin-Benson cycle, but also by the malate valve or NTRC (bright orange). Secondly, the FTR system is providing electrons to TRX (deep orange). Additionally, the nitrogen assimilation pathway and the cyclic electron transport rely on the reducing ability of FDX. Reduced TRX and NTRC proteins facilitate the reversible redox activation of their distinct target enzymes (deep purple). Exclusively, in vitro as well as in vivo confirmed targets are shown. These include enzymes involved in different biosynthesis pathways (AGPase, ATPase, CHLM, Mg-chelatase), the oxidative stress system (2-Cys PRX, MSR), the cyclic electron transport (PGR), the Calvin-Benson cycle (FBPase, PRK, Rubisco activase, SBPase), and the malate valve (NADP$^+$-MDH). The function of the proposed interaction between the TRX and NTRC systems, as well as the reduction pathway of TRXz and its role as redox regulator in general are not yet fully resolved (dotted red arrows). Whether TRX have an activating or inhibiting effect on the PGR target (bright purple) is also not clear so far (see above for Abbreviations)

lack of one of them might be complemented by the other, while the combined deficiency of both systems would cause the severe impairment of the redox regulation of these targets, hence with strong consequences on plant phenotype. In support of this notion, Nikkanen et al. (2016) showed the interaction of NTRC *in planta* with well-established TRX targets such as FBPase. However, in vitro assay analyses showed that NTRC is unable to reduce FBPase (Ojeda et al. 2017), suggesting that the effect of NTRC on the redox regulation of this enzyme might be indirect. In a recent report it was shown that the content of 2-Cys PRX affects the phenotype of *Arabidopsis* mutants devoid of NTRC, so that decreased contents of 2-Cys PRX alleviated the phenotype while increased contents had the opposite effect (Perez-

Ruiz et al. 2017). Based on these results it was proposed that the NTRC-dependent redox balance of 2-Cys PRX modulates the redox state of the pool of plastid TRX and, consequently, the redox regulation of the TRX targets. These results provide an explanation for the indirect participation of NTRC in the redox regulation of the TRX-dependent processes of the chloroplast.

3 The Extraplastidial Thioredoxin Systems

While functions of chloroplastic TRX have been extensively studied, extraplastidial TRX systems are much less characterized in plants. Among the around 40 TRX found in *Arabidopsis*, at least 20 of them are extraplastidial isoforms, based on the absence of a putative plastidial targeting sequence (Meyer et al. 2005, 2012). However, for some of them, the respective localization is still unexplored. It is now clearly established that functional TRX systems are present in most cellular compartments, including mitochondria, cytosol, nucleus, and endomembrane systems (see Fig. 4). In contrast to the large number of extraplastidial TRX, only two NTR genes (*NTRA* and *NTRB* in *Arabidopsis*) are found in the genome of most higher plants, suggesting multiple subcellular localizations of these isoforms. Consistently, NTR isoforms were found dual or even triple targeted in cytosol, mitochondria, and nucleus in *Arabidopsis* and pea (Laloi et al. 2001; Serrato and Cejudo 2003; Reichheld et al. 2005; Marchal et al. 2014).

3.1 Cytosolic Thioredoxin System

Cytosolic TRX are composed of h-type (h for heterotrophic) and other type TRX. TRXh isoforms have been mostly characterized in *Arabidopsis* but orthologues are found in other plants including rice and poplar (Meyer et al. 2006). All TRXh isoforms consist of a single TRX domain harboring a WCG(P)C active site and, according to their respective clustering on phylogenetic trees, have been divided into three different subtypes (I, II, III). TRX from type I (TRXh1, h3, h4, h5) are mainly cytosolic and typically reduced by NTR (Rivera-Madrid et al. 1995). Type II (h2, h7, h8) and type III (h9, h10, atypical CxxS1 and 2) TRXh are less characterized isoforms. While predominantly located in the cytosol, TRXh2, h8, and h9 are also located in the endomembrane system, presumably through the fact that they harbor myristoylated/palmitoylated residues in their N-terminal extensions (see below) (Traverso et al. 2013). Remarkably, while generally reduced by NTR, some TRXh isoforms are also reduced by the alternative thiol reduction systems glutathione (GSH)/glutaredoxin (GRX) (Gelhaye et al. 2003; Reichheld et al. 2007; Koh et al. 2008; Meng et al. 2010). For example, in *Arabidopsis* and poplar, TRXh9 is reduced through a disulfide cascade mechanism involving the transient glutathionylation of a Cys located in the N-terminal extension of TRXh9 (Koh et al. 2008).

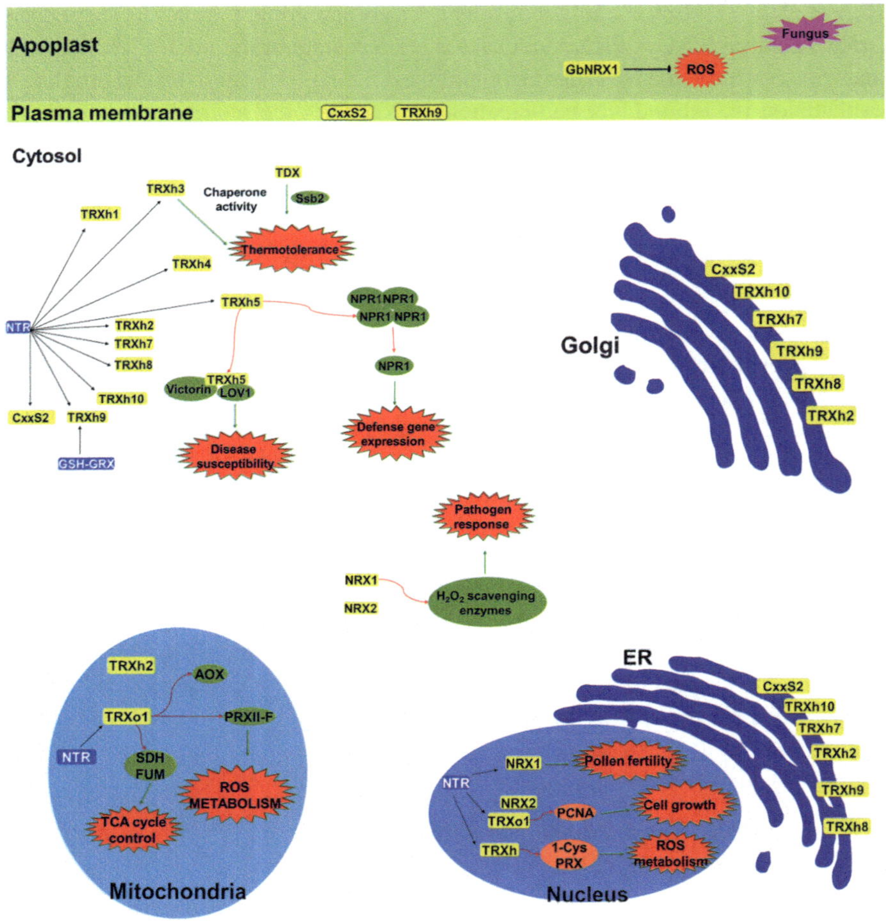

Fig. 4 Extraplastidial TRX systems in higher plants. Different plant TRX isoforms (yellow) are represented in their respective subcellular localization. When established, their respective reducers are represented (blue). Some key TRX target proteins (green) as well as the respective regulated functions (red spots) are also shown (see above for Abbreviations)

Only a few functions have been assigned to TRXh isoforms based on gene expression and genetic studies. Tissue-specific gene expression analyses of TRXh isoforms have revealed overlapping expression patterns in *Arabidopsis* (Reichheld et al. 2002; Belin et al. 2015). Overlapping gene expression between TRXh isoforms is likely explaining the lack of phenotype of most of the *trxh* knock-out mutants analyzed so far (our personal data). However, the characterization of a *ntra ntrb* double mutant deficient in the reduction of TRXh isoforms has revealed several plant development functions like pollen fitness, seed development, and plant growth (Reichheld et al. 2007). Moreover, it also shed light on cross talks acting between TRX and GRX systems in flower development and hormonal signaling (Reichheld et al. 2007; Marty et al. 2009; Bashandy et al. 2009, 2010).

Interestingly, the TRXh5 isoform is specifically involved in pathogen response pathways. TRXh5 is a major actor of plant immunity, regulating the oligomeric state of the non-pathogenesis-related protein expressor NPR1, a master regulator of the systemic acquired resistance (SAR). NPR1 normally exists in the cytosol as a thiol-bound oligomer. Upon pathogen challenge or treatment with salicylic acid, TRXh5 is upregulated (Laloi et al. 2004; Tada et al. 2008). TRXh5 then reduces NPR1 oligomers resulting in monomer release, translocation to the nucleus, and regulation of defense gene expression associated with both local and systemic resistance. Consequently, a *trxh5* knock-out mutant is impaired in establishing a SAR and is therefore more susceptible to pathogen infection (Tada et al. 2008). More recently, Kneeshaw et al. (2014) have shown that TRXh5 acts in the SAR signaling by selective NPR1 protein S-denitrosylation. TRXh acting as denitrosylase has been additionally shown in *ntra* mutants in which higher levels of S-nitrosylated proteins were found (Tada et al. 2008).

TRXh5 is also involved in the plant tolerance to the necrotrophic fungal pathogen *Cochliobolus victoriae.* The redox mechanism involves binding of victorin, the virulent effector of *C. victoriae*, to the first Cys residue in the TRXh5 active site (Sweat and Wolpert 2007; Lorang et al. 2012). Victorin binding to TRXh5 activates the locus orchestrating victorin effects protein LOV1, an *Arabidopsis* susceptibility protein, and elicits a resistance-like response that confers disease susceptibility. Lorang et al. (2012) proposed that victorin mimics a conventional pathogen virulence effector and confers virulence to *C. victoriae* solely because it incites plant defense mechanisms.

Specifically, in self-incompatible species, TRXh isoforms (called THL in *Brassica napus*) have been involved in recognition and rejection of self-incompatible pollen through redox regulation of a key actor (S-locus receptor kinase) of the self-incompatible signaling pathway (Bower et al. 1996; Cabrillac et al. 2001; Ivanov and Gaude 2009). While this specific regulatory mechanism has been questioned (Yamamoto and Nasrallah 2009), works in other self-incompatible species (i.e., *Solanaceae* and *Poaceae*) have also established a role of TRXh isoforms in self-incompatible pollen recognition, while acting by other ways (Li et al. 1996; Juárez-Díaz et al. 2006).

A function of cytosolic TRX isoforms in thermotolerance has been suggested. When overexpressed in plants, TRXh3 and the multidomain TRX tetratricoredoxin (TDX) enhance heat stress tolerance due to their chaperone activities (Park et al. 2009; Lee et al. 2009). Interestingly, TDX was shown to interact with the single-stranded DNA binding protein Ssb2, a yeast heat-shock protein 70 chaperone, suggesting that the TDX-dependent heat stress tolerance might be related to a co-chaperone activity (Vignols et al. 2003). Of note, none of the *trxh3* and *tdx* knock-out mutants does exhibit a heat stress sensitive phenotype, suggesting that this function is compensated by other proteins sharing chaperone activities.

3.2 Nuclear Thioredoxin System

Although several TRX isoforms have been identified in the nucleus, evidence of their functions in the nucleus is still scarce. Nucleoredoxins (NRX) are multidomain proteins consisting of two or three TRX-like domains fused to a Cys-rich C-terminal domain. In *Zea mays* and *Arabidopsis*, characterized NRX isoforms share a dual cytosolic/nuclear localization and harbor putative nuclear localization signals (Laughner et al. 1998; Marchal et al. 2014). On the other hand, a cotton NRX orthologue (GbNRX1) was preferentially found in the secretory pathway and the apoplast (see below) (Li et al. 2016). Both NRX isoforms (NRX1 and 2) found in *Arabidopsis* exhibit disulfide reduction activities. While NRX1 can be reduced by NTR in vitro, constituting a functional TRX reduction pathway in the nucleus (Marchal et al. 2014), the NRX2 physiological reducer is still unknown. Genetic studies have assigned a function of NRX1 in plant fertility. It is required for pollen tube navigation through the pistil, guiding it to the ovule (Qin et al. 2009; Marchal et al. 2014). Still it is not yet established, if this function is dependent on the nuclear or the cytosolic localization of NRX1 and what is the pathway involved.

Recent works have involved NRX1 in plant immunity responses (Kneeshaw et al. 2017). NRX1 gene expression is responsive to salicylic acid in *Arabidopsis* and poplar and is compromised in the SAR induced by pathogenic bacterial strains of *Pseudomonas syringae* (Kneeshaw et al. 2017). Upon pathogen trigger, NRX1 is induced and binds target enzymes of major hydrogen peroxide scavenging pathways, including catalases. Mutant *nrx1* plants displayed reduced catalase activity and are hypersensitive to oxidative stress. Remarkably, catalase is maintained in a reduced state by interaction with NRX1, a process necessary for its hydrogen peroxide scavenging activity. Thus, the works by Kneeshaw et al. (2017) conclude that hydrogen peroxide scavenging enzymes, like catalase, experience oxidative distress in reactive oxygen species (ROS)-induced environments and require reductive protection from NRX1 for optimal activity.

A pea orthologue of mitochondrial o-type TRX (PsTRXo1) shows a dual mitochondrial and nuclear localization (Martí et al. 2009). Recent works by Calderón et al. (2017) show that PsTRXo1 interacts with proliferating cell nuclear antigen (PCNA) and that its overexpression in cell suspension affects cell growth and cell cycle progression. While coinciding with an upregulation of PCNA protein, the effect of PsTRXo1 expression on cell growth is not yet fully established (Calderón et al. 2017).

TRXh isoforms accumulate in the nuclear compartment in tissues subjected to oxidative stress or genotoxic conditions (Serrato et al. 2001; Serrato and Cejudo 2003; Sarkar et al. 2005; Pulido et al. 2009) or during seed germination in which the NTR-TRXh system serves as a source of reducing power to regenerate 1-Cys (Pulido et al. 2009).

3.3 Mitochondrial Thioredoxin System

Although TRX and TRX reductase activities were reported in plant mitochondrial extracts two decades ago (Konrad et al. 1996; Banze and Follmann 2000), the first mitochondrial TRX system has been identified in *Arabidopsis* by Laloi et al. (2001).

In *Arabidopsis*, two NTR isoforms are localized in the mitochondria (NTRA and NTRB). Both of them reduce TRXo1, one of the two o-type TRX isoforms in *Arabidopsis*, which is targeted to the mitochondrial matrix through a cleavable N-terminal signal (Laloi et al. 2001; Reichheld et al. 2005). Mitochondrial localization of TRXo1 was later confirmed in other plants including poplar and pea (Gelhaye et al. 2004; Martí et al. 2009). Moreover, in poplar and *Arabidopsis*, the TRXh2 isoform has been identified in mitochondria, although its submitochondrial localizations have not been addressed yet (Gelhaye et al. 2004; Meng et al. 2010).

Proteomic affinity approaches have been used to identify potential mitochondrial TRX using purified mitochondrial protein extracts (Balmer et al. 2004; Martí et al. 2009; Marchand et al. 2010; Yoshida et al. 2013). These approaches have been highly valuable to identify putative TRX mitochondrial targets and open the field for further validation. Nevertheless, a few enzymatic and genetic evidences have further validated those interactions yet. In pea, TRXo1 has been shown to be an efficient reducer of the PRXII-F, suggesting that TRXo is involved in mitochondrial ROS metabolism (Barranco-Medina et al. 2008). Pea TRXh2 and *Arabidopsis* TRXo1 were also shown to affect in vitro the dimerization state of the alternative oxidase (AOX) by reducing a heterodimeric disulfide (Cys78–Cys78), and further restoring the AOX activity (Gelhaye et al. 2004; Yoshida et al. 2013). Nevertheless, evidence is lacking to clearly favor the interpretation of a regulatory function over that of a maintenance function, i.e., to avoid Cys over-oxidation upon ROS exposure. Moreover, genetic evidences are lacking to support this redox regulation in vivo.

Both proteomic and in vitro works have pinpointed a potent TRX-dependent regulation of tricarboxylic acid (TCA) cycle enzymes (Balmer et al. 2004; Martí et al. 2009; Yoshida et al. 2013; Schmidtmann et al. 2014). Thiol switching of mitochondrial citrate synthase and isocitrate dehydrogenase activity were reported in vitro (Schmidtmann et al. 2014; Yoshida and Hisabori 2014), while thiol switching of malate dehydrogenase activity was found unchanged (Yoshida and Hisabori 2016a; Huang et al. 2017). In order to further address the function of TRX on the TCA cycle in vivo, Daloso et al. (2015) have characterized *Arabidopsis* knock-out mutants for mitochondrial TRXo1 (*trxo1*) and mitochondrial NTR (double mutant *ntra ntrb*). Both mutants were affected in mitochondrial metabolic fingerprints and flux patterns, suggesting a role of the matrix TRX system in regulating metabolism in vivo. In a complementary approach, Daloso et al. (2015) evaluated in vitro activities of TCA cycle and associated enzymes in mutants deficient in mitochondrial TRX. Among TCA cycle enzyme activities affected in the mutants, both succinate dehydrogenase (SDH) and fumarase (FUM) were proposed to be deactivated by reduced TRXo1, suggesting that TRXo is acting as a direct regulator of carbon flow through the TCA cycle (Daloso et al. 2015).

3.4 Endomembrane Thioredoxin System

While TRXh were previously described as cytosolic isoforms, more recent works have established alternative cellular localizations for some members of this family.

In *Arabidopsis*, Traverso et al. (2013) have reexamined the subcellular localizations of members of the subtype II and III. By using a combination of predictive software and in vitro assays, they showed that members of the subtype II (TRXh2, h7, h8) and III (TRXh9, h10, atypical CxxS2) are subjected to N-terminal (Gly2 and Cys4) fatty acid acylations (N-myristoylation and S-palmitoylation). The role of these posttranslational modifications is to target proteins to membranes. Accordingly, TRX::GFP fusions revealed that myristoylation targeted proteins to the endomembrane system including the endoplasmic reticulum (ER) and Golgi, and that partitioning between this membrane compartment and the cytosol correlated with the catalytic efficiency of the N-myristoyltransferase acting at the N-terminus of the TRX. Moreover, a palmitoylatable Cys residue (only present in subtype III) flanking the N-myristoylation site was crucial to target proteins to the plasma membrane (Traverso et al. 2013). Independently, Meng et al. (2010) showed that N-myristoylation and S-palmitoylation of TRXh9 at Gly2/Cys4 residues is associated with the capacity of the protein to move from cell-to-cell, suggesting a role of this TRXh isoform in intercellular communication and cell growth.

In spite of the presence of distinct TRXh isoforms in the endomembrane system, their role is still unknown. While they might sustain other thiol oxidoreductases (i.e., protein disulfide isomerases, ER oxidases, etc.) in disulfide reduction/isomerization in the ER, no putative target proteins have been identified yet in these compartments. Concerning their reduction pathway, neither putative endomembrane nor plasma membrane NTR have been identified so far. Alternatively, GSH of the ER could constitute a physiological reducer of endomembrane TRX, as shown for subtype III isoform TRXh9 in poplar (Gelhaye et al. 2003; Koh et al. 2008). Of note, an atypical TRXh member of subtype III (CxxS2) is acting as a protein disulfide isomerase in vitro, which might be physiologically relevant (Serrato et al. 2008).

Finally, an NRX1 isoform (GbNRX1) has recently been involved in the apoplast immune response to the pathogenic fungus *Verticillium dahliae*. Here, GbNRX1 functions in apoplastic ROS scavenging after the ROS burst that occurs upon recognition of the pathogen (Li et al. 2016).

4 Conclusions and Perspectives

In this review, we have discussed emerging topics of higher plant TRX. It is becoming clear that TRX form an elaborate network in plants being important to integrate metabolism, stress responses, development, and gene expression in a fluctuating environment. The chloroplast contains light- and NADPH-dependent TRX systems, which are combined via the redox balance of 2-Cys PRX to allow a cooperative control of chloroplast functions and photoautotrophic growth in higher plants. Extraplastidial TRX have been found to play critical and integrative roles in controlling pathogen defense and abiotic stress tolerance of plants interfacing with other signals such as ROS and phytohormones. Additional emerging fields in TRX functions are the regulation of plant development and the interaction of TRX with transcription factors and genetic control elements in the nucleus. Future studies are required to examine the interorganellar cross talk between TRX at the cellular level and to explore further in vivo functions *in planta*.

References

Arscott LD, Gromer S, Schirmer RH et al (1997) The mechanism of thioredoxin reductase from human placenta is similar to the mechanisms of lipoamide dehydrogenase and glutathione reductase and is distinct from the mechanism of thioredoxin reductase from Escherichia coli. Proc Natl Acad Sci U S A 94:3621–3626

Arsova B, Hoja U, Wimmelbacher M et al (2010) Plastidial thioredoxin z interacts with two fructokinase-like proteins in a thiol-dependent manner: evidence for an essential role in chloroplast development in Arabidopsis and Nicotiana benthamiana. Plant Cell 22:1498–1515

Ballicora MA, Frueauf JB, Fu Y et al (2000) Activation of the potato tuber ADP-glucose pyrophosphorylase by thioredoxin. J Biol Chem 275:1315–1320

Balmer Y, Vensel WH, Tanaka CK et al (2004) Thioredoxin links redox to the regulation of fundamental processes of plant mitochondria. Proc Natl Acad Sci U S A 101:2642–2647

Banze M, Follmann H (2000) Organelle-specific NADPH thioredoxin reductase in plant mitochondria. J Plant Physiol 156:126–129

Barranco-Medina S, Krell T, Bernier-Villamor L et al (2008) Hexameric oligomerization of mitochondrial peroxiredoxin PrxIIF and formation of an ultrahigh affinity complex with its electron donor thioredoxin Trx-o. J Exp Bot 59:3259–3269

Bartsch S, Monnet J, Selbach K, Quigley F, Gray J, von Wettstein D, Reinbothe S, Reinbothe C (2008) Three thioredoxin targets in the inner envelope membrane of chloroplasts function in protein import and chlorophyll metabolism. Proc Natl Acad Sci U S A 105:4933–4938

Bashandy T, Taconnat L, Renou J-P et al (2009) Accumulation of flavonoids in an ntra ntrb mutant leads to tolerance to UV-C. Mol Plant 2:249–258

Bashandy T, Guilleminot J, Vernoux T et al (2010) Interplay between the NADP-linked thioredoxin and glutathione systems in Arabidopsis auxin signaling. Plant Cell 22:376–391

Belin C, Bashandy T, Cela J et al (2015) A comprehensive study of thiol reduction gene expression under stress conditions in Arabidopsis thaliana. Plant Cell Environ 38:299–314

Benitez-Alfonso Y, Cilia M, San Roman A et al (2009) Control of Arabidopsis meristem development by thioredoxin-dependent regulation of intercellular transport. Proc Natl Acad Sci U S A 106:3615–3620

Bernal-Bayard P, Hervas M, Cejudo FJ et al (2012) Electron transfer pathways and dynamics of chloroplast NADPH-dependent thioredoxin reductase C (NTRC). J Biol Chem 287:33865–33872

Bernal-Bayard P, Ojeda V, Hervas M et al (2014) Molecular recognition in the interaction of chloroplast 2-Cys peroxiredoxin with NADPH-thioredoxin reductase C (NTRC) and thioredoxin x. FEBS Lett 588:4342–4347

Biteau B, Labarre J, Toledano MB (2003) ATP-dependent reduction of cysteine-sulphinic acid by S. cerevisiae sulphiredoxin. Nature 425:980–984

Bohrer A-S, Massot V, Innocenti G et al (2012) New insights into the reduction systems of plastidial thioredoxins point out the unique properties of thioredoxin z from Arabidopsis. J Exp Bot 63:6315–6323

Bower MS, Matias DD, Fernandes-Carvalho E (1996) Two members of the thioredoxin-h family interact with the kinase domain of a Brassica S locus receptor kinase. Plant Cell 8:1641–1650

Breazale VD, Buchanan BB, Wolosiuk RA (1978) Chloroplast sedoheptulose 1,7-bisphosphatase: evidence for regulation by the ferredoxin/thioredoxin system. Z Naturforsch 33c:521–528

Buchanan BB (1980) Role of light in the regulation of chloroplast enzymes. Annu Rev Plant Physiol 31:341–374

Cabrillac D, Cock JM, Dumas C et al (2001) The S-locus receptor kinase is inhibited by thioredoxins and activated by pollen coat proteins. Nature 410:220–223
Calderón A, Ortiz-Espín A, Iglesias-Fernández R et al (2017) Thioredoxin (Trxo1) interacts with proliferating cell nuclear antigen (PCNA) and its overexpression affects the growth of tobacco cell culture. Redox Biol 11:688–700
Carrillo LR, Froehlich JE, Cruz JA et al (2016) Multi-level regulation of the chloroplast ATP synthase: the chloroplast NADPH thioredoxin reductase C (NTRC) is required for redox modulation specifically under low irradiance. Plant J 87:654–663
Cejudo FJ, Ferrandez J, Cano B et al (2012) The function of the NADPH thioredoxin reductase C-2-Cys peroxiredoxin system in plastid redox regulation and signalling. FEBS Lett 586:2974–2980
Chae HB, Moon JC, Shin MR et al (2013) Thioredoxin reductase type C (NTRC) orchestrates enhanced thermotolerance to Arabidopsis by its redox-dependent holdase chaperone function. Mol Plant 6:323–336
Chibani K, Tarrago L, Schürmann P et al (2011) Biochemical properties of poplar thioredoxin z. FEBS Lett 585:1077–1081
Collin V, Issakidis-Bourguet E, Marchand C et al (2003) The Arabidopsis plastidial thioredoxins: new functions and new insights into specificity. J Biol Chem 278:23747–23752
Collin V, Lamkemeyer P, Miginiac-Maslow M et al (2004) Characterization of plastidial thioredoxins from Arabidopsis belonging to the new y-type. Plant Physiol 136:4088–4095
Courteille A, Vesa S, Sanz-Barrio R et al (2013) Thioredoxin m4 controls photosynthetic alternative electron pathways in Arabidopsis. Plant Physiol 161:508–520
Da Q, Wang P, Wang M et al (2017) Thioredoxin and NADPH-dependent thioredoxin reductase C regulation of tetrapyrrole biosynthesis. Plant Physiol 175:652–666
Daloso DM, Müller K, Obata T et al (2015) Thioredoxin, a master regulator of the tricarboxylic acid cycle in plant mitochondria. Proc Natl Acad Sci U S A 112:E1392–E1400
Dangoor I, Peled-Zehavi H, Levitan A et al (2009) A small family of chloroplast atypical thioredoxins. Plant Physiol 149:1240–1250
Eliyahu E, Rog I, Inbal D et al (2015) ACHT4-driven oxidation of APS1 attenuates starch synthesis under low light intensity in Arabidopsis plants. Proc Natl Acad Sci U S A 112:12876–12881
Entus R, Poling M, Herrmann KM (2002) Redox regulation of Arabidopsis 3-deoxy-D-arabino-heptulosonate 7-phosphate synthase. Plant Physiol 129:1866–1871
Geigenberger P, Fernie A (2014) Metabolic control of redox and redox control of metabolism in plants. Antioxid Redox Signal 21:1389–1421
Geigenberger P, Kolbe A, Tiessen A (2005) Redox regulation of carbon storage and partitioning in response to light and sugars. J Exp Bot 56:1469–1479
Geigenberger P, Thormählen I, Daloso DM et al (2017) The unprecedented versatility of the plant thioredoxin system. Trends Plant Sci 22:249–262
Gelhaye E, Rouhier N, Jacquot JP (2003) Evidence for a subgroup of thioredoxin h that requires GSH/Grx for its reduction. FEBS Lett 555:443–448
Gelhaye E, Rouhier N, Gérard J et al (2004) A specific form of thioredoxin h occurs in plant mitochondria and regulates the alternative oxidase. Proc Natl Acad Sci U S A 101:14545–14550
Gould SB, Waller RF, McFadden GI (2008) Plastid evolution. Annu Rev Plant Biol 59:491–517
Hertle AP, Blunder T, Wunder T et al (2013) PGRL1 is the elusive ferredoxin-plastoquinone reductase in photosynthetic cyclic electron flow. Mol Cell 49:511–523
Huang J, Niazi AK, Young D et al (2017) Self-protection of Arabidopsis cytosolic malate dehydrogenase against oxidative stress. J Exp Bot. https://doi.org/10.1093/jxb/erx396
Ikegami A, Yoshimura N, Motohashi K et al (2007) The CHLI1 subunit of Arabidopsis thaliana magnesium chelatase is a target protein of the chloroplast thioredoxin. J Biol Chem 282:19282–19291
Ishiga Y, Ishiga T, Wangdi T et al (2012) NTRC and chloroplast-generated reactive oxygen species regulate Pseudomonas syringae pv. tomato disease development in tomato and Arabidopsis. Mol Plant-Microbe Interact 25:294–306

Ishiga Y, Ishiga T, Ikeda Y et al (2016) NADPH-dependent thioredoxin reductase C plays a role in nonhost disease resistance against Pseudomonas syringae pathogens by regulating chloroplast-generated reactive oxygen species. Peer J 4:e1938
Ivanov R, Gaude T (2009) Endocytosis and endosomal regulation of the S-receptor kinase during the self-incompatibility response in Brassica oleracea. Plant Cell 21:2107–2117
Jacquot JP, Rivera-Madrid R, Marinho P et al (1994) Arabidopsis thaliana NAPHP thioredoxin reductase. cDNA characterization and expression of the recombinant protein in Escherichia coli. J Mol Biol 235:1357–1363
Juárez-Díaz JA, McClure B, Vázquez-Santana S et al (2006) A novel thioredoxin h is secreted in Nicotiana alata and reduces S-RNase in vitro. J Biol Chem 281:3418–3424
Kirchsteiger K, Pulido P, Gonzalez M et al (2009) NADPH thioredoxin reductase C controls the redox status of chloroplast 2-Cys peroxiredoxins in Arabidopsis thaliana. Mol Plant 2:298–307
Kirchsteiger K, Ferrandez J, Pascual MB et al (2012) NADPH thioredoxin reductase C is localized in plastids of photosynthetic and nonphotosynthetic tissues and is involved in lateral root formation in Arabidopsis. Plant Cell 24:1534–1548
Kneeshaw S, Gelineau S, Tada Y et al (2014) Selective protein denitrosylation activity of thioredoxin-h5 modulates plant immunity. Mol Cell 56:153–162
Kneeshaw S, Keyani R, Delorme-Hinoux V et al (2017) Nucleoredoxin guards against oxidative stress by protecting antioxidant enzymes. Proc Natl Acad Sci U S A 114:8414–8419
Koh CS, Navrot N, Didierjean C et al (2008) An atypical catalytic mechanism involving three cysteines of thioredoxin. J Biol Chem 283:23062–23072
Konrad A, Banze M, Follmann H (1996) Mitochondria of plant leaves contain two thioredoxins. Completion of the thioredoxin profile of higher plants. J Plant Physiol 149:317–321
Laloi C, Rayapuram N, Chartier Y et al (2001) Identification and characterization of a mitochondrial thioredoxin system in plants. Proc Natl Acad Sci U S A 98:14144–14149
Laloi C, Mestres-ortega D, Marco Y et al (2004) The Arabidopsis cytosolic thioredoxin h5 gene induction by oxidative stress and its W-box-mediated response to pathogen elicitor. Plant Physiol 134:1006–1016
Laughner BJ, Sehnke PC, Ferl RJ (1998) A novel nuclear member of the thioredoxin superfamily. Plant Physiol 118:987–996
Laugier E, Tarrago L, Courteille A et al (2013) Involvement of thioredoxin y2 in the preservation of leaf methionine sulfoxide reductase capacity and growth under high light. Plant Cell Environ 36:670–682
Lee JR, Lee SS, Jang HH et al (2009) Heat-shock dependent oligomeric status alters the function of a plant-specific thioredoxin-like protein, AtTDX. Proc Natl Acad Sci U S A 106:5978–5983
Lemaire SD, Michelet L, Zaffagnini M et al (2007) Thioredoxins in chloroplasts. Curr Genet 51:343–365
Lennartz K, Plücken H, Seidler A et al (2001) HCF164 encodes a thioredoxin-like protein involved in the biogenesis of the cytochrome b_6f complex in Arabidopsis. Plant Cell 13:2539–2551
Lepistö A, Kangasjarvi S, Luomala EM et al (2009) Chloroplast NADPH-thioredoxin reductase interacts with photoperiodic development in Arabidopsis. Plant Physiol 149:1261–1276
Lepistö A, Pakula E, Toivola J et al (2013) Deletion of chloroplast NADPH-dependent thioredoxin reductase results in inability to regulate starch synthesis and causes stunted growth under short-day photoperiods. J Exp Bot 64:3843–3854
Li X, Nield J, Hayman D et al (1996) A self-fertile mutant of Phalaris produces an S protein with reduced thioredoxin activity. Plant J 10:505–513
Li YB, Han LB, Wang HY et al (2016) The thioredoxin GbNRX1 plays a crucial role in homeostasis of apoplastic reactive oxygen species in response to Verticillium dahliae infection in cotton. Plant Physiol 170:2392–2406
Lichter A, Häberlein I (1998) A light-dependent redox signal participates in the regulation of ammonia fixation in chloroplasts of higher plants – ferredoxin: glutamate synthase is a thioredoxin-dependent enzyme. J Plant Physiol 153:83–90
Lorang J, Kidarsa T, Bradford CS et al (2012) Tricking the guard: exploiting plant defense for disease susceptibility. Science 338:659–662

Luo T, Fan T, Liu Y et al (2012) Thioredoxin redox regulates ATPase activity of magnesium chelatase CHLI subunit and modulates redox-mediated signaling in tetrapyrrole biosynthesis and homeostasis of reactive oxygen species in pea plants. Plant Physiol 159:118–130

Machida T, Ishibashi A, Kirino A et al (2012) Chloroplast NADPH-dependent thioredoxin reductase from Chlorella vulgaris alleviates environmental stresses in yeast together with 2-Cys peroxiredoxin. PLoS One 7:e45988

Marchal C, Delorme-Hinoux V, Bariat L et al (2014) NTR/NRX define a new thioredoxin system in the nucleus of Arabidopsis thaliana cells. Mol Plant 7:30–44

Marchand CH, Vanacker H, Collin V et al (2010) Thioredoxin targets in Arabidopsis roots. Proteomics 10:2418–2428

Marri L, Zaffagnini M, Collin V et al (2009) Prompt and easy activation by specific thioredoxins of calvin cycle enzymes of Arabidopsis thaliana associated in the GAPDH/CP12/PRK supramolecular complex. Mol Plant 2:259–269

Martí MC, Olmos E, Calvete JJ et al (2009) Mitochondrial and nuclear localization of a novel pea thioredoxin: identification of its mitochondrial target proteins. Plant Physiol 150:646–657

Marty L, Siala W, Schwarzländer M et al (2009) The NADPH-dependent thioredoxin system constitutes a functional backup for cytosolic glutathione reductase in Arabidopsis. Proc Natl Acad Sci U S A 106:9109–9114

McKinney DW, Buchanan BB, Wolosiuk RA (1978) Activation of chloroplast ATPase by reduced thioredoxin. Phytochemistry 17:794–795

Meng L, Wong JH, Feldman LJ et al (2010) A membrane-associated thioredoxin required for plant growth moves from cell to cell, suggestive of a role in intercellular communication. Proc Natl Acad Sci U S A 107:3900–3905

Meyer Y, Reichheld J-P, Vignols F (2005) Thioredoxins in Arabidopsis and other plants. Photosynth Res 86:419–433

Meyer Y, Riondet C, Constans L et al (2006) Evolution of redoxin genes in the green lineage. Photosynth Res 89:179–192

Meyer Y, Belin C, Delorme-Hinoux V et al (2012) Thioredoxin and glutaredoxin systems in plants: molecular mechanisms, crosstalks, and functional significance. Antioxid Redox Signal 17:1124–1160

Michalska J, Zauber H, Buchanan BB et al (2009) NTRC links built-in thioredoxin to light and sucrose in regulating starch synthesis in chloroplasts and amyloplasts. Proc Natl Acad Sci U S A 106:9908–9913

Mikkelsen R, Mutenda KE, Mant A et al (2005) α-Glucan, water dikinase (GWD): a plastidic enzyme with redox-regulated and coordinated catalytic activity and binding affinity. Proc Natl Acad Sci U S A 102:1785–1790

Moon JC, Jang HH, Chae HB et al (2006) The C-type Arabidopsis thioredoxin reductase ANTR-C acts as an electron donor to 2-Cys peroxiredoxins in chloroplasts. Biochem Biophys Res Commun 348:478–484

Motohashi K, Hisabori T (2006) HCF164 receives reducing equivalents from stromal thioredoxin across the thylakoid membrane and mediates reduction of target proteins in the thylakoid lumen. J Biol Chem 281:35039–35047

Najera VA, Gonzalez MC, Perez-Ruiz JM et al (2017) An event of alternative splicing affects the expression of the NTRC gene, encoding NADPH-thioredoxin reductase C, in seed plants. Plant Sci 258:21–28

Naranjo B, Diaz-Espejo A, Lindahl M et al (2016a) Type-f thioredoxins have a role in the short-term activation of carbon metabolism and their loss affects growth under short-day conditions in Arabidopsis thaliana. J Exp Bot 67:1951–1964

Naranjo B, Mignee C, Krieger-Liszkay A et al (2016b) The chloroplast NADPH thioredoxin reductase C, NTRC, controls non-photochemical quenching of light energy and photosynthetic electron transport in Arabidopsis. Plant Cell Environ 39:804–822

Navrot N, Collin V, Gualberto J et al (2006) Plant glutathione peroxidases are functional peroxiredoxins distributed in several subcellular compartments and regulated during biotic and abiotic stresses. Plant Physiol 142:1364–1379

Née G, Zaffagnini M, Trost P et al (2009) Redox regulation of chloroplastic glucose-6-phosphate dehydrogenase: a new role for f-type thioredoxin. FEBS Lett 583:2827–2832
Nikkanen L, Toivola J, Rintamäki E (2016) Crosstalk between chloroplast thioredoxin systems in regulation of photosynthesis. Plant Cell Environ 39:804–822
Ojeda V, Perez-Ruiz JM, Gonzalez MC et al (2017) NADPH thioredoxin reductase C and thioredoxins act concertedly in seedling development. Plant Physiol 174:1436–1448
Okegawa Y, Motohashi K (2015) Chloroplastic thioredoxin m functions as a major regulator of Calvin cycle enzymes during photosynthesis in vivo. Plant J 84:900–913
Park SK, Jung YJ, Lee JR et al (2009) Heat-shock and redox-dependent functional switching of an h-type Arabidopsis thioredoxin from a disulfide reductase to a molecular chaperone. Plant Physiol 150:552–561
Pascual MB, Mata-Cabana A, Florencio FJ et al (2010) Overoxidation of 2-Cys peroxiredoxin in prokaryotes: cyanobacterial 2-Cys peroxiredoxins sensitive to oxidative stress. J Biol Chem 285:34485–34492
Pascual MB, Mata-Cabana A, Florencio FJ et al (2011) A comparative analysis of the NADPH thioredoxin reductase C-2-Cys peroxiredoxin system from plants and cyanobacteria. Plant Physiol 155:1806–1816
Perez-Ruiz JM, Cejudo FJ (2009) A proposed reaction mechanism for rice NADPH thioredoxin reductase C, an enzyme with protein disulfide reductase activity. FEBS Lett 583:1399–1402
Perez-Ruiz JM, Spinola MC, Kirchsteiger K et al (2006) Rice NTRC is a high-efficiency redox system for chloroplast protection against oxidative damage. Plant Cell 18:2356–2368
Perez-Ruiz JM, Guinea M, Puerto-Galan L et al (2014) NADPH thioredoxin reductase C is involved in redox regulation of the Mg-chelatase I subunit in Arabidopsis thaliana chloroplasts. Mol Plant 7:1252–1255
Perez-Ruiz JM, Naranjo B, Ojeda V et al (2017) NTRC-dependent redox balance of 2-Cys peroxiredoxins is needed for optimal function of the photosynthetic apparatus. Proc Natl Acad Sci U S A 114:12069–12074
Perkins A, Nelson KJ, Parsonage D et al (2015) Peroxiredoxins: guardians against oxidative stress and modulators of peroxide signaling. Trends Biochem Sci 40:435–445
Puerto-Galan L, Perez-Ruiz JM, Ferrandez J et al (2013) Overoxidation of chloroplast 2-Cys peroxiredoxins: balancing toxic and signaling activities of hydrogen peroxide. Front Plant Sci 4:310
Puerto-Galan L, Perez-Ruiz JM, Guinea M et al (2015) The contribution of NADPH thioredoxin reductase C (NTRC) and sulfiredoxin to 2-Cys peroxiredoxin overoxidation in Arabidopsis thaliana chloroplasts. J Exp Bot 66:2957–2966
Pulido P, Cazalis R, Cejudo FJ (2009) An antioxidant redox system in the nucleus of wheat seed cells suffering oxidative stress. Plant J 57:132–145
Pulido P, Spinola MC, Kirchsteiger K et al (2010) Functional analysis of the pathways for 2-Cys peroxiredoxin reduction in Arabidopsis thaliana chloroplasts. J Exp Bot 61:4043–4054
Qin Y, Leydon AR, Manziello A et al (2009) Penetration of the stigma and style elicits a novel transcriptome in pollen tubes, pointing to genes critical for growth in a pistil. PLoS Genet 8: e1000621
Reichheld J-P, Mestres-Ortega D, Laloi C et al (2002) The multigenic family of thioredoxin h in Arabidopsis thaliana: specific expression and stress response. Plant Physiol Biochem 40:685–690
Reichheld J-P, Meyer E, Khafif M et al (2005) AtNTRB is the major mitochondrial thioredoxin reductase in Arabidopsis thaliana. FEBS Lett 579:337–342
Reichheld J-P, Khafif M, Riondet C et al (2007) Inactivation of thioredoxin reductases reveals a complex interplay between thioredoxin and glutathione pathways in Arabidopsis development. Plant Cell 19:1851–1865
Rey P, Pruvot G, Becuwe N et al (1998) A novel thioredoxin-like protein located in the chloroplast is induced by water deficit in Solanum tuberosum L. plants. Plant J 13:97–107

Rey P, Cuine S, Eymery F et al (2005) Analysis of the proteins targeted by CDSP32, a plastidic thioredoxin participating in oxidative stress responses. Plant J 41:31–42
Richter AS, Peter E, Rothbart M et al (2013) Posttranslational influence of NADPH-dependent thioredoxin reductase C on enzymes in tetrapyrrole synthesis. Plant Physiol 162:63–73
Rivera-Madrid R, Mestres D, Marinho P et al (1995) Evidence for five divergent thioredoxin h sequences in Arabidopsis thaliana. Proc Natl Acad Sci U S A 92:5620–5624
Sahrawy M, Hecht V, Lopez-Jaramillo J et al (1996) Intron position as an evolutionary marker of thioredoxins and thioredoxin domains. J Mol Evol 42:422–431
Sanchez-Riego AM, Mata-Cabana A, Galmozzi CV et al (2016) NADPH-thioredoxin reductase C mediates the response to oxidative stress and thermotolerance in the cyanobacterium Anabaena sp. PCC7120. Front Microbiol 7:1283
Sanz-Barrio R, Corral-Martinez P, Ancin M et al (2013) Overexpression of plastidial thioredoxin f leads to enhanced starch accumulation in tobacco leaves. Plant Biotechnol J 11:618–627
Sarkar N, Lemaire S, Wu-Scharf D et al (2005) Functional specialization of Chlamydomonas reinhardtii cytosolic thioredoxin h1 in the response to alkylation induced DNA damage. Eukaryot Cell 4:262–273
Sasaki Y, Kozaki A, Hatano M (1997) Link between light and fatty acid synthesis: thioredoxin-linked reductive activation of plastidic acetyl-CoA carboxylase. Proc Natl Acad Sci U S A 94:11096–11101
Scheibe R, Anderson LE (1981) Dark modulation of NADP-dependent malate dehydrogenase and glucose-6-phosphate dehydrogenase in the chloroplast. Biochim Biophys Acta 636:58–64
Schmidtmann E, König AC, Orwat A et al (2014) Redox regulation of Arabidopsis mitochondrial citrate synthase. Mol Plant 7:156–169
Schürmann P, Buchanan BB (2008) The ferredoxin/thioredoxin system of oxygenic photosynthesis. Antioxid Redox Signal 10:1235–1274
Serrato AJ, Cejudo FJ (2003) Type-h thioredoxins accumulate in the nucleus of developing wheat seed tissues suffering oxidative stress. Planta 217:392–399
Serrato AJ, Crespo JL, Florencio FJ et al (2001) Characterization of two thioredoxins h with predominant localization in the nucleus of aleurone and scutellum cells of germinating wheat seeds. Plant Mol Biol 46:361–371
Serrato AJ, Perez-Ruiz JM, Cejudo FJ (2002) Cloning of thioredoxin h reductase and characterization of the thioredoxin reductase-thioredoxin h system from wheat. Biochem J 367:491–497
Serrato AJ, Perez-Ruiz JM, Spinola MC et al (2004) A novel NADPH thioredoxin reductase, localized in the chloroplast, which deficiency causes hypersensitivity to abiotic stress in Arabidopsis thaliana. J Biol Chem 279:43821–43827
Serrato AJ, Guilleminot J, Meyer Y et al (2008) AtCXXS: atypical members of the Arabidopsis thaliana thioredoxin h family with a remarkably high disulfide isomerase activity. Physiol Plant 133:611–622
Seung D, Thalmann M, Sparla F et al (2013) Arabidopsis thaliana AMY3 is a unique redox-regulated chloroplastic α-amylase. J Biol Chem 288:33620–33633
Silver DM, Silva LP, Issakidis-Bourguet E et al (2013) Insight into the redox regulation of the phosphoglucan phosphatase SEX4 involved in starch degradation. FEBS J 280:538–548
Skryhan K, Cuesta-Seijo JA, Nielsen MM et al (2015) The role of cysteine residues in redox regulation and protein stability of Arabidopsis thaliana starch synthase 1. PLoS One 10: e0136997
Sparla F, Costa A, Lo Schiavo F et al (2006) Redox regulation of a novel plastid-targeted beta-amylase of Arabidopsis. Plant Physiol 141:840–850
Sweat TA, Wolpert TJ (2007) Thioredoxin h5 is required for victorin sensitivity mediated by a CC-NBS-LRR gene in Arabidopsis. Plant Cell 19:673–687
Tada Y, Spoel SH, Pajerowska-Mukhtar K et al (2008) Plant immunity requires conformational changes of NPR1 via S-nitrosylation and thioredoxins. Science 321:952–956
Tarrago L, Laugier E, Zaffagnini M et al (2010) Plant thioredoxin CDSP32 regenerates 1-Cys methionine sulfoxide reductase B activity through the direct reduction of sulfenic acid. J Biol Chem 285:14964–14972

Thormählen I, Ruber J, von Roepenack-Lahaye E et al (2013) Inactivation of thioredoxin f1 leads to decreased light activation of ADP-glucose pyrophosphorylase and altered diurnal starch turnover in leaves of Arabidopsis thaliana. Plant Cell Environ 36:16–29

Thormählen I, Meitzel T, Groysman J et al (2015) Thioredoxin f1 and NADPH-dependent thioredoxin reductase C have overlapping functions in regulating photosynthetic metabolism and plant growth in response to varying light conditions. Plant Physiol 169:1766–1786

Thormählen I, Zupok A, Rescher J et al (2017) Thioredoxins play a crucial role in dynamic acclimation of photosynthesis in fluctuating light. Mol Plant 10:168–182

Toivola J, Nikkanen L, Dahlstrom KM et al (2013) Overexpression of chloroplast NADPH-dependent thioredoxin reductase in Arabidopsis enhances leaf growth and elucidates in vivo function of reductase and thioredoxin domains. Front Plant Sci 4:389

Traverso JA, Micalella C, Martinez A et al (2013) Roles of N-terminal fatty acid acylations in membrane compartment partitioning: Arabidopsis h-type thioredoxins as a case study. Plant Cell 25:1056–1077

Valerio C, Costa A, Marri L et al (2011) Thioredoxin-regulated β-amylase (BAM1) triggers diurnal starch degradation in guard cells, and in mesophyll cells under osmotic stress. J Exp Bot 62:545–555

Vieira Dos Santos C, Laugier E, Tarrago L et al (2007) Specificity of thioredoxins and glutaredoxins as electron donors to two distinct classes of Arabidopsis plastidial methionine sulfoxide reductases B. FEBS Lett 581:4371–4376

Vignols F, Mouaheb N, Thomas D et al (2003) Redox control of Hsp70-Co-chaperone interaction revealed by expression of a thioredoxin-like Arabidopsis protein. J Biol Chem 278:4516–4523

Wang P, Liu J, Liu B et al (2013) Evidence for a role of chloroplastic m-type thioredoxins in the biogenesis of photosystem II in Arabidopsis. Plant Physiol 163:1710–1728

Williams CH, Arscott LD, Muller S et al (2000) Thioredoxin reductase two modes of catalysis have evolved. Eur J Biochem 267:6110–6117

Wimmelbacher M, Börnke F (2014) Redox activity of thioredoxin z and fructokinase-like protein 1 is dispensable for autotrophic growth of Arabidopsis thaliana. J Exp Bot 65:2405–2413

Wolosiuk RA, Buchanan BB (1977) Thioredoxin and glutathione regulate photosynthesis in chloroplasts. Nature 266:565–567

Wolosiuk RA, Buchanan BB, Crawford NA (1977) Regulation of NADP-malate dehydrogenase by the light-actuated ferredoxin/thioredoxin system of chloroplasts. FEBS Lett 81:253–258

Wood ZA, Poole LB, Karplus PA (2003) Peroxiredoxin evolution and the regulation of hydrogen peroxide signaling. Science 300:650–653

Wulff RP, Lundqvist J, Rutsdottir G et al (2011) The activity of barley NADPH-dependent thioredoxin reductase C is independent of the oligomeric state of the protein: tetrameric structure determined by cryo-electron microscopy. Biochemist 50:3713–3723

Yamamoto M, Nasrallah JB (2009) In planta assessment of the role of thioredoxin h proteins in the regulation of S-locus receptor kinase signaling in transgenic Arabidopsis. Plant Physiol 163:1387–1395

Yamaryo Y, Motohashi K, Takamiya K et al (2006) In vitro reconstitution of monogalactosyldiacylglycerol (MGDG) synthase regulation by thioredoxin. FEBS Lett 580:4086–4090

Yoshida K, Hisabori T (2014) Mitochondrial isocitrate dehydrogenase is inactivated upon oxidation and reactivated by thioredoxin-dependent reduction in Arabidopsis. Front Environ Sci 2:38

Yoshida K, Hisabori T (2016a) Adenine nucleotide-dependent and redox-independent control of mitochondrial malate dehydrogenase activity in Arabidopsis thaliana. Biochim Biophys Acta 1857:810–818

Yoshida K, Hisabori T (2016b) Two distinct redox cascades cooperatively regulate chloroplast functions and sustain plant viability. Proc Natl Acad Sci U S A 113:E3967–E3976

Yoshida K, Noguchi K, Motohashi K et al (2013) Systematic exploration of thioredoxin target proteins in plant mitochondria. Plant Cell Physiol 54:875–892

Yoshida K, Hara S, Hisabori T (2015) Thioredoxin selectivity for thiol-based redox regulation of target proteins in chloroplasts. J Biol Chem 290:14278–14288

From Aspartate to Ethylene: Central Role of N, C, and S Shuttles by Aminotransferases During Biosynthesis of a Major Plant Growth Hormone

E. Le Deunff

Contents

E. Le Deunff (✉)
Normandie Université, UNICAEN, ICORE, Caen, France
e-mail: erwan.ledeunff@unicaen.fr

Progress in Botany (2019) 80: 253–294, DOI 10.1007/124_2018_17,
Published online: 28 February 2018

Abstract Ethylene biosynthesis originates from three amino acids: aspartate, cysteine, and methionine. In the aspartate-derived amino acid pathway, the ethylene pathway requires no less than seven aminotransferases that connect the metabolisms of nitrogen (N), sulfur (S), and carbon (C). Aminotransferases are fundamental enzymes in plants involved in N, S, and C shuttling through their implication in amino acid biosynthesis and catabolism. The role of these enzymes in the biosynthesis of hormones such as ethylene and auxins (IAA and PAA) is frequently overlooked. The functioning of aminotransferases is dependent on an essential cofactor: pyridoxal-5′-phosphate (PLP). This phosphorylated form of vitamin B_6 is synthesized from glutamine, the first product of N assimilation produced by the GS/GOGAT cycle after reduction of nitrate and the glyceraldehyde 3-phosphate (G3P) and ribose 5-phosphate (5RP) provided by the glycolytic and pentose phosphate pathways, respectively. Here we review the recent progress in characterization of the aspartate-derived metabolic pathway with a particular focus on methionine biosynthesis and its salvage pathway (Yang cycle) related to ethylene and polyamine biosynthesis. Emphasis is placed on the key role of aminotransferases in regulating these pathways and their relation with aromatic amino acid biosynthesis and catabolism. Indeed, the promiscuity of certain aminotransferases extends their catalytic function and gives them a key role in the metabolism of ethylene, IAA, and aromatic amino acids. In this respect, recent studies have identified specific aminotransferases as being the main targets involved in the root morphogenetic program in response to environmental cues, nutrient availability, and energy status. Thus, genetically engineered plants for some aminotransferases, such as ACC synthase and tryptophan aminotransferase, demonstrate a great potential to produce crop species with enhanced exploratory root growth and a better nitrogen use efficiency. How the network of aminotransferases is involved in nitrogen-sensing systems such as plant glutamate receptors, TOR, and GCN2 kinases is now becoming a fundamental issue. The use of specific and nonspecific inhibitors of the catalytic activity of certain aminotransferases should help future pharmacological and genetic approaches to unravel their role in N, S, and C sensory systems.

1 Introduction

Although N fertilization is one of the major components in the success of the green revolution and plant growth and productivity, the basis of the profound remodeling induced by nitrogen in nitrogen (N), sulfur (S), and carbon (C) allocation between shoots and roots remains always elusive (Good et al. 2004; Hirel et al. 2007; Masclaux-Daubresse et al. 2010). Thus, it is well known that N deficiency induces proliferation of roots at the expense of shoots whereas excess of N induced the opposite situation (Scheible et al. 1997). However, the basis of these physiological and molecular switches is still not understood (Scheible et al. 1997; Le Ny et al. 2013). For example, conventional approaches to study root functional and structural responses based on nitrate availability under homogeneous and heterogeneous N supply conditions are hampered by the fact that there is a very strong coordination and integration of absorption and assimilation of N with C fixation at primary metabolism level (Wang et al. 2003; Scheible et al. 2004; Bi et al. 2007; Nero et al. 2009; Bussell et al. 2013). NO_3^- uptake and its utilization by plants to match the N demand are sensitively regulated by internal N-sensing mechanisms that control functional and structural responses via a myriad of signaling molecules. Although many candidate systems for N detection have been proposed such as TOR (Target Of Rapamycin) and GCN2 (*G*eneral *C*ontrol *N*on-derepressible 2) kinases as well as GLR (Glutamate Like Receptor) receptors, most approaches for testing these N sensory systems ignore the involvement of hormones such as ethylene and IAA (Lam et al. 2006; Castilho et al. 2014; Xiong and Sheen 2014). However, compared to nitrate, ethylene and IAA play a major role in root growth and development because they act at very low concentrations (nM and μM) and at very short-term durations (min to hours) on root and shoot development (Robinson 2005; Le Deunff et al. 2016). Moreover, their interactions are mostly involved in the primary root and lateral root (LR) development as well as root hair growth (Stepanova et al. 2007; Muday et al. 2012; Hu et al. 2017). Interestingly, plant hormones such as ethylene, indole-3-acetic acid (IAA), and phenylacetic acid (PAA) are derived from L-methionine (Met), L-tryptophan (Trp), and L-phenylalanine (Phe) amino acids, respectively (Giovanelli et al. 1985, Stepanova et al. 2008; Sugawara et al. 2015). This confers to these plant hormones a specific place in N, S, and C primary metabolism in relation with growth during the plant life cycle. Among these hormones, ethylene occupies a special place since it is a volatile molecule involved in alarm, stress, and senescence (Abeles et al. 1992), but is also a powerful natural anesthetic (Baluska et al. 2016). Moreover, the first precursor of ethylene, methionine, is an essential amino acid at the intersection of the metabolism of N, S, and C. Its second precursor, S-adenosyl-L-methionine (AdoMet), is a fundamental cofactor involved in plant transmethylation reactions that is used by folates (THF) as a relay molecule to extend their capacities for methylation (Cossins 2000; Lu 2000; Gerelova et al. 2017). Finally, the direct precursor of ethylene, ACC, is a non-proteinogenic amino acid whose production is exquisitely regulated (Wang et al. 2002; Stepanova et al. 2007) and has a potential role as an agonist on plant glutamate receptors (Le Deunff and Lecourt 2016). Finally, ethylene acts on specific receptors and some components of its signaling pathway such as the transcription factors ethylene insensitive 3 (EIN3) and its homologue EIN3-LIKE 1 (EIL1) play a major role in the regulation of auxin biosynthetic genes (He et al. 2011) and responses to glucose through the hexokinase1 (HXK1) C sensor (Yanagisawa et al. 2003; Yoo et al. 2008).

So far, emphasis has been placed on the regulation of biosynthesis and signaling cascades of both ethylene and auxins and their interactions on root and shoots growth (Muday et al. 2012; Hu et al. 2017). However, a major challenge for the next few years is to address the precise molecular mechanisms underlying the complex signaling network that governs regulation of the metabolic interconnections between mineral nutrition, amino acid biosynthesis, and the production of plant growth regulators. In this aim, an alternative approach consists of stating that the biosynthesis of hormones such as ethylene and IAA is highly regulated through the aminotransferases network in primary metabolism (Le Deunff et al. 2016). Indeed, aminotransferases are the key enzymes that ensure N, S, and C shuttlings but are also essential in the biosynthesis of ethylene and auxin precursors Met, Trp, and Phe, and in the production of major cofactors such as vitamins B_9 (folates) and B_6.

This review focuses on the metabolic interconnections between ethylene biosynthesis and nitrogen metabolism and, to a lesser extent, sulfur metabolism. Because the focus is mainly on the network of PLP-dependent aminotransferases during ethylene and methionine biosynthesis, ethylene signaling is less discussed.

2 A Highly Regulated Pathway for the Biosynthesis of Methionine and S-Adenosylmethionine (AdoMet): The Ethylene Precursors

In plants, AA biosynthetic pathways can be organized depending on the origin of their carbon skeleton coming from glycolysis, the citric acid cycle, and the oxidative pentose phosphate pathway (Fig. 1; Coruzzi and Last 2000; Stitt et al. 2002). The carbon skeleton of methionine is mainly synthesized *via* a branch of aspartate pathway, but its sulfur atom is derived from cysteine and its methyl group from the β-carbon of serine. Other amino acids such as Thr, Ile, and Lys also originate from this aspartate-derived metabolic pathway (Azevedo et al. 2006; Jander and Joshi 2009). Due to the importance in human diets of these four amino acids (Met, Thr, Ile, and Lys), the biochemical regulatory mechanisms involved in the biosynthesis of aspartate-derived amino acids have been intensively studied. Thus, allosteric regulations of the branch-point enzymes by the pathway products have been determined by genetic and biochemical approaches and are summarized in Fig. 2. More details and explanations can be found in excellent reviews (Coruzzi and Last 2000; Azevedo et al. 2006; Jander and Joshi 2009; Galili 2011). Here we only focus on methionine biosynthesis that competes with threonine biosynthesis because these

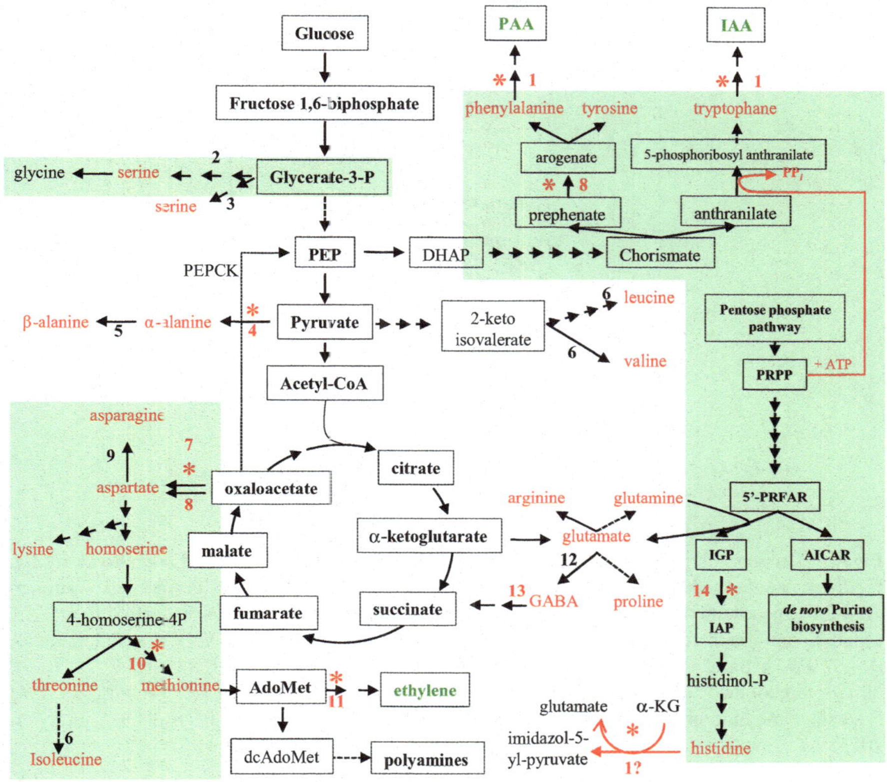

Fig. 1 Simplified overview of amino acid biosynthesis in plants. Asterisk indicates aminotransferases targeted by AVG and number represents PLP-dependent aminotransferases in the different biosynthetic pathways: **1**, tryptophan aminotransferase of *Arabidopsis* (TAA); **2**, 3-phosphoserine aminotransferase (PSAT) and ala-hydroxypyruvate aminotransferase (HA-AT); **3**, alanine-hydroxypyruvate aminotransferase (AH-AT); **4**, Alanine aminotransferase (AlaAT); **5**, β-alanine-pyruvate aminotransferase (BAPAT); **6**, branched-chain aminotransferase (BCAT); **7** and **8**, eukaryotic aspartate aminotransferase (ET-AAT) and prokaryotic type aspartate/prephenate aminotransferase (PT-AAT/PAT); **9**, asparagine synthase (ASN); **10**, cystathionine β-lyase (CBL); **11**, 1-aminocyclopropane-1-carboxylate synthase (ACS); **12**, glutamate decarboxylase (GAD); **13**, Gaba transaminases using either α-ketoglutarate (GABA-TK) or pyruvate (GABA-TP); **14**, histidinol-phosphate aminotransferase (HPA). Abbreviations used for metabolites are: *5′-PRFAR* phosphoribosyl formimino-5-aminoimidazole-′-carboxamide ribonucleotide, *IGP* imidazoleglycerolphosphate, *IAP* imidazoleacetolphosphate, *AICAR* 5-aminoimidazole-4-carboxamideribonucleotide, *α-KG* α-ketoglutarate, *IAA* indole-3-acetic acid, *PAA* phenylacetic acid, *SAM* S-Adenosylmethionine, *PEPCK* phospho*enol*pyruvate carboxykinase. Green boxes indicate plastidic location of enzymes. **?** indicates putative degradation pathway of histidine by tryptophan aminotransferase according to Brunke et al. (2014) in *C. glabrata* and *S. cerevisiae* (adapted from Coruzzi and Last 2000)

two branches of the aspartate pathway use initially a common substrate: *O*-phosphohomoserine (OPHS). It is now clearly established that in the methionine branch of the aspartate pathway all enzymes are present in the chloroplast. Therefore, chloroplasts are autonomous for *de novo* Met biosynthesis (Mills et al. 1980; Ravanel et al. 2004).

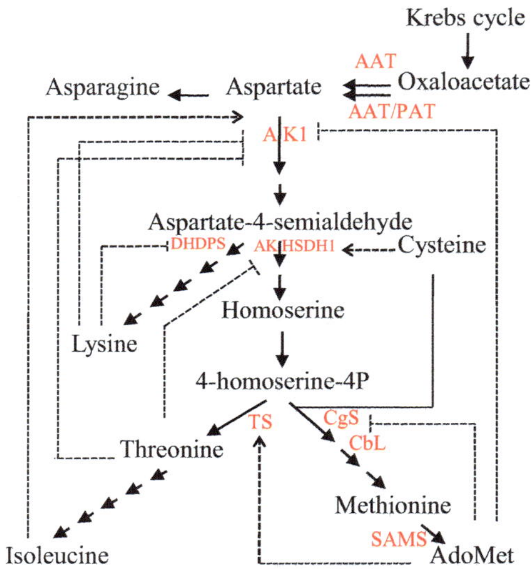

Fig. 2 Aspartate-derived amino acid pathway and allosteric regulations of the pathway by final products or intermediates. Positive allosteric regulation is indicated by a dashed arrow line and inhibition by a dashed T line. Enzymes submitted to allosteric regulation are indicated in red. *AAT* aspartate aminotransferase, *AAT/PAT* aspartate/prephenate aminotransferase, *AK* aspartate kinase, *AK-HDKS* aspartate kinase with homoserine dehydrogenase (HDKS) activity, *DHDPS* dihydrodipicolinate synthase, *TS* threonine synthase, *CgS* cystathionine γ-synthase, *SAMS* S-adenosylmethionine synthase (adapted from Coruzzi 2003; Jander and Joshi 2009; Sauter et al. 2013)

In this pathway, OPHS is converted to cystathionine *via* the cystathionine γ-synthase (CgS) whereas in the threonine branch pathway OPHS is converted to L-threonine *via* threonine synthase (TS, Fig. 2). Cystathionine γ-synthase and threonine synthase are rate-limiting steps in Met and Thr biosynthesis (Ravanel et al. 1998; Hesse et al. 2004). The activity of threonine synthase is allosterically activated by S-adenosyl-L-methionine (AdoMet), the end product of the methionine pathway (Curien et al. 1996, 1998; Laber et al. 1999; Zeh et al. 2002). In contrast, cystathionine γ-synthase is negatively regulated at posttranscriptional level by AdoMet that binds to the N-terminus of nascent cystathionine γ-synthase during translation (Fig. 2), reduces mRNA stability, and stops its translational activity (Chiba et al. 1999). Therefore, AdoMet levels exerted a feedback control on its own production that in turn partly modulates Met biosynthesis. AdoMet acts also in concert with lysine as an allosteric inhibitor on the activity of three monofunctional aspartate kinases (AK) involved in the first step of the aspartate-derived amino acid pathway (Fig. 2). Cysteine, used as substrate by CgS enzyme for providing the sulfur atom to methionine, also regulates the pathway by stimulating the aspartate kinase with homoserine dehydrogenase (AK-HSDH) activity (Jander and Joshi 2009). Indeed, allosteric co-regulations of CgS and AK-HSDH1 activities by AdoMet and cysteine, respectively, direct the flux from aspartate into Thr, Ile, and Met production.

2.1 Glutamine Involvement in the Methionine Pathway Is Hidden Behind the PLP Coenzyme Used by Aminotransferases

The three steps of methionine biosynthesis in the plastids from *O*-phosphohomoserine require the action of cystathionine γ-synthase (CgS), cystathionine β-lyase (CbL), and methionine synthase (MS), respectively (Fig. 3). Cystathionine γ-synthase and β-lyase are two enzymes that belong to the γ- and β-lyase subfamilies of the α family of aminotransferases (Christen and Mehta 2001). For their catalytic activity, they need pyridoxal-5′-phosphate (PLP), a derivative form of vitamin B_6 that is a versatile coenzyme. *De novo* PLP biosynthesis in the cytosol is catalyzed by pyridoxine synthase enzymes (PDX1 and PDX2) from glutamine, produced by the GS/GOGAT cycle after reduction of nitrate and glyceraldehyde 3-phosphate (G3P) and ribose 5-phosphate (5RP) provided by the glycolysis and pentose phosphate pathways, respectively (Tambasco-Studart et al. 2005; Fitzpatrick 2011; Colinas et al. 2016). PLP is also synthesized in chloroplasts from pyridoxamine 5′-phosphate (PMP) or pyridoxine 5′-phosphate (PNP) by pyridoxamine 5′-phosphate oxidase enzyme (Fitzpatrick 2011). It is not known how *de novo* PLP synthesis in cytosol modulates the PLP pool in the plastid and which plastidial carriers are involved in the transport of different forms of vitamin B_6 (Gerdes et al. 2012). Two olefinic compounds, DL-propargylglycine (PAG) and L-aminoethoxyvinylglycine (AVG), that act as suicide inhibitors of PLP-enzymes of the α family of aminotransferases (Lieberman 1979; Satoh and Yang 1989a) can inhibit in vitro and in vivo activities of CgS and CbL enzymes, respectively (Ravanel et al. 1998).

2.2 Serine Is Doubly Implicated in Methionine Biosynthesis

Serine is directly involved in the methionine pathway through cysteine biosynthesis (Romero et al. 2014). Cysteine provides a sulfur atom to methionine through the action of cystathionine γ-synthase (CgS) and comes from the reduction and assimilatory pathway of inorganic sulfate (SO_4^{2-}). The biosynthesis of cysteine requires serine as the amino acid skeleton-donor and bisulfide from sulfate reduction as a sulfur donor to form cysteine (Takahashi et al. 2011; Romero et al. 2014). In plants, the reduction of sulfate takes place in the chloroplast and produces sulfide, an inorganic anion of sulfur with the chemical formula S_2^-. However, in aqueous solution most sulfide ions are neutralized under the conjugate acid form of bisulfide: SH^-. The two consecutive reactions catalyzed by serine acyltransferase (SAT) and O-acetylserine(thiol)lyase (OSATL) produce cysteine (Fig. 3). SAT catalyzes the conversion of serine and acetyl-CoA into O-acetylserine (OAS) and acetyl-CoA-SH whereas OSATL catalyzes the final PLP-dependent conversion of bisulfide and OAS into cysteine and acetate (Bonner et al. 2005). These two enzymes can assemble in a hetero-oligomeric cysteine synthase complex formed by one SAT hexamer and two

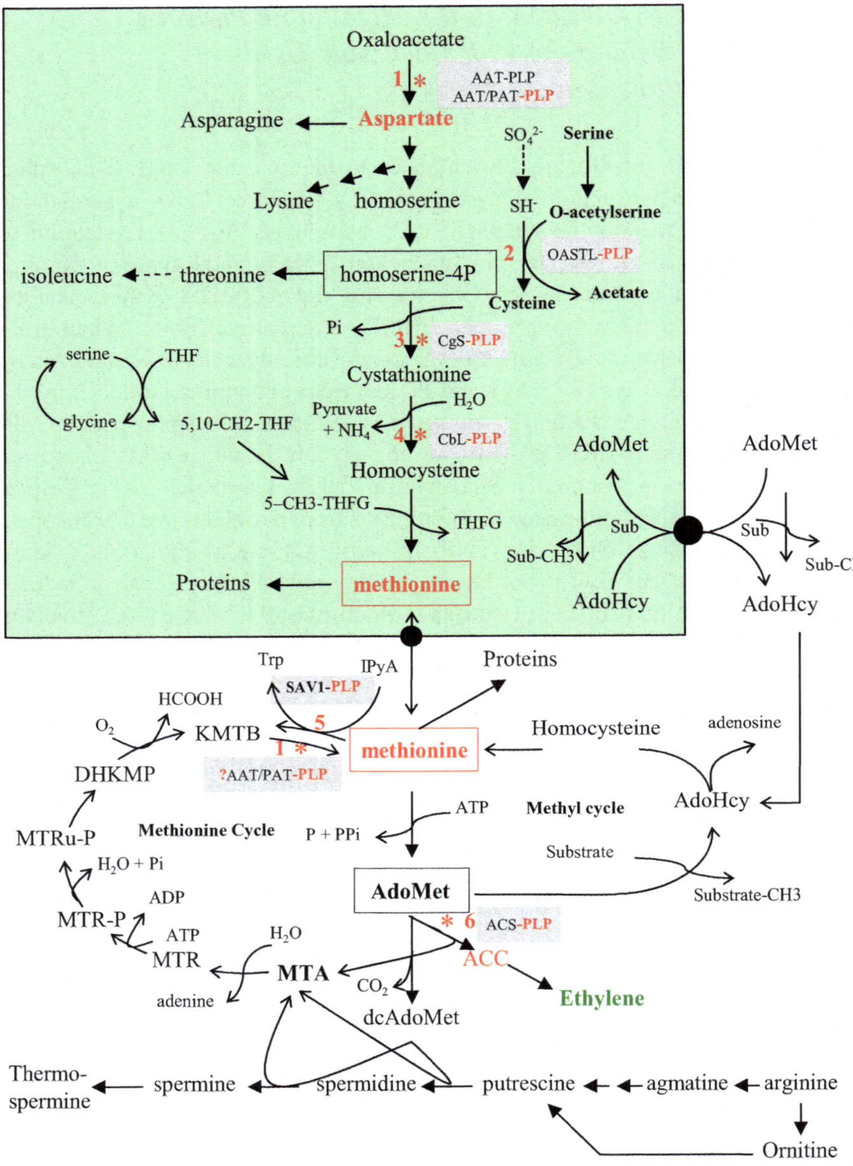

Fig. 3 Interconnections in biosynthetic pathways of methionine, polyamines, ethylene, and AdoMet and methionine salvage cycles in plants. Asterisk indicates aminotransferases targeted by AVG: **1**, eukaryotic aspartate aminotransferase (ET-AAT) and prokaryotic type aspartate/prephenate aminotransferase (PT-AAT/PAT); **2**, O-acetylserine(thiol)lyase (OASTL); **3**, cystathionine γ-lyase (CgL); **4**, cystathionine β-lyase (CbL); **5**, aminotransferase SAV1/ISS1; **6**, 1-aminocyclo-propane-1-carboxylate synthase (ACS). Abbreviations used for metabolites are: *AdoMet* S-adenosylmethionine, *dcAdoMet* decarboxylated S-adenosylmethionine, *SAH* S-adenosylhomocysteine, *MTA* 5′-methylthioadénosine, *MTR* 5′-methylthioribose, *MTR-P* 5′-methylthioribose-phosphate, *MTRu-P* 5′-methylthioribulose-phosphate, *DHKMP* 1,2-dihydro-3-oxo-5-methylthiopentene, *KMTB* 2-oxo-4-methylthiobutyrate; Green boxes indicate plastidic location of enzymes. **?** indicates putative aspartate aminotransferase according to Pommerrenig et al. (2011) (adapted from Ravanel et al. 2004; Pommerrenig et al. 2011; Sauter et al. 2013)

OASTL dimers requiring four molecules of PLP for functioning (Droux et al. 1998). This complex, first found in bacteria, plays a major role in plants for modulation of cysteine biosynthesis via bisulfide and OAS. Some authors consider that this complex could be a cellular sulfur sensor (Yi et al. 2010; Gerelova et al. 2017).

Serine is also indirectly involved in the last step of methionine biosynthesis through the use of a polyglutamate form of tetrahydrofolate (5-methyl-THF: 5-CH_3H_4PteGlu$_n$ with $n > 3$) as a methyl-group donor (Eichel et al. 1995). Thus, homocysteine (Hcy) methylation to form methionine is catalyzed by methionine synthase (MS). The methyl group given by 5-methyl-THF is provided by the β-carbon of serine after two consecutive reactions catalyzed in the plastids by serine hydroxymethyltransferase (THFT) and 5,10-methylenetetrahydrofolate reductase (MTHFR). THFT enzyme uses serine as a methylene group donor to convert tetrahydrofolate (H_4PteGlu$_n$) into methylene-THF (5,10-CH_2H_4PteGlu$_n$) and glycine (Fig. 3). Then, methylene-THF is converted to its reduced form 5-methyl-THF (5-CH_3H_4PteGlu$_n$) by MTHFR. The one-carbon group of 5-methyl-THF is used by methionine synthase (MS) to convert Hcy to methionine (Shah and Cossins 1970; Gorelova et al. 2017). Polyglutamate forms of THF are synthetized by folylpolyglutamate synthase (FPGS), an enzyme localized in mitochondria, cytosol, and plastids that catalyzes the attachment of glutamate tail to a THF molecule (Ravanel et al. 2001). In *Arabidopsis*, mutation of the AtDFB gene encoding FPGS enzyme showed a disruption in primary root growth by an alteration of the quiescent center of root meristem, suggesting that THF metabolism plays a major role in root cell proliferation (Srivastava et al. 2011; Reyes-Hernández et al. 2014). Therefore, serine is doubly required for methionine biosynthesis *via* the cysteine biosynthesis pathway and production of the polyglutamate form of 5-methyl-THF.

2.3 AdoMet Synthase Is a Fundamental Control Point of the Methionine Pathway

The last step of the methionine pathway is catalyzed by methionine adenosyltransferase (SAM or MAT), a cytosolic enzyme that catalyzes conversion of methionine and ATP into AdoMet. In the *Arabidopsis* genome, SAM is encoded by three genes (SAM1, SAM2, and SAM3) and their expression has been shown to be highly regulated at transcriptional (Peleman et al. 1989a, b; Boerjan et al. 1994; Chen et al. 2016) and posttranscriptional levels (Mao et al. 2015; Jin et al. 2017) in response to hormones, biotic and abiotic stress.

Recently, it has been shown in *Arabidopsis* that FERONIA (FER), a plasma membrane receptor-like kinase, may negatively regulate SAM (Mao et al. 2015). FERONIA receptor belongs to the *Cr*RLK (*Catharanthus roseus* Receptor Like

Kinase) family and is involved in the elongation of root hairs and leaf cells (Guo et al. 2009; Deslauriers and Larsen 2010; Duan et al. 2010). FER receptor ligand is the RALF peptide (Rapid Alkalinization Factor) that is able to bind to the FER receptor and trigger phosphorylation of many cytosolic proteins such as H^+-ATPase AHA2 involved in cell elongation (Haruta et al. 2014). Surprisingly, *fer* mutants display a dwarf phenotype and a significant increase in ethylene production and AdoMet contents. The dwarf phenotype of *fer* mutants was mimicked in transgenic plants over-expressing SAM whereas *sam1sam2* double mutant showed a wild-type phenotype suggesting that FER receptor might be involved in inhibition of SAM activity thereby reducing levels of AdoMet and ethylene. Further work is needed to determine mechanisms by which FER receptor inhibits SAM1 and SAM2 activities in responses to RALF peptide hormone and ethylene crosstalk.

SAM isoforms are also regulated at the posttranslational level by CPK28, a calcium-dependent protein kinase, that interacts with phosphorylated forms of SAM1, SAM2, and SAM3 to induce their degradation through the ubiquitin/26S proteasome pathway (Jin et al. 2017). At the biochemical level, *cpk8* mutants display an increase in AdoMet contents and SAM protein levels, whereas at the phenotypic level they present short hypocotyls and an increase in lignification caused by ethylene overproduction. Inhibiting ACC synthetase (ACS) activity with AVG treatment could restore the wild-type phenotype. Taken together these results demonstrate that the last step of the methionine biosynthetic pathway is highly regulated at the translational and posttranscriptional levels because AdoMet is a regulatory node molecule involved in plant development through trans-methylation reactions, crosstalk between ethylene and polyamine biosynthesis, and stress signaling responses.

3 Because of Its Vital Cellular Functions AdoMet Requires Two Salvage Pathways

In *Lemna paucicostata*, production of AdoMet consumes 80% of Met produced by the methionine pathway whereas 20% of Met is involved for protein synthesis (Giovanelli et al. 1985). AdoMet is one of the major cofactors used in nature with ATP and ensures vital biochemical functions in plant and animal cells (Lu 2000). Indeed, AdoMet is the methyl group donor during transmethylation reactions of lignins, chlorophylls, proteins, phospholipids, and nucleic acids biosynthesis, the precursor for biosynthesis of ethylene, polyamine, nicotianamine (siderophore), and biotin, and it is also involved in many transsulfuration reactions (Lu 2000). Since AdoMet is the end product of the methionine pathway and one of the major cofactors used in nature, the lack of recycling of the AdoMet methylthio-group and adenosine moiety reduces biosynthesis of ethylene and polyamines by a restriction in sulfur availability (Baur and Yang 1972; Bürstenbinder et al. 2007). From a literature survey, it appears that the cellular AdoMet homeostasis is controlled at several levels through: (1) allosteric and transcriptional regulations of the amino acid pathway derived from aspartate (Fig. 2), (2) translational and posttranscriptional regulation of methionine adenosyltransferase, (3) transmethylation reactions, (4) regeneration via the methyl cycle and Yang cycle, and (5) elevated shuttling of AdoMet into several biosynthetic pathway such as ethylene and polyamines (Bürstenbinder et al. 2007; Van de Poel et al. 2012). Because AdoMet is synthesized in the cytosol, its availability for plastidial reactions therefore requires a specific carrier, which is another regulatory point for AdoMet homeostasis (Fig. 3). In this respect, in *Arabidopsis* a chloroplastic AdoMet carrier was able to catalyze the unidirectional import of cytosolic AdoMet as well as the exchange between a cytosolic AdoMet and a chloroplastic AdoMet or S-adenosylhomocysteine (Ravanel et al. 2004). Therefore, because of its cellular importance, AdoMet requires two salvage pathways to regenerate its adenosine moiety, methylthio-group and methyl group used in methylation reactions.

3.1 A Methyl Cycle Serves to Regenerate the Methyl Group Issued from AdoMet

The S-adenosylhomocysteine (AdoHcy) issued from methylation reactions with AdoMet can be recycled to methionine from the homocysteine (Hcy) intermediate through the methyl or AdoMet cycle (Ravanel et al. 1998). AdoHcy issued from transmethylation reactions is converted into Hcy by AdoHcy hydrolase and Hcy is then regenerated to methionine by Met synthase (Fig. 3). Conversion of homocysteine into methionine can occur in mitochondria, chloroplasts, and cytosol (Clandinin and Cossins 1974; Shah and Cossins 1970; Eichel et al. 1995). The last two steps of this cycle are assumed to occur mainly in the cytosol (Hanson and Roje 2001; Roje et al. 2002a), since an AdoMet carrier is involved to ensure the exchanges between cytosolic and chloroplastic AdoMet or SAH (Ravanel et al. 2004). Met synthase catalyzes methionine formation and simultaneously allows regeneration of the methyl group of AdoMet *via* the use of a folic compound: 5-CH3H4PteGlu(n) (Cossins 1987; Roje et al. 2002b). The carbon of the methyl group provided by the folic compound originates from the β-carbon of serine (Fig. 3). The methyl cycle is fine-tuned by SAH levels, inhibiting folate compound biosynthesis through MTHFR (Jencks and Mathews 1987; Roje et al. 2002b) and by high levels of AdoMet that reduce methionine production (Crider et al. 2012). In summary, the methyl-group transfer catalyzed by AdoMet in methylation reactions is directly provided from serine through folates (vitamins B_9) that serve as donors or acceptors in one-carbon (C1 metabolism) transfer reactions. In other words, AdoMet is used by folates as a relay molecule to extend their capacities for methylation, probably because the methylthio-group of AdoMet is more reactive (Fig. 3). Therefore, the biosynthesis of methionine strongly depends on serine metabolism. In this respect, recent studies have demonstrated that the mutants of genes involved in the plastidial phosphorylated pathway of serine biosynthesis (*PPSB*) displayed arrested root development and exhibited a strong impairment of carbon and nitrogen metabolisms (Muños-Berthomeu et al. 2009, 2010; Cascales-Miñana et al. 2013). Furthermore, at the root morphological level, recent results in *Arabidopsis* showed that folic acid is essential in root organogenesis *via* root cell proliferation in meristems (Reyes-Hernández et al. 2014) and maturation of lateral root primordia upstream of auxin and independently of TOR kinase signaling (Ayala-Rodriguez et al. 2017; Li et al. 2017).

3.2 Ethylene and Polyamines Are By-Products Resulting from the Recycling of the Methylthio-Group of MTA Issued from AdoMet

5′-Methylthioadenosine (MTA), the by-product of the ACC synthase (ACS) and spermidine synthase (SPDS) reactions, is salvaged for the regeneration of the methylthio-group and adenine moiety through the methionine cycle, also known as the Yang or MTA cycle or methionine salvage pathway (Miyazaki and Yang 1987; Sekowska et al. 2004; Pommerrenig et al. 2011; Sauter et al. 2013). In plants, this cycle has been recently revised (Pommerrenig et al. 2011), and it is now composed of six reaction steps (Fig. 3) that involve successively: (1) the conversion of MTA into 5-methylthioribose (MTR) and adenine by MTA nucleosidase (MTN; Adams and Yang 1977; Wang et al. 1982; Rzewuski et al. 2007), (2) the phosphorylation of MTR into 5-methylthioribose-1-phosphate (MTR-1P) by the MTR kinase (MTK; Kushad et al. 1982; Sauter et al. 2004), (3) the isomerization of MTR-1P to 5-methylthioribulose-1-phosphate (MTRu-1P) by MTR-P isomerase (MTI; Pommerrenig et al. 2011), (4) then dehydratase-enolase-phosphatase (DEP) ensures the conversion of MTRu-1P into 1,2-dihydro-3-keto-5-methylthiopentene (DHKMP; Pommerrenig et al. 2011), (5) acireductone dioxygenase (ARD) in the presence of dioxygen catalyzes the conversion of DHKMP into 2-keto-4-methylthio butyrate (KMTB; Sauter et al. 2005; Bürstenbinder et al. 2007), and (6) finally transamination of KMTB to methionine is catalyzed by unknown aminotransferases (Kushad et al. 1983; Pommerrenig et al. 2011). In plants and many microorganisms, different unknown plastidial or cytosolic aspartate aminotransferases (AAT) are presumed to be involved in the last steps of the methionine regeneration pathway (Berger et al. 2003; Sekowska et al. 2004; Pommerrenig et al. 2011).

4 ACS and SAMDC Are Key Enzymes in Ethylene and Polyamine Biosynthetic Pathways and MTA Recycling

The biosynthesis of ethylene and polyamines are linked through the substrate AdoMet, which can be converted into decarboxylated S-adenosyl-L-methionine (dcAdoMet) and 1-aminocyclopropane-1-carboxylic acid (ACC), the ethylene precursor, through the action of S-adenosylmethionine decarboxylase (SAMDC) and 1-aminocyclopropane-1-carboxylate synthase (ACS), respectively (Fig. 3). In several physiological processes, ethylene and polyamines compete for their biosynthesis because these two metabolic pathways use initially a common substrate: AdoMet. Despite several studies that have revealed the competitive interaction between these two pathways during the processes of fruit ripening (Tassoni et al. 2000; Harpaz-Saad et al. 2012), root elongation (Kumar et al. 1996; Locke et al. 2000; Tassoni et al. 2000; Hummel et al. 2002), senescence (Pandey et al. 2000), and fruit or leaf abscission (Saftner 1989; Gil-Amado and Gomez-Jimenez 2012; Harpaz-Saad et al. 2012), little is known about the cross regulations of these two pathways. In fact, the partitioning of the AdoMet flux between these two pathways has never been precisely estimated because of the difficulties in measuring the contents of the different conjugated forms of ACC and polyamines that lead to underestimation of the flux in one or the other of these pathways.

4.1 SAMDC Is the Rate-Limiting Step for Polyamine Biosynthesis

Polyamine homeostasis is tightly regulated and achieved by the modulation of polyamine biosynthesis, conjugation, transport, and catabolism (Tiburcio et al. 2014). In the polyamine pathway AdoMet is converted to decarboxylated S-adenosyl-L-methionine (dcSAM) and MTA *via* SAMDC (Fig. 3). In *Arabidopsis* SAMDC genes belong to a small family of four genes that are not functionally redundant. Single and double mutants of SAMDC4 (*bud2-1*) and SAMDC1 (*samdc1*) increase root proliferation and growth by alteration of polyamine homeostasis (Kumar et al. 1996; Ge et al. 2006). Knockout of the SMDC4 gene affects growth and development, whereas alteration of the SMDC1 gene displays a more severe phenotype (Ge et al. 2006). The double mutant SMDC4-SMDC1 (*bud2-1-samdc1*) exhibited embryo lethality, probably through spermidine deficiency (Imai et al. 2004). SAMDC genes are regulated at the transcriptional level and induced by many abiotic stresses such as drought, chilling, heat, hypoxia, ozone, UV, and heavy metals (Alcázar et al. 2006). During tomato fruit ripening, it seems that no competition for AdoMet substrate occurs between SAMDC, ACS, and transmethylation reactions, suggesting that these three pathways can operate simultaneously (Van de Poel et al. 2012). This confirms studies where ethylene production was increased in transgenic plants over-expressing SAMDC (Mehta et al. 2002).

4.2 *ACS Aminotransferase Is the Rate-Limiting Step for Ethylene Biosynthesis*

In the ethylene pathway, AdoMet is converted by ACC synthase (ACS) into MTA and ACC, a non-proteinogenic amino acid, precursor of ethylene (Figs. 3 and 4). ACS is a PLP-dependent enzyme that belongs to the subgroup I of the α family of aminotransferases (Mehta et al. 1993; Christen and Mehta 2001). This subgroup contains major enzymes involved in carbon, nitrogen, and sulfur assimilation and shuttling such as ACC, aspartate, alanine, tyrosine, and aromatic aminotransferases (Mehta et al. 1993). X-ray structure analysis of the ACS protein revealed that amino acids surrounding the catalytic site share the same function as the aspartate aminotransferase (AAT) counterpart (Capitani et al. 1999; Yamagami et al. 2003). The ACS isoforms form homodimers and heterodimers and the active site of the dimers results

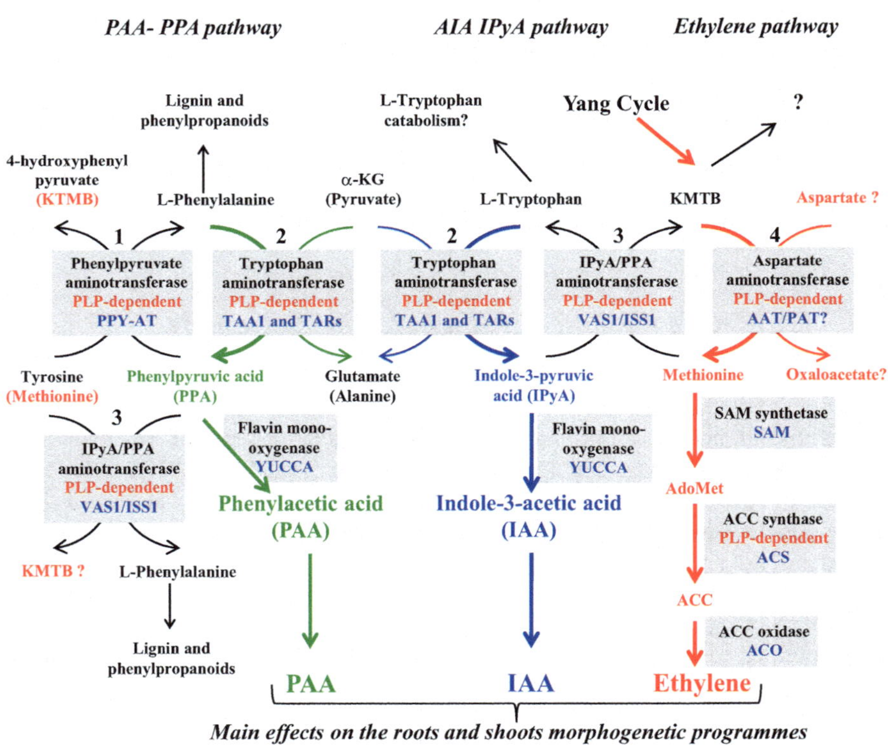

Fig. 4 Interconversion between aminotransferases belonging to the subgroup I of the α family involved in the metabolic pathways of auxins, ethylene, and aromatic amino acids (Phe, Tyr, Trp). **1**, phenylpyruvate aminotransferase (PPY-AT); **2**, tryptophan aminotransferase of *Arabidopsis* (TAA/TAR); **3**, IPyA/PPA aminotransferase SAV1/ISS1; **4**, eukaryotic aspartate aminotransferase (ET-AAT) and prokaryotic type aspartate/prephenate aminotransferase (PT-AAT/PAT). Abbreviations used for metabolite are: KTMB, 2-keto-4-methylthio butyrate; 2-KG, α-ketoglutarate (adapted from Zheng et al. 2013a; Yoo et al. 2013; Le Deunff et al. 2016; Pieck et al. 2015)

from the interaction of the two subunits with each other (Tarun and Theologis 1998; Tsuchisaka and Theologis 2004a). In *Arabidopsis*, there exist 12 genes encoding ACS but among them only eight genes (ACS2, 4–9, and 11) encode functional ACS (Bleecker and Kende 2000; Yamagami et al. 2003). Indeed, complementation of the *E. coli* aminotransferase mutant DL39 showed that ACS10 and ACS12 are aminotransferases with broad specificity for aspartate and aromatic amino acids whereas ACS1 forms inactive homodimers and ACS3 was a pseudogene (Tsuchisaka and Theologis 2004a). Based on protein sequences comparison, three subtypes of the eight functional ACS have been defined mainly based on the number of phosphorylation sites in their C-terminal sequence (Chae and Kieber 2005). These sites are involved in posttranslational regulation and modulate the stability and degradation of ACS proteins (Wang et al. 2004; Argueso et al. 2007; Hansen et al. 2009). Thus, the phosphorylation of ACS-type 1 (ACS1, ACS2, ACS6) are controlled by mitogen activated protein kinase (Tatsuki and Mori 2001; Liu and Zhang 2004) while the phosphorylation of ACS-type 2 (ACS4, ACS5, ACS8, ACS9, ACS11) are regulated by casein kinase (Tan and Xue 2014). The ACS-type 3 (ACS7) is a more stable protein because it has no C-terminal phosphorylation site but ubiquitin regulated its degradation via proteasome (Lyzenga et al. 2012).

Although ACS genes are strongly regulated at the transcriptional level depending on many developmental and environmental factors (Van der Straeten et al. 1992; Tsuchisaka and Theologis 2004b), to our knowledge only a few studies have compared the promoter sequences of ACS genes and search for common or specific responsive elements to transcription factors (Rodrigues-Pousada et al. 1993; Wand et al. 2005). As with many aminotransferases, it is likely that transcription of ACS genes is also controlled by more general mechanisms that affect the nutritional and energy status of the cell such as GCN2 and TOR kinase signaling systems (Xiong and Sheen 2015). Briefly, in plants GCN2 (*G*eneral *C*ontrol *N*on-derepressible 2) kinase is a cellular sensing system that allows overcoming amino acid starvation and stress conditions during growth (Hey et al. 2010; Zhang et al. 2003; Lageix et al. 2008; Castilho et al. 2014), whereas TOR (*T*arget *O*f *R*apamycin) kinase is a key regulator conserved in eukaryotic cells that controls nutrients and energy signaling topromote cell proliferation and growth (Loewith and Hall 2011; Xiong and Sheen 2014).

4.3 Cytosolic Levels of ACC Can Be Modulated in Many Ways

ACC produced by activity of the ACS isoforms is a neutral and non-proteinogenic amino acid that displays a structural analogy and molecular mass close to other signaling amino acids molecules such as GABA, α- and β-aminobutyric acids (α-ABA and β-ABA). If ACS is a rate-limiting step in ACC biosynthesis (Fig. 4), there are at least four different ways to modulate the excessive levels of ACC produced by ACS activation.

ACC Is Conjugated Under Different Forms as Many Signal Molecules The cellular levels of ACC can be decreased by conversion under different conjugated forms. Thus, ACC can be converted in tomato fruits by N-malonyltransferase (AMT) into 1-malonyl-ACC (MACC, Martin et al. 1995) and in *Arabidopsis* vegetative tissues by JA-aminosynthetase (JAR1, *J*asmonic *a*cid *r*esistance 1) into jasmonyl-ACC (JA-ACC; Staswick and Tiryaki 2004). ACC can also be conjugated to the reduced glutathione (GSH). GSH is a very abundant cellular tripeptide (γ-Glu-Cys-Gly) that serves as an essential antioxidant by scavenging reactive oxygen species during plant stresses (Noctor et al. 2011). γ-glutamyltranspeptidase (GGT) converted GSH and ACC into 1-(γ-glutamylamino)-ACC (GACC) in plant tissues (Amrhein et al. 1981).

ACC Can Be Stored in the Vacuole or Transported Long-Distance Another strategy developed by the plant for buffering an excess of ACC synthesis and ethylene production is ACC storage into the vacuole or long-distance translocation through the xylem (Bradford and Yang 1980; O'Neill 1997; Finlayson et al. 1999) and the phloem (Voesenek et al. 1990). Although ACC compartmentalization and storage into the vacuole have been studied for a long time, the characterization of genes encoding transporters of influx and efflux is not yet achieved (Bouzayen et al. 1991; Tophof et al. 1989; Saftner and Martin 1993). However, it has been recently demonstrated that ACC transport across the plasma membrane is provided by the lysine histidine transporter1 (LHT1). Indeed, neutral amino acids such as alanine and glycine have a competitive effect on the ACC-induced triple response. Likewise, *are2* mutants (*A*CC-*re*sistant 2) defective in LHT1 exhibit a dose-dependent response to ACC (Shin et al. 2015). Therefore, long-distance circulation and short-distance accumulation of ACC are dependent of amino acids transporters that have not been yet completely characterized.

ACC Is Catabolized by a PLP-Dependent Deaminase A recent study has shown that there exists in the *Arabidopsis* genome an ACC deaminase gene (*At*ACD1) encoding a protein homolog to bacterial and fungal ACD and previously identified as a D-cysteine desulfhydrase (D-CDes) (Riemenschneider et al. 2005). ACD1 is a PLP-dependent enzyme that belongs to the β family of aminotransferases (Christen and Mehta 2001) and converts ACC into α-ketobutyrate and ammonia, two important molecules involved in the GS/GOGAT cycle during N assimilation (McDonnell et al. 2009). The transgenic lines over-expressing and under-expressing the *At*ACD1 gene showed a significant modulation in the triple response of *Arabidopsis* etiolated seedlings. Likewise, antisense lines exhibited 70% reduction in ACD activity and a significant increase in ethylene production (McDonnell et al. 2009). These results strongly suggest that ACD is responsible for maintenance and control of ACC pools in vivo but the underlying regulatory mechanisms at the transcriptional and posttranscriptional levels of this gene remain to be discovered.

ACC Treatment Reduces Aspartate Levels by an Unknown Mechanism Seedling treatment by exogenous ACC induces a rapid decrease in primary and lateral root growth within minutes after application (Le et al. 2001; Swarup et al. 2007;

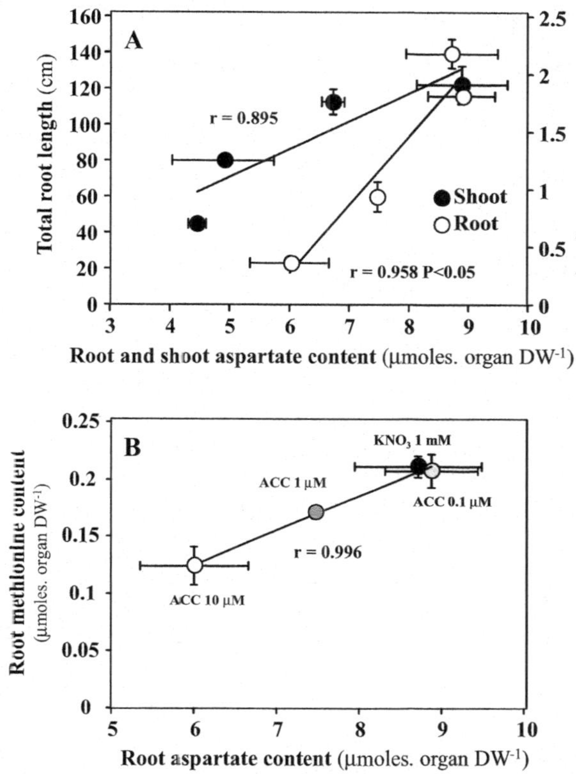

Fig. 5 (**a**) Relationships between the exploratory root length and shoot surface area with the aspartate contents in the roots and shoots of *Brassica napus* seedlings treated over 5 days on agar plates under homogeneous feeding of 1 mM $K^{15}NO_3$. (**b**) Relationship between aspartate and methionine contents in the roots. Values are mean ± SE of three or four repeats of four seedlings each (adapted from Lemaire et al. 2013)

Leblanc et al. 2008). Ultra performance liquid chromatography (UPLC) analyses of amino acid contents in root and shoot tissues of *B. napus* seedlings treated with increasing concentrations of ACC (from 0.1 to 10 µM) show that among amino acids, aspartate level is the most impaired and is positively correlated with changes in the root length and shoot surface area (Lemaire et al. 2013). In roots, a decrease in aspartate levels is also associated with a reduction in methionine (Fig. 5). Because ethylene biosynthesis depends directly on the aspartate-derived amino acid pathway, it is likely that ethylene can exert a feedback control on this pathway. Furthermore, aspartate issued from glycolysis and tricarboxylic acid (TCA) cycle plays a central role in C/N balance (Coruzzi 2003) and auxin catabolism (Ludwig-Müller 2009). How ACC or ethylene controls aspartate levels is a serious challenge in our understanding of the interaction between N, S, and C metabolisms. A combination of mutants of ethylene signaling and biosynthesis associated with pharmacological studies could be used to unravel this major mechanism involved in plant growth.

4.4 Are ACC and Polyamines Potential Ligands or Modulators of Plant Glutamate Receptors?

Because of their role as neurotransmitters or modulators in animal cells, it is not excluded that ACC, polyamines, GSH, and conjugated forms of ACC may also act on plant glutamate receptors. Assuming ACC acts as a ligand or modulator on the plant glutamate receptors (GLR), this would explain why levels of ACC are so controlled in plant cells (Le Deunff and Lecourt 2016).

The *Arabidopsis* genome encodes for 20 GLR genes with coding sequences close to the ionotropic glutamate receptors (iGluR) first identified and characterized in mammals (Price et al. 2012; Weiland et al. 2016). Analyses of DNA sequences of GLR have grouped these proteins into three clades (Davenport 2002; Chiu et al. 2002). Expression studies have shown that GLR genes were mainly expressed in roots but also in leaves and reproductive organs (Chiu et al. 2002). The structure of GLRs is organized into four different units comprised of an amino terminal domain (ATD), a ligand binding domain (LBD), a trans-membrane domain (TMD) formed by three complete trans-membrane domains (M1, M3, and M4) with one re-entrant loop (M2) that forms the ion channel, and a C-terminal tail. Plant GLRs probably form tetrameric channels selective for Na^+, K^+, and Ca^{2+} as mammalian iGLuR counterparts (Tapken and Hollmann 2008; Price et al. 2012). Heterologous expression in transfected human embryonic kidney and *Xenopus* oocytes associated with patch-clamp studies have shown that *At*GLR3.2, *At*GLR3.4, and *At*GLR1.4 can be gated by a broad spectrum of amino acids such as Asn, Ser, Gly, Met, Trp, Phe, Leu, Tyr, and Thr. Among these amino acids, Met was the most effective in *At*GLR1.4 receptor whereas Arg was the most effective antagonist of Met effect (Vincill et al. 2012, 2013; Tapken et al. 2013). Moreover, the Cys and GSH tripeptides were the most effective agonists for activating *At*GLR3.3 to suppress growth of the tomato bacterial pathogen *Pseudomonas syringae* in *Arabidopsis* leaves, suggesting that the methylthio-group common for both ligands could specifically interact with the receptor during the innate immunity response (Li et al. 2013).

If plants GLR are compared from an analogical perspective with their mammal's counterparts, very surprising results can be envisaged. Indeed, in mammals, the ethylene precursor ACC is a synthetic molecule called ACPC and frequently used as a partial agonist of iGluR on glycine binding sites (Nahum-Levy et al. 1999; Inanobe et al. 2005). Similarly, polyamines such as spermine and spermidine stabilize the formation of the LDB dimer in a tetrameric receptor and facilitate the attachment of ATD lobes by "gluing" them together (Mony et al. 2009). Because polyamines and ACC-ACP are directly produced downstream from AdoMet biosynthesis, further investigations are required because ACC and polyamines are involved in many plant responses to biotic and abiotic stresses (Le Deunff and Lecourt 2016). GLR are located in the plasma membrane but also in membranes of plastids and mitochondria (Weiland et al. 2016). This suggests that they can react to a broad spectrum of amino acids present in cytosol or circulating in the apoplast. In this regard, plant GLRs could act as an internal and external N sensor system and play a major role in the regulation of amino acids biosynthesis or their compartmentalization in different organelles in relation to the extracellular flows of amino acids.

5 Relationship Between Ethylene and N Metabolism: The Central Role of Aminotransferases

A close examination of Fig. 3 indicates that no less than six PLP-dependent aminotransferases, namely AAT/PAT, OASTL, CgS, CbL, ACS, and an unknown AAT, are required for ethylene biosynthesis and methionine salvage pathways (Fig. 3). Therefore, among the different plant amino acid pathways (Fig. 1), the methionine biosynthetic pathway needs the highest involvement of aminotransferases for its biosynthesis and is highly dependent on the availability in PLP cofactor (vitamin B_6). This ascertainment is exemplified in recent studies of pyridoxine synthase1 enzyme involved in PLP biosynthesis and encoded by two paralogous genes, namely PDX1.1 and PDX1.3 in *Arabidopsis* (Chen and Xiong 2009a, b; Boycheva et al. 2015). Disruption of either of these genes in single mutants results in vitamin B_6 deficiency and a differential reduction of ethylene production and auxin levels while disruption of both genes in double mutants is lethal for seedlings. At the phenotypic level, these mutants displayed impairment in root growth, reduction of meristem size, and altered root cell division and elongation (Chen and Xiong 2009a, b; Titiz et al. 2006). However, the single *pdx1.3* mutant is more impaired in root development than the *pdx1.1* mutant. These phenotypic differences are mainly explained by the presence of distinct regulatory elements in the upstream region of both genes that lead to a deficit in ethylene production and/or signaling. Thus, PDX1.1 expression is repressed by sucrose and its promoter possesses a sugar response element whereas PDX1.3 promoter contains ethylene and auxin response elements. Several evidences indicate that some components of the ethylene signaling pathway act upstream of the effect of auxin biosynthesis and transport (Růžička et al. 2007; Boycheva et al. 2015). Indeed, the root phenotype of the *pdx1.3* mutant is partially restored by ACC treatment but not by auxin treatment (Chen and Xiong 2009a, b; Boycheva et al. 2015). These results perfectly demonstrate that PLP availability plays a central role in ethylene biosynthesis and probably the methionine pathway as well as in phytohormone homeostasis. However, as previously mentioned it is not known how *de novo* PLP synthesis in cytosol or PLP salvage pathway modulate PLP pool in the plastid and which plastidial carriers are involved in the transport of different forms of vitamin B_6 (Gerdes et al. 2012).

5.1 Methionine and Tryptophan Interconversion by VAS1 and TAA Aminotransferases

For a long time, it was assumed that the unique role of methionine as a molecule in the cell was to produce proteins *via* its utilization in methionyl-tRNA formation as a substrate of aminoacyl-tRNA synthetase. However, it was demonstrated recently that the cytosolic PLP-dependent aminotransferase called VAS1/ISS1 (for re***v***ers***a***l of ***s****av3* phenotype and ***i***ndole ***s***evere ***s***ensitive1) catalyzes the conversion of methionine as amino acid donor and indole-3-pyruvic acid (IPA) as amino acceptor into L-tryptophan and 2-keto-4-methythiobutyrate (Zheng et al. 2013a; Pieck et al. 2015). IPA is the precursor of IAA (Fig. 4) that is involved in the most important route of plant IAA biosynthesis belonging to the tryptophan-dependent pathways (Stepanova et al. 2008, 2011; Mashiguchi et al. 2011; Won et al. 2011; Zhao 2014). Indeed, among 19 of naturally occurring amino acids examined, methionine was the most catalytically preferred amino acid donor followed by Phe with only 21% specificity relative to Met (Pieck et al. 2015). Therefore, VAS1/ISS1 catalyzes the opposing reaction of tryptophan aminotransferases TAA1 (Fig. 4) and tryptophan aminotransferases-related TAR1 and TAR2 (Stepanova et al. 2008). TAA PLP-dependent enzymes convert Trp by using α-ketoglutarate or pyruvate as organic acid donors, into IPA and glutamate or alanine, respectively (Fig. 4). In *Arabidopsis*, a *vas1-2* single mutant and *vas1-2sav3* double mutant showed a significant increase in IAA levels and a fivefold increase in the level of the ethylene precursor ACC (Zheng et al. 2013a). However, free L-methionine levels did not differ between mutant and WT, demonstrating that regulation of the methionine biosynthesis is controlled downstream by methionine synthase (MS) and methionine adenosyltransferase (SAM) (Bürstenbinder et al. 2007). Since the VAS1/ISS1 and TAA aminotransferases catalyze opposite reactions, it is assumed that VAS1 could counteract the production of IPA and methionine by subtracting a part of the IPA and methionine from IAA and methionine pathways, respectively (Zheng et al. 2013a). However, some evidence suggests that SAV1/ISS1 could be mainly involved either in Trp catabolism or in metabolism of Phe and Tyr (Pieck et al. 2015). Indeed, protein sequence analyses of SAV1/ISS1 with other plant aromatic aminotransferases (AroAT) showed that SAV1/ISS1 is a new subgroup of aromatic PLP-dependent enzymes belonging to the Iα aspartate family. This subgroup is distinct from Trp ATs and Tyr ATs previously described and could be involved in Phe and/or Tyr biosynthesis (Pieck et al. 2015). Unfortunately, these studies did not test whether SAV1/ISS1 could also catalyze the conversion of Met and phenylpyruvate into KMTB and Phe, or the conversion of Met and 4-hydroxyphenyl pyruvate into KTMB and Tyr (Fig. 4). The latter reactions would explain the role of SAV1/ISS1 in Phe and Tyr biosynthesis from methionine.

5.2 *Methionine and Phenylalanine Interconversion by PPY-AT in the New Phenylpyruvate Pathway*

Until now, the arogenate pathway was considered as the exclusive route for Tyr and Phe biosynthesis in plants (Cho et al. 2007; Maeda and Dudareva 2012). However, it was recently demonstrated that Phe could also be synthesized in cytosol *via* phenylpyruvate alternative pathway (Yoo et al. 2013; El-Azaz et al. 2016). This functional phenylpyruvate pathway was similar to most microorganisms and required two reaction steps. In the first reaction step, prephenate is converted to phenylpyruvate in plastids by arogenate dehydratases (ADT) that retain prephenate dehydratase activity (PDT) from a 22-amino acid region called PAC domain (for *P*DT activity *c*onferring *D*omain) conferring PDT activity to ADTs (El-Azaz et al. 2016). In the second reaction step, a cytosolic phenylpyruvate-aminotransferase (PPY-AT) catalyzes the interconversion of phenylpyruvate and tyrosine to 4-hydroxyphenylpyruvate and Phe (Yoo et al. 2013). As demonstrated PPY-AT could also convert phenylpyruvate and Met into Phe and 2-keto-4-methylthiobutyrate (Fig. 4) since PPY-AT is able to use Met as amino donor with only 13% specificity relative to tyrosine (Yoo et al. 2013). Therefore, the discovery of phenylpyruvate pathway links biosynthesis of phenylalanine with catabolism of tyrosine but also increases the complexity of the transaminases network involved in the regulation of aromatic amino acids and methionine (Fig. 4).

5.3 *In Search of the Last Missing Aminotransferase to Complete the Yang Cycle*

In plants and many microorganisms, due to their promiscuity, various aspartate aminotransferases (AAT) are presumed to be involved in the last step of the Yang cycle catalyzing the transamination of KMTB into methionine (Berger et al. 2003; Sekowska et al. 2004; Pommerrenig et al. 2011). In *Arabidopsis*, plastids contain two different AAT: a prokaryotic-type (PT-pAAT) and a eukaryotic-type (ET-pAAT) that are involved in the first step of the aspartate-derived amino acid pathway (de la Torre et al. 2006, 2014a, b). It is strongly suspected from analysis sequences that ET-pATT (ASP5) and PT-AAT could be implied in the last step of the Yang cycle (Pommerrenig et al. 2011). Moreover, it was recently discovered that plant PT-pAAT is bispecific since it also displays a prephenate aminotransferase activity (PAT) and catalyzes the conversion of glutamate and prephenate into arogenate and 2-oxoglutarate (Graindorge et al. 2010; Maeda et al. 2011). Therefore, it is not excluded that methylthio-group salvage of AdoMet by the Yang cycle in plastids could compete in its last step with the biosynthesis of Phe and Tyr (lignin and phenylpropanoids precursors) in the arogenate pathway catalyzed by the PAT activity of PT-pAAT (Figs. 1 and 4). In this respect, the PT-pAAT/PAT silenced plants show a severe reduction in growth as well as a decrease in chlorophyll and lignin biosynthesis (de la Torre et al. 2014a, b). Unfortunately, AdoMet and Met levels have not been measured so that it is difficult to evaluate the impact of this enzyme on the last step of the methionine pathway. Likewise, it is likely that the cytosolic aspartate aminotransferases ASP2 and ASP3 could also catalyze transamination of KMTB to methionine when the last step of Yang cycle occurs in the cytosol (Miesak and Coruzzi 2002). In summary, the functioning of AAT/PAT, SAV1/ISS1, and PPY-AT aminotransferases could be a major metabolic crossroad interconnecting the metabolism of methionine and aromatic amino acids (Trp, Phe, and Tyr) with biosynthesis of ethylene, IAA, PAA, and phenylpropanoids.

5.4 Is Tryptophan Aminotransferase Also Involved in Histidine Catabolism?

Although histidine catabolism is well established in animal cells, this pathway has not yet been established in plant cells (Hildebrandt et al. 2015). A surprising result has been recently obtained with the human fungal pathogen *Candida glabrata* and *Saccharomyces cerevisiae* that initialize His degradation *via* the aromatic aminotransferase ARO8. In yeast, ARO8 is known to be an AroAT that in vitro converts Phe, Tyr, and Trp as amino acid donor to phenylpyruvate, α-ketoglutarate, and pyruvate as amino acid acceptors (Iraqui et al. 1998; Urrestarazu et al. 1998). However, in both fungi ARO8 is ten-fold up-regulated by exogenous His and is also able to transfer the donor amino group of His to the α-ketoglutarate acceptor to produce imidazol-5-yl-pyruvate and glutamate (Brunke et al. 2014). Furthermore, heterologous expression of SAV1/ISS1 or TAA1 *Arabidopsis* genes can rescue the yeast mutant *Aro8Aro9* auxotroph for Phe and Tyr (Pieck et al. 2015). In *B. napus* seedlings, in vivo inhibition of TAA and ACS activities by root treatment with 10 μM AVG (Soeno et al. 2010) induced a 2.5-fold and 8-fold increase of His levels in the roots and shoots, respectively (Fig. 6). However, treatment with 1 mM glutamate was able to restore control levels of amino acids in roots and shoots with the exception of His (Le Deunff et al. 2018). In yeast, His starvation imposed by 3-aminotriazole induced the bZIP transcription factor GCN4 that initiates transcription of 539 genes such as genes encoding amino acids transporters and biosynthetic enzymes as well as genes involved in PLP biosynthesis (Niederberger et al. 1981; Natarajan et al. 2001; Hinnebusch 2005). In *Arabidopsis*, His homeostasis is crucial for root development since the *hap1* mutant of plastidial histidinol phosphate aminotransferases (*HPA*), the eighth enzyme of His biosynthesis, displays a defect in root meristem maintenance and reduction in root development (Mo et al. 2006). Taken together, these results suggest that TAA could be involved in His catabolism in plants and play a major role in the regulation of amino acid biosynthesis. Therefore, TAA/SAV1/ISS1 interconversion system could play a major role as a metabolic crossroad in the amino acid imbalance and changes in C/N ratio acting on N sensory systems such as the GCN4, GCN2, and TOR kinases signaling involved in plant nutrition.

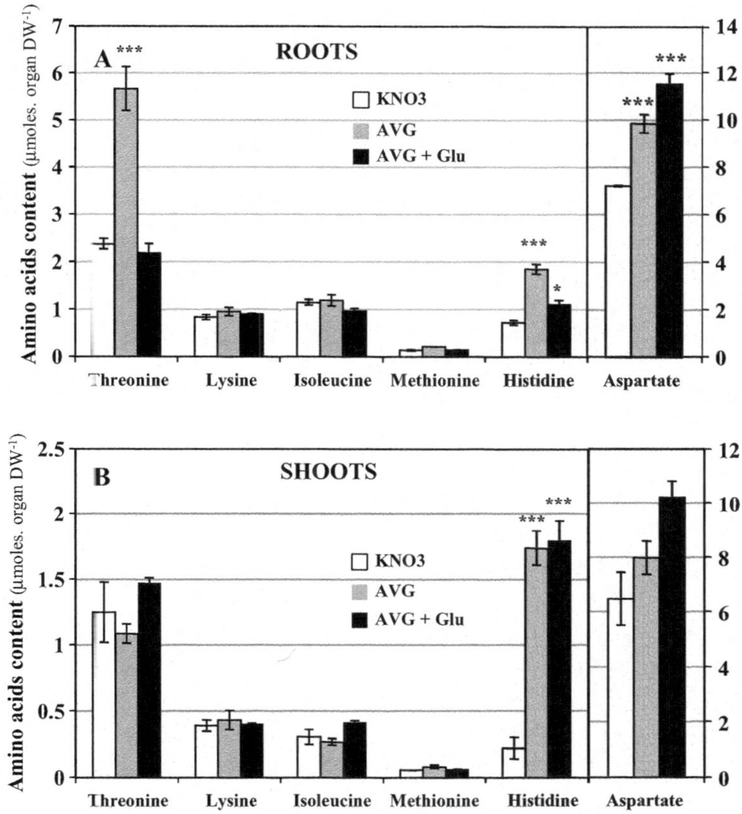

Fig. 6 Changes in free amino acid contents from aspartate-derived amino acid pathway induced by 10 µM AVG and 10 µM AVG + 1 mM Glu treatments in the root (**a**) and shoot tissues (**b**) of *Brassica napus* seedlings treated over 5 days on agar plates under homogeneous feeding of 1 mM KNO_3. Values are the average (±SE) of three agar plates ($n = 3$) of four seedlings each. Significant differences between control (1 mM KNO_3) and treatments are given for $*p < 0.05$; $**p < 0.01$; $***p < 0.005$; (t-test) (adapted from Le Deunff et al. 2018, submitted)

6 Use of Olefinic Glycine Inhibitors in Ethylene Biosynthesis: The Devious Trap

Since the 1970s, olefinic glycine analogues have been used as small bioactive molecules to inhibit specifically in vitro and in vivo PLP-dependent aminotransferases involved in the methionine biosynthetic pathway and ethylene biosynthesis (Rando 1974a, b; Berkowitz et al. 2006). In fact, this family of compounds acts as terminal inhibitors of subgroup I of aminotransferases belonging to the α family (Christen and Mehta 2001). The subgroup I contains ACC, aspartate, alanine, and aromatic (His, Phe, Tyr, Trp) aminotransferases (Christen and Mehta 2001). Recent findings have highlighted the non-specificity of these compounds in ACS and

TAA/TARs enzymes (Soeno et al. 2010). Due to the structural analogies of this family of inhibitors (Table 1) and the promiscuity of aminotransferases belonging to subgroup I, caution is now required when using these inhibitors to validate molecular studies with mutational approaches or transgenic plants on ethylene and IAA biosynthesis and signaling (Le Deunff and Lecourt 2016; Le Deunff et al. 2016). Moreover, cellular compartmentalization of their target enzymes in the peroxisomes, plastids, mitochondria, and cytosol probably modulate their effect in term of specificity, permeability, and stability during in vivo treatments. However the broad spectrum of action of these compounds remains a powerful tool (Le Deunff and Lecourt 2016). Indeed, they create an imbalance in the amino acid levels which makes it possible to test potential candidates for N sensory systems such as the TOR and GCN2 signaling pathways or to discover new targets involved in N detection and C/N ratio changes in primary metabolism (Le Deunff et al. 2018, submitted).

6.1 In Search for Highly Specific Inhibitors of ACS and TAA Aminotransferases

Because ethylene and auxin biosynthetic enzymes such as ACS and TAA are encoded by redundant genes in the *Arabidopsis* genome, the mutational approaches on single or multiple genes by reverse and forward genetics are often associated with plant lethality, poor growth, and sterility (Tsuchisaka et al. 2009; Stepanova et al. 2008). Chemical genetic or genomic approaches can overcome these limitations by selecting for small bioactive molecules able to inhibit specifically protein activity in any tissue and at any time during plant development and avoiding side effects (Zheng and Chan 2002; Blackwell and Zhao 2003; Robert et al. 2009). Recently, these chemical strategies have been used in *Arabidopsis* to find out new specific inhibitors of ACC synthase (Lin et al. 2010) and TAA enzymes (Soeno et al. 2010; Nakamura et al. 2016). Thus, more specific inhibitors of the ACS enzyme formed from a quinazolinone backbone (called compounds 9393, 9370, and 7303 compounds) have been discovered from their ability to suppress the constitutive triple response of ethylene overproducer mutant *eto1-4* (Lin et al. 2010). Similarly, 41 compounds derived from the 2-(aminooxy)-3-(naphthalen-2-yl)propanoic acid (KOK1169/AONP) backbone were able, in vivo and in vitro, to reduce TAA aminotransferase more specifically than L-α-aminooxy-phenylpropionic acid (AOPP) (Nakamura et al. 2016). Today, these compounds form a new class of specific TAA inhibitors designated as "pyruvamine." Therefore, it is reasonable to assume that the use of these specific and less specific inhibitors in differential transcriptomic studies should lead to the discovery of metabolic hubs in N and C metabolism or N sensory systems involved in the regulation of the root morphogenetic program.

Table 1 Origin and structure of amino acid analogs belonging to enol ether family able to inhibit catalysis reactions of PLP-dependent aminotransferases (adapted from Lieberman 1979; Berkowitz et al. 2006)

Synthetic and natural compounds	Structure of R group linked to Glycine	Origins	Target aminotransferases of the I α family	Authors
L-vinylglycine (VG)	CH_2=CH–	Synthetic	Aspartate amino transferase, ACC synthase	Rando (1974a, b) Cornell et al. (1984) Satoh and Yang (1989b) Capitani et al. (2005) McCarthy et al. (2001)
L-allylglycine	CH_2=CH–CH_2–	Synthetic	Glutamate decarboxylase, ACC synthase	Fisher and Davies (1976) Satoh and Yang (1989a)
Methoxyvinylglycine (MVG)	CH_3–O–CH=CH–	*Pseudomonas aeruginosa*	Aromatic acid amino transferase, β-cystathionine lyase	Scannell et al. (1972) Sahm et al. (1973) Pruess et al. (1974)
Aminoethoxyvinylglycine (AVG)	NH_2–CH_2–CH_2–O–CH=CH–	*Streptomyces* sp.	Ornithine aminotransferase, cystathionine β-lyase Tryptophan synthase, ACC synthetase, ACC deaminase	Satoh and Yang (1989a) Soeno et al. (2010) Capitani et al. (2002) Clausen et al. (1997) Huai et al. (2001) Penrose et al. (2001) Droux et al. (1995)
Rhizobitoxine	NH_2 $HOCH_2$–CH–CH_2–O–CH=CH–	*Rhizobium japonicun* *Pseudomonas andropogonis*	Cystathionine β-lyase, ACC synthetase	Giovanelli et al. (1971) Yasuta et al. (1999) Owens et al. (1971)

7 Are Aminotransferases Potential Targets for the Improvement of Nitrogen Use Efficiency (NUE)?

An alternative approach to address the metabolic interconnections between mineral nutrition and the production of plant growth regulators consists of proposing that the biosynthesis of hormones such as ethylene and IAA is highly regulated through the aminotransferase network in primary metabolism (Le Deunff et al. 2016; Le Deunff et al. 2018 submitted). Indeed, aminotransferases are the key enzymes that ensure N, S, and C shuttling through production of amino and organic acids but also biosynthesis of ethylene and auxins via Met, Trp, and Phe. Therefore, using mutants or nonspecific inhibitors of aminotransferases involved in ethylene and auxins biosynthesis can modulate the aminotransferase network and impact nitrogen use efficiency (NUE). It is likely that such approaches may disrupt N sensory systems such as TOR and GNC2 kinase signaling, the plant GCN4 transcription factor homolog, or GLR receptors. We present here some recent results in favor of this assumption.

7.1 Over-expression of the Tryptophan Aminotransferase Genes Improves Nitrogen Use Efficiency

In *Arabidopsis* the members of the Trp aminotransferases namely TAA1 (also known as WEI8, SAV3 and TIR2) and their related proteins TAR1 and TAR2 play critical role in plant development in response to stresses (Stepanova et al. 2008; Ma et al. 2014). Thus, the *Arabidopsis* double mutant *wei8tar2* displays a dwarf and bushy plant phenotype with reduced vasculature and abnormal flowers with complete sterility, whereas the triple mutant *wei8tar2tar1* is rootless, seedling-lethal, and impaired in embryo patterning (Stepanova et al. 2008). These mutants are insensitive to ethylene and ACC treatments because they lack the triple response. A recent study has shown that TAA1 is highly transcriptionally regulated by ARR transcription factors (A*rapidopsis r*esponse *r*egulator) that can bind after dimerization at two different sites. One site is present in the TAA1 promoter region whereas the other is found in the second intron of the gene sequence (Yan et al. 2017). Thus, transcriptional responses are modulated by a subset of ARRs implicated in binding on site 1 or 2. The ARRs biosynthesis is regulated by cytokinin, ethylene, light, and developmental signals. Especially, ethylene response is mediated by the EIN3 transcription factor, which is able to bind to the ARR12 protein and modulate TAA gene expression (Yan et al. 2017). These results explain previous data indicating that TAA1 transcription is activated by EIN3 (He et al. 2011). The major role of TAA1 aminotransferase in plant development was also demonstrated in maize (*Zea mays*) and rice (*Oryza sativa*) where the loss of function of two co-ortholog genes *Zm*VT2 (*V*anishing *T*assel2) and *Os*FIB (*Fi*sh *B*one) caused dramatic effects on vegetative and reproductive development associated with a reduction in IAA levels (Phillips et al. 2011; Yoshikawa et al. 2014).

Furthermore, it was found in *Arabidopsis* that under nitrogen limitation, aminotransferase TAR2 controls LR proliferation, suggesting that TAR2 is required for the reprogramming of root architecture in response to low nitrogen availability (Ma et al. 2014). Thus, transgenic plants over-expressing the TAR2 gene showed a significant increase in the number of LRs under low and high nitrogen conditions. In order to improve NUE in crop species, over-expression of the TAR2 gene has been recently engineered in wheat (*Triticum aestivum*). After a selection among the 12 TAR2 alleles present in the wheat genome, the best candidate *Ta*TAR2.1 gene expressed in different organs was over-expressed by using a constitutive promoter (Shao et al. 2017). The engineered plants grown under controlled-environment and field conditions showed an increase in LR-growth under low N supply conditions at the plantlet stage. At maturity and whatever the N supply levels applied, plants also displayed a significant increase in the biomass, plant height, spike number, and grain yield (Shao et al. 2017). Taken together, these results indicate that TAR2 genes show a potential for engineering crop plants for improving gain of yield under nitrogen-limiting conditions.

7.2 *Over-expression of Alanine Aminotransferase Genes Improves Nitrogen Use Efficiency*

Alanine aminotransferase (AlaAT) is a PLP-dependent aminotransferase that catalyzes the amino group transfer of alanine to α-ketoglutarate to form glutamate and pyruvate. This aminotransferase was used to engineer nitrogen use efficiency in maize, rice, and canola (Good et al. 2007; Shrawat et al. 2008). Over-expression of AlaAT from barley (*Hordeum vulgare*, *Hv*AlaAT) under the control of root-specific and constitutive promoters *Os*Ant1 and btg26 in rice and canola, respectively, induces a significant increase in NUE and root biomass in these crop species (Good et al. 2007; Shrawat et al. 2008). Especially, under N-limiting conditions, canola and rice plants exhibit increased biomass and yields. In field trials with 40% less applied N, canola maintains yields. The physiological reasons for this improvement remain elusive. It is assumed that the decrease in some amino acids in shoots such as Gln and Glu could alleviate the repression of these AA on nitrate uptake transporters and increase N uptake (Good et al. 2007). However, it is not excluded that the overproduction of pyruvate can play a major role in both gluconeogenesis and auxins biosynthesis (IAA and PAA). Indeed, auxins production needs pyruvate for TAA/TAR activities since pyruvate is one of the 2-oxoacid acceptors used for their functioning (Le Deunff et al. 2016). Similarly, phospho*enol*pyruvate carboxykinase PEPCK activity (catalysing the reversible conversion of oxaloacetate to phosphoenolpyryvate, see Fig. 1) is required for sink tissues metabolically active such as root, stem, and leaves (Malone et al. 2007). Unfortunately, no study of transcriptional regulation of AlaAT or PEPCK promoters has been done as with the TAA1 promoter in *Arabidopsis*. Therefore, it is difficult to know which endogenous and exogenous signals are involved in the regulation of AlaAT and PEPCK expression.

7.3 Could the Under-expression of ACC Synthase Genes Improve Nitrogen Use Efficiency?

Because a multigene family in *Arabidopsis* encodes ACS aminotransferases, *acs* mutants of this specie, mutated at multiple loci, has been engineered to understand the function and regulatory roles of these proteins (Tsuchisaka et al. 2009). Analyses of pentuple, hexuple, and octuple *acs* mutants demonstrated that in normal conditions ethylene acts as growth repressor in dark- or light-grown plants since pentuple and hexuple mutants are bushier and display significant greater height with the progressive decrease in ethylene production. Indeed, ethylene is known to down-regulate photosynthetic genes (Van Zhong et al. 2003). These *acs* mutants also exhibited a delayed flowering time, a less response to gravity, and enhanced susceptibility to the necrotrophic pathogen *Botrytis cinerea*. Among the mutants, growth of the octuple *acs* mutant is delayed during the initial stage of development and becomes taller and less bushy after 50 days of growth. Moreover, transcription analyses of *acs* mutants confirmed that there exists a relationship between expression of light signaling genes and ethylene biosynthesis. This study clearly highlights the importance of spatial and temporal combination between ACS isoforms in multiple ethylene-mediated physiological processes during growth and development. The multiple-locus *acs* mutants appear to be a valuable tool for deciphering how ethylene biosynthesis interacts with nitrogen metabolism in response to different levels of nitrate availability and how decrease in ethylene production affects the levels of Asp, Met, AdoMet, and PAs. In this respect, the crosstalk between nitrogen nutrition and the ethylene plant hormone in Col-0 and 20 natural accessions of *Arabidopsis* seedlings have been recently investigated (De Gernier et al. 2016). Comparison of the behavior of the 20 accessions behavior revealed that changes in root biomass and ethylene production were negatively correlated at 1 mM but positively correlated at 10 mM nitrate. Greater ethylene release and root biomass production under nitrogen limitation were mainly due to higher transcription levels in the roots of *ACS6* and of *ACO2* and *ACO4* genes, respectively. Taken together these studies indicate that ethylene modulates plant morphology and biomass allocation probably in relation to nitrate availability.

This conclusion is in line with recent findings demonstrating that endogenous glucose signals increase growth by promoting auxin signaling and by antagonizing ethylene signaling through the glucose sensor HXK1 (Moore et al. 2003). Indeed, glucose enhances the degradation of EIN3 and EIL1 (*E*IN3-*L*ike1) components of ethylene signaling through a proteasome-dependent mechanism controlled by HXK1 (Yanagisawa et al. 2003; Yoo et al. 2008). During primary root growth in *Arabidopsis*, EIN3 is known to exert a positive feedback loop on the expression of anthranilate synthase and TAA1/TAR2 genes involved in auxin biosynthesis whereas IAA enhances EIN3 stability by repressing EBF1/2(*E*IN3 *B*inding *F*-Box protein1)-mediated degradation of EIN3 (He et al. 2011). Although HXK1 and TOR kinase signaling systems in response to glucose seem to be mostly uncoupled more studies are required to understand possible interactions between both sensory systems (Sheen 2014).

Furthermore this conclusion is also corroborated by the overaccumulation of anthocyanin in ethylene signaling mutants *etr1*, *ein2*, *ein3/eil1*, and *rdh3* (root hair defective3) subjected to nitrogen deficiency, whereas exogenous application of ACC almost completely suppresses anthocyanin accumulation in Col-0 WT plants and ACC effect was attenuated in the ethylene signaling mutants (Wang et al. 2015). Anthocyanin is accumulated under low nitrate levels because nitrate relieves the repression of lateral organ boundary (LDB) transcription factors on anthocyanin biosynthetic genes (Rubin et al. 2009). The LDB genes also repress many other N-responsive genes such as nitrate reductase genes NIA2 and NIA2 and nitrate transporters *At*NRT1.1 (NPF6.3), *At*NRT2.1, and AtNRT2.5. Regulation by ethylene of nitrate responsive genes such as nitrate transporters is also demonstrated during N-induced nutritional stress. Indeed, rapid changes in nitrate supply such as transfers from high to low concentration and vice versa induce a burst of ethylene production in the roots associated with a differential expression of *At*NRT2.1 and *At*NRT1.1 (NPF6.3) nitrate transporters genes (Tian et al. 2009; Zheng et al. 2013b). In response to nitrate limitation, induction of *At*NRT2.1 plays a positive role on ethylene biosynthesis and signaling pathway but the ethylene signaling components such as EIN3/EIL1 induce in turn the repression of *At*NRT2.2 (Zheng et al. 2013b). In response to nitrate excess, expression of *At*NRT1.1 (NPF6.3) and *At*NRT2.1 genes is respectively up- and down-regulated (Tian et al. 2009). However, these responses are abolished in *etr1-3* and *ein2-1* ethylene receptor mutants, again demonstrating that components of ethylene signaling are probably involved in nitrate nutritional responses.

8 Conclusion

Ethylene biosynthesis from the amino acid aspartate needs at least seven different aminotransferases belonging to the subfamilies I and γ of the α family: AAT, AAT/PAT, CgS, CbL, Vas1/ISS1, ACS and an unknown AAT aminotransferase to complete definitively the Yang cycle. Five of these aminotransferases play a fundamental role in N, S and C shuttling in plant cells (namely CgS, CbL, ACS, AAT/PAT, VAS1/ISS1) and three of them (VAS1/ISS1, AAT/PAT, and an unknown AAT) connect the aspartate/methionine metabolism to aromatic amino acid biosynthesis and catabolism in a unique and major network of aminotransferases in the plastids.. Because of the promiscuous nature of the aminotransferases for their substrates, it is likely that different aminotransferases can be involved in the last step of the Yang cycle within plastids or the cytosol. The functioning of these aminotransferases depends directly on the availability of PLP that is synthesized from glutamine, the first amino acid produced by the glutamine synthetase in the nitrate reduction/assimilation pathway. Understanding how the genes encoding these aminotransferases and their cofactor are transcriptionally and translationally regulated is a major challenge in unraveling the interconnection between Trp, Met, and Phe amino acid biosynthesis, hormone production, and nitrogen metabolism. Involvement of N, S, and C sensory systems in the regulation of these aminotransferases such as hexokinase, glutamate like receptors, TOR, and GCN2 kinases seems highly probable.

References

Abeles FB, Morgan PW, Saltveit ME (1992) Ethylene in plant biology, 2nd edn. Academic Press, New York

Adams DO, Yang SF (1977) Methionine metabolism in apple tissue: implication of S-adenosylmethionine as an intermediate in the conversion of methionine to ethylene. Plant Physiol 60:892–896

Alcázar R, Marco F, Cuevas JC, Patron M, Ferrando A, Carrasco P, Tiburcio AF, Altabella T (2006) Involvement of polyamines in plant response to abiotic stress. Biotechnol Lett 28:1867–1876

Amrhein N, Schneebeck D, Skorupka H, Tophof S, Stöckigt J (1981) Identification of a major metabolite of the ethylene precursor 1-aminocyclopropane-1-carboxylic acid in higher plants. Naturwissenschaften 68(12):619–620

Argueso CT, Hansen M, Kieber JJ (2007) Regulation of ethylene biosynthesis. J Plant Growth Regul 262:92–105

Ayala-Rodriguez JA, Barrera-Ortiz S, Ruiz-Herrera LF, Lopez-Bucio J (2017) Folic acid orchestrates root development linking cell elongation with auxin response and acts independently of the Target of rapamycin signaling in *Arabidopsis thaliana*. Plant Sci 264:168–178

Azevedo RA, Lancien M, Lea PJ (2006) The aspartic acid metabolic pathway, an exciting and essential pathway in plants. Amino Acids 30:143–162

Baluska F, Yokawa K, Mancuso S, Baverstock K (2016) Understanding of anesthesia- why consciousness is essential for life and not based on genes. Commun Integr Biol 9(6):1–12. e1238118

Baur AH, Yang SF (1972) Methionine metabolism in apple tissue in relation to ethylene biosynthesis. Phytochemistry 11:3207–3214

Berger B, English S, Chan G, Knodel MH (2003) Methionine regeneration and aminotransferases in *Bacillus subtilis*, *Bacillus cereus*, and *Bacillus anthracis*. J Bacteriol 185:2418–2431

Berkowitz DB, Charrette BD, Karukurichi KR, McFadden JM (2006) α-Vinylic amino acids: occurrence, asymmetric synthesis, and biochemical mechanisms. Tetrahedron Asymmetry 17:869–882

Bi YM, Wang RL, Zhu T, Rothstein SJ (2007) Global transcription profiling reveals differential responses to chronic nitrogen stress and putative nitrogen regulatory components in *Arabidopsis*. BMC Genomics 8:281

Blackwell HE, Zhao Y (2003) Chemical genetic approaches to plant biology. Plant Physiol 133:448–455

Bleecker AB, Kende H (2000) Ethylene: a gaseous signal molecule in plants. Annu Rev Cell Dev Biol 16:1–18

Boerjan W, Bauw G, Van Montagu M, Inze D (1994) Distinct phenotypes generated by overexpression and suppression of S-adenosyl-L-methionine synthetase reveal developmental patterns of gene silencing in tobacco. Plant Cell 6:1401–1414

Bonner ER, Cahoon RE, Knapke SM, Jez JM (2005) Molecular basis of cysteine biosynthesis in plants: structural and functional analysis of O-acetylserine sulfhydrylase from *Arabidopsis thaliana*. J Biol Chem 280:38803–38813

Bouzayen M, Felix G, Latché A, Pech J-C, Boller T (1991) Iron: an essential cofactor for the conversion of 1-aminocyclopropane-1-carboxylic acid to ethylene. Planta 184:244–247

Boycheva S, Dominguez A, Rolcik J, Boller T, Fitzpatrick TB (2015) Consequences of a deficit in vitamin B6 biosynthesis *de novo* for hormone homeostasis and root development in *Arabidopsis*. Plant Physiol 167:102–117

Bradford KJ, Yang SF (1980) Xylem transport of 1-aminocyclopropane-1-carboxylic acid, an ethylene precursor, in waterlogged tomato plants. Plant Physiol 65:322–326

Brunke S, Seider K, Richter ME, Bremer-Streck S, Ramachandra S, Kiehntopf M, Brock M, Hube B (2014) Histidinedegradation via an aminotransferase increases the nutritional flexibility of *Candida glabrata*. Eukaryot Cell 13(6):758–765

Bürstenbinder K, Rzewuski G, Wirtz M, Hell R, Sauter M (2007) The role of methionine recycling for ethylene synthesis in *Arabidopsis*. Plant J 49:238–249

Bussell JD, Keech O, Fenske R, Smith SM (2013) Requirement for the plastidial oxidative pentose phosphate pathway for nitrate assimilation in *Arabidopsis*. Plant J 75:578–591

Capitani G, Hohenester E, Feng L, Storici P, Kirsch JF, Jansonius JN (1999) Structure of 1-aminocyclopropane-1-carboxylate synthase, a key enzyme in the biosynthesis of the plant hormone ethylene. J Mol Biol 294:745–756

Capitani G, McCarthy DL, Gut H, Grutter MG, Kirsch JF (2002) Apple 1-aminocyclopropane-1-carboxylate synthase in complex with the inhibitor L-aminoethoxyvinylglycine. Evidence for a ketimine intermediate. J Biol Chem 277:49735–49742

Capitani G, Tschopp M, Eliot AC, Kirsch JF, Grutter MG (2005) Structure of ACC synthase inactivated by the mechanism-based inhibitor L-vinylglycine. FEBS Lett 579:2458–2462

Cascales-Miñana B, Muños-Berthomeu J, Flores-Tomero M, Anoman AD, Pertusa J, Alaiz M, Osorio S, Fernie AR, Segura J, Ros R (2013) The phosphorylated pathway of serine biosynthesis is essential both for male gametophyte and embryo development and for root growth in *Arabidopsis*. Plant Cell 25:2084–2101

Castilho BA, Shanmugam R, Silva RC, Ramesh R, Himme BM, Sattleger E (2014) Keeping the eIF2 alpha kinase Gcn2 in check. Biochim Biophys Acta 1843:1948–1968

Chae HS, Kieber JJ (2005) Eto Brute? Role of ACS turnover in regulating ethylene biosynthesis. Trends Plant Sci 10:291–296

Chen H, Xiong L (2009a) Localized auxin biosynthesis and postembryonic root development in *Arabidopsis*. Plant Signal Behav 4:752–754

Chen H, Xiong L (2009b) The short-rooted vitamin B_6-deficient mutant pdx1 has impaired local auxin biosynthesis. Planta 229:1303–1310

Chen Y, Zou T, McCormick S (2016) S-adenosylmethionine synthetase 3 is important for pollen tube growth. Plant Physiol 172:244–253

Chiba Y, Ishikawa M, Kijima F, Tyson RH, Kim J, Yamamoto A, Nambara E, Leustek T, Wallsgrove RM, Naito S (1999) Evidence for autoregulation of cystathionine gamma-synthase mRNA stability in *Arabidopsis*. Science 286(5443):1371–1374

Chiu JC, Brenner ED, DeSalle R, Nitabach MN, Holmes TC, Coruzzi GM (2002) Phylogenetic and expression analysis of the glutamate-receptor–like gene family in *Arabidopsis thaliana*. Mol Biol Evol 19:1066–1082

Cho MH, Corea OR, Yang H, Bedgar DL, Laskar DD, Anterola AM, Moog-Anterola FA, Hood RL, Kohalmi SE, Bernards MA, Kang C, Davin LB, Lewis NG (2007) Phenylalanine biosynthesis in *Arabidopsis thaliana*. Identification and characterization of arogenate dehydratases. J Biol Chem 282:30827–30835

Christen P, Mehta PK (2001) From cofactor to enzymes. The molecular evolution of pyridoxal-5′-phosphate-dependent enzymes. Chem Rec 1:436–447

Clandinin MT, Cossins EA (1974) Methionine biosynthesis in isolated Pisum sativum mitochondria. Phytochemistry 13:585–591

Clausen T, Huber R, Messerschmidt A, Pohlenz HD, Laber B (1997) Slow-binding inhibition of *Escherichia coli* cystathionine β-lyase by L-aminoethoxyvinylglycine: a kinetic and X-ray study. Biochemist 36:12633–12643

Colinas M, Esenhut M, Tohge T, Pesquera M, Fernie AR, Weber APM, Fiztpatrick TB (2016) Balancing of B6 vitaminers is essential for plant development and metabolism in *Arabidopsis*. Plant Cell 28:439–453

Cornell NW, Zuurendonk PF, Kerich MJ, Straight CB (1984) Selective inhibition of alanine aminotransferase and aspartate aminotransferase in rat hepatocytes. Biochem J 220:707–716

Coruzzi GM (2003) Primary N-assimilation into amino acids in Arabidopsis. In: Somerville CR, Meyerowitz EM (eds) The *Arabidopsis* book. American Society of Plant Biologists, Rockville, pp 1–17

Coruzzi GM, Last RL (2000) Amino acids. In: Buchanan RB, Gruissem W, Jones R (eds) Biochemistry and molecular biology of plants. American Society of Plant Physiologists, Rockville, pp 358–410

Cossins EA (1987) In: Davies DD (ed) The biochemistry of plants, vol 11. Academic, San Diego, pp 317–353
Cossins EA (2000) The fascinating world of folate and one-carbon metabolism. Can J Bot 78:691
Crider KS, Yang TP, Berry RJ, Bailey LB (2012) Folate and DNA methylation: a review of molecular mechanisms and the evidence for folate's role. Adv Nutr 3:21–38
Curien G, Dumas R, Ravanel S, Douce R (1996) Characterization of an *Arabidopsis thaliana* cDNA encoding an S-adenosylmethionine-sensitive threonine synthase. FEBS Lett 390:85–90
Curien G, Job D, Douce R, Dumas R (1998) Allosteric activation of *Arabidopsis* threonine synthase by S-adenosylmethionine. Biochemist 31:13212–13221
Davenport R (2002) Glutamate receptors in plants. Ann Bot 90:549–557
De Gernier H, De Pessemier J, Xu J, Cristescu SM, Van Der Straeten D, Verbruggen N, Hermans C (2016) A comparative study of ethylene emanation upon nitrogen deficiency in natural accessions of *Arabidopsis thaliana*. Front Plant Sci 10(7):70
de la Torre F, De Santis L, Suárez M-F, Crespillo R, Cánovas FM (2006) Identification and functional analysis of a prokaryotic-type aspartate aminotransferase: implications for plant amino acid metabolism. Plant J 46:414–425
de la Torre F, El-Azaz J, Ávila C, Cánovas FM (2014a) Deciphering the role of aspartate and prephenate aminotransferase activities in plastid nitrogen metabolism. Plant Physiol 164(1):92–104
de la Torre F, Cañas RA, Pascual MB, Avila C, Cánovas FM (2014b) Plastidic aspartate aminotransferases and the biosynthesis of essential amino acids in plants. J Exp Bot 65(19):5527–5534
Deslauriers SD, Larsen PB (2010) FERONIA is a key modulator of brassinosteroid and ethylene responsiveness in *Arabidopsis* hypocotyls. Mol Plant 3:626–640
Droux M, Ravanel S, Douce R (1995) Methionine biosynthesis in higher plants. II. Purification and characterization of cystathionine β-lyase from spinach chloroplasts. Arch Biochem Biophys 316:585–595
Droux M, Ruffet ML, Douce R, Job D (1998) Interactions between serine acetyltransferase and O-acetylserine (thiol) lyase in higher plants–structural and kinetic properties of the free and bound enzymes. Eur J Biochem 255(1):235–245
Duan Q, Kita D, Li C, Cheung AY, Wu HM (2010) FERONIA receptorlikekinase regulates RHO GTPase signaling of root hair development. Proc Natl Acad Sci U S A 107:17821–17826
Eichel J, González JC, Hotze M, Matthews RG, Schröder J (1995) Vitamin-B12-independent methionine synthase from a higher plant (*Catharanthus Roseus*). Eur J Biochem 230:1053–1058
El-Azaz J, de la Torre F, Àvila C, Cánovas F (2016) Identification of a small protein domain present in all plant lineages that confers high prephenate dehydratase activity. Plant J 87:215–229
Finlayson SA, Lee I-J, Mullet JE, Morgan PW (1999) The mechanism of rhythmic ethylene production in sorghum. The role of phytochrome B and simulated shading. Plant Physiol 119:1083–1090
Fisher SK, Davies WE (1976) The effect of the convulsant allylglycine (2-amino-4-pentenoic acid) on the activity of glutamic acid decarboxylase and the concentration of GABA in different regions of guinea pig brain. Biochem Pharmacol 25(16):1881–1885
Fitzpatrick TB (2011) Vitamin B_6 in plants: more than meets the eye. In: Rebeille F, Douce R (eds) Advances in botanical research, vol 59. Elsevier, New York, pp 2–31
Galili G (2011) The aspartate-family pathway of plants: linking production of essential amino acids with energy and stress regulation. Plant Signal Behav 6:192–195
Ge C, Cui X, Wang Y, Hu Y, Fu Z, Zhang D, Cheng Z, Li J (2006) BUD2, encoding an S-adenosylmethionine decarboxylase, is required for *Arabidopsis* growth and development. Cell Res 16:446–456
Gerdes S, Lerma-Ortiz C, Frelin O, Seaver SMD, Henry CS, de Crécy-Lagard V, Hanson AD (2012) Plant B vitamin pathways and their comparmentation: a guide for the perplexed. J Exp Bot 63(15):5379–5395
Gerelova V, Ambach L, Rébeillé F, Stove C, Van Der Straeten D (2017) Folates in plants: research advances and progress in crop biofortification. *Frontiers in.* Plant Sci 5:21. https://doi.org/10.3389/fchem.2017.00021. eCollection 2017

Gil-Amado JA, Gomez-Jimenez MC (2012) Regulation of polyamine metabolism and biosynthetic gene expression during olive mature-fruit abscission. Planta 235:1221–1237
Giovanelli J, Owens LD, Mudd SH (1971) Mechanism of inhibition of spinach beta-cystathionase by rhizobitoxine. Biochim Biophys Acta 227(3):671–684
Giovanelli J, Mudd SH, Datko AH (1985) Quantitative analysis of pathways of methionine metabolism and their regulation in *Lemna*. Plant Physiol 78:555–560
Good AG, Shrawat AK, Muench DG (2004) Can less yield more? Is reducing nutrient input into the environment compatible with maintaining crop production? Trends Plant Sci 9:597–605
Good AG, Johnson SJ, De Pauw M, Carroll RT, Savidov N, Vidmar Jb LZ, Taylor G, Stroeher V (2007) Engineering nitrogen use efficiency with alanine aminotransferase. Can J Bot 85:252–262
Gorelova V, Ambach L, Rébeillé F, Stove C, Van Der Straeten D (2017) Folates in plants: research advances and progress in crop biofortification. Front Chem 5(2):21
Graindorge M, Giustini C, Jacomin AC, Kraut A, Curien G, Matringe M (2010) Identification of a plant gene encoding glutamate/aspartate-prephenate aminotransferase: the last homeless enzyme of aromatic amino acids biosynthesis. FEBS Lett 584:4357–4360
Guo H, Li L, Ye H, Yu X, Algreen A, Yin Y (2009) Three related receptor-like kinases are required for optimal cell elongation in *Arabidopsis thaliana*. Proc Natl Acad Sci U S A 106:7648–7653
Hansen M, Chae HS, Kieber JJ (2009) Regulation of ACS protein stability by cytokinin and brassinosteroid. Plant J 57:606–614
Hanson AD, Roje S (2001) One-carbon metabolismin higher plants. Annu Rev Plant Biol 52:119–137
Haruta M, Sabat G, Stecker K, Minkoff BB, Sussman MR (2014) A peptide hormone and its receptor protein kinase regulate plant cell expansion. Science 343(6169):408–411
Harpaz-Saad S, Yoon GM, Mattoo AK, Kieber JJ (2012) The formation of ACC and competition between polyamines and ethylene for SAM. In: The plant hormone ethylene, Annual plant reviews, vol 44. Wiley, Chichester, p 56
He W, Brumos J, Li H et al (2011) A small-molecule screen identifies L-Kynurenine as a competitive inhibitor of TAA1/TAR activity in ethylene-directed auxin biosynthesis and root growth in Arabidopsis. Plant Cell 23:3944–3960
Hesse H, Kreft O, Maimann S, Zeh M, Hoefgen R (2004) Current understanding of the regulation of methionine biosynthesis in plants. J Exp Bot 55:1799–1808
Hey SJ, Byrne E, Halford NG (2010) The interface between metabolic and stress signalling. Ann Bot 105 197–203
Hildebrandt TM, Nunes-Nesi A, Araújo WL, Braun H-P (2015) Amino acid catabolism in plants. Mol Plant 8:1563–1579
Hinnebusch AG (2005) Translational regulation of GCN4 and the general amino acid control of yeast. Annu Rev Microbiol 59:407–450
Hirel B, Le Gouis J, Ney B, Gallais A (2007) The challenge of improving nitrogen use efficiency in crop plants: towards a more central role forgenetic variability and quantitative genetics within integrated approaches. J Exp Bot 58:2369–2387
Hu Y, Vandenbussche F, Van Der Straeten D (2017) Regulation od seedling growth by ethylene and the ethylene-auxin crosstalk. Planta 245:467–489
Huai Q, Xia Y, Chen Y, Callahan B, Li N, Ke H (2001) Crystal structures of 1 amino-cyclopropane-1-carboxylate (ACC) synthase in complex with aminoethoxyvinylglycine and pyridoxal-5′-phosphate provide new insight into catalytic mechanisms. J Biol Chem 276:38210–38216
Hummel I, Couée I, El Amrani A, Martin-Tanguy J, Hennion F (2002) Involvement of polyamines in root development at low temperature in the subantarctic cruciferous species *Pringlea antiscorbutica*. J Exp Bot 53:1463–1473
Imai A, Matsuyama T, Hanzawa Y, Akiyama T, Tamaoki M, Saji H, Shirano Y, Kato T, Hayashi H, Shibata D, Tabat S, Komeda Y, Takahashi T (2004) Spermidine synthase genes are essential for survival *Arabidopsis*. Plant Physiol 135:1565–1573
Inanobe A, Furukawa H, Gouaux E (2005) Mechanism of partial agonist action at the NR1 subunit of NMDA receptors. Neuron 47:71–84

Iraqui I, Vissers S, Cartiaux M, Urrestarazu A (1998) Characterisation of Saccharomyces cerevisiae ARO8 and ARO9 genes encoding aromatic aminotransferases I and II reveals a new aminotransferase subfamily. Mol Gen Genet 257:238–248

Jander G, Joshi V (2009) Aspartate-derived amino acids biosynthesis in *Arabidopsis* thaliana. Arabidopsis Book 7:e0121. https://doi.org/10.1199/tab.0121

Jencks DA, Mathews RG (1987) Allosteric inhibition of methylenetetrahydrofolate reductase by adenosylmethionine. Effects of adenosylmethionine and NADPH on the equilibrium between active and inactive forms of the enzyme and on the kinetics of approach to equilibrium. J Biol Chem 262:2485–2493

Jin Y, Ye N, Zhu F, Li H, Wang J, Liwen J, Zhang J (2017) Calcium-dependent protein kinase CPK28 targets the methionine adenosyltransferases for degradation by the 26S proteasome and affects ethylene biosynthesis and lignin deposition in *Arabidopsis*. Plant J 90:304–318

Kumar A, Taylor MA, Mad Arif SA, Davies HV (1996) Patato plants expressing antisens and sense S-adenosylmethionine decarboxylase (SAMDC) transgenes show altered levels of polyamines and ethylene: antisens plants display abnormal phenotypes. Plant J 9:147–158

Kushad MM, Richardson DG, Ferro AJ (1982) 5-Methylthioribose kinase activity in plants. Biochem Biophys Res Commun 108:167–173

Kushad MM, Richardson DG, Ferro AJ (1983) Intermediates in the recycling of 5-methylthioribose to methionine in fruits. Plant Physiol 73:257–261

Laber B, Maurer W, Hanke C, Grafe S, Ehlert S, Messerschmidt A, Clausen T (1999) Characterization of recombinant *Arabidopsis thaliana* threonine synthase. Eur J Biochem 263:212–221

Lageix S, Lanet E, Pouch-Pélissier M-N, Espagnol M-C, Robaglia C, Deragon J-M, Pélissier T (2008) Arabidopsis eIF2A kinase GCN2 is essential for growth in stress conditions and is activated by wounding. BMC Plant Biol 8:134

Lam H-M, Chiao YA, Li M-W, Yung Y-K, Sang J (2006) Putative nitrogen sensing systems in plants. J Integr Plant Biol 43:873–888

Le Deunff E, Lecourt J (2016) Non-specificity of ethylene inhibitors: 'double edged' tools to find out new targets involved in the root morphogenetic programme. Plant Biol 18:353–361

Le Deunff E, Lecourt J, Malagoli P (2016) Fine-tuning of root elongation by ethylene: a tool to study dynamic structure-function relationships between root architecture and nitrate absorption. Ann Bot 118(4):607–620

Le J, Vandenbussche F, Van Der Straeten D, Verbelen J-P (2001) In the early response of Arabidopsis roots to ethylene, cell elongation is up-and downregulated and uncoupled from differentiation. Plant Physiol 125:519–522

Le Deunff E, Beauclair P, Leblanc A, Deleu C, Lecourt J (2018) Non-specificity of AVG inhibitor reveals the importance of aminotransferases network in the root morphogenetic program and nitrate absorption. Ann Bot (submitted MS/2018/)

Le Ny F, Leblanc A, Beauclair P, Deleu C, Le Deunff E (2013) In low transpiring conditions, nitrate and water fluxes for growth of *B. napus* plantlets correlate with changes in BnNrt2.1 and BnNrt1.1 nitrate transporters expression. Plant Signal Behav 8:e22902

Leblanc A, Renault H, Lecourt J, Etienne P, Deleu C, Le Deunff E (2008) Elongation changes of exploratory and root hair systems induced by AVG and ACC affect nitrate uptake and *BnNrt2.1* and *BnNrt1.1* gene expression in oil seed Rape. Plant Physiol 146:1028–1040

Lemaire L, Deleu C, Le Deunff E (2013) Modulation of ethylene biosynthesis by ACC and AIB reveals a structural and functional relationship between the $K^{15}NO_3$ uptake rate and root absorbing surfaces. J Exp Bot 64:2725–2737

Li F, Wang J, Ma C, Zhao Y, Wang Y, Hasi A, Qi Z (2013) Glutamate receptor-like channel3.3 is involved in mediating glutathionetriggered cytosolic calcium transients, transcriptional changes, and innate immunity responses in *Arabidopsis*. Plant Physiol 162:1497–1509

Li X, Cai W, Liu Y, Li H, Fu L, Liu Z, Xu L, Liu H, Xu T, Xiong Y (2017) Differential TOR activation and cell proliferation in *Arabidopsis* root and shoot apexes. Proc Natl Acad Sci U S A 114(10):2765–2770

Lieberman M (1979) Biosynthesis and action of ethylene. Annu Rev Plant Physiol 30:533–591

Lin L-C, Hsu J-E, Wang L-C (2010) Identification of novel inhibitors of 1-aminocyclopropane-1-carboxylic acid synthase by chemical screening in *Arabidopsis thaliana*. J Biol Chem 285:33445–33456

Liu Y, Zhang S (2004) Phosphorylation of 1-aminocyclopropane-1-carboxylic acid synthase by MPK6, a stress-responsive mitogen-activated protein kinase, induces ethylene biosynthesis in *Arabidopsis*. Plant Cell 16:3386–3399

Locke JM, Bryce JH, Morris PC (2000) Contrasting effects of ethylene perception and biosynthesis inhibitors on germination and seedling growth of barley (Hordeum vulgare L.) J Exp Bot 51:1843–1849

Loewith R, Hall MN (2011) Target of Rapamycin (TOR) in nutrient signaling and growth control. Genetics 189:1177–1201

Lu SC (2000) *S*-Adenosylmethionine. Int J Biochem Cell Biol 32:391–395

Ludwig-Müller J (2009) Auxin conjugates: their role for plant development and in the evolution of land plants. J Exp Bot 62:1757–1773

Lyzenga WJ, Booth JK, Stone SL (2012) The *Arabidopsis* RING-type E3 ligase XBAT32 mediates the proteasomal degradation of the ethylene biosynthetic enzyme, 1-aminocyclopropane-1-carboxylate synthase 7. Plant J 71:23–34

Ma W, Li J, Qu B, He X, Zhao X, Li B, Fu X, Tong Y (2014) Auxin biosynthetic gene TAR2 is involved in low nitrogen-mediated reprogramming of root architecture in *Arabidopsis*. Plant J 78:70–79

Maeda H, Dudareva N (2012) The shikimate pathway and aromatic amino acid biosynthesis in plants. Annu Rev Plant Biol 63:73–105

Maeda H, Yoo H, Dudareva N (2011) Prephenate aminotransferase directs plant phenylalanine biosynthesis *via* arogenate. Nature Chem Biol 7:19–21

Malone S, Chen Z-H, Bahram AR, Walker RP, Gray JE, Leegoog RC (2007) Phospho*enol*pyruvate carboxykinase in *Arabidopsis*: changes in gene expression, rotein and activity during vegetative and reproductive development. Plant Cell Physio 48(3):441–450

Mao D, Yu F, Li J, Van de Poel B, Tan D, Li J, Liu Y, Li X, Dong M, Chen L, Li D, Luan S (2015) FERONIA receptor kinase interacts with Sadenosylmethionine synthetase and suppresses S-adenosylmethionine production and ethylene biosynthesis in *Arabidopsis*. Plant Cell Environ 38:2566–2574

Martin MN, Cohen JD, Saftner RA (1995) A new 1-aminocyclopropane- 1-carboxylic acid-conjugating activity in tomato fruit. Plant Physiol 109(3):917–926

Masclaux-Daubresse C, Daniel-Vedele F, Dechorgnat J, Chardon F, Gaufichon L, Suziki A (2010) Nitrogen uptake, assimilation and remobilization in plants: challenges for sustainable and productive agriculture. Ann Bot 105:1141–1157

Mashiguchi K, Tanaka K, Sakai T, Sugawara S, Kawaide H et al (2011) The main auxin biosynthesis pathway in *Arabidopsis*. Proc Natl Acad Sci U S A 108:18512–18517

McCarthy DL, Capitani G, Feng L, Gruetter MG, Kirsch JF (2001) Glutamate 47 in 1-aminocyclopropane-1-carboxylate synthase is a major specificity determinant. Biochemist 40:12276–12284

McDonnell L, Plett JM, Andersson-Gunnerås S, Kozela C, Dugardeyn J, Van Der Straeten D, Glick BR, Sundberg B, Regan S (2009) Ethylene levels are regulated by a plant encoded 1-aminocyclopropane-1-carboxylic acid deaminase. Physiol Plant 136(1):94–109

Mehta PK, Hale TI, Christen P (1993) Aminotransferases: demonstration of homology and division into evolutionary subgroups. Eur J Biochem 214:549–561

Mehta RA, Cassol T, Li N, Ali N, Handa AK, Mattoo AK (2002) Engineered polyamine accumulation in tomato enhances phytonutrient content, juice quality, and vine life. Nat Biotechnol 20:613–618

Miesak B, Coruzzi GM (2002) Molecular and physiological analysis of *Arabidopsis mutants* defective in cytosolic or chloroplastic aspartate aminotransferase. Plant Physiol 129:650–660

Mills WR, Lea PJ, Miflin BJ (1980) Photosynthetic formation of the asparte family of amino acids in isolated chloroplasts. Plant Physiol 65:1166–1172

Miyazaki JH, Yang SF (1987) Metabolism of 5-methylthioribose to methionine. Plant Physiol 84:277–281

Mo X, Zhu Q, Li X et al (2006) The hpa1 mutant of *Arabidopsis* reveals a crucial role of histidinehomeostasis in root meristem maintenance. Plant Physiol 141:1425–1435

Mony L, Kew JN, Gunthorpe MJ, Paoletti P (2009) Allosteric modulators of NR2B-containing NMDA receptors: molecular mechanisms and therapeutic potential. Br J Pharmacol 157:1301–1317

Moore B, Zhou L, Rolland F, Hall Q, Cheng WH, Liu YX, Hwang I, Jones T, Sheen J (2003) Role of the *Arabidopsis* glucose sensor HXK1 in nutrient, light, and hormonal signaling. Science 300:332–336

Muday GK, Rahman A, Binder BM (2012) Auxin and ethylene: collaborators or competitors? Trends Plant Sci. 17:181–195

Muños-Berthomeu J, Cascales-Miñana B, Mulet JM, Bjora-Fernández P-RJ, Kuhn JM, Segura J, Ros R (2009) Plastidial glyceraldehyde-3-phosphate dehydrogenase deficiency leads to altered root development and affects the sugar and amino acid balancein Arabidopsis. Plant Physiol 151:541–558

Muños-Berthomeu J, Cascales-Miñana B, Alaiz M, Segura J, Ros R (2010) A critical role of plastidial glycolytic glyceraldehyde-3-phosphate dehydrogenase in the control of plant metabolism and development. Plant Signal Behav 5(1):67–69

Nahum-Levy R, Fossom LH, Skolnick P, Benveniste M (1999) Putative partial agonist 1-aminocyclopropanecarboxylic acid acts concurrently as a glycine-site agonist and a glutamate-site antagonist at N-methyl-D-aspartate receptors. Mol Pharmacol 56:1207–1218

Narukawa-Nara M, Nakamura A, Kikuzato K, Kakei Y, Sato A, Mitani Y, Yamasaki-Kokudo Y, Ishii T, Hayashi K-I, Asami T, Ogura T, Yoshida S, Fujioka S, Kamakura T, Kawatsu T, Tachikawa M, Soeno K, Shimada Y (2016) Aminooxy-naphthylpropionic acid and its derivatives are inhibitors of auxin biosynthesis targeting L-tryptophan aminotransferase: structure–activity relationships. Plant J 87:245–257

Natarajan K, Meyer MR, Jackson BM, Slade D, Roberts C, Hinnebusch AG, Marton MJ (2001) Transcriptional profiling shows that Gcn4p is a master regulator of gene expression during amino acid starvation in yeast. Mol Cell Biol 13:4347–4368

Nero D, Krouk G, Tranchina D, Coruzzi GM (2009) A system biology approach highlights a hormonal enhancer effect on regulation of genes in a nitrate responsive "biomodule". BMC Syst Biol 3:59

Niederberger P, Miozzari G, Hutter R (1981) Biological role of the general control of amino acid biosynthesis in Saccharomyces cerevisiae. Mol Cell Biol 1:584–593

Noctor G, Queval G, Mhamdi A, Chaouch S, Foyer CH (2011) Glutathione. Arabidopsis Book 9: e0142. doi/10.1199/tab.0142

O'Neill SD (1997) Pollination regulation of flower development. Annu Rev Plant Physiol Plant Mol Biol 48:547–574

Owens LD, Lieberman M, Kunishi AT (1971) Inhibition of ethylene production by rhizobitoxine. Plant Physiol 48:1–4

Pandey S, Ranade SA, Nagar PK, Kumar N (2000) Role of polyamines and ethylene as modulators of plant senescence. J Biosci 25:291–299

Peleman J, Boerjan W, Engler G, Seurinck J, Botterman J, Alliotte T, Van Montagu M, Inze D (1989a) Strong cellular preference in the expression of a housekeeping gene of *Arabidopsis thaliana* encoding Sadenosylmethionine synthetase. Plant Cell 1:81–93

Peleman J, Saito K, Cottyn B, Engler G, Seurinck J, Vanmontagu M, Inze D (1989b) Structure and expression analyses of the S-adenosylmethionine synthetase gene family in *Arabidopsis thaliana*. Gene 84:359–369

Penrose DM, Moffatt BA, Glick BR (2001) Determination of 1-aminocycopropane-1-carboxylic acid (ACC) to assess the effects of ACC deaminase-containing bacteria on roots of canola seedlings. Can J Microbiol 47(1):77–80

Phillips KA, Skirpan AL, Liu X, Christensen A, Slewinski TL, Hudson C, Barazesh S, Cohen JD, Malcomber S, McSteen P (2011) Vanishing tassel 2 encodes a grass-specific tryptophan aminotransferase required for vegetative and reproductive development in maize. Plant Cell 23:550–566

Pieck M, Yuan Y, Godfrey J, Fisher C, Zolj S, Vaughan D, Thomas N, Wu C, Ramos J, Lee N, Normanly J, Celenza JL (2015) Auxin and tryptophan homeostasis are facilitated by the ISS1/VAS1 aromatic aminotransferase in *Arabidopsis*. Genetics 201:185–199

Pommerrenig B, Feussner K, Zierer W, Rabinovych V, Klebl F, Feussner I, Sauer N (2011) Phloem-specific expression of Yang cycle genes and identification of novel Yang cycle enzymes in *Plantago* and *Arabidopsis*. Plant Cell 23:1904–1919

Price MB, Jelesko J, Okumoto S (2012) Glutamate receptor homologs in plants: functions and evolutionary origins. Front Plant Sci 3:235

Pruess DL, Scannell JP, Kellett M, Ax HA, Janecek J, Williams TH, Stempel A, Berger J (1974) Antimetabolites produced by microorganisms. X. L-2-Amino4-(2-aminoe- thoxy)-trans-3-butenoic acid. J Antibiot 27:229–233

Rando RR (1974a) Chemistry and enzymology of Kcat inhibitors. Science 185:320–324

Rando RR (1974b) Irreversible inhibition of aspartate aminotransferase by 2-amino-3-butenoic acid. Biochemist 13(19):3859–3863

Ravanel S, Gakière B, Job D, Douce R (1998) The specific features of methionine biosynthesis and metabolism in plants. Proc Natl Acad Sci U S A 95:7805–7812

Ravanel S, Cherest H, Jabrin S, Grunwald D, Surdin-Kerjan Y, Douce R, Rébeillé F (2001) Tetrahydrofolate biosynthesis in plants: molecular and functional characterization of dihydrofolate synthetase and three isoforms of folylpolyglutamate synthetase in Arabidopsis thaliana. Proc Natl Acad Sci U S A 98:15360–15365

Ravanel S, Block MA, Rippert P, Jabrin S, Curien G, Rébeillé F, Douce R (2004) Methionine metabolism in plants: chloroplasts are autonomous for *de novo* methionine synthesis and can import S-adenosylmethionine from the cytosol. J Biol Chem 279(21):22548–22557

Reyes-Hernández BJ, Srivastava AC, Ugartechea-Chirino Y, Shishkova S, Ramos-Parra PA, Lira-Ruan V, Díaz de la Garza RI, Dong G, Moon JC, Blancaflor EB, Dubrovsky JR (2014) The root indeterminacy-to-determinacy developmental switch is operated through a folate-dependent pathway in *Arabidopsis thaliana*. New Phytol 202:1223–1236

Riemenschneider A, Wegele R, Schmidt A, Papenbrock J (2005) Isolation and characterization of a d-cysteine desulhydrase protein from *Arabidopsis thaliana*. FEBS J 272:1291–1304

Robert S, Raikhel NV, Hicks G (2009) Powerful partners: arabidopsis and chemical genomics. Arabidopsis Book 7:e0109

Robinson D (2005) Integrated root responses to variations in nutrient supply. In: BassiriRad H (ed) Nutrient acquisition by plants: an ecological perspective. Springer, Berlin, Heidelberg, pp 43–61

Rodrigues-Pousada RA, De Rycke R, Dedonder A, Van Caeneghem W, Engler G, Van Montagu M, Van Der Straeten D (1993) The Arabidopsis 1-aminocyclopropane-1-carboxylatesynthase gene 1 is expressed during early development. Plant Cell 5:897–911

Roje S, Chan SY, Kaplan F, Raymond RK, Horne DW, Appling DR et al (2002a) Metabolic engineering in yeast demonstrates that s-adenosylmethionine controls flux through the methylene tetrahydrofolate reductase reaction in vivo. J Biol Chem 277:4056–4061

Roje S, Janave MT, Ziemak MJ, Hanson AD (2002b) Cloning and characterization of mitochondrial 5-formyltetrahydrofolate cycloligase from higher plants. J Biol Chem 277:42748–42754

Romero LC, Aroca MA, Laureano-Marin AM, Moreno I, Garcia I, Gotor C (2014) Cysteine and cysteine-related signaling pathways in *Arabidopsis thaliana*. Mol Plant 7(2):264–272

Rubin G, Tohge T, Matsuda F, Saito K, Scheible WR (2009) Members of the LBD family of transcription factors repress anthocyanin synthesis and affect additional nitrogen responses in *Arabidopsis*. Plant Cell 21:3567–3584

Růžička K, Ljung K, Vanneste S, Podhorská R, Beeckman T, Friml J, Benková E (2007) Ethylene regulates root growth through effects on auxin biosynthesis and transport-dependent auxin distribution. Plant Cell 19:2197–2212

Rzewuski G, Cornell KA, Rooney L, Bürstenbinder K, Wirtz M, Hell R, Sauter M (2007) OsMTN encodes a 5′-methylthioadenosine nucleosidase that is up-regulated during submergence-induced ethylene synthesis in rice (*Oryza sativa* L.) J Exp Bot 58:1505–1514

Saftner R (1989) Effects of organic amines on α-aminoisobutyric acid uptake into the vacuole and on ethylene production by tomato pericarp slices. Physiol Plant 75(4):485–491

Saftner RA, Martin MN (1993) Transport of 1-aminocyclopropane-1-carboxylic acid into isolated maize mesophyll vacuoles. Physiol Plant 87(4):535–543

Sahm U, Knobloch G, Wagner F (1973) Isolation and characterization of the methionine antagonist L-2-amino-4-methoxy-trans-3-butenoic acid from *Pseudomonas aeruginosa* grown on n-paraffin. J Antibiot 26:389–390

Satoh S, Yang SF (1989a) Inactivation of 1-aminocyclopropane-1-carboxylate synthase by L-vinylglycine as related to the mechanism-based inactivation of the enzyme by S-adenosyl-l-methionine. Plant Physiol 91:1036–1039

Satoh S, Yang SF (1989b) Specificity of S-adenosyl-L-methionine in the inactivation and the labeling of 1-aminocyclopropane-1-carboxylate synthase isolated from tomato fruits. Arch Biochem Biophys 271:107–112

Sauter M, Cornell KA, Beszteri S, Rzewuski G (2004) Functional analysis of methyl-thioribose kinase genes in plants. Plant Physiol 136:4061–4071

Sauter M, Lorbiecke R, Ouyang B, Pochapsky TC, Rzewuski G (2005) The immediate-early ethylene response gene OsARD1 encodes an acireductone dioxygenase involved in recycling of the ethylene precursor S-adenosylmethionine. Plant J 44:718–772

Sauter M, Moffatt B, Saechao MC, Helll R, Wirtz M (2013) Methionine salvage and S-adenosylmethionine: essential links between sulfur, ethylene and polyamine biosynthesis. Biochem J 451:145–154

Scannell JP, Pruess DL, Demny TC, Sello LH, Williams T, Stempel A (1972) Anti-metabolites produced by microorganisms. V. L-2-Amino-4- methoxy-trans-3-butenoic acid. J Antibiot 25:122–127

Scheible W, Lauerer M, Schulze E, Caboche M, Stitt M (1997) Accumulation of nitrate in the shoot acts as a signal to regulate shoot-root allocation in tobacco. Plant J 11:671–691

Scheible WR, Morcuende R, Czechowski T, Fritz C, Osuna D, Palacios-Rojas N, Schindelasch D, Thimm O, Udvardi MK, Stitt M (2004) Genome-wide reprogramming of primary and secondary metabolism, protein synthesis, cellular growth processes, and the regulatory infrastructure of *Arabidopsis* in response to nitrogen. Plant Physiol 136:2483–2499

Sekowska A, Dénervaud V, Ashida H, Michoud K, Haas D, Yokota A, Danchin A (2004) Bacterial variations on the methionine salvage pathway. BMC Microbiol 4:9

Shah S, Cossins E (1970) Pteroylglutamates and methionine biosynthesis in isolated chloroplasts. FEBS Lett 7:267–270

Shao A, Ma W, Zhao X, Hu M, He X, Teng W, Li H, Tong Y (2017) The auxin biosynthetic TRYPTOPHAN AMINOTRANSFERASE RELATED TaTAR2.1-3A increases grain yield of wheat. Plant Physiol 174(4):2274–2288

Sheen J (2014) Master regulators in plant Glucose signaling networks. J Plant Biol 57(2): 67–79

Shin K, Lee S, Song W-Y, Lee R-A, Lee I, Ha K, Ko J-C, Park S-K, Nam H-G, Lee Y (2015) Genetic identification of ACC-RESISTANT2 reveals involvement of LYSINE HISTIDINE TRANSPORTER1 in the uptake of 1-aminocyclopropane-1-carboxylic acid in Arabidopsis thaliana. Plant Cell Physiol 56(3):572–582

Shrawat AK, Carroll RT, DePauw M, Taylor GJ, Good AG (2008) Genetic engineering of improved nitrogen use efficiency in rice by the tissue-specific expression of alanine aminotransferase. Plant Biotechnol J 6:722–732

Soeno K, Goda H, Ishii T, Ogura T, Tachikawa T, Sasaki E, Yoshida S, Fujioka S, Asami T, Shimada Y (2010) Auxin biosynthesis inhibitors, identified by a genomics-based approach, provide insights into auxin biosynthesis. Plant Cell Physiol 51:524–536

Srivastava AC, Tang Y, de la Garza RI D, Blancaflor EB (2011) The plastidial folylpolyglutamate synthetase and root apical meristem maintenance. Plant Signal Behav 6:751–754

Staswick PE, Tiryaki I (2004) The oxylipin signal jasmonic acid is activated by an enzyme that conjugates it to isoleucine in *Arabidopsis*. Plant Cell 16(8):2117–2127

Stepanova AN, Yun J, Likhacheva AV, Alonso JM (2007) Multilevel interactions between ethylene and auxin in *Arabidopsis* roots. Plant Cell 19:2169–2185

Stepanova AN, Robertson-Hoyt J, Yun J, Benavente LM, Xie DY et al (2008) TAA1-mediated auxin biosynthesis is essential for hormone crosstalk and plant development. Cell 133:177–191

Stepanova AN, Yun J, Robles LM, Novak O, He W et al (2011) The *Arabidopsis* YUCCA1 flavin monooxygenase functions in the indole-3-pyruvic acid branch of auxin biosynthesis. Plant Cell 23:3961–3973

Stitt M, Müller C, Matt P, Gibon Y, Carillo P, Morcuende R, Scheieble WR, Krapp A (2002) Steps towards an integrated view of nitrogen metabolism. J Exp Bot 53(370):950–970

Sugawara S, Mashiguchi K, Tanaka K, Hishiyama S, Sakai T, Hanada K, Kinoshita-Tsujimura K, Yu H, Dai X, Takebayashi Y, Tajeda-Kamiya N, Kakimoto T, Kawaide H, Natsume M, Estelle M, Zhao Y, Hayashi K-I, Kamiya Y, Kasahara H (2015) Distinct characteristics of indole-3-acetic acid and phenylacetic acid, two common auxins in plants. Plant Cell Physiol 56(8):1641–1654

Swarup R, Perry P, Hagenbeek D, Van Der Straeten D, Beemster GT, Sandberg G, Bhalerao R, Ljung K, Bennett MJ (2007) Ethylene upregulates auxin biosynthesis in Arabidopsis seedlings to enhance inhibition of root cell elongation. Plant Cell 19:2186–2196

Takahashi H, Kopriva S, Giordano M, Saito K, Hell R (2011) Sulfur assimilation in photosynthetic organisms: molecular functions and regulations of transporters and assimilatory enzymes. Annu Rev Plant Biol 62:157–184

Tambasco-Studart M, Titiz O, Raschle T, Forster G, Amrhein N, Fitzpatrick TB (2005) Vitamin B_6 biosynthesis in higher plants. Proc Natl Acad Sci U S A 102:13687–13692

Tan ST, Xue HW (2014) Casein kinase 1 regulates ethylene synthesis by phosphorylating and promoting the turnover of ACS5. Cell Rep 9:1692–1702

Tapken D, Hollmann M (2008) *Arabidopsis thaliana* glutamate receptor ion channel function demonstrated by ion pore transplantation. J Mol Biol 383:36–48

Tapken D, Anschütz U, Liu L-H, Huelsken T, Seebohm G, Becker D, Hollmann M (2013) A plant homolog of animal glutamate receptors is an ion channel gated by multiple hydrophobic amino acids. Sci Signal 6:ra47

Tarun AS, Theologis A (1998) Complementation analysis of mutants of 1-aminocyclopropane 1-carboxylate synthase reveals the enzyme is a dimer with shared active sites. J Biol Chem 273:12509–12514

Tassoni A, van Buuren M, Franceschetti M, Fornalè S, Bagni N (2000) Polyamine content and metabolism in *Arabidopsis thaliana* and effect of spermidine on plant development. Plant Physiol Biochem 38:383–393

Tatsuki M, Mori H (2001) Phosphorylation of tomato 1-aminocyclopropane- 1-carboxylic acid synthase, LE-ACS2, at the C-terminal region. J Biol Chem 276:28051–28057

Tian QY, Sun P, Zhang WH (2009) Ethylene is involved in nitrate-dependent root growth and branching in *Arabidopsis thaliana*. New Phytol 184:918–931

Tiburcio AF, Altabella T, Bitrián M, Alcázar R (2014) The roles of polyamines during the lifespan of plants: from development to stress. Planta 240(1):1–18

Titiz O, Tambasco-Studart M, Warzych E, Apel K, Amrhein N, Laloi C, Fitzpatrick TB (2006) PDX1 is essential for vitamin B_6 biosynthesis, development and stress tolerance in *Arabidopsis*. Plant J 48:933–946

Tophof S, Martinoia E, Kaiser G, hartung W, Amrhein N (1989) Compartmentation and transport of 1-aminocyclopropane-1-carboxylic acid and N-manlonyl-1-aminocyclopropane-carboxylic acid in barley and wheat mesophyll-cells and protoplasts. Physiol Plant 75:333–339

Tsuchisaka A, Theologis A (2004a) Heterodimeric interactions among the 1-amino-cyclopropane-1-carboxylate synthase polypeptides encoded by the *Arabidopsis* gene family. Proc Natl Acad Sci U S A 101:2275–2280

Tsuchisaka A, Theologis A (2004b) Unique and overlapping expression patterns among the Arabidopsis 1-amino-cyclopropane- 1-carboxylate synthase gene family members. Plant Physiol 136:2982–3000

Tsuchisaka A, Yu G, Jin H, Alonso JM, Ecker JR, Zhang X, Gao S, Theologis A (2009) A combinatorial interplay among the 1-aminocyclopropane-1-carboxylate isoforms regulates ethylenebiosynthesis in Arabidopsis thaliana. Genetics 183(3):979–1003

Urrestarazu A, Vissers S, Iraqui I, Grenson M (1998) Phenylalanine- and tyrosine-auxotrophic mutants of Saccharomyces cerevisiae impaired in transamination. Mol Gen Genet 257:230–237

Van de Poel B, Bulens I, Markoula A, Hertog M, Dreesen R, Wirzt M, Vandoninck S, Oppermann Y, Keulemans J, Hell R, Waelkens E, De Prot MP, Sauter M, Nicolai BM, Geeraerd AH (2012) Targeted systems biology profiling of tomato fruit reveals coordination of the Yang cycle and a distinct regulation of ethylene biosynthesis during postclimacteric ripening. Plant Physiol 160:1498–1514

Van der Straeten D, odrigues-Pousada RA, Villarroel R, Hanley S, Goodman HM, Van Montagu M (1992) Cloning, genetic mapping, and expression analysis of an Arabidopsis thaliana gene that encodes 1-aminocyclopropane-1-carboxylate synthase. Proc Natl Acad Sci U S A 89:9969–9973

Van Zhong G, Jacqueline K, Burns JK (2003) Profiling ethylene-regulated gene expression in *Arabidopsis thaliana* by microarray analysis. Plant Mol Biol 53:117–131

Vincill ED, Bieck AM, Spalding EP (2012) Ca2+ conduction by an amino acid-gated ion channel related to glutamate receptors. Plant Physiol 159:40–46

Vincill ED, Clarin AE, Molenda JN, Spalding EP (2013) Interacting glutamate receptor-like proteins in phloem regulate lateral root initiation in *Arabidopsis*. Plant Cell 25:1304–1313

Voesenek LA, Harren FJ, Bögemann GM, Blom CW, Reuss J (1990) Ethylene production and petiole growth in rumex plants induced by soil waterlogging the application of a continuous flow system and a laser driven intracavity photoacoustic detection system. Plant Physiol 94:1071–1077

Wand NN, Shih M-C, Li N (2005) The GUS reporter-aided analysis of the promoter activities of Arabidopsis ACC synthase genes AtACS4, AtACS5, and AtACC7 induced by hormones and stresses. J Exp Bot 56(413):909–920

Wang SY, Adams DO, Lieberman M (1982) Recycling of 5′-methylthioadenosineribose carbon atoms into methionine in tomato tissue in relation to ethylene production. Plant Physiol 70:117–121

Wang KL, Li H, Ecker JR (2002) Ethylene biosynthesis and signaling networks. Plant Cell 14 (suppl.):S131–S151

Wang R, Okamoto M, Xing X, Crawford NM (2003) Microarray analysis of the nitrate response in *Arabidopsis* roots and shoots reveals over 1,000 rapidly responding genes and new linkages to glucose, trehalose-6-phosphate, iron, and sulfate metabolism. Plant Physiol 132:556–567

Wang KL, Yoshida H, Lurin C, Ecker JR (2004) Regulation of ethylene gas biosynthesis by the *Arabidopsis* ETO1 protein. Nature 428:945–950

Wang J, Wang Y, Yang J, Ma C, Zhang Y, Ge T, Qi Z, Kang Y (2015) *Arabidopsis* ROOT HAIR DEFECTIVE3 is involved in nitrogen starvation-induced anthocyanin accumulation. J Integr Plant Biol 57(8):708–721

Weiland M, Mancuso S, Baluska F (2016) Signalling via glutamate and GLRS in *Arabidopsis thaliana*. Funct Plant Biol 43:1–25

Won C, Shen X, Mashiguchi K, Zheng Z, Dai X et al (2011) Conversion of tryptophan to indole-3-acetic acid by tryptophan aminotransferase of *Arabidopsis* and YUCCAs in *Arabidopsis*. Proc Natl Acad Sci U S A 108:18518–18523

Xiong Y, Sheen J (2014) The role of Target of Rapamycin signaling networks in plant growth and metabolism. Plant Physiol 164:499–512

Xiong Y, Sheen J (2015) Novel links in the plant TOR kinase signaling network. Curr Opin Plant Biol 28:83–91

Yamagami T, Tsuchisaka A, Yamada K, Haddon WF, Harden LA, Theologis A (2003) Biochemical diversity among the 1-aminocyclopropane- 1-carboxylate synthase isozymes encoded by the *Arabidopsis* gene family. J Biol Chem 278:49102–49112

Yan Z, Liu X, Ljung K, Li S, Zhao W, Yang F, Wang M, Tao Y (2017) Type B response regulators act as central integrators in transcriptional control of the auxin biosynthesis enzyme TAA1. Plant Physiol 175(3):1438–1454

Yanagisawa S, Yoo S-D, Sheen J (2003) Differential regulation of EIN3 stability by glucose and ethylene signalling in plants. Nature 425:521–525

Yasuta T, Satoh S, Minamisawa K (1999) New assay for Rhizobitoxine based on inhibition of 1-aminocyclopropane-1-carboxylate synthase. Appl Environ Microbiol 65:849–852

Yi H, Galant A, Ravilious GE, Preuss ML, Jez JM (2010) Sensing sulfur conditions: simple to complex protein regulatory mechanisms in plant thiol metabolism. Mol Plant 3:269–279

Yoo SD, Cho YH, Tena G, Xiog Y, Sheen J (2008) Dual control of nuclear EIN3 by bifurcate MAPK cascades in C_2H_4 signalling. Nature 451:789–795

Yoo H, Widhalm JR, Qian Y, Maeda H, Cooper BR, Jannasch AS, Gonda I, Lewinsohn E, Rhodes D, Dudareva N (2013) An alternative pathway contributes to phenylalanine biosynthesis via a cytosolic tyrosine:phenylpyruvate aminotransferase. Nat Commun 4:2833

Yoshikawa T, Ito M, Sumikura T, Nakayama A, Nishimura T, Kitano H, Yamaguchi I, Koshiba T, Hibara KI, Nagato Y (2014) The rice FISH BONE gene encodes a tryptophan aminotransferase, which affects pleiotropic auxin-related processes. Plant J 78:927–936

Zeh M, Leggewie G, Hoefgen R, Hesse H (2002) Cloning and characterisation of a cDNA encoding a cobalamin-independent methionine synthase from potato (*Solanum tuberosum* L). Plant Mol Biol 48:255–265

Zhang Y, Dickinson JR, Paul MJ, Halford NG (2003) Molecular cloning of an *Arabidopsis* homologue of GCN2, a protein kinase involved in co-ordinated response to amino acid starvation. Planta 217:668–675

Zhao Y (2014) Auxin biosynthesis. Arabidopsis Book 12:e0173

Zheng XFS, Chan T-F (2002) Chemical genomics: a systematic approach in biological research and drug discovery. Curr Issues Mol Biol 4:33–43

Zheng Z, Guo Y, Novak O, Dai X, Zhao Y, Ljung K, Noel JP, Chory J (2013a) Coordination of auxin and ethylene biosynthesis by the aminotransferase VAS1. Nat Chem Biol 9(4):244–246

Zheng D, Han X, An Y, Guo H, Xia X, Yin W (2013b) The nitrate transporter NRT2. 1 functions in the ethylene response to nitrate deficiency in *Arabidopsis*. Plant Cell Environ 36:1328–1337

Understanding the Factors Influencing Stem Form with Modelling Tools

Robert Schneider

Contents

Abstract The shape of tree trunks has been studied in Forest Sciences for a long time, since the volume and biomass of the tree is directly related to its shape. This paper reviews the tools used to study the stem form, theories that explain the observed variations and some empirical observations of changes in stem form with climatic variables. A case study using balsam fir (*Abies balsamea*) data is used to highlight some modelling aspects. The ways to quantify stem form have evolved over time, and now it is mainly studied by either modelling the vertical growth distribution along the stem or with stem taper models. The results of the former will have implications in the fields of dendrochronology and wood properties and the latter is more important for estimating volume in national forest inventories or for

Communicated by Hans Pretzsch

R. Schneider (✉)
Chaire de recherche sur la forêt habitée, Département de Biologie, Université du Québec à Rimouski, Rimouski, QC, Canada
e-mail: Robert_Schneider@uqar.ca

Progress in Botany (2019) 80: 295–316, DOI 10.1007/124_2018_21,
Published online: 24 June 2018

tree-to-wood product conversion studies. More recently, allometric exponents, or the comparison of different allometric exponents, have also been used to gain insight on the effects of climate on stem form. These three modelling approaches are however empirically based. Several theories have been proposed to explain the effects of various factors on stem form, with the hydraulic and biomechanic theories being the most widely used, and often opposed. Both theories, when simplified to their simplest expression, underline the importance of crown dimensions in determining tree form. Nevertheless, these theories cannot explain all of the variations observed empirically. In the case study of balsam fir, climatic variables such as total summer precipitation and mean winter temperature are slightly more important in explaining tree taper, when compared to average wind speed. This signifies that the proposed theories, be it either hydraulic or biomechanic, should be hybridized with physiological processes in order to account for all the empirical observations.

1 Introduction

Foresters have long been interested in the shape of the main stem of a tree (Pressler 1865; Metzger 1893). The reason of this interest is that tree volume and biomass is dependant on tree form since the main stem accounts for over 60% of the total aboveground biomass of a tree with a diameter at breast height larger than 10 cm (Lambert et al. 2005). With green accounting and biomass estimation on large tracks of land, small variations in tree form can imply large biases (Weiskittel et al. 2015). Moreover, sampling height, or lack of homogeneity in sampling height, can strongly influence the past climate reconstruction when using tree ring chronologies due to variations in the vertical radial growth distribution (Autin et al. 2015).

The shape of the main stem can be quantified by its form or its taper. Stem form refers to geometrical shape of the stem. Generally, the tree is divided into three sections: the lower part of the tree that resembles a neiloid, the central part follows a paraboloid and the upper part which is often designated as a conoid. Stem taper on the other hand refers to the change in diameter between two points and the distance between these. Empirical results have long highlighted the relationship between crown size and stem taper (Larson 1963).

Besides the role of stem form in forest mensuration, forest ecology has always tried to understand to what degree taper is influenced by environmental factors. It is well known that auxins and other hormones such as indole-3-acetic acid (IAA), gibberellic acid and cytokinin initiate secondary growth in trees (Nilsson et al. 2008; Immanen et al. 2016). IAA is produced in the apical meristems (Zhao 2010) and diffused through the stem of the tree, thus initiating its radial growth.

Physiological processes can explain how the taper of a tree varies. However, these do not readily provide reasons why a tree would need to have more or less taper. The two most important theories used to explain such variations are based either on hydraulic or mechanical constraints. These two theories rely on the functions of the stem of the tree. It must ensure mechanical stability such that the tree does not buckle under its own weight or through stresses imposed by wind or snow. The stem must also transport metabolites produced in the leaves to the other parts of the tree, and water and nutrients from the roots up.

This review will start by exploring empirical methods to quantify changes in stem form (Sect. 3). The underlying hypothesis of the hydraulic (Sect. 4.1) and biomechanical (Sect. 4.2) theories will then be explored, followed by a review of previously published and new results on the effects of climatic variables on the shape of forest trees (Sect. 5).

2 Data Description

In several Sects. 3.1 and 4, data on balsam fir (*Abies balsamea*) is used and described here. A total of 3,019 trees were sampled in 656 plots across the managed forest of the province (Fig. 1), ranging from the temperate (45.2° N) to the boreal (51.2° N) forests and from a continental (79.2° W) to a maritime (64.5° W) climate.

In each ecological region of the province, the five most important ecotypes were retained (Robitaille et al. 2015) and five plots were randomly located in each ecotype. A forest inventory was carried out in a 400-m^2 plot to obtain stand characteristics. Stand density, basal area and dominant height were obtained from

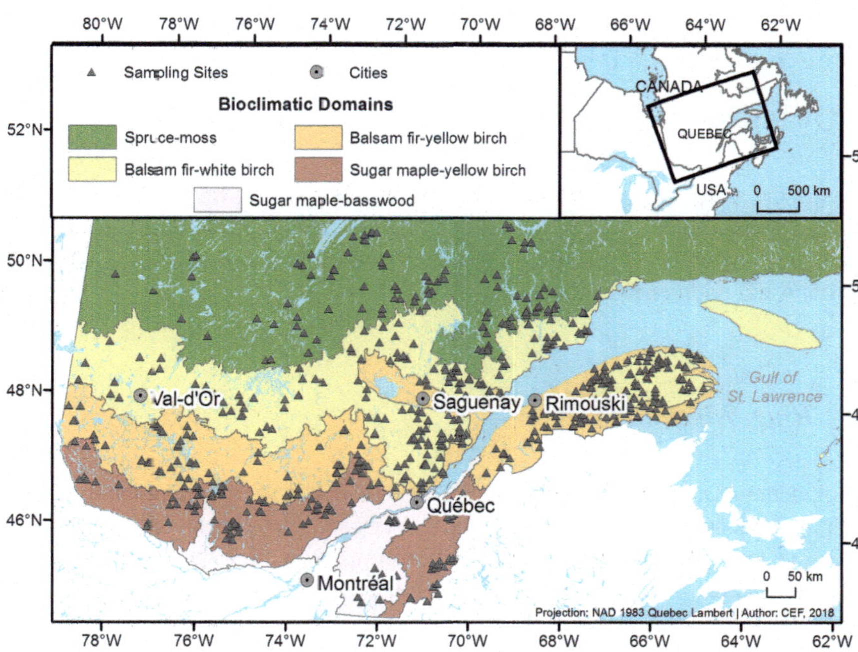

Fig. 1 Location of the plots across the province of Quebec, Canada

Table 1 Mean tree, stand and environmental characteristics for the sampled plots with balsam fir (standard deviations in parentheses)

Variable	Mean
Diameter at breast height (cm)	22.1 (4.7)
Total tree height (m)	16.7 (2.9)
Tree age (years)	72.9 (22.4)
Altitude (m)	421 (181)
Stand density (n/ha)	1,303 (520)
Stand basal area (m^2/ha)	33.0 (9.8)
Mean summer temperature (°C)	15.2 (1.1)
Mean winter temperature (°C)	−14.1 (2.3)
Total summer precipitation (mm)	1,061 (109)
Mean wind speed (m/s)	3.6 (1.8)
Mean wind standard deviation (m/s)	2.0 (0.9)

the diameter at breast height (dbh) measurements made on trees with a dbh greater or equal to 9 cm. Between 5 and 10 dominant or codominant trees were then randomly selected in a 1,964-m^2 plot with the same centre as the inventory plot. Each tree was then felled and disks taken at stump height (0.15 cm), 0.60, 1.00, breast height (1.3 m), 3.00 m and every 2 m thereafter until the top of the tree. The disks were sanded, and ring widths measured along four radii for the lower disks lower than 3.00 m and on two opposing radii on the other disks. For a complete description of sampling design, see Schneider et al. (2013) or Franceschini and Schneider (2014). Only the rings produced after 1970 were kept for the analysis, leading to a total of 1,082,125 ring widths. Data characteristics are available in Table 1.

3 Quantifying Stem Form

Through the years, several methods have been used to study the effect of different variables on the shape of a tree stem. Of these, three methods that have been used in recent years will be presented: the vertical profile of radial growth, stem taper equations and allometric exponents.

3.1 Ring Width Models

The diameter of a stem at a given point is the sum of the yearly radial increments at that height. Generally, the yearly increment is largest at crown base and close to the stump (e.g. see Larson 1963 for early observations in the first half of the twentieth century; Deleuze and Houllier 2002; Cortini et al. 2013). Differences in radial growth rates within a tree can lead to important biases when constructing long-term ring chronologies, if sampling of dead trees is not carried out at the same height as those of the living trees (Autin et al. 2015).

Vertical growth distribution models can thus be used to study the effects of various factors on tree development. These models can be divided into two groups, the first comprising a segmented regression (Courbet 1999; Courbet and Houllier 2002; Cortini et al. 2013) and the second using trigonometric functions (Deleuze and Houllier 2002). Here, the models proposed by Courbet and Houllier (2002) and Deleuze and Houllier (2002) are compared using balsam fir stem analysis data presented in Sect. 2.

The model by Courbet and Houllier (2002) (adapted for relative ring width, Eq. 1) and the ShX model from Deleuze and Houllier (2002) (Eq. 2) were retained:

$$\frac{\mathrm{rw}_h}{\mathrm{rw}_{1.3}} = b_1\left(b_2 \cdot x - x^2\right)^{b_3} \quad \text{for } x \leq x_1 \tag{1a}$$

$$\frac{\mathrm{rw}_h}{\mathrm{rw}_{1.3}} = b_4 + b_5 \cdot x \quad \text{for } x_1 < x \leq x_2 \tag{1b}$$

$$\frac{\mathrm{rw}_h}{\mathrm{rw}_{1.3}} = \frac{b_5}{\cos\left(b_6 + \frac{x-b_7}{H-b_7}\right)} \quad \text{for } x_2 < x \leq H \tag{1c}$$

$$\mathrm{ra}_h = x \cdot \frac{b_1 \sin h(b_2 x) + b_3 x \sin h(b_2(H-x))}{\sin h(b_2 H)} \tag{2}$$

where rw_h, $\mathrm{rw}_{1.3}$ and ra_h are the ring widths (rw) and areas (ra) at height h or 1.3 m, H the total tree height, x the distance from stem apex ($x = H - h$) and x_1 and x_2 the segmented regression break points. For Eq. (1), the number of parameters can be reduced to four by ensuring at each break point the continuity of the function as well as the first derivative. Each model was calibrated on each tree and year independently (i.e. 88,735 regressions). The standardized residuals were then plotted as a function of relative height (h/H, Fig. 2). It can be seen that there is a lot of difference between both tested models. The model by Courbet and Houllier (2002) (Eq. 1) shows very little bias, with a small over-prediction at the base of the tree (Fig. 2a) whereas the model by Deleuze and Houllier (2002) (Eq. 2) has a strong trend within the residuals (Fig. 2b). The model in Eq. (2) thus seems to be not flexible enough to follow the observed trends within the data. Similar results are observed (and not presented) if ra is used in Eq. (1), or $\frac{\mathrm{rw}_h}{\mathrm{rw}_{1.3}}$ in Eq. (2).

The advantage of using ring width/area models is that the effect of various climatic variables such as monthly precipitations and temperatures of the current and previous year can be used to describe how the radial growth is distributed along the main stem of a tree, in the same way the effect of current and past climate influences radial growth at breast height (e.g. as in Huang et al. 2010). The disadvantage of these models is that the form of the whole tree must be obtained by summing the radial increments at every point along the stem. The problem with such an approach is with error propagation (Temesgen et al. 2015). If the stem taper is needed, a taper equation must be used in order to eliminate this.

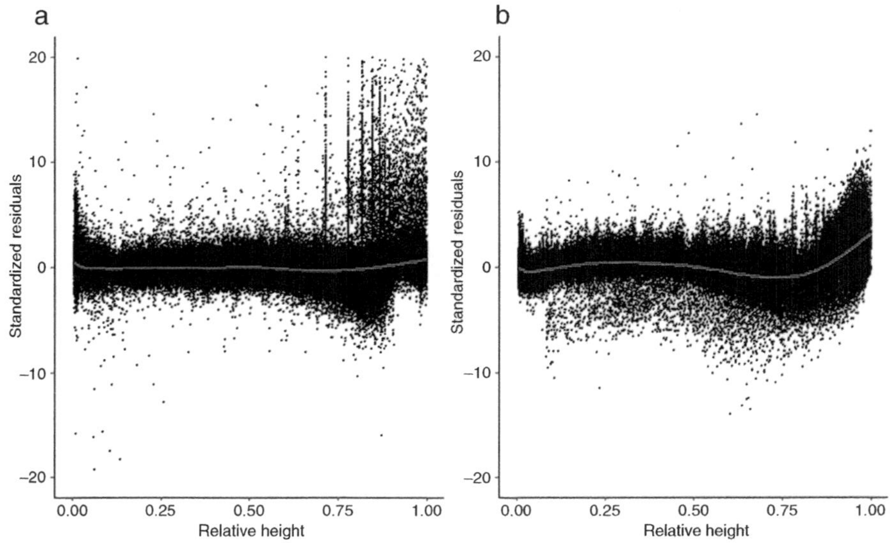

Fig. 2 Standardized residuals for the ring width (**a**, Eq. 1) and ring area (**b**, Eq. 2) plotted against relative height, defined as height of the disk (h) divided by total tree height (H). Smoothing function used to highlight trends in the residuals (blue line)

3.2 Stem Taper

Stem taper models are equations that predict the diameter at any height using the diameter at breast height and total height of the tree. The parameters of these models can be used to evaluate how the form of a tree changes with environmental variables (e.g. Schneider et al. 2018; Sharma and Parton 2009). These models can be classified into two large groups, the first one where the function is segmented into several parts (Max and Burkhart 1976) and the second where the diameter is predicted by one function. Segmented taper functions separate the tree into two or three sections, with a function for each section based on the form of the tree. The second group can be subdivided into three subgroups, whereby (1) the exponent varies with height within the tree in order to adjust to the changing shape of the stem (Kozak 1988), (2) the exponent is height invariant (Ormerod 1973) or (3) the equation is either a polynomial (Bruce et al. 1968) or trigonometric function (Thomas and Parresol 1991). By comparing different taper models, Rojo et al. (2005) and Li and Weiskittel (2010) concluded that variable exponent models performed the best.

A few variable exponent stem taper models are variations of the fixed exponent model proposed by Ormerod (1973):

$$d = D\left(\frac{H - h}{H - \mathrm{BH}}\right)^{\beta} \tag{3}$$

where BH is the breast height (i.e. height where D is measured), d is the stem diameter at height h, D the diameter at breast height, H the total tree height and β the fixed exponent. This model was then modified by Reed and Byrne (1985) in order to account for changes in stem form with tree slenderness, such that $\beta = 1 - \left(\frac{\frac{H}{D} - 30}{120}\right)$. Newberry and Burkhart (1986) further modified the taper equation to include crown characteristics (height to crown base, crown ratio) and site index in the model.

The variable exponent taper equation, whereby the exponent varies with the height within the tree, was first proposed by Kozak (1988) and Newnham (1988). Newnham (1988) isolated β (Eq. 4) and regressed it with measurements of d, h, D and H.

$$\beta = \frac{\ln\left(\frac{H-h}{H-\mathrm{BH}}\right)}{\ln\left(\frac{d}{D}\right)} \tag{4}$$

It was thus found that the exponent varies within a tree, in the same way the form of tree changes from a neiloid at the base of a tree to a paraboloid in the central part and a cone near the apex of a tree. Since then, several variable equations were proposed (e.g. Muhairwe 1999; Sharma and Oderwald 2001).

There are several statistical considerations that need attention when taper equations are fitted. The first of which is the correlation between measurements of a same tree, and the hierarchal nature of the data (e.g. several diameters of the stem, trees sampled within the same plot). These issues can easily be integrated by using a mixed-model framework (Garber and Maguire 2005). Furthermore, the variance is heterogeneous along the stem as consequence of the model formulation. For example, $d = D$ when $h = \mathrm{BH}$ and $d = 0$ when $h = H$. As a consequence, the error in $\widehat{d}$ is very small as h tends towards BH or H. These trends were integrated into a stem taper model (Schneider et al. 2013), with a variance function leading to a prediction error being very small at the top of the tree and at breast height (Fig. 3). This variance function has important consequences when the taper function is used to calculate the volume of a section of the tree. As illustrated in Fortin et al. (2013), the volume of the first log of white spruce tree with a diameter at breast height of 25 cm and a total height of 25 m (0.3–2.3 m) is 98.0 dm^3 with a standard deviation of 5.62 dm^3 and the volume of the second log (2.3–4.3 m) is 72.7 dm^3 with a standard deviation of 10.73 dm^3. The uncertainty of the volume of a section containing D is thus smaller than a section above BH, as the uncertainty in stem diameter around BH is smaller.

3.3 Comparing Allometric Exponents

An allometric relationship relates dimensions X and Y of an individual as:

$$Y = c \cdot X^{\alpha} \tag{5}$$

where c is a constant and α the allometric exponent. The exponent can be interpreted as the ratio of the relative growth of Y over the relative growth of X (Pretzsch et al. 2013). Theoretical values of α for various variables (e.g. D, H, tree volume, crown

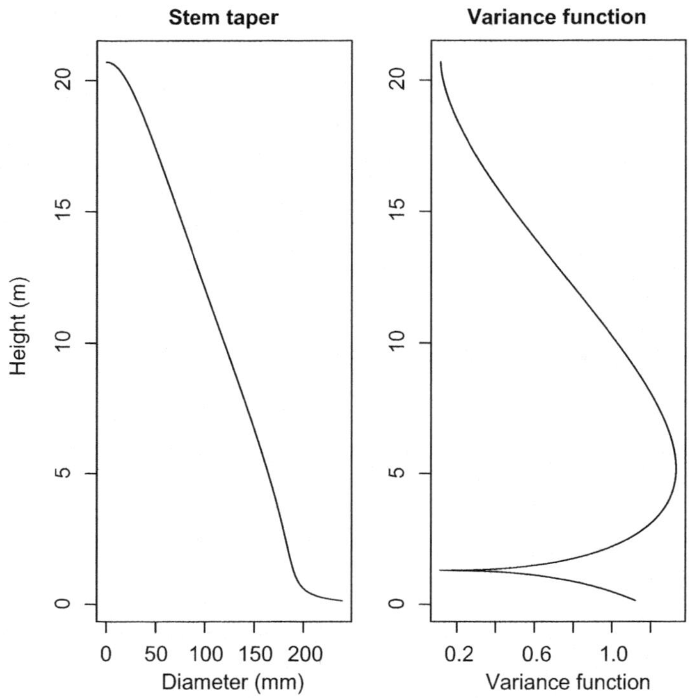

Fig. 3 Predicted stem taper and variance function for the taper equation from Schneider et al. (2013)

diameter, etc.) in the *X* and *Y* positions have been proposed by the metabolic scaling theory (West et al. 1999, 2009). Other studies have shown that divergences exist between the theoretical and observed exponent values (Duursma et al. 2010; Pretzsch et al. 2013; Franceschini and Schneider 2014). The divergences have been partially explained by modifying the paradigms of the metabolic scaling theory, such as varying the fractal dimension of the branching pattern (Smith et al. 2014) or adding climate dependencies such as variations in metabolic activity with temperature (Gillooly et al. 2001).

Even if the allometric exponents differ from theoretical values, they can still be used to gain insight on changes in stem form. For example, Franceschini et al. (2016) found that the *D* to volume and *D* to *H* allometric exponents are both influenced by growing season temperature for six North American species. If both allometric relationships vary in the same manner, then the changes in volume can be attributed to changes in height growth. On the other hand, if the *D* to volume allometric exponent increases more than the *D* to *H* allometric exponent, this would indicate that there are changes in stem form. This is what is observed for four of the six species. This thus implies that tree form changes with climate. This approach however cannot identify precisely what changes explain the observed shifts in tree form.

4 Theoretical Conception of Stem Form

Pressler (1865) suggested that the ring basal area growth of a stem is proportional to the amount of foliage above that point, implying that the growth is constant below the crown base and decreases from crown base to the tree apex. Metzger (1893) on the other hand conceived a tree as a cantilever beam that had to resist bending forces brought on by the wind on the crown. This led to the "*d*3" law whereby the distance from the crown centre is proportional to the diameter to the power of three at that point (distance $\propto d^3$) in order to respect an equal bending stress along the cantilever. Since then, there has been a lot of debate on why tree stems have the form they have. Larson (1963) identified four theories to explain stem form: nutritional, water conductance, hormonal and biomechanical. In the following section, theories explaining stem form based on both hydraulic and biomechanical approaches will be reviewed.

4.1 Hydraulic Approaches

One of the functions of tree stem is to transport metabolites and water between the roots and the foliage, and to a lesser extent store nutrients and metabolites. The stem of a tree was thus proposed to be a collection of pipes connecting root units to foliage units (Shinozaki et al. 1964):

$$A_s = \alpha_s A_f \tag{6}$$

where A_s and A_f are the cross-sectional areas of the pipes in the stem and foliage above the cross-section of the stem, respectfully and α_s the pipe model coefficient.

In order to model tree growth, or obtain the whole cross-sectional area of the tree, the concept of disused pipes is needed, whereby a pipe linking a root unit to leaf unit becomes disconnected when the leaf unit is shed. Several growth models were based on the pipe model theory to predict either stand average tree or cohort-average tree diameter and height growth (Valentine 1988; Osawa et al. 1991; Mäkelä 1997). Some models are more detailed and offer the possibility to simulate stem taper (Mäkelä 2002; Valentine et al. 2012) as well.

Taper is obtained dynamically (Mäkelä 2002). The foliage of each whorl is known by integrating a foliage distribution model (Kershaw and Maguire 2000; Schneider et al. 2011b). The disused pipe area due to foliage shedding can then be obtained by comparing the foliage in an interval along the main stem of year (t) and that of the previous year ($t - 1$). If the foliage increases between t and $t - 1$, A_s increases as per the pipe model. If it decreases, the difference in A_s between the 2 years becomes disused pipes. Validation results presented by Mäkelä (2002) indicate that simulated taper was very close to the observed taper.

Under this framework, stem taper is intimately related to both foliage shedding and changes in the foliage distribution. Several factors that influence foliage distributions have been identified. Both stand (i.e. site fertility and stand density) and tree level (i.e. tree size, age, and crown length) have been used to explain variations in foliage distribution (Mäkelä and Vanninen 2001; Weiskittel et al. 2009; Schneider et al. 2011b; Power et al. 2013; Williams et al. 2017). As illustrated in Schneider et al. (2011a), a top-heavy foliage distribution will lead to a high sapwood area at crown base whereas a bottom-heavy foliage distribution will have less sapwood area, if the total crown foliage biomass is the same. This change in sapwood area taper will lead to differences in stem taper. Nevertheless, this approach does not integrate sapwood to heartwood turnover rates, as sapwood will not immediately change into heartwood when the leaf is shed. The mechanisms underlying the turnover are not well known, and most approaches rely on results obtained from model simulations (e.g. Sievänen et al. 1997).

There has however been much debate on the validity of the pipe model theory (Kershaw and Maguire 2000; Kershaw 2001; Valentine 2001; Sattler and Comeau 2015), if one considers sapwood area equivalent to the used pipe area. There is a lot of evidence on the changes in sapwood conductivity in the radial direction (Domec et al. 2006; Fiora and Cescatti 2008; Chang et al. 2014; Landsberg et al. 2017), and might lead to ways of linking the theoretical pipes of the pipe model theory to sapwood area.

4.2 *Biomechanical Approaches*

Tree stems are also used to support the crown in order to have access to the light needed for photosynthesis. Biomechanics has thus also been used to explain tree form. Metzger's "*d*3" law was the first attempt to explain tree taper based on the representation of a tree as a cantilever with constant stress along the stem (Metzger 1893). Dean and Long (1986) used the same foundations to derive the diameter d of a tree at any point along the stem as given by:

$$d = \left[\frac{32\ W\ A_f\ L}{\pi\sigma}\right]^{\alpha_W} \tag{7}$$

where W is the wind pressure applied to the centre of the crown, L the distance between the point where diameter d is measured and the crown centre and α_W the exponent that varies between 1/3 and 1/2 depending how well the tree is rooted. Based on this model, the taper of a tree depends on the amount of foliage. With trees with more foliage, the diameter at the same distance from the crown centre will be larger in order maintain the structural integrity of the stem. Hence, stem taper should be influenced by the same factors as those that effect crown dimensions, as with the models based on the pipe model theory.

The constant stress cantilever model has however had some criticism (Niklas and Spatz 2000; Slater 2016), even though model simplifications might explain some of the deviations between the model developed by Niklas and Spatz (2000) and the cantilever model (Mattheck 2000). Moreover, the effect of wind on the movement of a tree is complex, where the branches act as a damping mass thus reducing mechanical loads on the stem (James et al. 2006; Eloy 2011). Nevertheless, models based on the "*d*3" law and its variants are able to reproduce observed variations in stem taper (Dean et al. 2002) and other ecological observations such as the self-thinning rule (Dean and Baldwin 1996). On a broader scale, plant biomechanics have been used to explain tree and plant development (e.g. King et al. 2009; Niklas 2013).

5 Empirical Evidences of Changes in Stem Form

5.1 Literature Review

The first studies on stem form were interested in determining the principal factors influencing stem form (see Larson 1963 for review), and very early the relation between crown size and taper was established (Pressler 1865). Trees with large crowns have more taper than trees with shorter crowns, whereby the addition of such variables in taper equations greatly increases their precision (Leites and Robinson 2004). Crown dimensions are not usually measured in national inventories, and thus equations have sought to include proxies for crown dimensions such as stand density and/or basal area. Thus, trees in denser stands will have less taper (Morris and Parker 1992; Muhairwe 1999; Sharma and Zhang 2004; Schneider et al. 2013), as their crowns are shorter (Ducey 2009). Other stand-level variables also influence the tree shape, such as site index (Newberry and Burkhart 1986; Schneider et al. 2013).

Silvicultural interventions can also play a role in determining tree form. Commercial thinning increases the growth of the tree at its base, thus increasing tree taper (Karlsson 2000; Mäkinen and Isomäki 2004). The effect of reducing stand density on stem form is however species dependant, and most probably site dependant. No changes were observed in the form of black spruce trees after commercial thinning, whereas changes were observed for jack pine (Goudiaby et al. 2012). In fact, radial growth of black spruce was constant throughout the stem, and maximum at tree base for jack pine. These observations were explained by the foliage biomass distribution. Finally, fertilization also changes stem taper (Groot et al. 1984; Wiklund et al. 1995; Karlsson 2000; Younger et al. 2008).

The effect of climatic effects on stem form, although studied very early (e.g. Schwarz 1899), has not seen a lot of research in recent years, even if these effects could have very important repercussions on reconstructing long-term dendrochonologies using either dead or subfossilized trees (Autin et al. 2015). Larger changes in tree form with climatic variables were observed for conifers when compared to broadleaved species (Franceschini et al. 2016). Changes in tree shape between ecozones are also observed across large tracks of land

(Klos et al. 2007; Schneider et al. 2013) and have been attributed to differences in climatic conditions.

Studies on the influence of different climatic variables on stem shape have shown that temperature and precipitations can both result in trees with different forms. Jack pine (*Pinus banksiana*) tree form was found to vary with the growing season degree-days (Morris and Forslund 1992). On the other hand, Nigh and Smith (2012) reported that mean annual precipitation and Julian date of the first frost were the best climatic variables to explain taper, where taper increased with higher values of these two variables. They hypothesized that the tree crowns were under the control of these variables, and thus that both climatic variables had an indirect effect on tree shape through changes in crown length. Differences in stand dynamics and wind patterns were also suggested as possible reasons for observing shifts in stem form. Lee et al. (2006) linked changes in the taper in the lower part of Korean red pine (*Pinus densiflora*) to the amount of precipitations late in the growing season whereas shifts in tree form in the upper part of the stem were related to temperature, precipitation and wind during the growing season. Crown dynamics, and their effect on tree biomechanics were used to explain these observations.

The effect of climatic variables varies between species. The variations in stem form with growing season temperature were found to be more important in conifers when compared to broadleaved species (Franceschini et al. 2016). This result was interpreted using differences in primary and secondary growth rates between these clades, as broadleaved species generally have less apical dominance than conifers (Millet 2012). A biomechanical explanation could also be put forth, whereby broadleaved species, due to their architecture, are less prone to wind pressure through the damping effect (James et al. 2006; Eloy 2011). Differences in stem form between species have also been linked to certain functional traits such as shade and waterlogging tolerance (Schneider et al. 2018). Thus, species with higher shade or waterlogging tolerances will decrease their taper, leading to an increase of up to 6% in tree volume when actual and projected climatic variables are compared. It was hypothesised that these observations were explained by hydraulic limitations.

Wind has been found to greatly effect tree form. Guyed trees show lower radial increment at breast height when compared to unguyed trees (Moore et al. 2014). In fact, trees under less wind stress demonstrate a shift in their allocation patterns (Meng et al. 2006), as guyed trees invest more in height growth than radial growth. In other words, radial growth at given point in the stem is closely related to the bending moment above that point (Meng et al. 2006). Gardiner et al. (2016) presents a review on the effects of wind on plant growth and adaptation through a biomechanical perspective.

5.2 Untangling the Effect of Wind and Climate on Stem Form

It is however difficult to untangle the confounding effects of wind and other climatic variables in order to evaluate which effect, wind or climate, is the most important in determining tree form. A reduced form of the taper equation applied by Schneider et al. (2018) is thus used (Eq. 8):

$$d_{ijk}^2 = \frac{1}{1 + e^{(\alpha + b_{i,1} + b_{ij,1})}} \times \mathrm{DBH}_{ij}^2 \times \left(\frac{HT_{ij} - h_{ijk}}{HT_{ij} - 1.3}\right) \times \left(\frac{h_{ijk}}{1.3}\right)^{2 - (\boldsymbol{x}_{ijk}\boldsymbol{\beta} + b_{i,2} + b_{ij,2})} + \varepsilon_{ijk}$$

$$\text{for } i = 1, 2, \ldots, q; j = 1, 2, \ldots, n_i \text{ and } k = 1, 2, \ldots, o_{ij}. \tag{8}$$

where:

- $\boldsymbol{\alpha}$ and $\boldsymbol{\beta}$ are a fixed effect or a vector of fixed effects parameters estimated by the regression, respectively.
- q, n and o_{ij} the total number of plots, the total number of trees and the number of disks in the tree i of plot j, respectively.
- i, j and k the plot, tree within plot and disk within tree within plot indices, respectively, such that
- $b_{i,1}$ and $b_{i,2}$ plot random effects where:

$$\boldsymbol{b}_i = (b_{i,1}, b_{i,2})^T \sim N_2(\boldsymbol{0}, \boldsymbol{G}_{plot})$$

$$\boldsymbol{G}_{\text{plot}} = \begin{bmatrix} \sigma_{\text{plot},1}^2 & \rho_{\text{plot}}\sigma_{\text{plot},1}\sigma_{\text{plot},2} \\ \rho_{\text{plot}}\sigma_{\text{plot},1}\sigma_{\text{plot},2} & \sigma_{\text{plot},2}^2 \end{bmatrix}$$

- $b_{ij,1}$ and $b_{ij,2}$ tree random effects where:

$$\boldsymbol{b}_{ij} = (b_{ij,1}, b_{ij,2})^T \sim N_2(\boldsymbol{0}, \boldsymbol{G}_{tree})$$

$$\boldsymbol{G}_{\text{tree}} = \begin{bmatrix} \sigma_{\text{tree},1}^2 & \rho_{\text{tree}}\sigma_{\text{tree},1}\sigma_{\text{tree},2} \\ \rho_{\text{tree}}\sigma_{\text{tree},1}\sigma_{\text{tree},2} & \sigma_{\text{tree},2}^2 \end{bmatrix}$$

- ε_{ijk} the residual error, where

$$\varepsilon_{ij} = (\varepsilon_{ij1}, \varepsilon_{ij2}, \ldots, \varepsilon_{ijk})^T \sim N_k\left(\boldsymbol{0}, \boldsymbol{\Psi}_{ij}^{1/2}\boldsymbol{\Lambda}_{ij}\boldsymbol{\Psi}_{ij}^{1/2}\right)$$

which depends on a correlation ($\boldsymbol{\Psi}$) and variance ($\boldsymbol{\Lambda}$) matrix. The complete description of the model is given in Schneider et al. (2013, 2018) and Fortin et al. (2013). The fixed effects of the model are decomposed into dendrometric, climatic and wind components (Eq. 9):

$$\boldsymbol{x}_{ijk}\boldsymbol{\beta} = \beta_0 + \beta_1\frac{h}{H} + \beta_2\log\left(\frac{h}{H}\right) + \boldsymbol{x}_{ijk}^{\text{dendro}}\boldsymbol{\beta}^{\text{dendro}} + \boldsymbol{x}_{ij}^{\text{climate}}\boldsymbol{\beta}^{\text{climate}} + \boldsymbol{x}_i^{\text{wind}}\boldsymbol{\beta}^{\text{wind}} \tag{9}$$

Wind metrics (mean wind speed and its standard deviation at an altitude of 30 m) for each of the 656 plots illustrated in Fig. 1 were obtained from the Canadian Wind

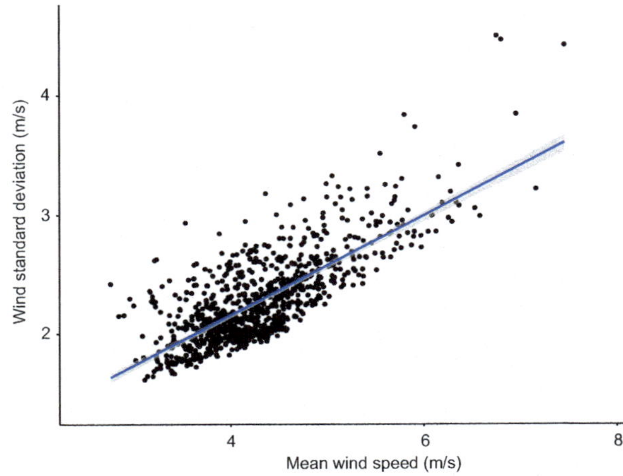

Fig. 4 Mean wind speed standard deviation versus mean wind speed for the 656 plots in Fig. 1

Atlas (Environment and Climate Change Canada, Government of Canada 2003) based on the wind model by Yu et al. (2006). There was however a very important correlation between mean wind speed and mean wind speed standard deviation ($R = 0.97$, Fig. 4), and thus the former was retained for the analysis (W: mean wind speed (m/s) in $\boldsymbol{x}_i^{\text{wind}}\boldsymbol{\beta}^{\text{wind}}$ term of Eq. 9). Furthermore, the 30-year prior to felling monthly average climatic variables for each plot were spatially interpolated with the BIOSIM software using the altitude and spatial coordinates of each plot (Régnière et al. 1995; Régnière and Saint-Amant 2008). Of all the potential variables available (i.e. mean winter, spring, summer and fall temperatures and total precipitations), only mean winter (twin, °C) and summer (tsum, °C) temperatures as well as total summer precipitations (psum, mm) as well as altitude (alt, m) were used in the $\boldsymbol{x}_{ij}^{\text{climate}}\boldsymbol{\beta}^{\text{climate}}$ term of Eq. (9), as these were the ones that were found to be statistically significant in predicting balsam fir stem taper (Schneider et al. 2018). Finally, D (mm), stand density (SD, stems/ha) and stand basal area (BA, m^2/ha) were kept in the list of potential candidate variables in the $\boldsymbol{x}_{ijk}^{\text{dendro}}\boldsymbol{\beta}^{\text{dendro}}$ term of Eq. (9).

Every combination using either all or some of the dendrometric, climatic and/or wind variables were tested, yielding 255 different models (i.e. from 8 models with only 1 covariate added to 1 model with all the covariates added). The three best models, with their AIC (Akaike information criterion), ΔAIC ($\Delta AIC_i = AIC_i - \min(AIC)$ for model i (Burnham and Anderson 2002)) and AIC weights ($AICwt_i = \exp(-0.5\ \Delta AIC_i)/\Sigma \exp(-0.5\ \Delta AIC_i)$ (Burnham and Anderson 2002)) are given in Table 2.

The AIC weights of each model can then be used to evaluate the relative importance of each variable (Burnham and Anderson 2002). The relative importance of a variable is equal to the sum of the AICwt over all the models where the variable in question is present. This method can thus evaluate the importance of each variable, even if it is not present in the best model (i.e. in the model with the lowest AIC). Diameter at breast height has the highest relative importance, very closely

Table 2 Variable exponent specifications, AIC, ΔAIC and AIC weights for the three best models for stem taper (Eq. 8)

Model	AIC	ΔAIC	AIC weights
1. $\boldsymbol{x}_{ijk}\boldsymbol{\beta} = \beta_0 + \beta_1\frac{h}{H} + \beta_2\log\left(\frac{h}{H}\right) + \beta_3 D + \beta_4\text{alt} + \beta_5\text{BA} + \beta_6\text{twin} + \beta_7\text{psum} + \beta_8 W$	626,303.9	0	0.227
2. $\boldsymbol{x}_{ijk}\boldsymbol{\beta} = \beta_0 + \beta_1\frac{h}{H} + \beta_2\log\left(\frac{h}{H}\right) + \beta_3 D + \beta_5\text{BA} + \beta_6\text{twin} + \beta_7\text{psum} + \beta_8 W + \beta_9\text{tsum}$	626,304.8	0.913	0.144
3. $\boldsymbol{x}_{ijk}\boldsymbol{\beta} = \beta_0 + \beta_1\frac{h}{H} + \beta_2\log\left(\frac{h}{H}\right) + \beta_3 D + \beta_4\text{alt} + \beta_{10}\text{SD} + \beta_5\text{BA} + \beta_6\text{twin} + \beta_7\text{psum} + \beta_{10} W$	626,304.9	1.038	0.135

h height within the tree, *H* total tree height, *D* diameter at breast height, *alt* altitude, *BA* stand basal area, *twin* mean winter temperatures, *psum* summer precipitations, *W* average wind speed, *tsum* mean summer precipitations, *SD* stand density

Table 3 Relative importance of each variable used in Eq. (9)

Variable	Category	Relative importance
Diameter at breast height	Dendrometric	1.000
Total summer precipitation	Climatic	0.999
Mean winter temperature	Climatic	0.999
Stand basal area	Dendrometric	0.973
Average wind speed	Wind	0.960
Altitude	Climatic	0.573
Mean summer temperature	Climatic	0.420
Stand density	Dendrometric	0.378

Table 4 Parameter estimates (standard errors in parentheses) for the climatic and wind variables of the three best models presented in Table 1

Parameter	Variable	Model 1	Model 2	Model 3
β_6	Mean winter temperature (twin)	0.0036 (0.0008)	0.0036 (0.0009)	0.0037 (0.0008)
β_7	Total summer precipitations (psum)	−0.0003 (0.0001)	−0.0002 (0.0001)	−0.0003 (0.0001)
β_7	Mean wind speed (W)	0.0026 (0.0009)	0.0027 (0.0009)	0.0026 (0.0009)
β_8	Mean summer temperature (tsum)		−0.0032 (0.0018)	

followed by total summer precipitations and mean winter temperature (Table 3). Average wind speed is ranked 5th, with still a very high relative importance. There is an important drop in relative importance after the average wind speed. This thus seems to indicate that even though wind stress is important, other physiological processes contribute slightly more to determining the shape of a tree, at the very least in a temperate to boreal climate.

The parameter estimates of the best three models based on AIC for mean winter temperature and mean wind speed are positive. The opposite is observed for the parameter estimates for total summer precipitations and mean summer temperatures, which are negative (Table 4). Parameter values are directly related to the shape of the stem (Schneider et al. 2013), with positive values indicating that the taper increases with the value of the parameter, and inversely for negative values. This implies that trees will have more taper with heavier wind loads and higher winter temperatures and will have less taper with higher summer precipitations and summer temperatures. The effect of wind follows previous observations, as presented earlier. However, the changes with the temperatures and precipitations are harder to interpret through a biomechanical approach. Stand density and/or basal area are also included in each model, and thus different stand dynamics are already accounted for, at least partially.

Trees are very resistant to cavitation (Cochard and Delzon 2013; Brown 2013). Nevertheless, they still have to ensure that cavitation is reduced in order to maintain sap conduction. Cavitation increases in hot and/or dry conditions (Rennenberg et al. 2006) or in cold conditions (Arias et al. 2017). If cavitation is important enough, it can lead to the death of the tree (Allen et al. 2010). Thus, trees must adapt themselves to reduce the effects of cavitation. The changes in stem form with the various climatic variables studied might be explained by such adaptations. For example, if a tree is growing in an area with cold winters, it might shift its radial growth distribution towards the top of the tree, in order to have more sapwood. This safety factor would thus produce trees with less taper. In the same manner, trees with a wetter climate will not have the need to produce as much conducting area as cavitation due to water stress is less likely. The trees would thus have more taper. These explanations remain hypothetical and would need to be further validated.

6 Conclusion

The debate has been ongoing in order to evaluate if tree shape is determined by either mechanical or hydraulic constraints. The stem of a tree has to ensure mechanical stability in order for the tree to support the crown all the while maintaining hydraulic conductivity. In conifers, the tracheid cells in the stem perform both functions whereas broadleaved species are composed of fibres and vessels, the prior contributing to biomechanical stability and the latter to sap conductivity. Some studies have found that hydraulic constraints are more important, while others reported that biomechanical theories better explain tree growth and form (Farnsworth and Gardingen 1995; Domec and Gartner 2002; Eloy et al. 2017).

Others have tried to propose optimization approaches between the different functions of a tree (Farnsworth and Niklas 1995), relying on the underlying objective that a tree optimizes its volume occupancy (West et al. 1999, 2009). Nevertheless, these approaches oppose hydraulic to biomechanical functions. These functions are however intimately related because often the same cells carry both functions out in conifers or both types of cells (i.e. conducting and support) are present in broadleaved species. Thus, optimization processes opposing hydraulic to biomechanical might not be necessary, since if one function is secured, the other should also be ensured.

Furthermore, the results presented here indicate that climatic variables are slightly more important in determining tree form, when compared to average wind speed. It could be that variations in wind are more important than wind speed per se. However, these two methods of quantifying wind could not be disentangled with the data on hand. Nevertheless, the results indicate that other processes must also be considered in developing theories of stem form. For hydraulic theories, there is a need to better understand the relationships between sapwood and used pipe area, or revisit the pipe model theory in order to better relate the state variables with measurable metrics found in a tree. In biomechanical approaches, some physiological considerations must be included, such as adaptations to cavitation and their implications on stem wood production. In other words, instead of opposing conductivity versus mechanical stability, theoretical models should hybridize with some types of acclimations, such as those proposed by Mäkelä (1999).

Acknowledgements I would like to thank Tony Franceschini for help finding some of the references and comments on an early draft of the manuscript. I would also like to thank Mélanie Desrochers of the Center for Forest Research (CEF-CFR) for the preparation of the map illustrated in Fig. 1. Finally, I would like to also thank the Forest Inventory Branch (Direction des inventaires forestiers) and the Forest Research Branch (Direction de la recherche forestière) of the Quebec Ministry of Forests, Wildlife and Parks for the stem analysis database.

References

Allen CD, Macalady AK, Chenchouni H et al (2010) A global overview of drought and heat-induced tree mortality reveals emerging climate change risks for forests. For Ecol Manag 259:660–684

Arias NS, Scholz FG, Goldstein G, Bucci SJ (2017) The cost of avoiding freezing in stems: trade-off between xylem resistance to cavitation and supercooling capacity in woody plants. Tree Physiol 37:1251–1262

Autin J, Gennaretti F, Arseneault D, Bégin Y (2015) Biases in RCS tree ring chronologies due to sampling heights of trees. Dendrochronologia 36:13–22

Brown HR (2013) The theory of the rise of sap in trees: some historical and conceptual remarks. Phys Perspect 15:320–358

Bruce D, Curtis RO, Vancoevering C (1968) Development of a system of taper and volume tables for red alder. For Sci 14:339–350

Burnham KP, Anderson DR (2002) Model selection and multimodel inference: a practical information-theoretic approach. Springer, New York

Chang X, Zhao W, He Z (2014) Radial pattern of sap flow and response to microclimate and soil moisture in Qinghai spruce (Picea crassifolia) in the upper Heihe River basin of arid northwestern China. Agric For Meteorol 187:14–21

Cochard H, Delzon S (2013) Hydraulic failure and repair are not routine in trees. Ann For Sci 70:659–661

Cortini F, Groot A, Filipescu CN (2013) Models of the longitudinal distribution of ring area as a function of tree and stand attributes for four major Canadian conifers. Ann For Sci 70:637–648

Courbet F (1999) A three-segmented model for the vertical distribution of annual ring area: application to *Cedrus atlantica* Manetti. For Ecol Manag 119:177–194

Courbet F, Houllier F (2002) Modelling the profile and internal structure of tree stem. Application to Cedrus atlantica (Manetti). Ann For Sci 59:63–80

Dean TJ, Baldwin VC (1996) The relationship between Reineke's stand-density index and physical stem mechanics. For Ecol Manag 81:25–34

Dean TJ, Long JN (1986) Validity of constant-stress and elastic-instability principles of stem formation in Pinus contorta and Trifolium pratense. Ann Bot 58:833–840

Dean TJ, Roberts S, Gilmore D et al (2002) An evaluation of the uniform stress hypothesis based on stem geometry in selected North American conifers. Trees 16:559–568

Deleuze C, Houllier F (2002) A flexible radial increment taper equation derived from a process-based carbon partitioning model. Ann For Sci 59:141–154

Domec JC, Gartner BL (2002) Age-and position-related changes in hydraulic versus mechanical dysfunction of xylem: inferring the design criteria for Douglas-fir wood structure. Tree Physiol 22:91–104

Domec J-C, Lachenbruch B, Meinzer FC (2006) Bordered pit structure and function determine spatial patterns of air-seeding thresholds in xylem of Douglas-fir (Pseudotsuga menziesii; Pinaceae) trees. Am J Bot 93:1588–1600

Ducey MJ (2009) Predicting crown size and shape from simple stand variables. J Sustain For 28:5–21

Duursma RA, Mäkelä A, Reid DEB et al (2010) Self-shading affects allometric scaling in trees. Funct Ecol 24:723–730

Eloy C (2011) Leonardo's rule, self-similarity, and wind-induced stresses in trees. Phys Rev Lett 107:258101

Eloy C, Fournier M, Lacointe A, Moulia B (2017) Wind loads and competition for light sculpt trees into self-similar structures. Nat Commun 8:1014

Environment and Climate Change Canada, Government of Canada (2003) Canadian Wind Energy Atlas. http://www.windatlas.ca/index-en.php. Accessed 28 Jan 2018

Farnsworth KD, Gardingen PR (1995) Allometric analysis of Sitka spruce branches: mechanical versus hydraulic design principles. Trees 10:1–12

Farnsworth KD, Niklas KJ (1995) Theories of optimization, form and function in branching architecture in plants. Funct Ecol 9:355–363

Fiora A, Cescatti A (2008) Vertical foliage distribution determines the radial pattern of sap flux density in Picea abies. Tree Physiol 28:1317–1323

Fortin M, Schneider R, Saucier J-P (2013) Volume and error variance estimation using integrated stem taper models. For Sci 59:345–358

Franceschini T, Schneider R (2014) Influence of shade tolerance and development stage on the allometry of ten temperate tree species. Oecologia 176:739–749

Franceschini T, Martin-Ducup O, Schneider R (2016) Allometric exponents as a tool to study the influence of climate on the trade-off between primary and secondary growth in major north-eastern American tree species. Ann Bot 117:551–563

Garber SM, Maguire DA (2005) Vertical trends in maximum branch diameter in two mixed-species spacing trials in the central Oregon Cascades. Can J For Res 35:295–307

Gardiner B, Berry P, Moulia B (2016) Wind impacts on plant growth, mechanics and damage. Plant Sci 245:94–118

Gillooly JF, Brown JH, West GB et al (2001) Effects of size and temperature on metabolic rate. Science 293:2248–2251

Goudiaby V, Brais S, Berninger F, Schneider R (2012) Vertical patterns in specific volume increment along stems of dominant jack pine (Pinus banksiana) and black spruce (Picea mariana) after thinning. Can J For Res 42:733–748

Groot A, Brown KM, Morrison IK, Barker JE (1984) A 10-year tree and stand response of jack pine to urea fertilization and low thinning. Can J For Res 14:44–50

Huang J, Tardif JC, Bergeron Y et al (2010) Radial growth response of four dominant boreal tree species to climate along a latitudinal gradient in the eastern Canadian boreal forest. Glob Change Biol 16:711–731

Immanen J, Nieminen K, Smolander O-P et al (2016) Cytokinin and auxin display distinct but interconnected distribution and signaling profiles to stimulate cambial activity. Curr Biol 26:1990–1997

James KR, Haritos N, Ades PK (2006) Mechanical stability of trees under dynamic loads. Am J Bot 93:1522–1530

Karlsson K (2000) Stem form and taper changes after thinning and nitrogen fertilization in Picea abies and Pinus sylvestris stands. Scand J For Res 15:621–632

Kershaw JA Jr (2001) Influence of vertical foliage structure on the distribution of stem cross-sectional area increment in western hemlock and balsam fir [For Sci 46(1): 86–94]. For Sci 47:117–118

Kershaw JA, Maguire DA (2000) Influence of vertical foliage structure on the distribution of stem cross-sectional area increment in western hemlock and balsam fir. For Sci 46:86–94

King DA, Davies SJ, Tan S, Nur Supardi MN (2009) Trees approach gravitational limits to height in tall lowland forests of Malaysia. Funct Ecol 23:284–291

Klos RJ, Wang GG, Dang Q-L, East EW (2007) Taper equations for five major commercial tree species in Manitoba, Canada. West J Appl For 22:163–170

Kozak A (1988) A variable-exponent taper equation. Can J For Res 18:1363–1368
Lambert MC, Ung CH, Raulier F (2005) Canadian national tree aboveground biomass equations. Can J For Res 35:1996–2018
Landsberg J, Waring R, Ryan M (2017) Water relations in tree physiology: where to from here? Tree Physiol 37:18–32
Larson PR (1963) Stem form development of forest trees. For Sci 9:a0001–a0042
Lee W-K, Biging GS, Son Y et al (2006) Geostatistical analysis of regional differences in stem taper form of Pinus densiflora in central Korea. Ecol Res 21:513–525
Leites LP, Robinson AP (2004) Improving taper equations of loblolly pine with crown dimensions in a mixed-effects modeling framework. For Sci 50:204–212
Li R, Weiskittel AR (2010) Comparison of model forms for estimating stem taper and volume in the primary conifer species of the North American Acadian Region. Ann For Sci 67:302
Mäkelä A (1997) A carbon balance model of growth and self-pruning in trees based on structural relationships. For Sci 43:7–24
Mäkelä A (1999) Acclimation in dynamic models based on structural relationships. Funct Ecol 13:145–156
Mäkelä A (2002) Derivation of stem taper from the pipe theory in a carbon balance framework. Tree Physiol 22:891–905
Mäkelä A, Vanninen P (2001) Vertical structure of Scots pine crowns in different age and size classes. Trees 15:385–392
Mäkinen H, Isomäki A (2004) Thinning intensity and long-term changes in increment and stem form of Norway spruce trees. For Ecol Manag 201:295–309
Mattheck C (2000) Comments on "Wind-induced stresses in cherry trees: evidence against the hypothesis of constant stress levels" by K.J. Niklas, H.-C. Spatz, Trees (2000) 14:230–237. Trees 15:63–63
Max TA, Burkhart HE (1976) Segmented polynomial regression applied to taper equations. For Sci 22:283–289
Meng SX, Lieffers VJ, Reid DEB et al (2006) Reducing stem bending increases the height growth of tall pines. J Exp Bot 57:3175–3182
Metzger K (1893) Der Wind als massgebender Faktor für das Wachstum der Bäume. Mündener Forstl Hefte 3:35–86
Millet J (2012) L'architecture des arbres des régions tempérées. Multimondes, Montreal
Moore JR, Cown DJ, Lee JR et al (2014) The influence of stem guying on radial growth, stem form and internal resin features in radiata pine. Trees 28:1197–1207
Morris DM, Forslund RR (1992) The relative importance of competition, microsite, and climate in controlling the stem taper and profile shape in jack pine. Can J For Res 22:1999–2003
Morris DM, Parker WH (1992) Variable-quality form in mature jack pine stands: quantification and relationship with environmental factors. Can J For Res 22:279–289
Muhairwe CK (1999) Taper equations for Eucalyptus pilularis and Eucalyptus grandis for the north coast in New South Wales, Australia. For Ecol Manag 113:251–269
Newberry JD, Burkhart HE (1986) Variable-form stem profile models for loblolly pine. Can J For Res 16:109–114
Newnham RM (1988) A variable-form taper equation. Information report PI-X-83, Petawawa National Forest Institute
Nigh G, Smith W (2012) Effect of climate on lodgepole pine stem taper in British Columbia, Canada. Forestry 85:579–587
Niklas KJ (2013) Biophysical and size-dependent perspectives on plant evolution. J Exp Bot 64:4817–4827
Niklas KJ, Spatz H-C (2000) Wind-induced stresses in cherry trees: evidence against the hypothesis of constant stress levels. Trees 14:230–237
Nilsson J, Karlberg A, Antti H et al (2008) Dissecting the molecular basis of the regulation of wood formation by auxin in hybrid aspen. Plant Cell 20:843–855
Ormerod DW (1973) A simple bole model. For Chron 49:136–138

Osawa A, Ishizuka M, Kanazawa Y (1991) A profile theory of tree growth. For Ecol Manag 41:33–63
Power H, Schneider R, Berninger F (2013) Understanding changes in black (Picea mariana) and white spruce (Picea glauca) foliage biomass and leaf area characteristics. Trees 28:345–357
Pressler MR (1865) Gesetz der Stammbildung und dessen forstwirthschaftliche Bedeutung insbesondere für den Waldbau höchsten reinertrago. Arnold Verlag, Leipzig
Pretzsch H, Dauber E, Biber P (2013) Species-specific and ontogeny-related stem allometry of European forest trees: evidence from extensive stem analyses. For Sci 59:290–302
Reed DD, Byrne JC (1985) A simple, variable form volume estimation system. For Chron 61:87–90
Régnière J, Saint-Amant R (2008) BioSIM 9: manuel de l'utilisateur. Centre de foresterie des Laurentides
Régnière J, Cooke BJ, Bergeron V (1995) BioSIM: un instrument informatique d'aide à la décision pour la planification saisonnière de la lutte antiparasitaire. Guide d'utilisation
Rennenberg H, Loreto F, Polle A et al (2006) Physiological responses of forest trees to heat and drought. Plant Biol 8:556–571
Robitaille A, Saucier J-P, Chabot M et al (2015) An approach for assessing suitability for forest management based on constraints of the physical environment at a regional scale. Can J For Res 45:529–539
Rojo A, Perales X, Sánchez-Rodríguez F et al (2005) Stem taper functions for maritime pine (Pinus pinaster Ait.) in Galicia (Northwestern Spain). Eur J For Res 124:177–186
Sattler DF, Comeau PG (2015) Crown allometry and application of the pipe model theory to white spruce (Picea glauca (Moench) Voss) and aspen (Populus tremuloides Michx.) in the western boreal forest of Canada. Can J For Res 46:262–273
Schneider R, Berninger F, Ung C-H et al (2011a) Within crown variation in the relationship between foliage biomass and sapwood area in jack pine. Tree Physiol 31:22–29
Schneider R, Fortin M, Berninger F et al (2011b) Modeling Jack pine (Pinus banksiana) foliage density distribution. For Sci 57:180–188
Schneider R, Fortin M, Saucier J-P (2013) Équations de défilement en forêt naturelle pour les principales essences commerciales du Québec. Mémoire de Recherche Forestière No. 168, Ministère des Ressources Naturelles, Direction de la Recherche Forestière, 34 pp
Schneider R, Franceschini T, Fortin M, Saucier J-P (2018) Climate-induced changes in the stem form of 5 North American tree species. For Ecol Manag (in press). https://www.sciencedirect.com/science/article/pii/S0378112717315876
Schwarz F (1899) Physiologische Untersuchungen über Dickenwachstum und Holzqualität von Pinus silvestris. P. Parey, Berlin
Sharma M, Oderwald RG (2001) Dimensionally compatible volume and taper equations. Can J For Res 31:797–803
Sharma M, Parton J (2009) Modeling stand density effects on taper for jack pine and black spruce plantations using dimensional analysis. For Sci 55:268–282
Sharma M, Zhang SY (2004) Variable-exponent taper equations for jack pine, black spruce, and balsam fir in eastern Canada. For Ecol Manag 198:39–53
Shinozaki K, Kyoji Y, Hozumi K, Kira T (1964) A quantitative analysis of plant form – pipe model theory I. Basic analyses. Jpn J Ecol 14:97–105
Sievänen R, Nikinmaa E, Perttunen J (1997) Evaluation of importance of sapwood senescence on tree growth using the model Lignum. Silva Fenn 31:329–340
Slater D (2016) An argument against the axiom of uniform stress being applicable to trees. Arboric J 38:143–164
Smith DD, Sperry JS, Enquist BJ et al (2014) Deviation from symmetrically self-similar branching in trees predicts altered hydraulics, mechanics, light interception and metabolic scaling. New Phytol 201:217–229
Temesgen H, Affleck D, Poudel K et al (2015) A review of the challenges and opportunities in estimating above ground forest biomass using tree-level models. Scand J For Res 30:326–335
Thomas CE, Parresol BR (1991) Simple, flexible, trigonometric taper equations. Can J For Res 21:1132–1137

Valentine HT (1988) A carbon-balance model of stand growth: a derivation employing pipe-model theory and the self-thinning rule. Ann Bot 62:389–396

Valentine HT (2001) Influence of vertical foliage structure on the distribution of stem cross-sectional area increment in western hemlock and balsam fir. For Sci 47:115–116

Valentine HT, Mäkelä A, Green EJ et al (2012) Models relating stem growth to crown length dynamics: application to loblolly pine and Norway spruce. Trees 26:469–478

Weiskittel AR, Kershaw JA Jr, Hofmeyer PV, Seymour RS (2009) Species differences in total and vertical distribution of branch- and tree-level leaf area for the five primary conifer species in Maine, USA. For Ecol Manag 258:1695–1703

Weiskittel AR, MacFarlane DW, Radtke PJ et al (2015) A call to improve methods for estimating tree biomass for regional and national assessments. J For 113:414–424

West GB, Brown JH, Enquist BJ (1999) A general model for the structure and allometry of plant vascular systems. Nature 400:664–667

West GB, Enquist BJ, Brown JH (2009) A general quantitative theory of forest structure and dynamics. Proc Natl Acad Sci 106:7040–7045

Wiklund K, Konöpka B, Nilsson L-O (1995) Stem form and growth in Picea abies (L.) karst, in response to water and mineral nutrient availability. Scand J For Res 10:326–332

Williams GM, Nelson AS, Affleck DLR (2017) Vertical distribution of foliar biomass in western larch (Larix occidentalis). Can J For Res 48:42–57

Younger NL, Temesgen H, Garber SM (2008) Taper and volume responses of Douglas-fir to sulfur treatments for control of Swiss needle cast in the Coast Range of Oregon. West J Appl For 23:142–148

Yu W, Benoit R, Girard C et al (2006) Wind Energy Simulation Toolkit (WEST): a wind mapping system for use by the wind-energy industry. Wind Eng 30:15–33

Zhao Y (2010) Auxin biosynthesis and its role in plant development. Annu Rev Plant Biol 61:49–64

Northern Forest Trees Under Increasing Atmospheric Humidity

Elina Oksanen, Jenna Lihavainen, Markku Keinänen, Sarita Keski-Saari, Sari Kontunen-Soppela, Arne Sellin, and Anu Sõber

Contents

Abstract Several climate change scenarios predict increasing precipitation for northern latitudes and several other regions in the globe, leading to increase in atmospheric water vapour content [expressed as increased relative humidity (RH) or as decreased water vapour pressure deficit (VPD)] and environmental wetness. Plants are known to be sensitive to high humidity (low VPD) as indicated by changes in stomatal function, transpiration, mineral nutrient uptake, growth, development, sugar metabolism and leaf epicuticular wax composition in several species and studies. To understand the impact of increasing humidity (lower VPD) on forest ecosystems, a long-term and large-scale field experiment (FAHM, Free Air

E. Oksanen (✉), J. Lihavainen, M. Keinänen, S. Keski-Saari, and S. Kontunen-Soppela
Department of Environmental and Biological Sciences, University of Eastern Finland, Joensuu, Finland
e-mail: elina.oksanen@uef.fi

A. Sellin and A. Sõber
Institute of Ecology and Earth Sciences, University of Tartu, Tartu, Estonia

Progress in Botany (2019) 80: 317–336, DOI 10.1007/124_2017_15,
Published online: 30 January 2018

Humidity Manipulation) is conducted in Estonia with silver birch and hybrid aspen, representing widespread deciduous species in northern Europe. The experiment revealed that high humidity is an important climatic factor, causing reduction in growth rate of the aboveground parts, leaf biomass and area, bud size, sap flux and earlier bud break and delayed leaf fall. In the belowground parts, root biomass (fine-root biomass in particular) was increased although the root compartment was exposed to anaerobic conditions, increased soil pH and soil water potential, with lowered soil respiration and altered microbial and fungal communities. In the leaves, high humidity shifted the metabolism towards nonstructural carbohydrates, antioxidants and phenolic compounds, while nutrient (N and P) content, photosynthesis, dark respiration, hydraulic conductance and the density of glandular trichomes were reduced. The change in the chemical composition of the surface wax layer resulted in lower hydrophobicity, exposing the leaves for fungal pathogen attacks. In the stem, N and P content in wood and number of living parenchyma cells were increased, while the stem wood density was reduced and wood chemistry altered. Reduction in the growth rates under high humidity was more pronounced in hybrid aspen than in silver birch. Birch showed more efficient adjustment of leaf and root morphology, hydraulic architecture, primary carbon and nitrogen metabolism and leaf surface properties mitigating the negative impacts of high humidity. At the ecosystem level, high humidity altered soil processes (moisture content, microbiota, nitrification), with potential to affect competition and species composition. A complementary laboratory study with birch demonstrated that additional N supply can counteract the effects of low VPD on cellular metabolism and leaf surface wax quality.

Keywords Aspen, Birch, Forest ecosystem, Growth, Leaf epicuticular waxes, Metabolism, Nitrogen, Nutrients, Relative humidity (RH), Sap flux, Soil processes, Trichomes, Vapour pressure deficit (VPD)

1 Introduction

Several climate change scenarios predict increasing precipitation at the northern latitudes leading to increase in atmospheric water vapour content and wetter conditions (IPCC 2013). The predicted rise in annual precipitation in northern Europe varies from 4 to 34% for the time period of 2071–2100, depending on the scenario (Jacob et al. 2014). In addition to soil water availability, plants are affected by the relative humidity of the surrounding air (often measured as relative humidity, RH, or as water vapour pressure deficit, VPD). Dai (2006) showed a slightly positive trend in relative humidity (RH) in the northern hemisphere over the recent decades (1976–2004). Besides the northern European area, an increase in RH has been reported over oceans and regionally in Africa, the Middle East, India and Asia (China), whereas decreasing trends were reported in the low and midlatitude areas (central Europe, the USA, South America) (Simmons et al. 2010; IPCC 2013). The regional and temporal differences in changes can be large both in magnitude and direction, as indicated, e.g. by Fatichi et al. (2015), reporting a declining trend in RH from January to June but an increasing trend in RH from July to August over the recent decades 1981–2011 in central Europe (Switzerland). The increase in RH is closely associated with increasing precipitation and cloud cover, with decreasing atmospheric pressure (Boé and Terray 2014; Fatichi et al. 2015) and the whole hydrological cycle (Willett et al. 2007). Atmospheric water vapour has an important role as a greenhouse gas absorbing solar heat and holding it in the troposphere, with a positive feedback on the global warming (Dessler et al. 2013). VPD is closely related to plant productivity; VPD explains more of the variation in the primary production of vegetation than soil temperature, air temperature or soil moisture does (Eamus et al. 2013; Fu and Shen 2016). Therefore, air humidity is an important climatic variable that should be taken into account when estimating the climate change impacts on plant production (Zhang et al. 2017a).

Info Box

There are several definitions and units describing the atmospheric water vapour content, i.e. air humidity. *Relative humidity (RH, %)* and *vapour pressure deficit (VPD, Pa)* are the most commonly used definitions. *RH* describes the actual water vapour content relative to maximum water vapour content of air at a certain temperature. Relative humidity of 100% means that the maximum water vapour holding capacity of air is reached, and at this saturation point (also called a dew point), water condensates onto solid surfaces or into fog and clouds and precipitates. Humidity can be expressed as *VPD*, describing the difference between the actual water vapour pressure and the maximum vapour pressure at the saturation point at a certain temperature. Increasing RH decreases VPD and atmospheric evaporative demand, which drives water evaporation from soil and plants, i.e. evapotranspiration. Evapotranspiration rate depends also on temperature, solar radiation and wind speed and is closely related to plant functioning. *Absolute humidity* is the actual water vapour content of air, whereas *specific humidity* is the ratio of the mass of water vapour to the total mass of air. Specific humidity follows closely Clausius-Clapeyron equation: the maximum water vapour capacity of air increases exponentially with temperature. Therefore, it is predicted that specific humidity will increase in the future due to global warming, especially at the northern latitudes (IPCC 2013).

High RH = low VPD

Low RH = high VPD

2 High Humidity Affects Physiology, Biochemistry and the Structure of Plants

2.1 Stomata

Stomata are key players in regulating the water and carbon fluxes of plants. For maintaining photosynthesis, plants need to keep their stomata open to facilitate carbon dioxide (CO_2) influx to the mesophyll. However, at the same time, water vapour evaporates through the open pores during the transpiration process. Increasing RH decreases the evaporative demand by reducing the VPD gradient between leaf interior and surrounding air. Low VPD promotes stomatal opening and facilitates CO_2 influx to the leaf mesophyll, which can enhance photosynthesis (e.g. Dai et al. 1992) together with the relatively small loss of water.

Apart from VPD and leaf turgor, stomata respond to changes in light, CO_2 concentration and metabolic signals (Shimazaki et al. 2007; Sellin et al. 2010; Niglas et al. 2017). Plant responses to changing VPD are tightly associated with stomatal regulation and abscisic acid (ABA) signalling (Thioune et al. 2017). It has been shown that ABA levels are low and stomatal development may be disturbed by low VPD in growth chamber or greenhouse conditions: plants produce large stomata that are unable to close in response to normal closing stimuli such as darkness, high VPD or ABA (Torre et al. 2003; Fanourakis et al. 2013; Aliniaeifard and van Meeteren 2013). On the contrary, in field conditions, trees grown under decreased VPD exhibit higher stomatal sensitivity to changing VPD (Li and Li 2014; Niglas et al. 2015), indicating the complex nature of stomatal reactions. It appears to be the daily fluctuation of VPD that maintains stomatal responsiveness to closing stimuli (Arve et al. 2017).

2.2 Transpiration and Mineral Nutrient Uptake

Although reduced transpiration rate under low VPD can ameliorate the water status of plants and protect them from the water deficit, it may hinder mineral nutrient uptake by the mass flow of water (Barber 1995; Tinker and Nye 2001). The mass flow of water brings mineral nutrients and other solutes in contact with the root system enabling their uptake. Therefore, the influence of air humidity on plants and mineral nutrient uptake can depend on the nutrient and water status of the soil (Osonubi and Davies 1980).

Several experiments have demonstrated a positive relationship between the transpiration rate and mineral nutrition of plants (McDonald et al. 2002; Cernusak et al. 2011). The decrease of calcium (Ca) and magnesium (Mg) concentrations in leaves and fruits is the most commonly reported effect of high humidity (Armstrong and Kirkby 1979; Adams and Holder 1992; Suzuki et al. 2015), but also the concentrations of important macronutrients, nitrogen (N), phosphorus (P) and potassium (K), have positive association with the transpiration rate (McDonald et al. 2002; Novák and Vidovič 2003; Cernusak et al. 2011).

The transpiration rate affects leaf temperature and energy budget by evaporative cooling. Restrained transpiration cooling under low VPD can raise leaf temperature (Bakker 1991). On the other hand, misting or fogging is used to reduce transpiration rate and leaf temperature to promote the production of crops in greenhouse cultivation (Suzuki et al. 2015; Leyva et al. 2015; Zhang et al. 2017b). Besides the climatic factors (radiation, temperature, wind, VPD), water evaporation from plants depends on soil water content, plant structure (canopy density, leaf area, hydraulic architecture), cuticle properties and stomatal conductance (Monteith 1995; Sellin and Lubenets 2010). The water fluxes and related uptake of solutes can also be promoted by water channel proteins, namely aquaporins, located in the plasma membranes of root and leaf mesophyll cells (Kaldenhoff et al. 1998). The expression of aquaporin-coding genes was affected by high VPD in several species such as *Arabidopsis thaliana* (Levin et al. 2009) and hybrid poplar (*Populus trichocarpa* × *deltoides*) (Laur and Hacke 2013). Larger endodermal cells, thicker root endodermis and higher xylem exudation rate in plants grown under low VPD compared with high VPD can reflect the increased abundance or activity of aquaporins (Claverie et al. 2016; Kholová et al. 2016).

2.3 Growth and Development

Different plant species and genotypes vary in their responses to changing VPD. There is evident genotypic variation in stomatal and transpiration responses to VPD in beans (*Phaseolus*), maize (*Zea mays*) and wheat (*Triticum aestivum*) (Ryan et al. 2016; Claverie et al. 2016; Medina et al. 2017). The differences in water use characteristics will eventually manifest in the growth and development of plants. Accordingly, low VPD has been shown to affect plant growth and leaf size as well as leaf development in several herbaceous and woody species varying from positive effects (e.g. Leuschner 2002; Lendzion and Leuschner 2008) to negative impact (Holder and Cockshull 1990). In some cases, high humidity may enhance the growth rate at the beginning of the experiment (Armstrong and Kirkby 1979).

Air humidity affects the leaf morphology by modifying mesophyll structure, cell expansion and the formation of leaf veins. Cell expansion and cell division are often promoted under low VPD due to the high turgor of leaf cells (Leuschner 2002; Carins Murphy et al. 2014). Expansin and extensin genes, which are related to cell wall loosening and expansion/elongation of plant organs, are affected by changing VPD in soybean (*Glycine max*) leaves (Devi et al. 2015a). Larger leaf area in woodland herbaceous species in low than high VPD was a consequence of promoted epidermal cell expansion (Leuschner 2002). On the other hand, promoted cell division seemed to explain large leaf area observed in myrtle beech (*Nothofagus cunninghamii*) under low VPD (Hovenden et al. 2012). Leaf expansion can decrease the density of stomata, leaf veins and trichomes (Carins Murphy et al. 2014; Shibuya et al. 2015). A study on wheat revealed that stomatal density was reduced under low nighttime VPD only in the leaf adaxial epidermis (Claverie et al. 2016).

An anatomical study showed that the leaves of some plant species were thicker and the fractional air space of leaf mesophyll was generally larger under low than high VPD (Leuschner 2002) promoting the diffusion of water and CO_2 in the leaf. Severe disorders in leaf and fruit formation have been reported, for example, in cucumber (*Cucumis sativus*) plants, due to calcium deficiency elicited by high humidity (Staub and Navazio 1993). High air humidity may enhance root growth, as seen in studies with root and tuber crops, such as potato (*Solanum tuberosum*) (Wheeler et al. 1989) and peanut (*Arachis hypogaea*) (Mortley et al. 2000). Although low nighttime VPD does not affect the growth of wheat, it affects the hydraulic properties of roots and leaves (Claverie et al. 2016).

2.4 Biochemical and Molecular Changes

Most air humidity studies have focused on growth responses, stomatal regulation, morphological changes and plant nutrition, but very few have focused on biochemical or molecular changes (Koch et al. 2006; Devi et al. 2015b; Georgii et al. 2017). Low VPD was found to alter the epicuticular wax layer composition in *Brassica oleracea* (Koch et al. 2006), which may affect cuticular permeability to water and leaf wettability. However, in the same experiment, two other species, *Eucalyptus gunnii* and *Tropaeolum majus*, were unaffected. There is evidence that short-term exposure to different VPD conditions affects the gene expression patterns both in the roots and the shoot (Levin et al. 2009; Devi et al. 2015b) and starch and sucrose metabolism in the leaves (Shirke and Pathre 2004). The changes in plant metabolome, downstream from transcription and translation, are the end results of altered transcription and enzyme activities and thus the basis of plant phenotypic responses to changing environment.

2.5 Interaction with Abiotic Stresses

The studies on the interaction between VPD and environmental stress factors are scarce; some recent studies have focused on VPD effects under drought, heat and salt stresses. Low VPD increases cuticular permeability and impairs stomatal functioning, exposing the plants to excess water loss and desiccation (Fanourakis et al. 2016). The stomatal acclimation to VPD conditions modifies the tree responses to drought and temperature (Marchin et al. 2016). Rise in air temperature increases VPD, and accordingly, VPD plays an important role in heat stress, especially through hormonal signalling (Georgii et al. 2017). The high VPD during heat stress was the major driver of metabolic responses in *Arabidopsis thaliana* (Georgii et al. 2017). It was also shown that low VPD reduced the shift in the metabolite profile that occurred under drought or heat stress. Transcriptomic analysis under combined drought and heat stress demonstrates that low VPD enhances the heat-specific gene responses and alleviates drought-related responses. This is in accordance with the reduced transpirational cooling under low VPD. High relative humidity can mitigate the negative effects of salt stress and improve the ion homeostasis, as seen in mangrove trees (*Rhizophora apiculata* and *R. stylosa*), melon (*Cucumis melo*) cultivars, *Jatropha curcas* and cucumber (Ball et al. 1997; An et al. 2002; Rodrigues et al. 2016; Shibuya et al. 2017).

3 High Humidity Experiments with Forest Trees and Forest Ecosystems

While the responses of trees to increasing temperature and CO_2 are already well documented (e.g. Saxe et al. 2001; Ainsworth and Long 2005; Leakey et al. 2009), the impact of changes in atmospheric humidity on forest trees and ecosystems is much less studied. The increase in daytime VPD is shown to affect transpiration and forest net primary production more than the temperature (Eamus et al. 2013). Evidence is accumulating that also decreasing tendency in VPD during the growing season will have an impact on the growth of northern forests, evapotranspiration, ecosystem functions and global carbon dynamics.

Leaves of forest trees are frequently wetted by rain, dew or fog. The condensation of water onto the leaf surface impairs stomatal functioning, gas exchange and photosynthesis (Ishibashi and Terashima 1995). High humidity and prolonged water retention on the leaf surface promote fungal pathogen attacks (Sturrock et al. 2011). In addition, a high influx of phytotoxic air pollutants through the open stomata under low VPD may have harmful impact on trees. For example, in sugar maple (*Acer saccharum*) and white ash (*Fraxinus americana*), photosynthesis was more repressed when sulphur dioxide (SO_2) and ozone (O_3) fumigations were conducted under low VPD compared with high VPD (Carlson 1979). Similarly, SO_2 fumigation under low VPD caused stronger growth inhibition, leaf necrosis and leaf abscission in paper birch (*Betula papyrifera*) compared with fumigation under high VPD (Norby and Kozlowski 1982). Thus, increasing air humidity due to climate change can increase the susceptibility of trees to fungal diseases and air pollutants.

Forest trees can experience low or high VPD over long time periods: for days, weeks or months. Therefore, long-term experiments are required to study the ecophysiological responses to VPD changes. A special experimental facility for Free Air Humidity Manipulation (FAHM) was established in Estonia to investigate the long-term effects of increasing air humidity on tree performance and on the functioning of the deciduous forest ecosystem (Kupper et al. 2011). The first studies on the effects of low VPD with several tree species showed a general decrease in leaf and canopy transpiration rates (Meinzer 2003; Hölscher et al. 2005; Kupper et al.

2011), affecting soil microbial activity, ectomycorrhizal community as well as water, carbon and nutrient cycling due to higher soil water reserves (Hansen et al. 2013; Parts et al. 2013; Kukumägi et al. 2014).

3.1 Free Air Humidity Manipulation (FAHM)

Free Air Humidity Manipulation (FAHM) infrastructure was established in a former agricultural field in 2006–2007 to study the effects of artificially decreased VPD on northern forest ecosystems. The experimental site is located at Rõka in the Järvselja Experimental Forest District in south-eastern Estonia (58°14′N, 27°18′E, altitude 40–48 m) (Fig. 1). Estonia lies in between the temperate continental and maritime climate zones, and the vegetation belongs to the boreo-nemoral zone. The soil type in the experimental site is endogleyic planosol (Hansen et al. 2013). Average temperature is 17°C and −6.7°C in July and January, respectively. The growing season lasts around 180 days from mid-April to October.

The experimental site comprises three control and three humidified plots (Fig. 1). Two fast-growing deciduous tree species, silver birch (*Betula pendula* Roth) and hybrid aspen (*Populus tremula* × *tremuloides* Michx.), are used as model species in the FAHM experiment. Half of each plot (diameter 14 m) is planted with silver birch and the other half with hybrid aspen. Silver birch is widely distributed in Eurasia (Kellomäki and Wang 2001; Possen et al. 2011). It is a pioneer species and economically and ecologically the most important deciduous tree species in

Fig. 1 Layout of the Free Air Humidity Manipulation (FAHM) field site and experimental plot, located in Estonia (green area in the map). Photo was taken in summer 2012 after silver birches were harvested at the end of the year 2011. In the humidified experimental plots (H), air humidity is elevated over ambient level (C) by a misting system. Experimental plots are split between two tree species, silver birch and hybrid aspen, and between two understory vegetation types, diverse forest species and pioneer grass species. For a more detailed description of the experimental design and the misting system, see Kupper et al. (2011) and Tullus et al. (2012)

Scandinavia and large areas of Eurasia. *Betula* species are among the key species in the responses of northern ecosystems to climate change. The genome of silver birch is now assembled (Salojärvi et al. 2017), and it is an attractive target for modern genomics-based biotechnology, phenotyping and research on resistance traits and adaptation to climate change. Hybrid aspen is the fastest-growing deciduous tree in the boreal zone, and it is regarded as one of the most promising species for short-rotation forestry in the northern Europe (Tullus et al. 2012).

In the humidified plots, RH is elevated with a misting pipe system producing fine water droplets (diameter ~10 μm), leading to on average 7% higher RH in the long term compared with the ambient level (Fig. 1). The droplet size was optimized to minimize leaf wetting and temperature decline in the humidified plots. Misting was started in 2008 and has been applied throughout the growing seasons between May and October until today (Kupper et al. 2011; Tullus et al. 2012). The first generation of trees was harvested at the end of 2011 (birch) and 2012 (aspen). The second generation of birches was planted on the site in 2012 and new aspen saplings regenerated naturally from the stump sprouts.

3.2 Ecophysiological Responses in the FAHM Experiment

The results from FAHM experiment demonstrate slower tree growth under low VPD (Fig. 2) both in silver birch and hybrid aspen (Sellin et al. 2017a; Tullus et al. 2017). However, the radial growth of birch stems was less affected in comparison with the height growth, leading to lower slenderness (i.e. increased ratio of stem width to height) under high humidity. Lower slenderness is in line with the carbon allocation to water-conducting vascular tissues at the expense of photosynthetic leaf area and increasing proportion of the living parenchyma cells and rays in stem wood relative to dead xylem elements. These morphological and anatomical changes in the stem wood tissues favour radial growth to provide mechanical stability and compensate for the reduced hydraulic conductance but increase the maintenance respiration costs (due to the increased number of living parenchyma cells) (Sellin et al. 2017a). The decline of leaf hydraulic conductance was demonstrated for silver birch growing under low VPD (Sellin et al. 2013). Air humidification reduced the hydraulic conductance (Sellin et al. 2017b) and changed the xylem anatomy also in hybrid aspen (Jasińska et al. 2015). Decreased hydraulic capacity becomes a disadvantageous factor for the trees grown under low VPD, limiting leaf water supply and reducing canopy conductance at least under the conditions of soil water deficit (Niglas et al. 2014). The effects of elevated humidity on the chemical properties of the stem wood varied between the tree species (Tullus et al. 2014). In aspen, cellulose content was higher and hemicellulose content was lower, whereas in birch, cellulose content was lower under high humidity (Tullus et al. 2014).

Naturally, there was a year-to-year variation in growth responses, due to the prevailing weather conditions. Humidification reduced the growth rate of aspen and birch trees during normal rainy seasons (Kupper et al. 2011; Tullus et al. 2012; Sellin

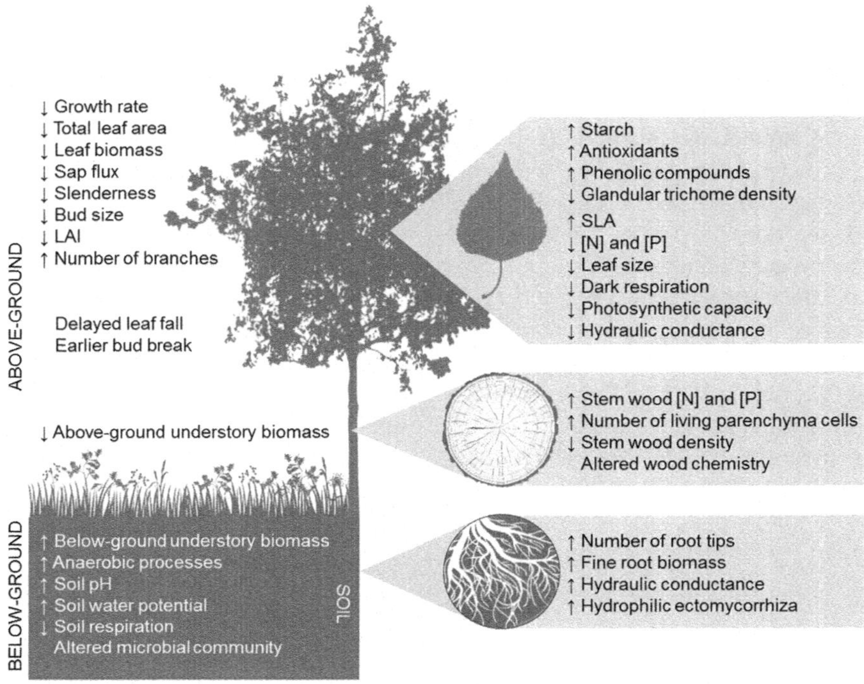

Fig. 2 An overview of forest tree and ecosystem level responses to long-term air humidification treatment in the FAHM field experiment. Arrows mark the direction of responses (↓ lower/decrease, ↑ higher/increase) to elevated humidity. *LAI* leaf area index, *SLA* specific leaf area

et al. 2013). However, humidification tended to alleviate drought stress during a dry season, particularly at the end of summer 2011 (Niglas et al. 2014). The reduction in growth rate and biomass production in response to humidification was more pronounced in aspen than in birch (Rosenvald et al. 2014). The slower growth rate under elevated humidity was confirmed in the second-generation stump sprouts of hybrid aspen (Rohula et al. 2017; Sellin et al. 2017b). In addition, the growth and biomass partitioning of the understory vegetation responded to humidification treatment, but the responses varied based on study year and tree species (Torga et al. 2017). In the birch stands, the aboveground biomass of the understory vegetation was lower and the belowground biomass higher under humidification in 2010 (Torga et al. 2017). The competitive status of trees affected their growth responses to humidification treatment (Tullus et al. 2017) indicating potential impacts of decreasing VPD on forest species composition and forestry practices in the future.

Reduced leaf photosynthetic capacity was detected for silver birch (Sellin et al. 2013), along with the decline in foliar P and N contents and nutrient imbalance (P:N ratio in particular). Humidification treatment reduced the concentrations of N and P mostly in the upper canopy (Sellin et al. 2015) indicating changes in the nutrient allocation. On the other hand, higher content of N and P was detected in the stem wood under low VPD which may have resulted from the enhanced resorption of nutrients and the higher number of the living parenchyma cells (Tullus et al. 2014). Impaired leaf nutrient status and soil hypoxia caused by increased soil wetness led to the retardation of foliar development (leaf area and mass), which was clearly demonstrated at leaf, tree and stand levels in both species, having a direct negative impact on productivity (Sellin et al. 2017a). High humidity reduced bud size and primordia growth in silver birch, but these changes did not translate into reduced leaf area in the current-year shoots (Kukk et al. 2015). Nevertheless, changes in leaf morphogenesis in response to high humidity can have long-term effects on the crown development, including a trend towards increased branchiness (Sellin et al. 2015). Humidification stimulated bud break in aspen (Kukk et al. 2015) and delayed leaf fall in birch (Godbold et al. 2014). The prolonged leaf lifespan enables trees to photosynthesize and to acquire and resorb mineral nutrients for a longer time, which may compensate for the reduced photosynthetic capacity and mineral nutrient status.

The root systems of birch trees showed structural and morphological adjustments in response to high air humidity (Rosenvald et al. 2014; Godbold et al. 2014). Increased root hydraulic capacity (Sellin et al. 2015), greater fine-root biomass, higher root-specific area and higher turnover rate may be considered to mitigate low nutrient gaining due to reduced transpirational flux under elevated humidity. Acclimation of root morphological traits of silver birch in humidified plots depends on the type of the understory vegetation (Parts et al. 2013). For example, specific root length and density were very high in the humidified plots with the early-successional grass understory. Silver birch displayed a better acclimation capacity to high humidity conditions than hybrid aspen did. One of the explanations for this could be the natural variation in silver birch stands and thereby higher plasticity compared with monoclonal hybrid aspens (Rosenvald et al. 2014).

3.3 Metabolic Responses in the FAHM Experiment

Leaves play a key role in plant responses to variable environmental conditions. The foliar metabolite profiles of trees grown under ambient and reduced VPD condition were analysed by gas chromatography-mass spectrometry (GC-MS) (Lihavainen et al. 2016a). There was a shift in the metabolite profiles of aspen and birch leaves in response to humidification treatment. The changes in metabolite levels were subtle, and there were significant differences in 28 metabolites in birch and 6 metabolites in aspen (Lihavainen et al. 2016a). Metabolite responses were largely similar in both species but stronger in birch. The differences appeared in carbohydrates, amino acids, ascorbate metabolites and secondary metabolites. The high starch content in the leaves indicated an increased carbon storage activity at the expense of biomass accumulation, sugar transport and respiration processes. This was supported by the reduced dark respiration rate of the second-generation stump sprouts of hybrid aspen in the humidified plots (Rohula et al. 2017).

Changes in N assimilation were observed in amino acid profiles, in particular as a lower glutamine-to-glutamate ratio in the low VPD conditions, reflecting a reduced ammonium assimilation rate. Carbon was allocated to the biosynthesis of secondary compounds through the phenylpropanoid pathway leading to increased production of quinic acid derivatives, catechins, hemiterpenoid glycoside, quercetin glycosides and other phenolic glycosides. The increased production of these secondary compounds indicates the activation of defence mechanisms against a stress situation, i.e. nutrient deficiency and soil hypoxia in this case (Sellin et al. 2015). Many secondary compounds act as antioxidants against oxidative stress under unfavourable environmental conditions. In addition to the phenolic compounds, elevated levels of ascorbate metabolites and alpha-tocopherol were observed in the leaves under low VPD treatment indicating the development of antioxidative defence.

At the ecosystem level, high content of nonstructural carbohydrates and phenolics in the leaf litter is known to slow down the decomposition rate (e.g. Tuchman et al. 2003). In 2010, tree leaf litter flux was lower under aspen stands in the humidified plots than in the control plots (Torga et al. 2017). Therefore, increasing air humidity may affect the carbon and nutrient cycling and eventually the primary production of the northern forest ecosystems.

3.4 Laboratory Studies Confirming the Field Results

For the mechanistic understanding, it is necessary to compare and verify the results from the field experiments with the more strictly controlled laboratory studies, where the environmental 'noise' (e.g. change in weather conditions) is minimized. According to the FAHM experiment in the field, high humidity promoted branching and reduced the slenderness of silver birch plants and altered N allocation pattern among plant organs and within crown (Lihavainen et al. 2016b). The foliar concentrations of starch, quercetin glycosides and glutamate were higher in silver birch plants growing under low VPD than under high VPD in a short-term chamber experiment, while the levels of shikimic acid, ribonic acid, amino acids alanine and valine, several citric acid cycle intermediates as well as glutamine-to-glutamate ratio were lower (Lihavainen et al. 2016b).

The chamber study revealed several differences in the responses between the seedlings and the larger trees and/or experimental systems. For example, enhanced growth appeared to be an initial response in the chamber experiment, accompanied by the efficient recycling of N, and enhanced root and stem radial growth (Lihavainen et al. 2016b), which could compensate for the reduced transpiration flux and nutrient acquisition under high humidity. Mineral nutrient uptake under low VPD could be enhanced by the higher number of root tips, larger absorptive root area and transpiring leaf area (Kupper et al. 2017). In the chamber study, carbon accumulation in the form of poly- and oligosaccharides in low VPD-grown leaves was indicated as increased raffinose content, which may reflect impaired phloem loading process (Lihavainen et al. 2016b). The development of antioxidative defence under low VPD, on the other hand, was obvious only in the long-term field experiment (Lihavainen et al. 2016a).

Because the acquisition and translocation of N through transpiration flux were reduced in the field conditions, the chamber experiment was designed to study the impact of N availability on the metabolism in more detail using two N fertilization levels under low VPD conditions (Lihavainen et al. 2016b). Additional N fertilization mitigated the effects of low VPD on leaf metabolism. For example, high N supply maintained the glutamine-to-glutamate ratio at the same levels as in the plants under high VPD with moderate N fertilization. The chamber experiment verified that metabolic responses to low VPD were mainly driven by the limitation in N availability.

The chamber environment is also suitable for molecular studies. The gene expression studies with birches indicated very complex genetic regulation system related to N metabolism (Lihavainen et al. 2016b). The leaf mRNA levels of the key enzymes, nitrate reductase (*BpNIA*) and nitrite reductase (*BpNiR*), were upregulated in the low VPD conditions, while the N concentration of the leaves was reduced. The complex regulation pathways may involve feedback inhibition of gene expressions and the key role of enzyme activators such as molybdenum (Mo), which was at high level in these low VPD plants. Further, enzyme levels do not always correlate directly with the corresponding gene expression.

3.5 The Key Role of Leaf Surface

The surface of the leaves contains an epicuticular wax layer that is acting as a protective barrier between the leaf tissue and the surrounding environment. The chemical properties and structure of the cuticular wax layer are reactive to many environmental and biotic factors, such as UV radiation, air pollutants, herbivores and pathogens and air humidity. The cuticular wax layer of silver birch leaves is characterized by a high amount of dammarane and epoxydammarane triterpenoid aglycones, flavonoid aglycones and very long-chain aliphatic hydrocarbons, forming a strong hydrophobic layer (Lihavainen et al. 2017). Leaves of many plant species are covered by different kinds of trichomes. Silver birch leaves are densely covered with glandular trichomes that store and excrete lipophilic compounds to the leaf surface and provide chemical defence. Glandular trichomes show the highest metabolic activity in the young developing leaves (Valkama et al. 2004).

Both FAHM and laboratory studies (described above) indicated that the glandular trichome density is smaller under low VPD (Lihavainen et al. 2017), correlating with the amount of the major triterpenoid and flavonoid aglycones. There was some variation in the trichome responses between the adaxial and abaxial surfaces of the leaf, indicating differences in the function of the two leaf sides. The chamber experiment revealed that the chemical properties of cuticular waxes are affected by low VPD in silver birch although the total wax content may remain unaffected (Lihavainen et al. 2017). The low content of alkanes and the low proportion of highly hydrophobic flavonoids under the low VPD conditions reduce the hydrophobicity of the cuticular wax layer. The lower hydrophobicity of wax layer increases the leaf wettability, which in turn affects water retention on the leaf surface. Leaf wetting exposes the leaves for fungal pathogen attacks. Susceptibility to herbivores and pathogens is further enhanced by the lower glandular trichome density, because the trichomes are exuding defence compounds to the leaf surface. In the first study years in the FAHM experiment, fungal damages were more frequent in the birch stands in the humidified plots than in the plots with ambient VPD (Sellin et al. 2017a). The chamber study also revealed that high N availability can mitigate the effects of low VPD on cuticular wax chemistry by affecting the biosynthesis of wax compounds and genetic regulation of some key enzymes in the triterpenoid pathway.

4 Future Perspectives

The results obtained by FAHM experiment described above indicate that atmospheric humidity is a highly important climatic factor that affects growth, leaf and root morphology, stomatal function, water relations, stem wood properties and mineral nutrition of northern forest trees, as demonstrated with silver birch and hybrid aspen. These results are in line with the previous results with different plant species but give a more comprehensive picture of responses at the whole-tree and ecosystem level. The species-specific responses under low VPD were indicated as a more pronounced growth reduction in hybrid aspen compared with silver birch. Birch showed more efficient adjustment of the leaf and root morphology and hydraulic architecture that could compensate for the reduced transpirational flux, mineral nutrient acquisition and photosynthetic capacity under decreased VPD. Low VPD triggered the readjustment of primary carbon and nitrogen metabolism and altered the leaf surface properties. The experimental knowledge of responses and acclimation capacity of birch to several climate change factors, and the recent sequencing of silver birch genome (Salojärvi et al. 2017) with several candidate genes for adaptation to different environmental variables (such as temperature and precipitation), makes silver birch an attractive model for forest biotechnology and breeding. During the last decades, *Populus* species have been extensively studied for acclimation and adaptation, for example, for dry and wet conditions (Street et al. 2006). Both species are considered typical pioneer species, able to quickly colonize disturbed sites (e.g. Worrell 1995).

The FAHM study showed the importance of long-term experiments, because the responses to low VPD were dependent on the prevailing weather conditions. Further, several indirect effects of low VPD through altered soil processes (moisture, microbiota, nitrification) and the competitive status of trees became apparent. High levels of antioxidants and phenolic compounds under low VPD indicate that trees employed defence against stress factors such as nutrient limitation, soil hypoxia and exposure to fungal attacks. The high contents of carbohydrates and phenolics in the green leaves and leaf litter with the reduced foliage biomass and mineral nutrient content can further affect the decomposition rates and carbon and nutrient cycling of the forest ecosystems.

Besides the field experiments, it is important to conduct complementary laboratory studies. In this study, laboratory studies demonstrated that additional N supply can mitigate the effects of low VPD on cellular metabolism and leaf surface waxes in silver birch. Concluding from above, rising air humidity has negative impact on the productivity and functioning of forest trees, with a potential effect on species composition in the future. Increasing temperature and atmospheric CO_2 concentration are expected to increase the forest productivity in northern Europe due to longer growing season and higher carbon assimilation rate. However, the growth stimulation can be smaller than predicted if the climate change is accompanied by rising atmospheric humidity.

Acknowledgements The authors thank Jaak Sõber and Priit Kupper for the technical constructions, coordination and management of the FAHM experiment.

References

Adams P, Holder R (1992) Effects of humidity, Ca and salinity on the accumulation of dry matter and Ca by the leaves and fruit of tomato (*Lycopersicon esculentum*). J. Hortic Sci 67:137–142

Ainsworth EA, Long SP (2005) What have we learned from 15 years of free-air CO2 enrichment (FACE)? A meta-analytic review of the responses of photosynthesis, canopy properties and plant production to rising CO2. New Phytol 165(2):351–372

Aliniaeifard S, van Meeteren U (2013) Can prolonged exposure to low VPD disturb the ABA signalling in stomatal guard cells? J Exp Bot 64:3551–3566

An P, Inanaga S, Lux A, Li XJ, Ali MEK, Matsui T, Sugimoto Y (2002) Effects of salinity and relative humidity on two melon cultivars differing to salt tolerance. Biol Plant 45:409–415

Armstrong M, Kirkby E (1979) The influence of humidity on the mineral composition of tomato plants with special reference to calcium distribution. Plant Soil 52:427–435

Arve LE, Kruse OMO, Tanino KK, Olsen JE, Futsæther C, Torre S (2017) Daily changes in VPD during leaf development in high air humidity increase the stomatal responsiveness to darkness and dry air. J Plant Physiol 211:63–69

Bakker JC (1991) Analysis of humidity effects on growth and production of glasshouse fruit vegetables. Dissertation, Landbouwuniversiteit, Wageningen

Ball M, Cochrane M, Rawson H (1997) Growth and water use of the mangroves *Rhizophora apiculata* and *R. stylosa* in response to salinity and humidity under ambient and elevated concentrations of atmospheric CO_2. Plant Cell Environ 20:1158–1166

Barber SA (1995) Soil nutrient bioavailability: a mechanistic approach. Wiley, Chichester

Boé J, Terray L (2014) Land–sea contrast, soil-atmosphere and cloud-temperature interactions: interplays and roles in future summer European climate change. Clim Dyn 42:683–699

Carins Murphy MR, Jordan GJ, Brodribb TJ (2014) Acclimation to humidity modifies the link between leaf size and the density of veins and stomata. Plant Cell Environ 37:124–131

Carlson RW (1979) Reduction in the photosynthetic rate of *Acer*, *Quercus* and *Fraxinus* species caused by sulphur dioxide and ozone. Environ Pollut 18(2):159–170

Cernusak LA, Winter K, Turner BL (2011) Transpiration modulates phosphorus acquisition in tropical tree seedlings. Tree Phys 31:878–885

Claverie E, Schoppach R, Sadok W (2016) Nighttime evaporative demand induces plasticity in leaf and root hydraulic traits. Physiol Plant 158(4):402–413

Dai A (2006) Recent climatology, variability, and trends in global surface humidity. J Clim 19:3589–3606

Dai Z, Edwards GE, Ku MS (1992) Control of photosynthesis and stomatal conductance in *Ricinus communis* L. (castor bean) by leaf to air vapour pressure deficit. Plant Physiol 99:1426–1434

Dessler AE, Schoeberl MR, Wang T, Davis SM, Rosenlof KH (2013) Stratospheric water vapour feedback. Proc Natl Acad Sci 110:18087–18091

Devi MJ, Taliercio EW, Sinclair TR (2015a) Leaf expansion of soybean subjected to high and low atmospheric vapour pressure deficits. J Exp Bot 66:1845–1850

Devi MJ, Sinclair TR, Taliercio E (2015b) Comparisons of the effects of elevated vapour pressure deficit on gene expression in leaves among two fast-wilting and a slow-wilting soybean. PLoS One 10:e0139134

Eamus D, Boulain N, Cleverly J, Breshears DD (2013) Global change-type drought-induced tree mortality: vapour pressure deficit is more important than temperature per se in causing decline in tree health. Ecol Evol 3(8):2711–2729

Fanourakis D, Heuvelink E, Carvalho SM (2013) A comprehensive analysis of the physiological and anatomical components involved in higher water loss rates after leaf development at high humidity. J Plant Physiol 170:890–898

Fanourakis D, Bouranis D, Giday H, Carvalho DR, Nejad AR, Ottosen CO (2016) Improving stomatal functioning at elevated growth air humidity: a review. J Plant Physiol 207:51–60

Fatichi S, Molnar P, Mastrotheodoros T, Burlando P (2015) Diurnal and seasonal changes in near-surface humidity in a complex orography. J Geophys Res Atmos 120:2358–2374

Fu G, Shen ZX (2016) Environmental humidity regulates effects of experimental warming on vegetation index and biomass production in an alpine meadow of the Northern Tibet. PLoS One 11(10):e0165643

Georgii E, Jin M, Zhao J et al (2017) Relationships between drought, heat and air humidity responses revealed by transcriptome-metabolome co-analysis. BMC Plant Biol 17(1):120

Godbold D, Tullus A, Kupper P, Sõber J, Ostonen I, Godbold JA, Lukac M, Ahmed IU, Smith AR (2014) Elevated atmospheric CO2 and humidity delay leaf fall in *Betula pendula*, but not in *Alnus glutinosa* or *Populus tremula* × *tremuloides*. Ann For Sci 71:831–842

Hansen R, Mander Ü, Soosaar K, Maddison M, Lõhmus K, Kupper P, Kanal A, Sõber J (2013) Greenhouse gas fluxes in an open air humidity manipulation experiment. Landsc Ecol 28:637–649

Holder R, Cockshull K (1990) Effects of humidity on the growth and yield of glasshouse tomatoes. J Hortic Sci 65:31–39

Hölscher D, Koch O, Korn S, Leuschner C (2005) Sap flux of five co-occurring tree species in a temperate broad-leaved forest during seasonal soil drought. Trees 19(6):628–637

Hovenden MJ, Vander Schoor JK, Osanai Y (2012) Relative humidity has dramatic impacts on leaf morphology but little effect on stomatal index or density in *Nothofagus cunninghamii* (Nothofagaceae). Aust J Bot 60:700–706

IPCC (2013) Climate change 2013: the physical science basis. In: Stocker TF, Qin D, Plattner G-K, Tignor M, Allen SK, Boschung J, Nauels A, Xia Y, Bex V, Midgley PM (eds) Contribution of Working Group I to the fifth assessment report of the Intergovernmental Panel on Climate Change. Cambridge University Press, Cambridge

Ishibashi M, Terashima I (1995) Effects of continuous leaf wetness on photosynthesis: adverse aspects of rainfall. Plant Cell Environ 18:431–438

Jacob D, Petersen J, Eggert B et al (2014) EURO-CORDEX: new high-resolution climate change projections for European impact research. Reg Environ Chang 14(2):563–578

Jasińska AK, Alber M, Tullus A, Rahi M, Sellin A (2015) Impact of elevated atmospheric humidity on anatomical and hydraulic traits of xylem in hybrid aspen. Funct Plant Biol 42(6):565–578

Kaldenhoff R, Grote K, Zhu J, Zimmermann U (1998) Significance of plasmalemma aquaporins for water transport in *Arabidopsis thaliana*. Plant J 14:121–128

Kellomäki S, Wang KY (2001) Growth and resource use of birch seedlings under elevated carbon dioxide and temperature. Ann Bot 87(5):669–682

Kholová J, Zindy P, Malayee S et al (2016) Component traits of plant water use are modulated by vapour pressure deficit in pearl millet (*Pennisetum glaucum* (L.) R. Br.) Funct Plant Biol 43(5):423–437

Koch K, Hartmann KD, Schreiber L, Barthlott W, Neinhuis C (2006) Influences of air humidity during the cultivation of plants on wax chemical composition, morphology and leaf surface wettability. Environ Exp Bot 56:1–9

Kukk M, Räim O, Tulva I, Sõber J, Lõhmus K, Sõber A (2015) Elevated air humidity modulates bud size and the frequency of bud break in fast-growing deciduous trees: silver birch (*Betula pendula* Roth.) and hybrid aspen (*Populus tremula* L. × *P. tremuloides* Michx.) Trees 29(5):1381–1393

Kukumägi M, Ostonen I, Kupper P et al (2014) The effects of elevated atmospheric humidity on soil respiration components in a young silver birch forest. Agric For Meteorol 194:167–174

Kupper P, Sõber J, Sellin A et al (2011) An experimental facility for free air humidity manipulation (FAHM) can alter water flux through deciduous tree canopy. Environ Exp Bot 72(3):432–438

Kupper P, Rohula G, Inno L, Ostonen I, Sellin A, Sõber A (2017) Impact of high daytime air humidity on nutrient uptake and night-time water flux in silver birch, a boreal forest tree species. Reg Environ Chang 17:2149–2157

Laur J, Hacke UG (2013) Transpirational demand affects aquaporin expression in poplar roots. J Exp Bot 64 2283–2293

Leakey AD, Ainsworth EA, Bernacchi CJ, Rogers A, Long SP, Ort DR (2009) Elevated CO_2 effects on plant carbon, nitrogen, and water relations: six important lessons from FACE. J Exp Bot 60(10):2859–2876

Lendzion J, Leuschner C (2008) Growth of European beech (Fagus sylvatica L.) saplings is limited by elevated atmospheric vapour pressure deficits. For Ecol Manag 256:648–655

Leuschner C (2002) Air humidity as an ecological factor for woodland herbs: leaf water status, nutrient uptake, leaf anatomy, and productivity of eight species grown at low or high VPD levels. Flora 197:262–274

Levin M, Resnick N, Rosianskey Y, Kolotilin I, Wininger S, Lemcoff JH, Cohen S, Galili G, Koltai H, Kapulnik Y (2009) Transcriptional profiling of *Arabidopsis thaliana* plants' response to low relative humidity suggests a shoot–root communication. Plant Sci 177:450–459

Leyva R, Constán-Aguilar C, Sánchez-Rodríguez E, Romero-Gámez M, Soriano T (2015) Cooling systems in screenhouses: effect on microclimate, productivity and plant response in a tomato crop. Biosyst Eng 129:100–111

Li J, Li X (2014) Response of stomatal conductance of two tree species to vapour pressure deficit in three climate zones. J Arid Land 6:771–781

Lihavainen J, Keinänen M, Keski-Saari S, Kontunen-Soppela S, Sõber A, Oksanen E (2016a) Artificially decreased vapour pressure deficit in field conditions modifies foliar metabolite profiles in birch and aspen. J Exp Bot 67(14):4367–4378

Lihavainen J, Ahonen V, Keski-Saari S, Kontunen-Soppela S, Oksanen E, Keinänen M (2016b) Low vapour pressure deficit affects nitrogen nutrition and foliar metabolites in silver birch. J Exp Bot 67(14):4353–4365

Lihavainen J, Ahonen V, Keski-Saari S, Sõber A, Oksanen E, Keinänen M (2017) Low vapour pressure deficit reduces glandular trichome density and modifies the chemical composition of cuticular waxes in silver birch leaves. Tree Physiol 37:1166–1181

Marchin RM, Broadhead AA, Bostic LE, Dunn RR, Hoffmann WA (2016) Stomatal acclimation to vapour pressure deficit doubles transpiration of small tree seedlings with warming. Plant Cell Environ 39(10):2221–2234

McDonald EP, Erickson JE, Kruger EL (2002) Research note: can decreased transpiration limit plant nitrogen acquisition in elevated CO_2? Funct Plant Biol 29:1115–1120

Medina V, Teran JCBM, Gepts P, Gilbert ME (2017) Low stomatal sensitivity to vapour pressure deficit in irrigated common, lima and tepary beans. Field Crop Res 206:128–137

Meinzer FC (2003) Functional convergence in plant responses to the environment. Oecologia 134(1):1–11

Monteith J (1995) Accommodation between transpiring vegetation and the convective boundary layer. J Hydrol 166:251–263

Mortley D, Bonsi C, Loretan P, Hill W, Morris C (2000) High relative humidity increases yield, harvest index, flowering, and gynophore growth of hydroponically grown peanut plants. Hortic Sci 35:46–48

Niglas A, Kupper P, Tullus A, Sellin A (2014) Responses of sap flow, leaf gas exchange and growth of hybrid aspen to elevated atmospheric humidity under field conditions. AoB Plants 6:plu021

Niglas A, Alber M, Suur K, Jasińska AK, Kupper P, Sellin A (2015) Does increased air humidity affect stomatal morphology and functioning in hybrid aspen? Botany 93:243–250

Niglas A, Papp K, Sękiewicz M, Sellin A (2017) Short-term effects of light quality on leaf gas exchange and hydraulic properties of silver birch (*Betula pendula*). Tree Phys 37:1218–1228

Norby RJ, Kozlowski TT (1982) The role of stomata in sensitivity of *Betula papyrifera* seedlings to SO2 at different humidities. Oecologia 53(1):34–39

Novák V, Vidovič J (2003) Transpiration and nutrient uptake dynamics in maize (*Zea mays* L.) Ecol Model 166:99–107

Osonubi O, Davies W (1980) The influence of plant water stress on stomatal control of gas exchange at different levels of atmospheric humidity. Oecologia 46:1–6

Parts K, Tedersoo L, Lõhmus K, Kupper P, Rosenvald K, Sõber A, Ostonen I (2013) Increased air humidity and understory composition shape short root traits and the colonizing ectomycorrhizal fungal community in silver birch stands. For Ecol Manag 310:720–728

Possen BJ, Oksanen E, Rousi M et al (2011) Adaptability of birch (*Betula pendula* Roth) and aspen (*Populus tremula* L.) genotypes to different soil moisture conditions. For Ecol Manag 262(8):1387–1399

Rodrigues CR, Silveira JA, Viégas RA, Moura RM, Aragão RM, Silva EN (2016) Combined effects of high relative humidity and K+ supply mitigates damage caused by salt stress on growth, photosynthesis and ion homeostasis in *J. curcas* plants. Agric Water Manag 163:255–262

Rohula G, Tulva I, Tullus A, Sõber A, Kupper P (2017) Endogenous regulation of night-time water relations in hybrid aspen grown at ambient and elevated air humidity. Reg Environ Chang 17:2169–2178

Rosenvald K, Tullus A, Ostonen I, Uri V, Kupper P, Aosaar J, Varik M, Sõber J, Niglas A, Hansen R (2014) The effect of elevated air humidity on young silver birch and hybrid aspen biomass allocation and accumulation–Acclimation mechanisms and capacity. For Ecol Manag 330:252–260

Ryan AC, Dodd IC, Rothwell SA, Jones R, Tardieu F, Draye X, Davies WJ (2016) Gravimetric phenotyping of whole plant transpiration responses to atmospheric vapour pressure deficit identifies genotypic variation in water use efficiency. Plant Sci 251:101–109

Salojärvi J, Smolander OP, Nieminen K et al (2017) Genome sequencing and population genomic analyses provide insights into the adaptive landscape of silver birch. Nat Genet 49(6):904–912

Saxe H, Cannell MG, Johnsen Ø, Ryan MG, Vourlitis G (2001) Tree and forest functioning in response to global warming. New Phytol 149(3):369–399

Sellin A, Lubenets K (2010) Variation of transpiration within a canopy of silver birch: effect of canopy position and daily versus nightly water loss. Ecohydrology 3:467–477

Sellin A, Eensalu E, Niglas A (2010) Is distribution of hydraulic constraints within tree crowns reflected in photosynthetic water-use efficiency? An example of *Betula pendula*. Ecol Res 25:173–183

Sellin A, Tullus A, Niglas A, Õunapuu E, Karusion A, Lõhmus K (2013) Humidity-driven changes in growth rate, photosynthetic capacity, hydraulic properties and other functional traits in silver birch (*Betula pendula*). Ecol Res 28:523–535

Sellin A, Rosenvald K, Õunapuu-Pikas E, Tullus A, Ostonen I, Lõhmus K (2015) Elevated air humidity affects hydraulic traits and tree size but not biomass allocation in young silver birches (*Betula pendula*). Front Plant Sci 6:860

Sellin A, Alber M, Keinänen M et al (2017a) Growth of northern deciduous trees under increasing atmospheric humidity: possible mechanisms behind the growth retardation. Reg Environ Chang 17:2135–2148

Sellin A, Alber M, Kupper P (2017b) Increasing air humidity influences hydraulic efficiency but not functional vulnerability of xylem in hybrid aspen. J Plant Physiol 219:28–36

Shibuya T, Itagaki K, Ueyama S, Hirai N, Endo R (2015) Atmospheric Humidity influences oviposition rate of *Tetranychus urticae* (Acari: Tetranychidae) through morphological responses of host *Cucumis sativus* leaves. J Econ Entomol 109(1):255–258

Shibuya T, Kano K, Endo R, Kitaya Y (2017) Effects of the interaction between vapour-pressure deficit and salinity on growth and photosynthesis of *Cucumis sativus* seedlings under different CO_2 concentrations. Photosynthetica. https://doi.org/10.1007/s11099-017-0746-8

Shimazaki K, Doi M, Assmann SM, Kinoshita T (2007) Light regulation of stomatal movement. Annu Rev Plant Biol 58:219–247

Shirke PA, Pathre UV (2004) Influence of leaf-to-air vapour pressure deficit (VPD) on the biochemistry and physiology of photosynthesis in *Prosopis juliflora*. J Exp Bot 55:2111–2120

Simmons A, Willett K, Jones P, Thorne P, Dee D (2010) Low-frequency variations in surface atmospheric humidity, temperature, and precipitation: inferences from reanalyses and monthly gridded observational data sets. J Geophys Res Atmos 115:D01110

Staub JE, Navazio JP (1993) Temperature and humidity affect pillowy fruit disorder in cucumber. Hortic Sci 28:822–823

Street NR, Skogstrom O, Sjodin A et al (2006) The genetics and genomics of the drought response in *Populus*. Plant J 48:321–341

Sturrock R, Frankel S, Brown A, Hennon P, Kliejunas J, Lewis K, Worrall J, Woods A (2011) Climate change and forest diseases. Plant Pathol 60:133–149

Suzuki M, Umeda H, Matsuo S, Kawasaki Y, Ahn D, Hamamoto H, Iwasaki Y (2015) Effects of relative humidity and nutrient supply on growth and nutrient uptake in greenhouse tomato production. Sci Hortic 187:44–49

Thioune EH, McCarthy J, Gallagher T, Osborne B (2017) A humidity shock leads to rapid, temperature dependent changes in coffee leaf physiology and gene expression. Tree Phys 37 (3):367–379

Tinker BP, Nye PH (2001) Solute movement in rhizosphere. Oxford University Press, Oxford, p 444

Torga R, Mander Ü, Soosaar K et al (2017) Weather extremes and tree species shape soil greenhouse gas fluxes in an experimental fast-growing deciduous forest of air humidity manipulation. Ecol Eng 106:369–377

Torre S, Fjeld T, Gislerød HR, Moe R (2003) Leaf anatomy and stomatal morphology of greenhouse roses grown at moderate or high air humidity. J Am Soc Hortic Sci 128:598–602

Tuchman NC, Wahtera KA, Wetzel RG, Teeri JA (2003) Elevated atmospheric CO2 alters leaf litter quality for stream ecosystems: an in situ leaf decomposition study. Hydrobiologia 495:203–211

Tullus A, Kupper P, Sellin A, Parts L, Sõber J, Tullus T, Lõhmus K, Sõber A, Tullus H (2012) Climate change at northern latitudes: rising atmospheric humidity decreases transpiration, N-uptake and growth rate of hybrid aspen. PLoS One 7:e42648

Tullus A, Sellin A, Kupper P et al (2014) Increasing air humidity – a climate trend predicted for northern latitudes – alters the chemical composition of stemwood in silver birch and hybrid aspen. Silva Fennica 48(4):1107

Tullus A, Kupper P, Kaasik A et al (2017) The competitive status of trees determines their responsiveness to increasing atmospheric humidity – a climate trend predicted for northern latitudes. Glob Chang Biol 23(5):1961–1974

Valkama E, Salminen JP, Koricheva J, Pihlaja K (2004) Changes in leaf trichomes and epicuticular flavonoids during leaf development in three birch taxa. Ann Bot 94(2):233–242

Wheeler R, Tibbitts T, Fitzpatrick A (1989) Potato growth in response to relative humidity. Hortic Sci 24:482–484

Willett KM, Gillett NP, Jones PD, Thorne PW (2007) Attribution of observed surface humidity changes to human influence. Nature 449:710–712

Worrell R (1995) European aspen (*Populus tremula* L.) – a review with particular reference to Scotland I. Distribution, ecology and genetic variation. Forestry 68:93–105

Zhang P, Zhang J, Chen M (2017a) Economic impacts of climate change on agriculture: the importance of additional climatic variables other than temperature and precipitation. J Environ Econ Manag 83:8–31

Zhang D, Du Q, Zhang Z, Jiao X, Song X, Li J (2017b) Vapour pressure deficit control in relation to water transport and water productivity in greenhouse tomato production during summer. Sci Rep 7:srep43461

Printed by Printforce, the Netherlands